Molecular Drug Properties

Edited by
Raimund Mannhold

Methods and Principles in Medicinal Chemistry

Edited by R. Mannhold, H. Kubinyi, G. Folkers

Previous Volumes of this Series:

G. Cruciani (ed.)

Molecular Interaction Fields

Vol. 27

2006, ISBN 978-3-527-31087-6

M. Hamacher, K. Marcus, K. Stühler, A. van Hall, B. Warscheid, H. E. Meyer (eds.)

Proteomics in Drug Research

Vol. 28

2006, ISBN 978-3-527-31226-9

D. J. Triggle, M. Gopalakrishnan, D. Rampe, W. Zheng (eds.)

Voltage-Gated Ion Channels as Drug Targets

Vol. 29

2006, ISBN 978-3-527-31258-0

D. Rognan (ed.)

Ligand Design for G Protein-coupled Receptors

Vol. 30

2006, ISBN 978-3-527-31284-9

D. A. Smith, H. van de Waterbeemd, D. K. Walker

Pharmacokinetics and Metabolism in Drug Design, 2nd Ed.

Vol. 31

2006, ISBN 978-3-527-31368-6

T. Langer, R. D. Hofmann (eds.)

Pharmacophores and Pharmacophore Searches

Vol. 32

2006, ISBN 978-3-527-31250-4

E. Francotte, W. Lindner (eds.)

Chirality in Drug Research

Vol. 33

2006, ISBN 978-3-527-31076-0

W. Jahnke, D. A. Erlanson (eds.)

Fragment-based Approaches in Drug Discovery

Vol. 34

2006, ISBN 978-3-527-31291-7

J. Hüser (ed.)

High-Throughput Screening in Drug Discovery

Vol. 35

2006, ISBN 978-3-527-31283-2

K. Wanner, G. Höfner (eds.)

Mass Spectrometry in Medicinal Chemistry

Vol. 36

2007, ISBN 978-3-527-31456-0

Molecular Drug Properties

Measurement and Prediction

Edited by
Raimund Mannhold

WILEY-VCH Verlag GmbH & Co. KGaA

Series Editors

Prof. Dr. Raimund Mannhold
Molecular Drug Research Group
Heinrich-Heine-Universität
Universitätsstrasse 1
40225 Düsseldorf
Germany
mannhold@uni-duesseldorf.de

Prof. Dr. Hugo Kubinyi
Donnersbergstrasse 9
67256 Weisenheim am Sand
Germany
kubinyi@t-online.de

Prof. Dr. Gerd Folkers
Collegium Helveticum
STW/ETH Zurich
8092 Zurich
Switzerland
folkers@collegium.ethz.ch

Volume Editor

Prof. Dr. Raimund Mannhold
Molecular Drug Research Group
Heinrich-Heine-Universität
Universitätsstrasse 1
40225 Düsseldorf
Germany
mannhold@uni-duesseldorf.de

Cover Illustration

Molecular lipophilicity potentials for an extended, more lipophilic and a folded, less lipophilic conformer of verapamil are shown ($\Delta \log P_{MLP} = 0.6$). Violet regions: higher lipophilicity; blue regions: medium lipophilicity; yellow regions: weakly polar; red regions: strongly polar (Preparation of this graph by Pierre-Alain Carrupt is gratefully acknowledged.)

Library of Congress Card No.:
applied for

British Library Cataloguing-in-Publication Data
A catalogue record for this book is available from the British Library.

Bibliographic information published by the Deutsche Nationalbibliothek
Die Deutsche Nationalbibliothek lists this publication in the Deutsche Nationalbibliografie; detailed bibliographic data are available in the Internet at <http://dnb.d-nb.de>.

Composition SNP Best-set Typesetter Ltd., Hong Kong

Printing Betz-Druck GmbH, Darmstadt

Bookbinding Litges & Dopf GmbH, Heppenheim

Cover Design Grafik-Design Schulz, Fußgönheim

Printed in the Federal Republic of Germany
Printed on acid-free paper

ISBN 978-3-527-31755-4

Dedicated with love
to my wife Barbara
and my daughter Marion

Contents

Molecular Drug Properties. Measurement and Prediction. R. Mannhold (Ed.)

ISBN: 978-3-527-31755-4

List of Contributors

Alex Avdeef
*p*ION INC
5 Constitution Way
Woburn, MA 01801
USA

Jonas Boström
AstraZeneca R&D
Department of Lead Generation
43183 Mölndal
Sweden

Giulia Caron
CASMedChem laboratory
Dipartimento di Scienza
Tecnologia del Farmaco
Università di Torino
Via P. Giuria 9
10125 Torino
Italy

Pierre-Alain Carrupt
Unit of Pharmacochemistry
School of Pharmaceutical Sciences
University of Geneva,
University of Lausanne
Quai Ernest-Ansermet 30
1211 Geneva 4
Switzerland

Giuseppe Ermondi
CASMedChem laboratory
Dipartimento di Scienza
Tecnologia del Farmaco
Università di Torino
Via P. Giuria 9
10125 Torino
Italy

Peter Ertl
Novartis Institutes
for Biouedical Research
4002 Basel
Switzerland

Bernard Faller
Novartis Pharma AG
Lichtstrasse 35
4056 Basel
Switzerland

Andreas Frank
Institute for Organic Chemistry and Biochemistry
Technical University Munich
Lichtenbergstrasse 4
85747 Garching
Germany

Molecular Drug Properties. Measurement and Prediction. R. Mannhold (Ed.)

ISBN: 978-3-527-31755-4

Alexandra Galland
Unit of Pharmacochemistry
School of Pharmaceutical Sciences
University of Geneva, University of Lausanne
Quai Ernest-Ansermet 30
1211 Geneva 4
Switzerland

J Andrew Grant
AstraZeneca Pharmaceuticals Mereside
DECS Global Compound Sciences
Alderley Park,
Cheshire SK10 4TG
UK

Davy Guillarme
Laboratory of Analytical
Pharmaceutical Chemistry
School of Pharmaceutical Sciences
University of Geneva,
University of Lausanne
Boulevard d'Ivoy 20
1211 Geneva 4
Switzerland

Yveline Henchoz
Unit of Pharmacochemistry
School of Pharmaceutical Sciences
University of Geneva,
University of Lausanne
Quai Ernest-Ansermet 30
1211 Geneva 4
Switzerland

Ovidiu Ivanciuc
Sealy Center for Structural Biology
and Molecular Biophysics
Departments of Biochemistry
and Molecular Biology
University of Texas Medical Branch
301 University Boulevard
Galveston, TX 77555-0857
USA

Horst Kessler
Institute for Organic Chemistry and Biochemistry
Technical University Munich
Lichtenbergstrasse 4
85747 Garching
Germany

Andreas Klamt
COSMO*logic* GmbH & Co. KG
Burscheider Str. 515
51381 Leverkusen
Germany

Institute of Physical and Theoretical Chemistry
University of Regensburg
93040 Regensburg
Germany

Christopher A. Lipinski
Scientific Advisor
Melior Discovery
10 Connshire Drive
Waterford, CT 06385-4122
USA

Franco Lombardo
Novartis Institute for
Biomedical Research
250 Massachusetts Avenue
Cambridge, MA 02139
USA

Burkhard Luy
Institute for Organic Chemistry and Biochemistry
Technical University Munich
Lichtenbergstraße 4
85747 Garching
Germany

Raimund Mannhold
Molecular Drug Research Group
Heinrich-Heine-Universität
Universitätsstraße 1
40225 Düsselsorf
Germany

Sophie Martel
Unit of Pharmacochemistry
School of Pharmaceutical Sciences
University of Geneva,
University of Lausanne
Quai Ernest-Ansermet 30
1211 Geneva 4
Switzerland

Sorel Muresan
AstraZeneca R&D
Computational Chemistry
431 83 Mölndal
Sweden

Claude Ostermann
Nycomed GmbH
Byk-Gulden-Str. 2
78467 Konstanz
Germany

Alessandro Pedretti
Istituto di Chimica Farmaceutica
Facoltà di Farmacia
Università di Milano
Via Mangiagalli 25
20131 Milano
Italy

Gennadiy I. Poda
Pfizer Global R & D
700 Chesterfield Parkway West
Mail Zone BB2C
Chesterfield, MO 63017
USA

Oleg Raevsky
Department of Computer-Aided
Molecular Design
Institute of Physiologically
Active Compounds
Russian Academy of Sciences
Severnii proezd, 1
142432, Chernogolovka,
Moscow region
Russia

Serge Rudaz
Laboratory of Analytical
Pharmaceutical Chemistry
School of Pharmaceutical Sciences
University of Geneva,
University of Lausanne
Boulevard d'Ivoy 20
1211 Geneva 4
Switzerland

Jens Sadowski
AstraZeneca
Lead Generation KJ257
43183 Mölndal
Sweden

Marina Shalaeva
Pfizer Global Research
and Development
Groton Laboratories
Groton, CT 06340
USA

Brian J Smith
The Walter and Eliza Hall
Institute of Medical Research
Department of Structural Biology
1G Royal Parade, Parkville,
Victoria 3050
Australia

Bernard Testa
Pharmacy Department
University Hospital Centre
CHUV-BH 04
46 Rue du Bugnon
1011 Lausanne
Switzerland

Igor Tetko
GSF – National Research Centre
for Environment and Health
Institute for Bioinformatics
(MIPS)
Ingolstädter Landstraße 1
85764 Neuherberg
Germany

Suzanne Tilton
Novartis Institute
for Biomedical Research
250 Massachusetts Avenue
Cambridge, MA 02139
USA

Jean-Luc Veuthey
Laboratory of Analytical
Pharmaceutical Chemistry
School of Pharmaceutical Sciences
University of Geneva,
University of Lausanne
Boulevard d'Ivoy 20
1211 Geneva 4
Switzerland

Giulio Vistoli
Istituto di Chimica Farmaceutica
Facoltà di Farmacia
Università di Milano
Via Mangiagalli 25
20131 Milano
Italy

Han van de Waterbeemd
AstraZeneca
DECS – Gobal Compound Sciences
Mereside 50S39
Macclesfield
Cheshire SK10 4TG
UK

Preface

Despite enormous investments in pharmaceutical research and development, the number of approved drugs has declined in recent years. The attrition of compounds under development is dramatically high. Safety, insufficient efficacy and, to some extent, absorption, distribution, metabolism, excretion and toxicity (ADMET) problems are the responsible factors. Formerly, drugs were discovered by testing compounds synthesized in time-consuming multistep processes against a battery of *in vivo* biological screens. Promising compounds were then further tested in development, where their pharmacokinetic (PK) properties, metabolism and potential toxicity were investigated. Adverse findings were often made at this stage and projects were re-started to find another clinical candidate. Drug discovery has undergone a dramatic change over the last two decades due to a methodological revolution including combinatorial chemistry, high-throughput screening and *in silico* methods, which greatly increased the speed of the process of drug finding and development.

More recently, the bottleneck of drug research has shifted from hit-and-lead discovery to lead optimization, and more specifically to PK lead optimization. Some major reasons are (i) the imperative to reduce as much as feasible the extremely costly rate of attrition prevailing in preclinical and clinical phases, and (ii) more stringent concerns for safety. The testing of ADME properties is now done much earlier, i.e. before a decision is taken to evaluate a compound in the clinic.

As the capacity for biological screening and chemical synthesis has dramatically increased, so have the demands for large quantities of early information on ADME data. The physicochemical properties of a drug have an important impact on its PK and metabolic fate in the body, and so a good understanding of these properties, coupled with their measurement and prediction, are crucial for a successful drug discovery programme.

The present volume is dedicated to the measurement and the prediction of key physicochemical drug properties with relevance for their biological behavior including ionization and H-bonding, solubility, lipophilicity as well as three-dimensional structure and conformation. Potentials and limitations of the relevant techniques for measuring and calculating physicochemical properties of drugs are critically discussed and comprehensively exemplified in 17 chapters from 35 distinguished authors, from both academia and the pharmaceutical industry.

Molecular Drug Properties. Measurement and Prediction. R. Mannhold (Ed.)

ISBN: 978-3-527-31755-4

We are indebted to all authors for their well-elaborated chapters, and we want to express our gratitude to Dr Andreas Sendtko and Dr Frank Weinreich from Wiley-VCH for their valuable contributions to this volume and the ongoing support of our series *Methods and Principles in Medicinal Chemistry*.

Raimund Mannhold, *Düsseldorf* August 2007
Hugo Kubinyi, *Weisenheim am Sand*
Gerd Folkers, *Zürich*

A Personal Foreword

Several editors of previous volumes in this series lised the platform of the Personal Foreword to reflect routes and contents of their scientific lives and in particular to appreciate the invaluable support by rewarded colleagues. It is a pleasure for me to continue this tradition.

After the study of pharmaceutical sciences in Frankfurt/Main I joined the Department of Clinical Physiology at the Heinrich-Heine-Universität Düsseldorf to start my PhD work dedicated to pharmacological studies of the calcium channel blocker verapamil under the supervision of Raimund Kaufmann. He was a very liberal scientific teacher and he allowed me to fine-tune the contents of my PhD work according to my personal preferences.

Frequent contacts with the manufacturer of verapamil, the Knoll company in Ludwigshafen, enabled an intense communication with Hugo Kubinyi, working at that time as a medicinal chemist for Knoll. As a consequence of frequent fruitful discussions with Hugo I included quantitative structure–activity relationship (QSAR) studies on verapamil congeners in my PhD work and continued working in the QSAR field till the present.

Two Dutch colleagues and friends have strongly influenced me since the early 1980s. I first met Roelof Rekker, one of the fathers of log *P* calculation approaches, on the occasion of one of the famous Noordwijkerhout meetings. Roelof fascinated me with his elegant lipophilicity studies. After years of fruitful cooperation I had the privilege to coauthor with him our booklet "Calculation of Drug Lipophilicity" updating the Σf system, the first fragmental approach for lipophilicity calculation.

My first personal contact to Henk Timmerman happened on the wonderful island of Capri during a symposium on pharmaceutical sciences. Henk Timmerman headed one of the largest and most important departments of Medicinal Chemistry in European academia. It was very impressive to face his views on our research field, and his integrated and straightforward way to guide research projects. For several years I collaborated with his group and, as an added bonus, became a great fan of Amsterdam.

In the early 1990s, I founded the book series *Methods and Principles in Medicinal Chemistry* with Verlag Chemie; Henk Timmerman and Povl Krogsgaard Larsen joined me on the initial board of series editors. Hugo Kubinyi followed Povl Krogsgaard Larsen after the first three volumes were released. Henk contributed

Molecular Drug Properties. Measurement and Prediction. R. Mannhold (Ed.)

ISBN: 978-3-527-31755-4

to the series very intensely and successfully for many years, and I want to thank him for the times of coediting this book series. When retiring from the chair of Medicinal Chemistry at the Vrije Universiteit of Amsterdam, he forwarded his work in the series to Gerd Folkers from ETH, Zurich.

In the late 1990s another fruitful and pleasant cooperation arose in Perugia, Italy, with the chemometric group of Sergio Clementi and Gabriele Cruciani, two guys with excellent skills and scientific enthusiasm. Since 1997 I have spent weeks up to months each year in Perugia for joint projects on three-dimensional (3D) QSAR and virtual screening studies. Fortunately, these stays also enable a further specialization in Italian food and wine.

The present volume is dedicated to the measurement and the prediction of key physicochemical drug properties with relevance for their biological behavior, including ionization and H-bonding, solubility, lipophilicity as well as 3D structure and conformation.

In the *Introductory section*, Bernard Testa, Giulio Vistoli and Alessandro Pedretti give us "A Fresh Look at Molecular Structure and Properties", which are key concepts in drug design, but may not mean the same to all medicinal chemists. This chapter serves as a general opening, and invites readers to stand back and reflect on the information contained in chemical compounds and on its description. The authors base their approach on a discrimination between the "core features" and the physicochemical properties of a compound.

Han van de Waterbemd focuses on "Physicochemical Properties in Drug Profiling". These properties play a key role in drug metabolism and pharmacokinetics (DMPK). Their measurement and prediction is relatively easy compared to DMPK and safety properties, where biological factors come into play. However, the latter depend to some extent on physicochemical properties as they dictate the degree of access to biological systems. The change in work practice towards high-throughput screening (HTS) in biology using combinatorial libraries has also increased the demands on more physicochemical and absorption, distribution, metabolism and excretion (ADME) data. Han's chapter reviews the key physicochemical properties, both how they can be measured as well as how they can be calculated in some cases.

Alex Avdeef opens the section on *Electronic Properties* considering "Drug Ionization and Physicochemical Profiling". The ionization constant tells the pharmaceutical scientist to what degree the molecule is charged in solution at a particular pH. This is important to know, since the charge state of the molecule strongly influences its other physicochemical properties. After an in-depth discussion of the accurate determination of ionization constants, Alex focuses on three physicochemical properties where the ionization constant relates to a critical distribution or transport function: (i) octanol–water and liposome–water partitioning, (ii) solubility, and (iii) permeability.

Ovidiu Ivanciuc describes the computation of "Electrotopological State (E-state) Indices" from the molecular graph and their application in drug design. The E-state encodes at the atomic level information regarding electronic state and topological accessibility. Computing of E-state indices is based exclusively on the molecular topology and it can be done efficiently for large chemical libraries. Comparative QSAR models from a large variety of descriptors show that the E-state indices are often selected in the best QSAR models.

"Polar Surface Area" (PSA) is the topic of Peter Ertl's chapter. PSA has been shown to provide very good correlations with intestinal absorption, blood–brain barrier penetration and several other drug characteristics. It has also been effectively used to characterize drug-likeness during virtual screening and combinatorial library design. The descriptor seems to encode an optimal combination of H-bonding features, molecular polarity and solubility properties. PSA can be easily and rapidly calculated as a sum of fragment contributions using only the molecular connectivity of a structure.

Lastly, Oleg Raevsky discusses "H-bonding Parameterization in QSAR and Drug Design". Studies based on direct thermodynamic parameters of H-bonding and exact 3D structures of H-bonding complexes have essentially improved our understanding of solvation and specific intermolecular interactions. These studies consider the structure of liquid water, new X-ray data for specific H-bonding complexes, partitioning in water–solvent–air systems, a refinement in the PSA approach, improvement of GRID potentials, and calculation schemes of optimum H-bonding potential values for any concrete H-bonding atoms. Oleg exemplifies the successful application of direct H-bonding descriptors in QSAR and drug design.

Conformational Aspects are covered in the next section. First, Jens Sadowski discusses automatic "Three-dimensional Structure Generation" as a fundamental operation in computational chemistry. It has become a standard procedure in molecular modeling and appropriate software has been available for many years. Several of the most common concepts as well as their strengths and limitations are shown in detail. An evaluation study of the two most commonly used programs, CONCORD and CORINA, indicates their general applicability for robust, fast and automatic 3D structure generation. Within the limitation of single conformation generation, reasonable rates of reproducing experimental geometries and other quality criteria are reached. For many applications, the obtained 3D structures are good enough to be used without any further optimization.

Then, Jonas Boström and Andrew Grant review "Exploiting Ligand Conformations in Drug Design". Section 1 gives a theoretical outline of the problems and presents details of various implementations of computer codes to perform conformational analysis. Section 2 describes calculations illustrative of the current accuracy in generating the conformation of a ligand when bound to proteins (the bioactive conformer) by comparisons to crystallographically observed data. The final section concludes by presenting some practical applications of using knowledge of molecular conformation in actual drug discovery projects.

Finally, Burkhard Luy, Andreas Frank and Horst Kessler discuss "Conformational Analysis of Drugs by Nuclear Magnetic Resonance Spectroscopy". The determination and refinement of molecular conformations comprehends three main methods: distance geometry (DG), molecular dynamics (MD) and simulated annealing (SA). In principle, it is possible to exclusively make use of DG, MD or

SA, but normally it is strongly suggested to combine these methods in order to obtain robust and reliable structural models. Only when the results of different methods match should a 3D structure be presented. There are various ways of combining the described techniques and the procedural methods may differ depending on what kind of molecules are investigated. In this chapter, the authors give instructions on how to obtain reliable structural models.

Solubility is a fundamental characteristic of drug candidates. In synthetic chemistry, low solubility can be problematic for homogeneous reactions, and in preclinical experimental studies, low solubility may produce experimental errors or precipitation.

First, Chris Lipinski debates "Experimental Approaches to Aqueous and Dimethylsulfoxide Solubility". The emphasis is on the discovery stage as opposed to the development stage. The reader will find numerous generalizations and rules-of-thumb relating to solubility in a drug discovery setting. The solubility of drugs in water is important for oral drug absorption. Drug solubility in dimethylsulfoxide (DMSO) is important in the biological testing of a compound formatted as a DMSO stock solution. Solubility in aqueous media and DMSO is discussed in the context of both similarities and differences.

Then, Andreas Klamt and Brian Smith discuss the "Challenge of Drug Solubility Prediction". While standard models have emerged for log *P*, no such convergence can be observed for log *S*, probably due to its inherent nonlinear character. Thus, nonlinear models are required, but it is questionable whether neural network techniques will ever yield reliable models, because the number of good quality data required will be of the order of hundreds of thousands. In the authors' view, the best way is to make use of the fundamental laws of physical chemistry and thermodynamics as much as possible. Using the supercooled state of the drug as intermediate state, and splitting log *S* into one smaller contribution arising from the free energy of fusion and a large contribution from the solubility of the supercooled drug, appear to be the only sensible way for reasonable calculation.

A quite comprehensive section concerns *Lipophilicity*, one of the most informative physicochemical properties in medicinal chemistry and since long successfully used in QSAR studies.

"Chemical Nature and Biological Relevance of Lipophilicity" are the topics of the starting chapter by Giulia Caron and Giuseppe Ermondi. Sections on chemical concepts to understand the significance of lipophilicity, lipophilicity systems, the determination of log *P* and a general factorization of lipophilicity are dedicated to reflect the chemical nature of lipophilicity. In the second part, the biological relevance of lipophilicity is exemplified for membrane permeation, receptor affinity and the control of undesired human ether-a-go-go-related gene activity.

Pierre-Alain Carrupt and colleagues review "Chromatographic Approaches for Measuring Log *P*". They present a brief overview of the main features of reversed-phase liquid chromatography (isocratic condition and gradient elution) and capillary electrophoresis (microemulsion electrokinetic chromatography, microemulsion electrokinetic chromatography and liposome/vesicular electrokinetic chromatography) methods used for lipophilicity determination of neutral compounds or the neutral form of ionizable compounds. Relationships between lipophilicity and retention parameters obtained by reversed-phase liquid chromatography methods using isocratic or gradient condition are reviewed. Advantages and limitations of the two approaches are also pointed out and general guidelines to determine partition coefficients in 1-octanol–water are proposed. Finally, recent data on lipophilicity determination by capillary electrophoresis of neutral compounds and neutral form of ionizable compounds are reviewed.

Raimund Mannhold and Claude Ostermann describe the "Prediction of Log *P* with Substructure-based Methods". Substructure-based methods are either fragmental (use fragments and apply correction factors) or atom based (use atom types and do not apply correction rules). Significant electronic interactions are comprised within one fragment; this is a prime advantage of using fragments. On the other hand, fragmentation can be arbitrary and missing fragments may prevent calculation. An advantage of atom-based methods is that ambiguities are avoided; a shortcoming is the failure to deal with long-range interactions. The predictive power of six substructure-based methods is compared via a benchmarking set of 284 drugs.

Igor Tetko and Gennadyi Poda focus on the "Prediction of Log *P* with Property-based Methods", which are either based on 3D structure representation including empirical approaches, quantum chemical semiempirical calculations, continuum solvation models, molecular dynamics calculations, molecular lipophilicity potential calculations, and lattice energy calculations, or on topological descriptors using graph molecular connectivity or E-state descriptors. Tetko and Poda used the same dataset of 284 drugs, and showed best predictivity for A_S+logP and ALOGPS methods, based on topological descriptors.

Finally, Franco Lombardo and colleagues consider "The Good, the Bad and the Ugly of Distribution Coefficients". The question of "how" and "what" log *D* values we use in our daily work is an important one. Sections on log *D* versus log *P*, issues and automation in the determination of log *D*, pH-partition theory and ion-pairing, and on computational approaches for log *D* are dedicated to answer this question in detail. Computational approaches for log *D* might tempt medicinal chemists to use routinely a computed value as a surrogate of measured values. However, "good" practice should be to determine at least a few values for representative compounds and continue monitoring the performance of computation with additional determinations alongside the medicinal chemistry work.

Physicochemical properties guide *Drug- and lead-likeness* in a dedicated manner. In the concluding chapter, Sorel Muresan and Jens Sadowski discuss simple calculated compound properties and related aspects in this context. The presence or absence of specific chemical features as well as their correlation with each other and with biological potency are of high importance for success in selecting starting points for lead generation and in guiding chemical optimization. A number of important concepts such as property ranges, chemical substructure filters, ligand efficiency and drug-likeness as a classification problem are discussed, and some of them are finally demonstrated in an example of how to select compounds for acquisition.

It was an outstanding experience to plan, organize and realize this book, and to work with such a distinguished group of contributors. I hope that the readers will enjoy the work they did. I won new friends during this book project, one of which is Pierre-Alain Carrupt. He prepared the cover graphics, which represents the molecular lipophilicity potentials for my "PhD molecule" verapamil in its extended and folded conformation.

This is already the 37th volume in our series on *Methods and Principles in Medicinal Chemistry* which started in 1993 with a volume on *QSAR: Hansch Analysis and Related Approaches*, written by Hugo Kubinyi. An average release of roughly three volumes per year indicates the increasing appreciation of the series in the MedChem world. I want to express my sincere thanks to my editor friends Hugo Kubinyi and Gerd Folkers for their continuous and precious contributions to the steady development of our series.

Finally I want to acknowledge the pleasant collaboration with Dr Andreas Sendtko and Dr Frank Weinreich from Wiley-VCH during all steps of editing this volume.

Raimund Mannhold, *Düsseldorf*
August 2007

Part I
Introduction

1
A Fresh Look at Molecular Structure and Properties

Bernard Testa, Giulio Vistoli, and Alessandro Pedretti

Abbreviations

α_1-AR	α_1-adrenoceptors
ADME	absorption, distribution, metabolism and excretion
MC	Monte Carlo
MD	molecular dynamics
MEP	molecular electrostatic potential
MIF	molecular interaction field
PCA	principal component analysis
PSA	polar surface area
QSAR	quantitative structure–activity relationship
SAR	structure–activity relationship
SAS	solvent accessible surface

1.1
Introduction

Molecular structure and properties are key concepts in drug design, but they may not mean the same to all medicinal chemists, not to mention other researchers involved in drug discovery and development such as biochemists, pharmacologists and toxicologists (see Chapter 2). It is therefore the merit of this book to offer a rationalization of these concepts with a view to advocating their value and clarifying their use.

One of the sources of the fuzziness surrounding these concepts may well be the implicit assumption in structure–activity relationship (SAR) studies that molecular structure contains (i.e. encodes) the information on the biological activity of a given compound. Such an assumption cannot be incorrect, since this would imply the fallacy of SAR studies. However, the assumption becomes misleading if not properly qualified to the effect that the molecular structure of a given compound contains only part of the information on its bioactivity. Indeed, what the structure of a compound encodes is information about the molecular features accounting

Molecular Drug Properties. Measurement and Prediction. R. Mannhold (Ed.)

ISBN: 978-3-527-31755-4

for its recognition by a biological system. Such a recognition obviously occurs at the molecular level – the biological components which "recognize" the compound being bio(macro)molecular entities or complexes such as membranes, transporters, enzymes, receptors or polynucleosides. The mutual recognition and interaction of bioactive compound and biochemical entity translates into the formation of a functional complex which triggers the cascade of biochemical events that leads to the observed biological response [1–3].

As far as SARs are concerned, the outcome of processes such as "recognition" and "functional response" need to be formalized for incorporation into mathematical models or simulations. The same is true for "molecular structure", which remains an abstract concept until expressed formally and in quantitative terms. This is what medicinal chemists and their biological colleagues have achieved, as formalized in Table 1.1. Indeed, SAR studies, in general, and quantitative SAR (QSAR) studies, in particular, can be subdivided into four components [4]. First, we find the biological systems themselves, be they functional proteins, molecular machines, membranes, organelles, cells, tissues, organs, organisms, populations or even ecosystems. Second, there are the molecular compounds that interact with these biological systems, be they hits, lead candidates, drug candidates, drugs, agrochemicals, toxins, pollutants and more generally any type of bioactive compounds; in (Q)SAR studies, these compounds are described by their molecular features (i.e. their structure and properties). The third component in (Q)SAR studies are the responses produced by a biological system when interacting with bioactive compounds; here again, a description in the form of pharmacokinetic, pharmacological or toxicological descriptors is necessary. As for the last component, we find mathematical models or simulations which describe how the biological response varies with variations in the molecular structure of bioactive

Tab. 1.1 The four components of SAR and QSAR studies (modified from Ref. [4]).

Component	*Definition*	*Description in SARs*
(A) Biological systems	any biological entity, from a functional protein to an ecosystem	virtual (*in silico*) 3D models; mathematical models
(B) Bioactive compounds	e.g. hits, lead candidates, drug candidates, drugs, toxins, agrochemicals, pollutants	molecular features (i.e. their structure and properties)
(C) Biological responses	the response of A when exposed to B	pharmacological or toxicological descriptors
(D) Mathematical models or simulations	virtual or mathematical models of how variations in C change with variations in the molecular structure of B	variations in C = variations in the values of the descriptors; variations in B = variations in the molecular features of the bioactive compounds

compounds. As is well known to medicinal chemists, the usual statement ". . . how the biological response varies with the structure of bioactive compounds" is a simplifying shortcut.

This book focuses on molecular features and properties, their meaning, measurement, computation, and encoding into parameters and descriptors. The present chapter serves as a general opening, and invites readers to stand back and reflect on the information contained in chemical compounds and on our description of it. We base our approach on a discrimination between the "core features" of a molecule/compound and the physicochemical properties of a compound.

1.2 Core Features: The Molecular "Genotype"

1.2.1 The Argument

In our view, the core features of a molecule are the constant (unchangeable) ones, i.e. those features whose change necessarily implies a transformation into another molecule. This view is somewhat analogous with the genome, since unless they are clones different multicellular organisms necessarily have different genotypes. For this reason we use the term molecular "genotype" to describe the ensemble of the molecular core features.

As shown in Fig. 1.1, the constant features of a molecule/compound are the number and nature of its atoms (its composition), the connectivity of its atoms

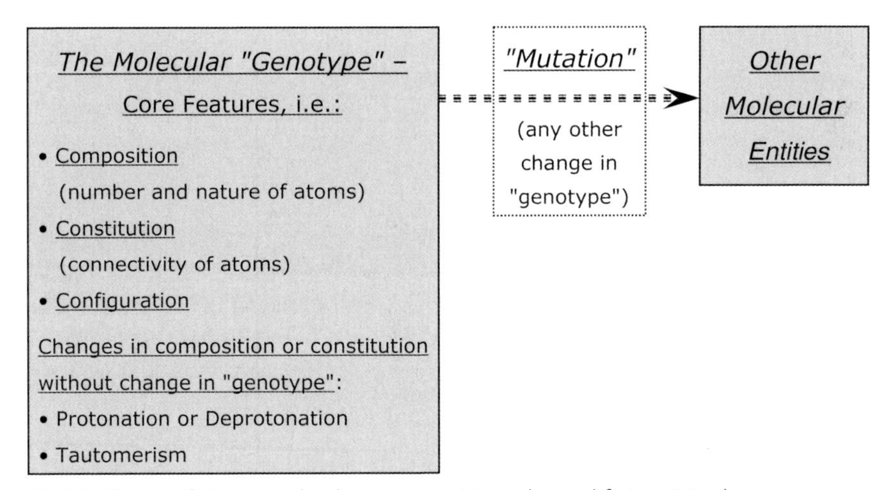

Fig. 1.1 The core features (molecular "genotype") of a molecule/compound are presented here. Attention is drawn to the fact that changes in composition, constitution (connectivity) and configuration (stereochemical features) implies a "mutation" to another molecule/compound. The exceptions are ionization and tautomerism, which are not defined as implying a "mutation" of the "genotype".

(its constitution), and its absolute configuration. Indeed, any change (i.e. "mutation") in composition, constitution or configuration yields another molecule/compound, i.e. a derivative/analog, a constitutional isomer or a stereoisomer.

Note, however, that the above scheme needs further qualification. First and strictly speaking, protonation and deprotonation involve a change in composition and connectivity, but they are reversible processes whose equilibrium is a condition-dependent property. Nevertheless, the low energy barrier and reversibility of the process lead us to view a base and its conjugated acid as two states of the same molecular "genotype". As for tautomerism, it involves a low-energy change in connectivity, again with a condition-dependent equilibrium. Again, two tautomers can be considered as two distinct states of the same compound. A further and more general proviso is the fact that our entire argument is limited to covalent bonds, with the consequence that an ion and its counterion are considered as two separate molecular entities.

1.2.2 Encoding the Molecular "Genotype"

Can various components of the core features be encoded in a form suitable for SAR investigations? Interestingly, the answer is clearly a positive one.

- Composition is partly encoded in molecular weight – a parameter sometimes used.
- Topological indices are used to describe some components of connectivity. A more complete description is afforded by unidimensional codes (linear line notations) such as SMILES. Connectivity plus explicit attention to valence electrons is afforded by the electrotopological indices (see Chapter 4).
- Configuration is described by the *R*- and *S*-descriptors for enantiomerism, and the *E*- and *Z*-descriptors for geometric π-diastereomerism [5].

1.3 Observable and Computable Properties: The Molecular "Phenotype"

1.3.1 Overview

The phenotype of an organism is its huge repertoire of observable properties. This phenotype is the visible expression of the organism' genotype, but is also controlled by the organism's life history and environment. That is to say that a given genotype can translate into a large variety of potential phenotypes – a "phenotype space" [6].

In close analogy with this biological definition, we will designate as molecular "phenotype" the ensemble of observable and computable properties of a chemical entity. These indeed are the observable expression of the core features of the compound and like a biological phenotype they are influenced by the environment, here the molecular environment. There is a major difference, however, since compounds have no life history, but as we shall see in the last part of this chapter, compounds have a "property space" just like organisms have a phenotype space.

Energy interaction between a probe and a compound is necessary for molecular properties to be observed. As a result, properties can be categorized according to the nature of the probe used to observe them. Properties revealed by low-energy interactions are schematized in Fig. 1.2, which outlines that:

- Spectral properties arise through interactions with electromagnetic radiation.
- Some pharmacologically important properties such as pK_a, tautomeric equilibrium, conformational behavior, solubility and partitioning are temperature and solvent dependent.
- Interactions between a vast number of identical molecules give rise to such solid- or liquid-state properties as melting point and boiling point.
- Interaction with (recognition by) biomolecules triggers the cascade that leads to a biological response (see above).

The approach we follow below in surveying molecular properties is a different one based on their interdependence and the progressive emergence of biologically relevant properties (Fig. 1.3).

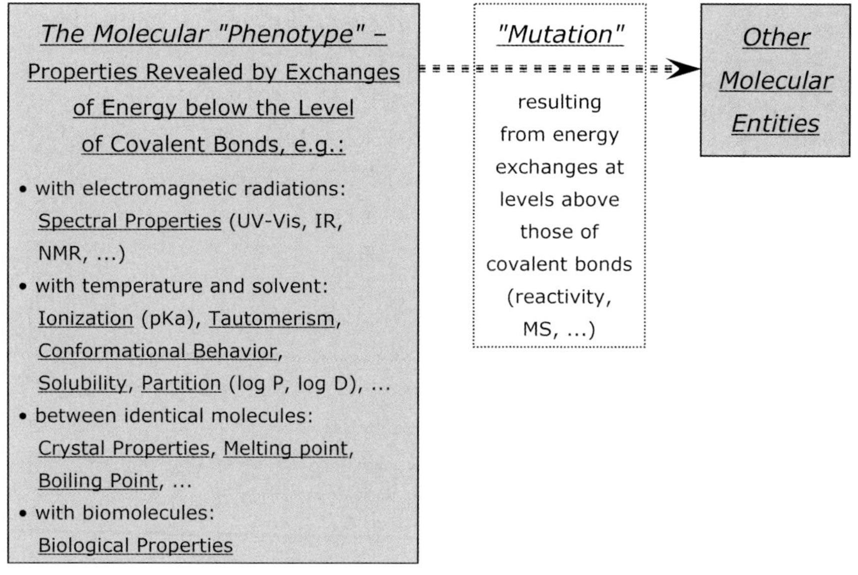

Fig. 1.2 Properties revealed by low-energy exchanges belong to the molecular "phenotype", as exemplified here. This is contrasted with some other chemical properties (e.g. reactivity) which involve the cleavage and/or formation of covalent bonds, and thus imply a "mutation" of the "genotype". UV, ultraviolet; IR, infrared; NMR, nuclear magnetic resonance; MS, mass spectroscopy.

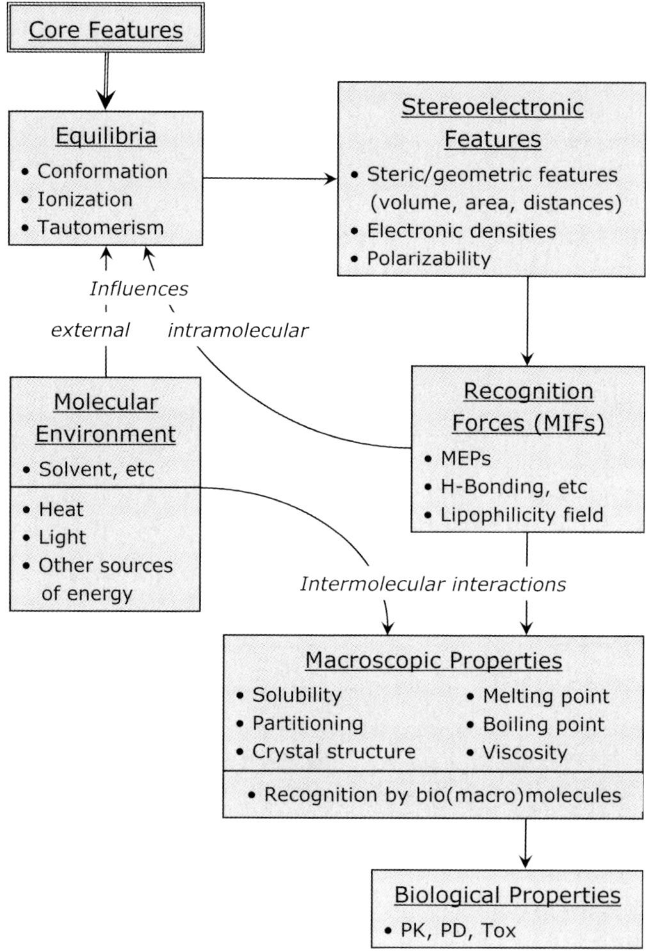

Fig. 1.3 A survey of molecular properties based on their interdependence and the progressive emergence of biologically relevant properties. See text for further details. MIFs, molecular interaction fields; MEPs, molecular electrostatic potentials; PK, pharmacokinetic(s); PD, pharmacodynamic(s).

1.3.2
Equilibria

A two-dimensional (2D) molecule is a simplified abstraction because molecules have a three-dimensional (3D) form and shape. Furthermore, form and shape fluctuate, making them four-dimensional (4D) objects. Some molecular entities may be extremely flexible, others rather rigid, but a totally rigid molecule exists only at 0 K.

A major fluctuation is the conformational behavior of molecular entities, as discussed explicitly in Chapter 9, but also in Chapters 7 and 8. Other equilibria, already mentioned above, are ionization and tautomerism. The former is the most important as far as drug research is concerned and it is discussed extensively in Chapter 3.

1.3.3 Stereoelectronic Features

The form and shape of a molecule (i.e. its steric and geometric features) derive directly from the molecular "genotype", but they cannot be observed without a probe. Furthermore, they vary with the conformational, ionization and tautomeric state of the compound. Thus, the computed molecular volume can vary by around 10% as a function of conformation. The same is true of the molecular surface area, whereas the key (i.e. pharmacophoric) intramolecular distances can vary much more.

A similar argument can be made for electronic features such as electron density, polarization and polarizability. These are critically dependent on the ionization state of the molecule, but the conformational state is also highly influential. One highly approximate yet useful reflection of electron density is afforded by the polar surface area (PSA), a measure of the extent of polar (hydrophilic) regions on a molecular surface (see Chapter 5).

1.3.4 Recognition Forces and Molecular Interaction Fields (MIFs)

The stereoelectronic features produce actions at a distance by the agency of the recognition forces they create. These forces are the hydrophobic effect, and the capacity to enter ionic bonds, van der Waals interactions and H-bonding interactions. The most convenient and informative assessment of such recognition forces is afforded by computation in the form of MIFs, e.g. lipophilicity fields, hydrophobicity fields, molecular electrostatic potentials (MEPs) and H-bonding fields (see Chapter 6) [7–10].

Like the stereoelectronic features that generate them, the MIFs are highly sensitive to the conformational and ionization state of the molecule. However, they in turn have a marked intramolecular influence on the conformational and ionization equilibria of the compound. It is the agency of the MIFs that closes the circle of influences from molecular states to stereoelectronic features to MIFs (Fig. 1.3).

1.3.5 Macroscopic Properties

As shown in Fig. 1.3, MIFs account not only for intramolecular effects, but also for intermolecular interactions, allowing macroscopic properties to emerge. The interactions of a chemical with a solvent reveal such pharmacologically essential properties as solubility (Chapters 10 and 11) and partitioning/lipophilicity (Chapters 12–16). The interactions between a large number of identical molecules translate into solid-state properties (including melting point and solubility) or liquid-state properties such as viscosity and boiling point. Note that these macroscopic properties are also influenced by energy influx, both directly and indirectly (via equilibria).

As the same types of intermolecular forces are involved, there is no qualitative difference between solute–solvent interactions and the recognition of a compound by a bio(macro)molecular compound.

Having explained the origin of the adaptable (condition-dependent) character of molecular properties, we now turn to illustrations of this phenomenon. Indeed, stating the variable nature of molecular properties is not sufficient to appreciate its significance in drug design and SAR studies.

1.4 Molecular Properties and their Adaptability: The Property Space of Molecular Entities

1.4.1 Overview

The concept of property space, which was coined to quantitatively describe the phenomena in social sciences [11, 12], has found many applications in computational chemistry to characterize chemical space, i.e. the range in structure and properties covered by a large collection of different compounds [13]. The usual methods to approach a quantitative description of chemical space is first to calculate a number of molecular descriptors for each compound and then to use multivariate analyses such as principal component analysis (PCA) to build a multidimensional hyperspace where each compound is characterized by a single set of coordinates.

Whereas this approach has proven very successful in comparing chemical libraries and designing combichem series, it nevertheless is based on the assumption that the molecular properties being computed are discrete, invariant ones [14]. This assumption derives from the restrictions imposed by the handling of huge databases, but like many assumptions it tends to fade in the background and be taken as fact. Yet as chemistry progresses, so does our understanding of molecular structure taken in its broadest sense, i.e. the mutual interdependence between geometric features and physicochemical properties.

The growing computational power available to researchers proves an invaluable tool to investigate the dynamic profile of molecules. Molecular dynamics (MD) and Monte Carlo (MC) simulations have thus become pivotal techniques to explore the dynamic dimension of physicochemical properties [1]. Furthermore, the powerful computational methods based in particular on MIFs [7–10] allow some physicochemical properties to be computed for each conformer (e.g. virtual log P), suggesting that to the conformational space there must correspond a property space covering the ensemble of all possible conformer-dependent property values.

The concept of property space is progressively being used to gain a deeper understanding of the dynamic behavior of a single compound in different media (as we illustrate below with acetylcholine, see Section 1.4.2) or bound to biological targets (the carnosine–carnosinase complex, see Section 1.4.3), but it can be used also with a set of compounds to derive fertile descriptors for dynamic QSAR analyses (4D QSAR, see Section 1.4.4).

In this dynamic vision, a molecular property can be described by either (i) an average value or (ii) descriptors defining its property space. The average value of a property, and especially a weighted average, contains more information than a conformer-specific value (even if it is that of the lowest-energy or hypothetical bioactive conformer). However, this average value does not yield information on the property space itself. To this end, one should use descriptors specifying the property range and distribution in relation to conformational changes and other property profiles.

A property space can be defined using two classes of descriptors. The first class includes descriptors quantifying the variability (spread) of values; their range is probably the most intuitive one. The second class of descriptors relates the dynamic behavior of a given property with other geometric or physicochemical properties. Such correlations can reveal if and how two molecular properties change in a coherent manner.

The relations between physicochemical properties and geometric descriptors describe the ability of a physicochemical property to fluctuate when the 3D geometry fluctuates. These relationships also lead to the concept of molecular sensitivity, since there will be sensitive molecules whose property values are markedly influenced by small geometric changes and insensitive molecules whose properties change little even during major geometric fluctuations. We can assume that molecular sensitivity may affect biological properties, as the latter are in themselves dynamic properties whose emergence will depend on the ability of a molecule to fit into and interact with an active site. Furthermore, molecular sensitivity and adaptability appear as two sides of the same coin, since sensitive molecules will need only small conformational changes to adapt their properties to the environment.

1.4.2
The Versatile Behavior of Acetylcholine

Our first exploration of property space was focused on acetylcholine. This molecule was chosen for its interesting structure, major biological role, and the abundant data available on its conformational properties [15]. The behavior of acetylcholine was analyzed by MD simulations in vacuum, in isotropic media (water and chloroform) [16] and in an anisotropic medium, i.e. a membrane model [17]. Hydrated *n*-octanol (1 mol water/4 mol octanol) was also used to represent a medium structurally intermediate between a membrane and the isotropic solvents [17].

The conformational profile of acetylcholine depended on the τ_2 and τ_3 dihedral angles since τ_1 and τ_4 remained constant during all monitored simulations (Fig. 1.4). It was found that acetylcholine assumes seven low-energy conformations

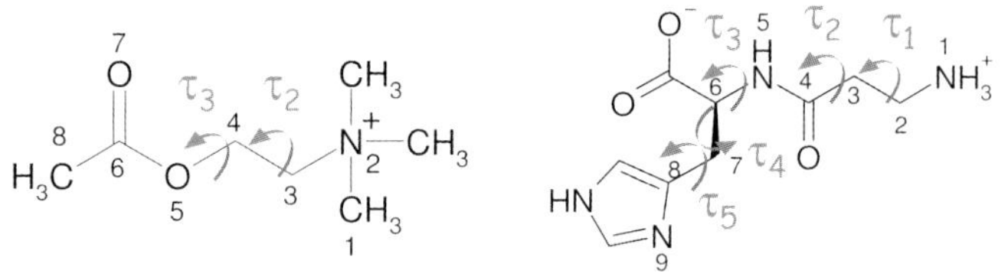

τ_1 = C1-N2-C3-C4
τ_2 = N2-C3-C4-O5
τ_3 = C3-C4-O5-C6
τ_4 = C4-O5-C6-O7

τ_1 = N1-C2-C3-C4
τ_2 = C2-C3-C4-N5
τ_3 = C4-N5-C6-C7
τ_4 = N5-C6-C7-C8
τ_5 = C6-C7-C8-N9

Fig. 1.4 Relevant dihedral angles in acetylcholine (left) and carnosine (right).

(i.e. the full-extended forms, τ_2=**t** and τ_3=**t**, and three pairs of chiral conformational clusters **+g+g/−g−g**; **+gt/−gt** and **t+g/t−g**), which can be clustered in folded (if τ_2 assumes synclinal conformations) and extended forms (if τ_2 is in antiperiplanar geometry). Thus, the conformational profile of acetylcholine strongly depends on τ_2, since τ_3 shows no clear preference in the range 60–300°. Clearly, the extended conformers were poorly populated in a vacuum, presumably due to intramolecular attractions between the cationic head and the electron-rich oxygen atoms. The proportion of extended conformers markedly increased in the isotropic solvents (as seen in Table 1.2) even if their increase seems due mainly to the physical presence of the solvent (i.e. friction and shielding effect) rather than to its specific physicochemical properties (i.e. polarity, H-bonding). In other words, solvent polarity does not appear to significantly affect the conformational profile of acetylcholine.

Notwithstanding this, Table 1.2 clearly shows that the behavior of acetylcholine reflected the physicochemical properties of the simulated media by adapting its property space. This is particularly evident when examining the lipophilicity averages, since the polarity of acetylcholine increased in all media compared to vacuum; although the differences between the mean log P values were small, they were significant as assessed by their 99.9% confidence limits.

The adaptability of acetylcholine appears even more evident when comparing the log P averages per conformational cluster (Fig. 1.5), which were markedly influenced by the isotropic media. Thus, all averages were lower in water than *in vacuo*, while in chloroform they assumed intermediate values, suggesting that acetylcholine can adjust its lipophilicity behavior by selecting the most suitable conformers within each conformational cluster rather than by modifying its conformational profile. The effects of water and chloroform are easily interpretable in terms of polarity and friction, but in a solvent such as octanol whose size is comparable to its own, the solute minimized steric repulsion by mimicking the shape of the solvent. In octanol, the extended conformers of acetylcholine successfully mimicked the preferred zig-zag conformation of the solvent. It is very intriguing

Tab. 1.2 Limits, ranges and mean values ±99.9% confidence limits of the molecular properties of acetylcholine conformers generated during MD simulations.

Property	***Medium*** [1]				
	Vacuum ($\varepsilon = 1$)	***Chloroform***	***Water***	***Octanol***	***Membrane***
SAS ($Å^2$)	343 to 377 34 358 ± 0.21	336 to 376 40 356 ± 0.25	341 to 378 37 361 ± 0.30	335 to 374 39 358 ± 0.42	337 to 371 34 354 ± 0.30
PSA ($Å^2$)	24.2 to 44.0 20.0 35.0 ± 0.12	28.5 to 50.4 21.9 40.1 ± 0.16	24.4 to 44.8 20.4 37.8 ± 0.11	32.0 to 51.1 19.1 42.7 ± 0.20	30.1 to 49.3 19.2 40.7 ± 0.14
Log P_{oct} [3]	−2.53 to −2.15 0.38 −2.34 ± 0.0026	−2.53 to −2.19 0.34 −2.36 ± 0.0026	−2.55 to −2.20 0.35 −2.42 ± 0.0026	−2.52 to −2.24 0.28 −2.40 ± 0.0030	−2.51 to −2.23 0.28 −2.39 ± 0.0030
Dipole moment	5.51 to 10.1 4.50 7.78 ± 0.035	7.43 to 9.54 2.07 8.40 ± 0.016	7.80 to 9.71 1.91 8.88 ± 0.014	7.63 to 9.45 1.88 8.67 ± 0.020	7.56 to 9.40 1.84 8.66 ± 0.019
Extended geometries (%)	6.4	19.7	16.7	22.8	0

1 In each box, the first line shows the limits (minimum to maximum value), the second line the range and the third line the mean ± 99.9% confidence limits (*t*-test). The compiled results are from [16, 17].
2 Distance (in Å) between (N^+) and (OC)CH_3.
3 "Virtual" log *P* calculated by the molecular lipophilicity potential.

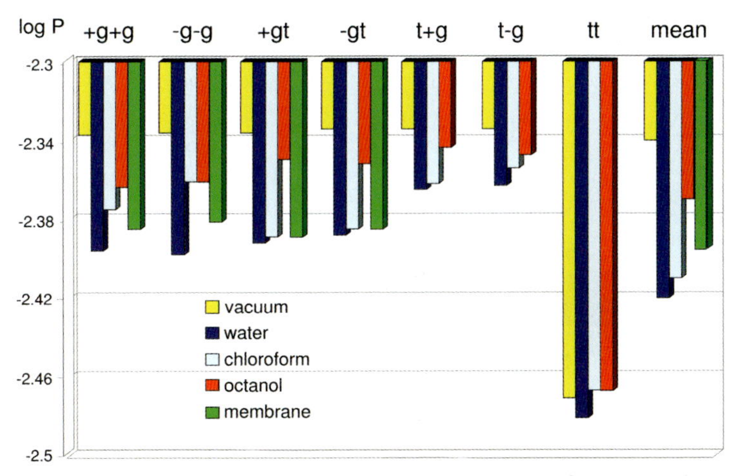

Fig. 1.5 Medium effects on average log *P* values for each conformational cluster.

to note that acetylcholine can modulate the properties of its fully extended conformers in an apparently contrasting way, selecting conformers that are simultaneously the most extended ones to better mimic the shape of the solvent and the most lipophilic ones to preserve an intermediate polarity. This suggests that conformational space and property spaces are quite independent, and that each cluster of conformers spans most of the property space of acetylcholine.

Conversely, in a membrane model, acetylcholine showed mean log *P* values very similar to those exhibited in water. This was due to the compound remaining in the vicinity of the polar phospholipid heads, but the disappearance of extended forms decreased the average log *P* value somewhat. This suggests that an anisotropic environment can heavily modify the conformational profile of a solute, thus selecting the conformational clusters more suitable for optimal interactions. In other words, isotropic media select the conformers, whereas anisotropic media select the conformational clusters. The difference in conformational behavior in isotropic versus anisotropic environments can be explained considering that the physicochemical effects induced by an isotropic medium are homogeneously uniform around the solute so that all conformers are equally influenced by them. In contrast, the physicochemical effects induced by an anisotropic medium are not homogeneously distributed and only some conformational clusters can adapt to them.

Taken globally, the results show a remarkable adaptability of acetylcholine which can be justified considering both its intrinsic flexibility, and the fact that its intramolecular interactions are not very strong and that almost all media can compete with them. Such adaptability finds a noteworthy implication in significant pairwise correlations between physicochemical properties and geometrical descriptors as well as among physicochemical properties. Thus, Fig. 1.6 shows the revealing 3D

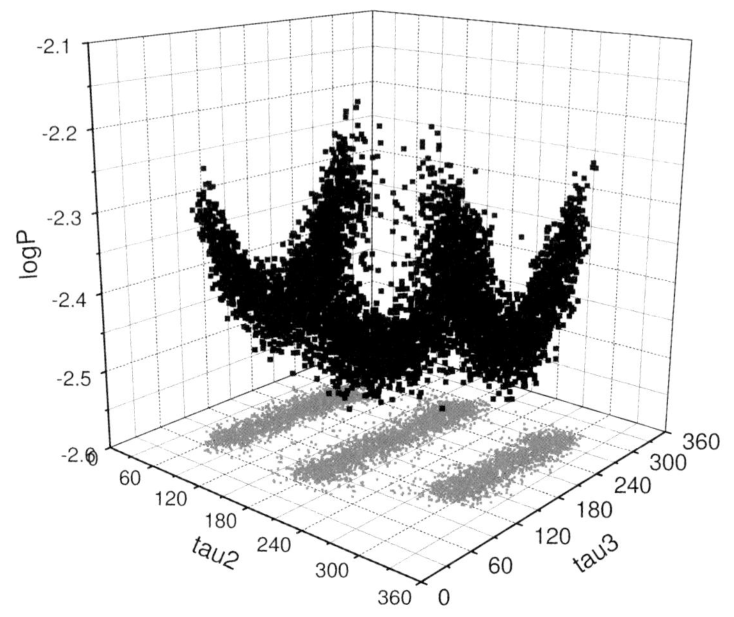

Fig. 1.6 Three-dimensional plot of τ_2 and τ_3 versus virtual log *P* as obtained from MD simulation of acetylcholine in water. Reproduced from Ref. [16] with kind permission of American Chemical Society 2005.

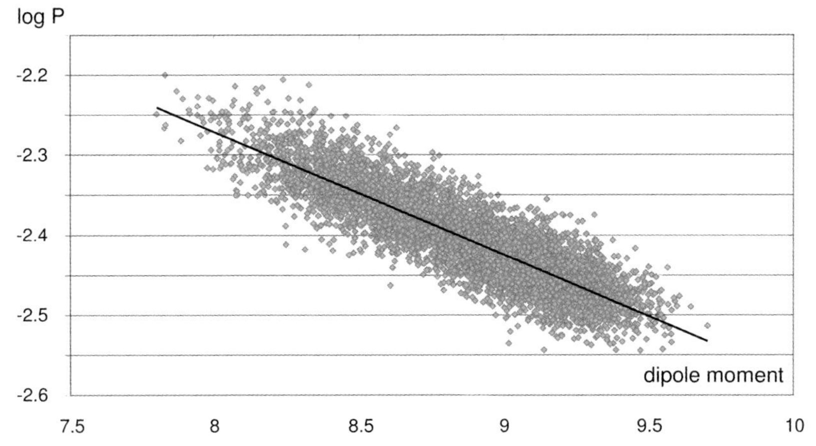

Fig. 1.7 Correlation between virtual log *P* (calculated with the molecular lipophilicity potential) and the dipole moment (r^2=0.76) as obtained from MD simulation of acetylcholine in water. Reproduced from Ref. [16] with kind permission of American Chemical Society 2005.

plots of virtual log *P* versus τ_2 and τ_3 as obtained from MD simulation in water (but all media gave fully comparable plots). Here, lipophilicity was not influenced by variations of τ_2, since the same range was covered for each of the three classes of conformers (i.e. with τ_2=**−g** or **+g** or **t**), while it was highly sensitive to variations in τ_3, with the most lipophilic conformers having τ_3=*gauche* and the most hydrophilic ones having τ_3=*trans*, suggesting that the main variations in log *P* are due to the accessibility of the ester moiety.

Among the correlations between physicochemical properties, the most noteworthy one was between dipole moment and log *P* (e.g. in water, see Fig. 1.7). Clearly, a higher dipole moment implies a greater hydrophilicity, but the fact that the two parameters correlate despite their different nature can be seen as a mutual validation of the respective algorithms used to calculate them.

1.4.3 The Carnosine–Carnosinase Complex

The second example of property space applications concerns the dipeptide carnosine (β-alanine-L-histidine, see Fig. 1.4) which represents the archetype of a series of histidine-containing dipeptides whose full physiological role remains poorly understood despite extensive studies in recent years [18–20]. Carnosine is synthesized by carnosine synthetase and hydrolyzed by dipeptidases (also called carnosinases) which belong to the metalloproteases [21].

The dynamic profile of carnosine was investigated by comparing MD simulations in isotropic solvents (i.e. water and chloroform) with simulation of the compound bound to serum carnosinase (CN1) [22]. This enzyme is characterized by its distribution in plasma and brain, and its ability to hydrolyze also anserine and homocarnosine [23]. The conformational profile of carnosine can be defined by five torsion angles (i.e. τ_1–τ_5) [24]. The first two angles concern the β-alanine residue, while τ_3, τ_4 and τ_5 involve the L-histidine residue. In fact, τ_1, τ_2 and τ_3 remained constant during the simulations in isotropic solvents (i.e. water and chloroform) due to the strong intramolecular ion-pair which heavily influences the behavior of carnosine in its zwitterionic form. In this case, the variability in conformational and property spaces was almost totally due to the orientation of the imidazole moiety, since the simulated solvents were not able to break the intramolecular salt bridge. Specifically, the β-alanine residue was constantly rigidified in water and chloroform, with τ_1 = **+g**, τ_2 = **−g**, and τ_3 = **−g**. In contrast, the L-histidine residue assumed four different conformers depending on τ_4 and τ_5 (i.e. **t+g**, **t−g**, **−g+g** and **−g−g**).

When comparing the conformational profile of carnosine in isotropic solvents and bound to carnosinase, a contrasting behavior is apparent. Indeed, Fig. 1.8(a) clearly shows that the β-alanine residue is more flexible in the enzyme-bound complex than in isotropic solvents, while the L-histidine residue appears constrained by interactions with carnosinase (Fig. 1.8b). This discrepancy can be explained considering the interaction pattern binding carnosine to the enzyme (Fig. 1.9). Thus, the polar residues lining the catalytic site of carnosinase (including the key zinc ions) can successfully compete with the intramolecular ionic bond in carnosine, while the L-histidine residue must retain an accessible conformation which optimizes the contacts of imidazole with the enzyme. These results confirm that an isotropic solvent is unable to heavily modify the conformational profile of a solute, while an anisotropic medium (including a protein, which is also a structured anisotropic medium) can do it. Interestingly, the membrane model reduced the conformational space of acetylcholine, while its specific recognition interactions with carnosinase partially enlarged that of carnosine.

The marked rigidity of carnosine reflected in its property spaces (Table 1.3) deserves some considerations. The high polarity of carnosine is illustrated by the fact that its PSA in all media is about 50% of its SAS, whereas in acetylcholine PSA is about 10% of the SAS. It is worth observing that, despite the rigidity which characterizes the conformational profile of carnosine, this molecule can modulate its physicochemical properties according to the polarity of the medium, as seen in its lipophilicity space. Indeed, carnosine in water shows the lowest log P average, but the highest lipophilicity average when bound to the enzyme, probably due to the marked accessibility of the imidazole ring into the catalytic site. This profile confirms the results observed in isotropic media, that the contribution to lipophilicity of the β-alanine residue is nearly constant – the variability being mainly due to the accessibility of the histidine moiety. Similarly, the most polar group in acetylcholine (i.e. its ammonium head) gave a quite constant contribution, variability in lipophilicity being due to the accessibility of the ester moiety. This suggests that molecules can modulate the physicochemical profile of highly polar groups only with great difficulty. The marked accessibility of the imidazole ring of bound carnosine finds convincing confirmation in the average SAS and PSA values, which are highest in the carnosinase-bound form. Finally, carnosine showed the most narrow property ranges when bound to the enzyme, although

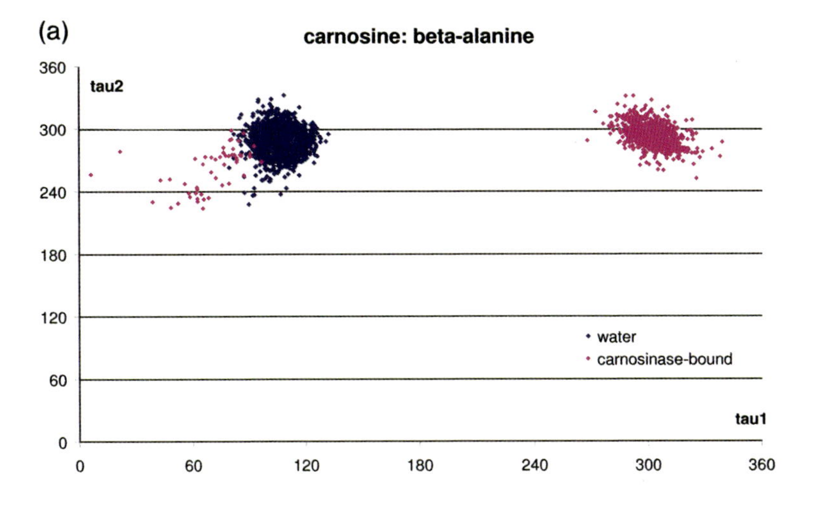

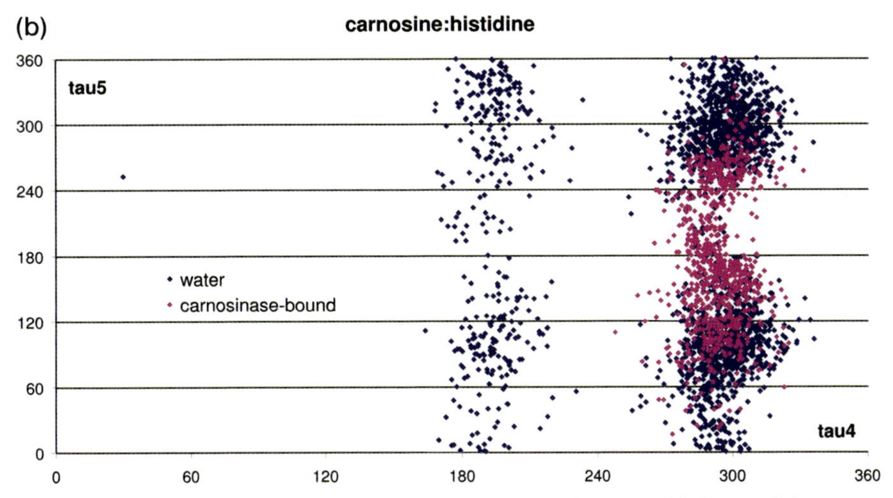

Fig. 1.8 The conformational behavior of the two amino acyl residues of carnosine (τ_2 versus τ_3 plot) as simulated for 5 ns in water (in blue) or when bound to carnosinase (in pink). (a) Conformational behavior of the β-alanine residue (τ_1 versus τ_2 plot). (b) Conformational behavior of the histidine residue (τ_4 versus τ_5 plot).

its conformational space was then markedly enlarged. This suggests that a ligand must assume well-defined property profiles to optimize its recognition by and binding to an enzyme and that this adaptation is only partly explained by a mere conformational fit. In other words, conformational space and property spaces are only partly correlated.

Taken together, our MD simulations of acetylcholine and carnosine emphasize the marked difference between them. Indeed, acetylcholine is representative of a sensitive molecule whose physicochemical and structural properties can vary in a coherent manner, aptly adapting themselves to the simulated media. Conversely,

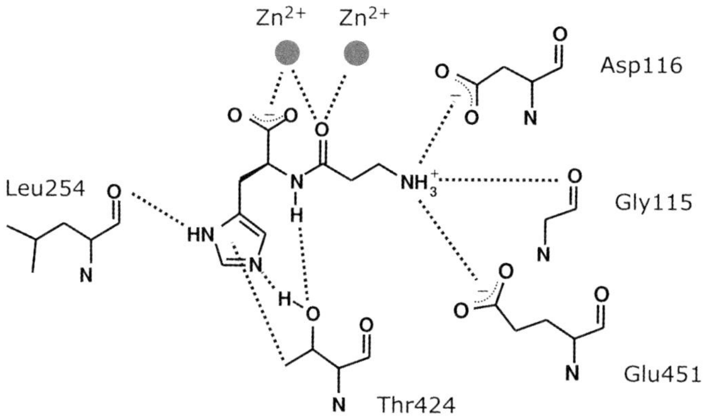

Fig. 1.9 Bidimensional representation of the interaction pattern between carnosinase and its substrate carnosine. The model shows how the enzyme recognizes (binds) the ammonium group, the carboxylate group and the unsubstituted imidazole ring. The amido bond is simultaneously bound for recognition and polarized for catalysis. Reproduced from Ref. [22] with kind permission of American Chemical Society 2005.

Tab. 1.3 Limits, ranges and mean values ± 99.9% confidence limits of the molecular properties of carnosine conformers generated during MD simulations.

Property	*Medium*[1]		
	Water	*Chloroform*	*Carnosinase*
SAS ($Å^2$)	395 to 434	336 to 376	410 to 446
	39	40	36
	417 ± 0.23	356 ± 0.25	427 ± 0.28
PSA ($Å^2$)	180 to 226	177 to 221	192 to 234
	47	44	42
	203 ± 0.33	201 ± 0.33	214 ± 0.27
Log P_{oct}[2]	−4.57 to −3.98	−4.49 to −3.85	−4.45 to −3.90
	0.59	0.66	0.55
	−4.28 ± 0.0043	−4.20 ± 0.0044	−4.17 ± 0.0042

1 In each box, the first line shows the limits (minimum to maximum value), the second line the range and the third line the mean ± 99.9% confidence limits (*t*-test).
2 "Virtual" log *P* calculated by the molecular lipophilicity potential.

carnosine appears markedly rigidified by an intramolecular ionic bridge which influences both its conformational space (which is frozen in few conformations) and its property spaces, as evidenced by their narrow ranges and insignificant pairwise correlations. Nevertheless, MD simulations revealed that carnosine could also adjust its physicochemical properties to the simulated medium, suggesting that the conformational space is easier to constrain than the property spaces, which indeed conserve a significant elasticity even in very constrained molecules. In other words, some physicochemical adaptability to the molecular environment is retained even in rather rigid compounds. Such molecular adaptability can clearly influence biological activity and molecular descriptors accounting for adaptability might find fertile applications in QSAR as described below.

1.4.4 Property Space and Dynamic QSAR Analyses

Our third example illustrates the use in QSAR analyses of parameters describing the property range and distribution in relation to conformational changes and other property profiles. As previously stated, a property space can be defined using two classes of descriptors, i.e. the distribution of property values and the relations between properties. Thus, the relations between geometric descriptors and physicochemical properties describe the ability of a physicochemical property to fluctuate when the 3D geometry fluctuates. These relations lead to the concept of molecular sensitivity, since there will be sensitive molecules whose property values are markedly influenced by small geometric changes and insensitive molecules whose properties change little even during major geometric fluctuations.

From a mathematical point of view, such correlations may be analyzed by considering their regression coefficients. However, using regression coefficients as independent variables may lead to mathematical dead-ends. We thus looked for a descriptor of property space that would be both informative and simple to use. The descriptor we propose and evaluate is the amplitude of variation of a given physicochemical property for a given variation in molecular geometry. If we consider a physicochemical property X for which conformer-specific values can be computed (e.g. dipole moment, polar surface area, virtual log P, etc.), its pairwise sensitivity value ($Pairwise\ Sensitivity_{X,Gij}$) for two given conformers (i, j) and a given geometric descriptor G (e.g. an intramolecular distance, a torsion angle, etc.) can be defined as the ratio between the absolute value of the difference of X and the corresponding absolute value of the difference in G:

$$Pairwise\ Sensitivity_{X,Gij} = \frac{|X_i - X_j|}{|G_i - G_j|} \quad (1)$$

The global sensitivity ($Sensitivity_{X,Gij}$) will be the average of the pairwise sensitivities computed for all possible pairs of N conformers, i.e. for $N(N-1)$ pairs.

For any given physicochemical property of a molecule, one can calculate several sensitivity values according to the geometric descriptors being used. However, when investigating a set of heterogeneous compounds, a geometric descriptor applicable to all molecules must be selected. To this end, the root mean square deviation of atomic coordinates aptly describes geometric differences between pairs of conformers as a function of their atomic positions.

The objective of this example [25] was to examine whether and how range and sensitivity can be successfully used as descriptors of the space of relevant physicochemical properties, and correlated with affinities and receptor subtype selectivities for a heterogeneous set of ligands of α_1-adrenoceptors (α_1-ARs) taken from the literature [26] and characterized by their large differences in binding affinities. The conformational space was explored using a MC procedure, and the properties considered were dipole moment, lipophilicity, polar area and surface area.

A search for relations between affinity data (pK_i) and descriptors of property spaces (range and sensitivity) failed to uncover any significant correlation (all r values <0.5). This result was expected and understandable, since affinity depends on the ligand ability to assume well-defined property values – a type of information not encoded in range and sensitivity. In contrast, significant correlations were found between some receptor selectivities and some property space descriptors. Indeed, ΔpK_{a-b} and ΔpK_{a-d} yielded significant correlations ($r > 0.7$) with log P, PSA and SAS ranges, whereas ΔpK_{b-d} yielded no correlation whatsoever ($r < 0.1$).

A clear trend was also apparent among the physicochemical properties, since the lipophilicity range yielded the best correlations for both ΔpK_{a-b} and ΔpK_{a-d}, while the dipole space yielded the lowest. Interestingly, all significant correlation coefficients were positive, implying that α_1-AR selectivities are mainly proportional to variations in physicochemical properties, as expressed mainly by range.

The above observations may imply that the ability to selectively interact with the α_{1a}-AR is encoded in property space descriptors and especially in the lipophilicity space, whereas selective interaction with the α_{1b}-AR is only partially encoded in property space descriptors and α_{1d}-AR selectivity not at all. To verify the above hypothesis, we recalculated regressions coefficients between ΔpK_{a-b} selectivity and property space parameters, removing the strongly selective α_{1b}-AR ligands. This indeed produced a slight increase (about 0.05–0.10) in all correlation coefficients between property spaces and ΔpK_{a-b}. The best correlation, i.e. between {*range*_log P} and ΔpK_{a-b}, is shown in Eq. 2 and Fig. 1.10:

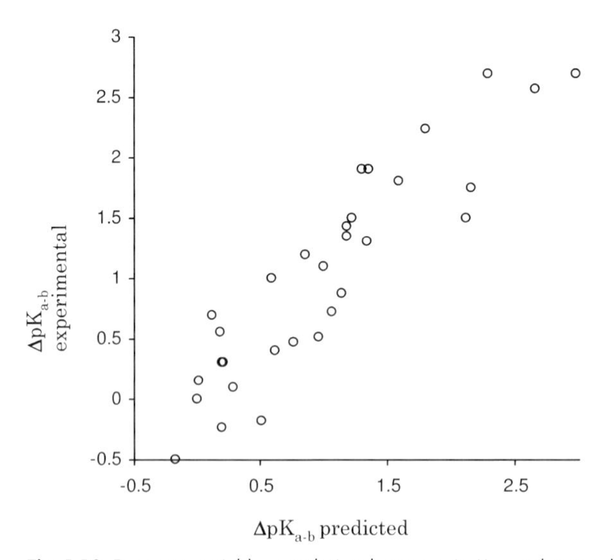

Fig. 1.10 Best one-variable correlation between ΔpK_{a-b} and *range*_log P (Eq. 2).

$$\Delta pK_{a-b} = 1.49(\pm 0.12)\{range\_\log P\} - 0.12(\pm 0.13) \quad (2)$$

$$n = 32, r^2 = 0.79, q^2 = 0.78, s = 0.41$$

Clearly, this equation cannot take into account α_{1b}-selective ligands (i.e. with $\Delta pK_{a-b} < 0$). Indeed, a hypothetical molecule with an impossibly low $\{range\_\log P\}$ of 0 would be predicted to have a ΔpK_{a-b} equal to –0.12. Nevertheless, the goodness-of-fit of this equation is remarkable considering the heterogeneous nature of the ligands and its high q^2 value (i.e. good predictive power) obtained with a single independent variable.

Given the absence of correlation between the sensitivity and range descriptors, we also examined whether a two-variable equation would improve on Eq. (2). As shown by Eq. (3), the inclusion of two independent variables in the same equations improved their predictive capacity:

$$\Delta pK_{a-b} = 1.61(\pm 0.13)\{range\_\log P\} + 0.34(\pm 0.04)\{sensitivity\_\log P\} - 0.76(\pm 0.19) \quad (3)$$

$$n = 32, r^2 = 0.84, q^2 = 0.83, s = 0.38\ F = 74.38$$

Compared to Eq. (2), Eq. (3) shows a slight statistical improvement. Also, is has a better predictability for α_{1b} selective ligands, since a hypothetical molecule with very low $\{range\_\log P\}$ and $\{sensitivity\_\log P\}$ values would be predicted to have a ΔpK_{a-b} equal to –0.76.

From a methodological viewpoint, our results suggest that range and sensitivity are useful descriptors of property spaces and can parameterize the capacity of a given molecule to span broad conformational and property spaces. In other words, range and sensitivity appear as promising descriptors of the dynamic behavior of a molecule. Their application to other dynamic QSAR studies [in particular, absorption, distribution, metabolism and excretion (ADME) behavior] is under investigation.

1.5 Conclusions

As the present and other chapters show, the usual way medicinal chemists describe molecular structure and properties is necessarily limited and partial. Indeed, it deals with 4D structures (3D geometry plus conformation) in molecular graphics, 3D and 2D representations on paper, 1D strings in codes, and 0D points in chemical spaces. However, as we have tried to show here, molecules are 3D objects whose shape and the various molecular fields they generate vary in space and time, effectively making them *N*-dimensional objects. The dynamic complexity of molecules arises from their interactions with energy fields and with neighboring molecules, to such extent that a fully isolated molecule is unobservable, the concept of it being a mere abstraction. However, beyond the grasp of this paradigm lies the challenge of expressing and using our ever-increasing wealth of molecular information in manners suitable for higher-level QSARs yet to be conceived. We hope the present book may contribute to such a progress.

References

1 Testa, B., Kier, L. B., Carrupt, P. A. A systems approach to molecular structure, intermolecular recognition, and emergence–dissolvence in medicinal research. *Med. Res. Rev.* **1997**, *17*, 303–326.

2 Testa, B. Drugs as chemical messages: molecular structure, biological context, and structure–activity relationships. *Med. Chem. Res.* **1997**, *7*, 340–365.

3 Testa, B., Kier, L. B., Bojarski, A. J. Molecules and meaning: how do molecules become biochemical signals? *SEED Electronic J.* **2002**, *2*, 84–101. http://www.library.utoronto.ca/see/SEED/Vol2-2/2-2%20resolved/Testa.pdf.

4 Testa, B., Vistoli, G., Pedretti, A. Musings on ADME predictions and structure–activity relations. *Chem. Biodivers.* **2005**, *2*, 1411–1427.

5 Testa, B. *Principles of Organic Stereochemistry*, Dekker, New York, NY, **1979**.

6 Kirschner, M. W., Gerhart, J. C. *The Plausibility of Life*, Yale University Press, New Haven, CT, **2005**.

7 Testa, B., Carrupt, P. A., Gaillard, P., Billois, F., Weber, P. Lipophilicity in molecular modeling. *Pharm. Res.* **1996**, *13*, 335–343.

8 Carrupt, P. A., Testa, B., Gaillard, P. Computational approaches to lipophilicity: Methods and applications. *Rev. Computat. Chem.* **1997**, *11*, 241–315.

9 Carrupt, P. A., El Tayar, N., Karlén, A., Testa, B. Value and limits of molecular electrostatic potentials for characterizing drug–biosystem interactions. *Methods Enzymol.* **1991**, *203*, 638–677.

10 Rey, S., Caron, G., Ermondi, G., Gaillard, P., Pagliara, A., Carrupt, P. A., Testa, B. Development of molecular hydrogen bonding potentials (MHBPs) and their application to structure–permeation relations. *J. Comput.-Aided Mol. Des.* **2001**, *19*, 521–535.

11 Lazarsfeld, P. F., Barton, H. A. Qualitative measurement in the social sciences: classifications, typologies, and indices. In *The Policy Sciences*, Lerner, D., Lasswell, H. D. (eds.), Stanford University Press, Stanford, CA, **1951**, pp. 151–192.

12 Merschrod, K. A property-space perspective for interaction terms of ordinal variables. *Quality and Quantity*, **1982**, *16*, 549–558

13 Dobson, C. M. Chemical space and biology. *Nature* **2004**, *432*, 824–828.

14 Feher, M., Schmidt, J. M. Property distributions: differences between drugs, natural products, and molecules from combinatorial chemistry. *J. Chem. Inf. Comput. Sci.* **2003**, *43*, 218–227.

15 Vistoli, G., Pedretti, A., Villa, L., Testa, B. The solute–solvent system: solvent constraints on the conformational dynamics of acetylcholine. *J. Am. Chem. Soc.* **2002**, *124*, 7472–7480.

16 Vistoli, G., Pedretti, A., Villa, L., Testa, B. Solvent constraints on the property space of acetylcholine. I. Isotropic solvents. *J. Med. Chem.* **2005**, *48*, 1759–1767.

17 Vistoli, G., Pedretti, A., Testa, B. Solvent constraints on the property space of acetylcholine. II. Ordered media. *J. Med. Chem.* **2005**, *48*, 6926–6935.

18 Guiotto, A., Calderan, A., Ruzza, P., Borin, G. Carnosine and carnosine-related antioxidants: a review. *Curr. Med. Chem.* **2005**, *12*, 2293–2315.

19 Bauer, K. Carnosine and homocarnosine, the forgotten, enigmatic peptides of the brain. *Neurochem. Res.* **2005**, *30*, 1339–1345.

20 Aldini, G., Facino, R. M., Beretta, G., Carini, M. Carnosine and related dipeptides as quenchers of reactive carbonyl species: from structural studies to therapeutic perspectives. *Biofactors* **2005**, *24*, 77–87.

21 Pegova, A., Abe, H., Boldyrev, A. Hydrolysis of carnosine and related compounds by mammalian carnosinases. *Comp. Biochem. Physiol. B Biochem. Mol. Biol.* **2000**, *127*, 443–446.

22 Vistoli, G., Pedretti, A., Cattaneo, M., Aldini, G., Testa B. Homology modeling of human serum carnosinase, a potential medicinal target, and MD simulations of its allosteric activation by citrate. *J. Med. Chem.* **2006**, *49*, 3269–3277.

23 Teufel, M., Saudek, V., Ledig, J. P., Bernhardt, A., Boularand, S., Carreau, A., Cairns, N. J., Carter, C., Cowley, D. J., Duverger, D., Ganzhorn, A. J., Guenet, C., Heintzelmann, B., Laucher, V., Sauvage, C., Smirnova, T. Sequence identification and characterization of human carnosinase and a closely related non-specific dipeptidase. *J. Biol. Chem.* **2003**, *278*, 6521–6531.

24 Dies, R. P., Baran, E. J. A density functional study of some physical properties of carnosine (*N*-β-alanyl-L-histidine). *J. Mol. Struct. Theochem.* **2003**, *621*, 245–251.

25 Vistoli, G., Pedretti, A., Villa, L., Testa, B. Range and sensitivity as descriptors of molecular property spaces in dynamic QSAR analyses. *J. Med. Chem.* **2005**, *48*, 4947–4952.

26 Bremner, J. B., Coban, B., Griffith, R., Groenewoud, K. M., Yates, B. F. Ligand design for alpha$_1$ adrenoceptor subtype selective antagonists. *Bioorg. Med. Chem.* **2000**, *8*, 201–214.

2
Physicochemical Properties in Drug Profiling

Han van de Waterbeemd

Abbreviations

ADME	absorption, distribution, metabolism and excretion
BBB	blood–brain barrier
BCS	Biopharmaceutics Classification Scheme
BMC	biopartitioning micellar chromatography
Caco-2	adenocarcinoma cell line derived from human colon
CNS	central nervous system
DMPK	drug metabolism and pharmacokinetics
FaSSIF	fasted-state simulated artificial intestinal fluid
HB	H-bonding
HDM	hexadecane membranes
HSA	human serum albumin
HTS	high-throughput screening
IAM	immobilized artificial membrane
ILC	immobilized liposome chromatography
MAD	maximum absorbable dose
MEKC	micellar electrokinetic chromatography
PAMPA	parallel artificial membrane permeation assay
PBPK	physiologically-based pharmacokinetic modeling
P-gp	P-glycoprotein
PK	pharmacokinetic(s)
PPB	plasma protein binding
PSA	polar surface area ($Å^2$)
QSAR	quantitative structure–activity relationship
SPR	surface plasmon resonance

Symbols

A_D	cross-sectional area ($Å^2$)
Clog P	calculated logarithm of the octanol–water partition coefficient (for neutral species)

Molecular Drug Properties. Measurement and Prediction. R. Mannhold (Ed.)

ISBN: 978-3-527-31755-4

D	distribution coefficient (often in octanol–water)
$diff(\log P^{N-I})$	difference between $\log P^N$ and $\log P^I$
$\Delta\log P$	difference between $\log P$ in octanol–water and alkane–water
k_a	transintestinal rate absorption constant (min^{-1})
K_a	dissociation constant
Elog D	experimental log D based on a high-performance liquid chromatography method
log D	logarithm of the distribution coefficient, usually in octanol–water at pH 7.4
$\log D^{7.4}$	logarithm of the distribution coefficient, in octanol–water at pH 7.4
log P	logarithm of the partition coefficient, usually in octanol–water (for neutral species)
$\log P^I$	logarithm of the partition coefficient of a given compound in its fully ionized form, usually in octanol–water
$\log P^N$	logarithm of the partition coefficient of a given compound in its neutral form, usually in octanol–water
MW	molecular weight (Da)
P	partition coefficient (often in octanol–water)
P_{app}	permeability constant measured in Caco-2 or PAMPA assay ($cm\,min^{-1}$)
pK_a	ionization constant in water
PPB%	percentage plasma protein binding
S	solubility ($mg\,mL^{-1}$)
SITT	small intestinal transit time (4.5 h = 270 min)
SIWV	small intestinal water volume (250 mL)
V	volume (mL or L)
V_{dss}	volume of distribution at steady state ($L\,kg^{-1}$)

2.1 Introduction

An important part of the optimization process of potential leads to candidates suitable for clinical trials is the detailed study of the absorption, distribution, metabolism and excretion (ADME) characteristics of the most promising compounds. Experience has shown that physicochemical properties play a key role in drug metabolism and pharmacokinetics (DMPK) [1–5]. In 1995, 2000 and 2004 specialized but very well attended meetings were held to discuss the role of log P and other physicochemical properties in drug research and lead profiling, and the reader is referred to the various proceedings for highly recommended reading on this subject [4, 6, 7].

The molecular structure is at the basis of physicochemical, DMPK, as well as safety/toxicity properties, as outlined in Fig. 2.1. Measurement and prediction of

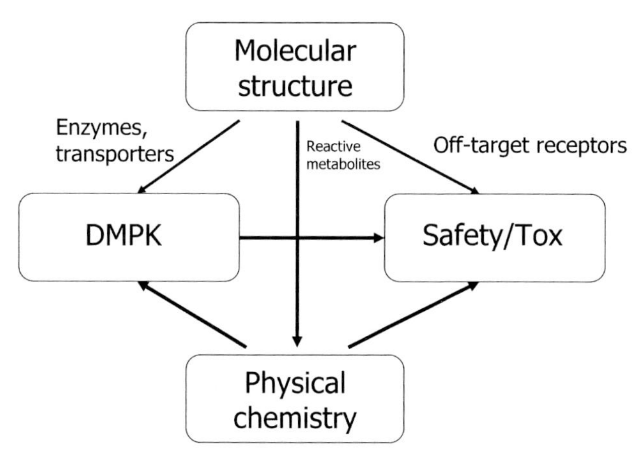

Fig. 2.1 Dependency of DMPK and safety/toxicity properties on structural and physicochemical properties.

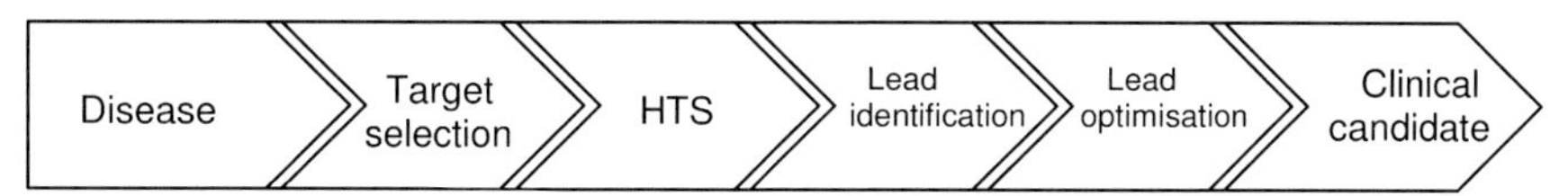

Fig. 2.2 The drug discovery process.

physicochemical properties is relatively easy compared to DMPK and safety properties, where biological factors come into play. However, DMPK and toxicity properties depend to a certain extent on the physicochemical properties of the compounds as these dictate the degree of access to biological systems such as enzymes and transporters.

The change in work practice towards high-throughput screening (HTS) in biology using combinatorial libraries has also increased the demands on more physicochemical and ADME data. There has been an increasing interest in physicochemical hits and leads profiling in recent years, using both *in vitro* and *in silico* approaches [8–11]. This chapter will review the key physicochemical properties, both how they can be measured as well as how they can be calculated in some cases. Chemical stability [12] is beyond the scope of this chapter, but is obviously important for a successful drug candidate.

The need and precision of a particular physicochemical property for decision making in a drug discovery project depends on the stage in the drug discovery process (see Fig. 2.2). Whilst calculated simple filters may be sufficient in library design, more experimental data are required in lead optimization. Striking the right balance between computational and experimental predictions is an important challenge in cost-efficient and successful drug discovery.

Physicochemical properties are considerably interrelated as visualized in Fig. 2.3. The medicinal chemist should bear in mind that modifying one often means

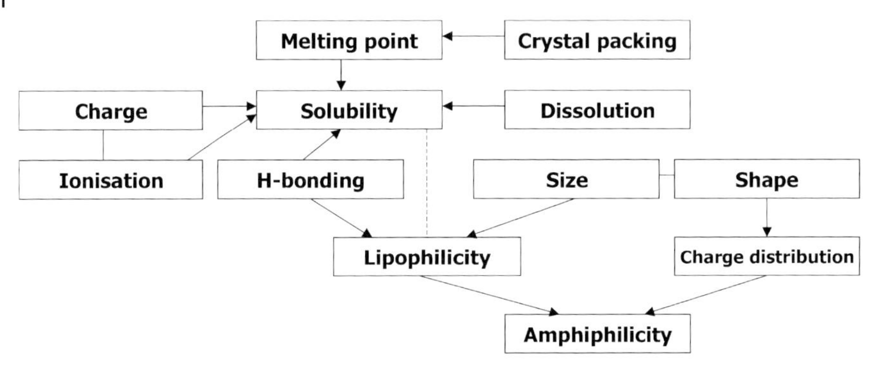

Fig. 2.3 Dependencies between various physicochemical properties.

also changes in other physicochemical properties, and hence indirectly influencing the DMPK and safety profile of the compound.

2.2
Physicochemical Properties and Pharmacokinetics

2.2.1
DMPK

The study of DMPK has changed from a descriptive to a much more predictive science [3]. This is driven by great progress in bioanalytics, development of *in vitro* assays and *in silico* modeling/simulation, and a much better basic understanding of the processes. Thus, and fortunately, ADME-related attrition has lowered from around 40% in 1990 to around 10% in 2005 [13].

2.2.2
Lipophilicity – Permeability – Absorption

As an example of the role of physicochemical properties in DMPK, the properties relevant to oral absorption are described in Fig. 2.4. It is important to note that these properties are not independent, but are closely related to each other. Oral absorption is the percentage of drug taken up from the gastrointestinal lumen into the portal vein blood. The processes involved are a combination of physical chemistry and biological (transporters, metabolizing enzymes). The transfer process through a membrane without any biological component is often called permeability. It can be mimicked in an artificial membrane such as the parallel artificial membrane permeation assay (PAMPA) set-up (see Section 2.8.1). However, *in vivo* permeability cannot be measured in isolation from biological events. All so-called *in vitro* measures for permeability are nothing else than different types of lipophilicity measures. In plotting oral absorption (percentage or fraction) against any

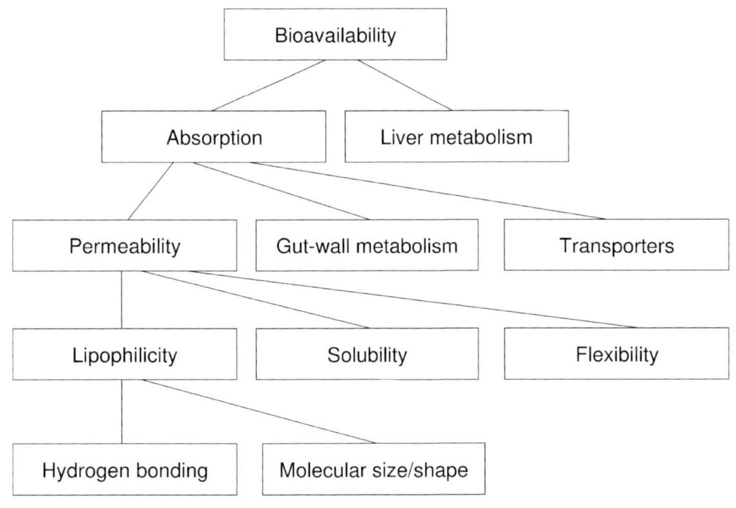

Fig. 2.4 Importance of physical chemistry properties on permeability, absorption and bioavailability [16]. (With kind permission of Elsevier.)

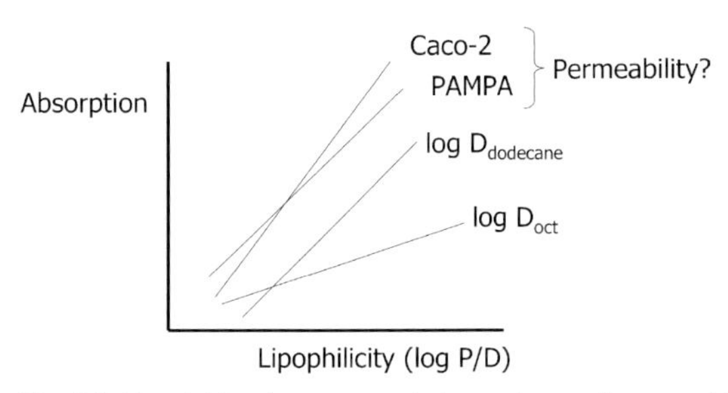

Fig. 2.5 Trendships between oral absorption and permeability/lipophilicity. In reality these relationships are most likely sigmoidal, i.e. more complex than these trends indicate.

"permeability" or lipophilicity scale (see Fig. 2.5) one observes a trend indicating that higher permeability or lipophilicity leads to better absorption. Often a plateau is observed too, indicating that such relationships are in fact nonlinear and can be approached by a sigmoidal function. Several lipophilicity scales can be related to each other via a Collander (Eq. 1) or an extended Collander relationship (Eq. 2) by adding a parameter for the difference in H-bonding (HB) between the two solvent systems. The equivalent for relating, for example, PAMPA scales to each other or PAMPA with Caco-2 has also been published [14, 15].

$$\log P_1 = a \log P_2 + b \tag{1}$$

$$\log P_1 = p \log P_2 + q\mathrm{HB} + r \tag{2}$$

Instead of using surrogate measures for oral absorption with a lipophilicity or permeability assay *in vitro*, oral absorption can also be estimated *in silico* by using human oral absorption data from the literature [16]. This data is rather sparse because oral absorption is not systematically measured in clinical trials. The data is also skewed towards high absorption compounds. In addition, interindividual variability is important (around 15%). Of course absorption can also be dose and formulation dependent. Therefore, early estimates are only rough guides to get the ballpark right.

2.2.3
Estimation of Volume of Distribution from Physical Chemistry

The distribution of a drug in the body is largely driven by its physicochemical properties and in part for some compounds by the contribution of transporter proteins [17]. By using the Oie–Tozer equation and estimates for ionization (pK_a), plasma protein binding (PPB) and lipophilicity (log $D^{7.4}$) quite robust predictions for the volume of distribution at steady state (V_{dss}), often within 2-fold of the observed value, can be made [18].

2.2.4
PPB and Physicochemical Properties

The percentage of binding to plasma proteins (PPB%) is an important factor in PK and is determinant in the actual dosage regimen (frequency), but not important for the daily dose size [3]. The daily dose is determined by the required free or unbound concentration of drug required for efficacy [3]. Lipophilicity is a major driver to PPB% [19, 20]. The effect of the presence of negative (acids) or positive (bases) charges has different impacts on binding to human serum albumin (HSA), as negatively charged compounds bind more strongly to HSA than would be expected from the lipophilicity of the ionized species at pH 7.4 [19, 20] (see Fig. 2.6).

2.3
Dissolution and Solubility

Each cellular membrane can be considered as a combination of a physicochemical and biological barrier to drug transport. Poor physicochemical properties may sometimes be overcome by an active transport mechanism. Before any absorption can take place at all, the first important properties to consider are dissolution and solubility [21]. Many cases of solubility-limited absorption have been reported and therefore solubility is now seen as a property to be addressed at the early stages of drug discovery. Only compound in solution is available for permeation across the gastrointestinal membrane. Solubility has long been recognized as a limiting factor in the absorption process leading to the implementation of high-throughput solubility screens in early stages of drug design [22–26]. Excessive lipophilicity is a common cause of poor solubility and can lead to erratic and incomplete absorp-

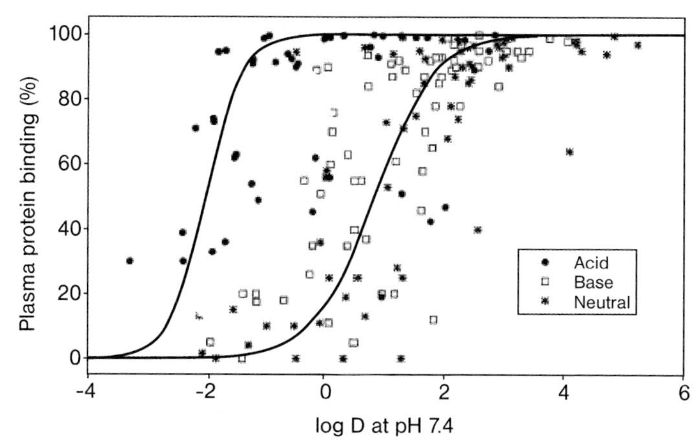

Fig. 2.6 Trendships between percentage human PPB (hPPB%) and octanol–water log $D^{7.4}$ [20]. Note the around 2 log units downshift of the sigmoidal relationship for acids as compared to neutrals and basics. (With kind permission of Springer-Kluwer.)

Tab. 2.1 Desired solubility (μg mL^{-1}) needed for expected doses [26].

Dose (mg kg^{-1})	*Permeability*		
	High	*Medium*	*Low*
0.1	1	5	21
1	10	52	210
10	100	520	2100

tion following oral administration. Estimates of desired solubility for good oral absorption depend on the permeability of the compound and the required dose, as illustrated in Table 2.1 [26]. The incorporation of an ionizable center, such as an amine or similar function, into a template can bring a number of benefits including water solubility.

The concept of maximum absorbable dose (MAD) relates drug absorption to solubility via [27, 28]:

$$\mathrm{MAD} = S \times k_a \times \mathrm{SIWV} \times \mathrm{SITT} \qquad (3)$$

where S = solubility (mg mL^{-1}) at pH 6.5, k_a = transintestinal absorption rate constant (min^{-1}), SIWV = small intestinal water volume (mL), assumed to be around 250 mL, and SITT = small intestinal transit time (min), assumed to be 4.5 h = 270 min.

Dissolution testing has been used as a prognostic tool for oral drug absorption [29]. A Biopharmaceutics Classification Scheme (BCS) has been proposed under which drugs can be categorized into four groups according to their solubility and permeability properties [30]. As both permeability as well as solubility can be further dissected into more fundamental properties, it has been argued that the principal properties are not solubility and permeability, but rather molecular size and H-bonding [31]. The BCS has been adopted as a regulatory guidance for bioequivalence studies.

2.3.1 Calculated Solubility

As a key first step towards oral absorption, considerable effort went into the development of computational solubility prediction [32–39]. However, partly due to a lack of large sets of experimental data measured under identical conditions, today's methods are not robust enough for reliable predictions [40]. Further fine-tuning of the models can be expected now high-throughput data has become available to construct such models. Models will be approximate since they do not take into account the effect of crystal packing, ionic force, type of buffer, temperature, etc. Solubility is typically measured in an aqueous buffer only partly mimicking the physiological state. More expensive fasted-state simulated artificial intestinal fluid (FaSSIF) solutions have been used to measure solubility, which in some cases appear to give better predictions in physiologically based pharmacokinetic (PBPK) modeling than solubility data using a simpler aqueous buffer [41].

2.4 Ionization (pK_a)

It was assumed for a long time that molecules can only cross a membrane in their neutral form. This dogma, based on the pH-partition theory, has been challenged [42, 43]. Using cyclic voltammetry it was demonstrated that compounds in their ionized form pass into organic phases and might well cross membranes in this ionized form [44].

The importance of drug ionization using cell-based methods such as Caco-2 in the *in vitro* prediction of *in vivo* absorption was discussed [45]. It was observed that when the apical pH used in Caco-2 studies was lowered from 7.4 to 6.0 a better correlation was obtained with *in vivo* data, demonstrating that careful selection of experimental conditions *in vitro* is crucial to produce a reliable model. Studies with Caco-2 monolayers also suggested that the ionic species might contribute considerably to overall drug transport [46].

Various ways that a charged compound may cross a membrane by a "passive" mechanism have been described [42]. These include transport as ion (*trans*- and/or paracellular), ion-pair or protein-assisted (using the outer surface of a protein spanning a membrane).

Therefore a continued interest exists in the role of pK_a in oral absorption, which often is related to its effect on lipophilicity and solubility. Medicinal chemists can modulate these properties through structural modifications [47]. Various methods to measure pK_a values have been developed [47–50] and considerable databases are now available.

The difference between the log P of a given compound in its neutral form (log P^N) and its fully ionized form (log P^I) has been termed *diff*(log P^{N-I}) and contains series-specific information, and expresses the influence of ionization on the intermolecular forces and intramolecular interactions of a solute [44, 51, 52].

2.4.1 Calculated pK_a

A number of approaches to predict ionization based on structure have been published (for a review, see [53]) and some of these are commercially available. Predictions tend to be good for structures with already known and measured functional groups. However, predictions can be poor for new innovative structures. Nevertheless, pK_a predictions can still be used to drive a project in the desired direction and the rank order of the compounds is often correct. More recently training algorithms have also become available which use in-house data to improve the predictions. This is obviously the way forward.

2.5 Molecular Size and Shape

Molecular size can be a further limiting factor in oral absorption [54]. The Lipinski "Rule-of-5" proposes an upper limit of molecular weight (MW) of 500 as acceptable for orally absorbed compounds [25]. High-MW compounds tend to undergo biliary excretion. Size and shape parameters are generally not measured, but rather calculated. A measured property is the so-called cross-sectional area, which is obtained from surface activity measurements [55].

2.5.1 Calculated Size Descriptors

MW is often taken as the size descriptor of choice, while it is easy to calculate and is in the chemist's mind. However, other size and shape properties are equally simple to calculate, and may offer a better guide to estimate potential for permeability. Thus far no systematic work has been reported investigating this in detail. Cross-sectional area A_D obtained from surface activity measurements have been reported as a useful size descriptor to discriminate compounds which can access the brain ($A_D < 80\,Å^2$) of those that are too large to cross the blood–brain barrier (BBB) [55]. Similar studies have been performed to define a cut-off for oral absorption [56].

2.6 H-bonding

Molecular size and H-bonding have been unraveled as the two major components of log P or log D [57–59]. It was found that H-bonding capacity of a drug solute correlates reasonably well with passive diffusion. Δlog P, the difference between octanol–water and alkane–water partitioning, was suggested as a good measure for solute H-bonding [58, 60, 61]. However, this involves tedious experimental work and it appeared that calculated descriptors for H-bonding could most conveniently be assessed, in particular also for virtual compounds.

2.6.1 Calculated H-bonding descriptors

Considerable interest is focused on the calculation of H-bonding capability in the design of combinatorial libraries, for assessing the potential for oral absorption and permeability [16, 62–65]. A number of different descriptors for H-bonding have been discussed [66], one of the simplest being the count of the number of H-bond forming atoms [67].

A simple measure of H-bonding capacity, originally proposed by Van de Waterbeemd and Kansy, is the polar surface area (PSA), defined as the sum of the fractional contributions to surface area of all nitrogen and oxygen atoms and hydrogens attached to these [68]. PSA was used to predict passage of the BBB [69–71], flux across a Caco-2 monolayer [72] and human intestinal absorption [73, 74]. The physical explanation is that polar groups are involved in desolvation when they move from an aqueous extracellular environment to the more lipophilic interior of membranes. PSA thus represents, at least part of, the energy involved in membrane transport. PSA is dependent on the conformation and the original method [68] is based on a single minimum energy conformation. Others [73] have taken into account conformational flexibility and coined a dynamic PSA, in which a Boltzmann-weighted average PSA is computed. However, it was demonstrated that PSA calculated for a single minimum energy conformation is in most cases sufficient to produce a sigmoidal relationship to intestinal absorption, differing very little from the dynamic PSA described above [74]. A fast calculation of PSA as a sum of fragment-based contributions has been published [75], allowing use of these calculations for large datasets such as combinatorial or virtual libraries. The sigmoidal relationship can be described by $A\% = 100/[1 + (\mathrm{PSA}/\mathrm{PSA}_{50})^{\gamma}]$, where $A\%$ is percentage of orally absorbed drug, PSA_{50} is the PSA at 50% absorption level and γ is a regression coefficient [76].

Poorly absorbed compounds have been identified as those with a PSA > 140 Å^2. Considering more compounds, considerable more scatter was found around the sigmoidal curve observed for a smaller set of compounds [74]. This is partly due to the fact that many compounds do not show simple passive diffusion only, but are affected by active carriers, efflux mechanisms involving P-glycoprotein (P-gp) and other transporter proteins, and gut wall metabolism. These factors also contribute to the considerable inter-individual variability of human oral absorption data. A further refinement in the PSA approach is expected to come from taking into account the strength of the H-bonds, which in principle already is the basis of the HYBOT approach [63–65].

2.7 Lipophilicity

Octanol–water partition (log *P*) and distribution (log *D*) coefficients are widely used to make estimates for membrane penetration and permeability, including gastro-intestinal absorption [77, 78], BBB crossing [60, 69] and correlations to pharmacokinetic properties [1]. The two major components of lipophilicity are molecular size and H-bonding [57], which each have been discussed above (see Sections 2.5 and 2.6).

According to published International Union of Pure and Applied Chemistry recommendations the terms hydrophobicity and lipophilicity are best described as follows [79]:

- *Hydrophobicity* is the association of nonpolar groups or molecules in an aqueous environment which arises from the tendency of water to exclude nonpolar molecules
- *Lipophilicity* represents the affinity of a molecule or a moiety for a lipophilic environment. It is commonly measured by its distribution behavior in a biphasic system, either liquid–liquid (e.g. partition coefficient in 1-octanol–water) or solid–liquid (retention on reversed-phase high-performance liquid chromatography or thin-layer chromatography system).

The *intrinsic lipophilicity* (*P*) of a compound refers only to the equilibrium of the unionized (neutral) drug between the aqueous phase and the organic phase. It follows that the remaining part of the overall equilibrium, i.e. the concentration of ionized drug in the aqueous phase, is also of great importance in the overall observed partition ratio. This in turn depends on the pH of the aqueous phase and the acidity or basicity (pK_a) of the charged function. The overall ratio of drug, ionized and unionized, between the phases has been described as the *distribution coefficient* (*D*), to distinguish it from the intrinsic lipophilicity (P). The term has become widely used in recent years to describe, in a single term, the *effective (or net) lipophilicity* of a compound at a given pH taking into account both its intrinsic lipophilicity and its degree of ionization. The distribution coefficient (*D*) for a monoprotic acid (HA) is defined as:

$$D = [\mathrm{HA}]_{\mathrm{organic}}/([\mathrm{HA}]_{\mathrm{aqueous}} + [\mathrm{A}^-]_{\mathrm{aqueous}}) \quad (4)$$

where [HA] and [A⁻] represent the concentrations of the acid in its unionized and dissociated (ionized) states, respectively. The ionization of the compound in water is defined by its dissociation constant (K_a) as:

$$K_a = [H^+][A^-]/[HA] \quad (5)$$

sometimes referred to as the Henderson–Hasselbalch relationship. The combination of Eqs. (4)–(6) gives the pH-distribution (or "pH-partition") relationship:

$$D = P/(1 + \{K_a/[H^+]\}) \quad (6)$$

more commonly expressed for monoprotic organic *acids* in the form:

$$\log(\{P/D\} - 1) = \text{pH} - \text{p}K_a \quad (7)$$

or

$$\log D = \log P - \log(1 + 10^{\text{pH} - \text{p}K_a}) \quad (8)$$

For monoprotic organic *bases* (BH^+ dissociating to B) the corresponding relationships are:

$$\log(\{P/D\} - 1) = \text{p}K_a - \text{pH} \quad (9)$$

or:

$$\log D = \log P - \log(1 + 10^{\text{p}K_a - \text{pH}}) \quad (10)$$

From these equations it is possible to predict the effective lipophilicity (log *D*) of an acidic or basic compound at any pH value. The data required in order to use the relationship in this way are the intrinsic lipophilicity (log *P*), the dissociation constant (pK_a) and the pH of the aqueous phase. The overall effect of these relationships is the effective lipophilicity of a compound, at physiological pH, is approximately the log *P* value minus one unit of lipophilicity, for every unit of pH the pK_a value is below (for acids) and above (for bases) pH 7.4. Obviously for compounds with multifunctional ionizable groups the relationship between log *P* and log *D*, as well as log *D* as a function of pH become more complex [65, 68, 70]. For diprotic molecules there are already 12 different possible shapes of log *D*–pH plots.

Traditional octanol–water distribution coefficients are still widely used in quantitative structure–activity relationship (QSAR) and in ADME/PK studies. However, alternative solvent systems have been proposed [80]. To cover the variability in biophysical characteristics of different membrane types a set of four solvents has been suggested, sometimes called the "critical quartet" [81]. The 1,2-dichloroethane–water system has been promoted as a good alternative to alkane–water due to its far better dissolution properties [82, 83], but may find little application because of its carcinogenic properties.

Several approaches for higher throughput lipophilicity measurements have been developed in the pharmaceutical industry [50] including automated shake-plate methods [84] and immobilized artificial membranes [85]. A convenient method to measure octanol–water partitioning is based on potentiometric titration, called the pH method [86]. Methods based on chromatography are also widely used, e.g. chromatographic hydrophobicity indices measured on immobilized artificial membranes (IAM) [19, 87]. Another chromatography-based method is called Elog *D* giving log *D* values comparable to shake-flask data [88].

2.7.1 Calculated log *P* and log *D*

A number of rather comprehensive reviews on lipophilicity estimation have been published and are recommended for further reading [89–91]. Due to its key importance, a continued interest is seen to develop good log *P* estimation programs [82–94]. Most log *P* approaches are limited due to a lack of parameterization of certain fragments. For the widely used CLOGP program (Daylight/Biobyte computer program for the calculation of log *P*), a version making estimates for missing fragments has become available [95].

With only few exceptions, most log *P* programs refer to the octanol–water system. Based on Rekker's fragmental constant approach, a log *P* calculation for aliphatic hydrocarbon–water partitioning has been reported [96]. Another more recent approach to alkane–water log *P* and log *D* is based on the program VolSurf [97]. It is believed that these values may offer a better predictor for uptake in the brain. The group of Abraham investigated many other solvent systems and derived equations to predict log *P* from structure for these solvent systems, which are also commercially available [94].

Log *D* predictions are more difficult as most approaches rely on the combination of estimated log *P* and estimated pK_a. Obviously, this can lead to error accumulation and errors of 2 log units or more can be found. Some algorithms, however, are designed to learn from experimental data so that the predictions improve over time. An interesting approach is also the combination of a commercial log *D* predictor with proprietary descriptors using a Bayesian neural network approach [98].

2.8 Permeability

An overview of permeability assays is presented in Table 2.2. As discussed earlier in this chapter, these permeability scales are correlated to each other as well as the various lipophilicity scales via extended Collander equations.

2.8.1 Artificial Membranes and PAMPA

When screening for absorption by passive membrane permeability, artificial membranes have the advantage of offering a highly reproducible and high-throughput system. Artificial membranes have been compared to Caco-2 cells and for passive

Tab. 2.2 *In vitro* models for membrane permeability.

Permeability model	***Reference***
Solvent–water partitioning	
octanol–water distribution	52
Chromatography	
IAM	99–103
ILC	104
MEKC	105
BMC	106
Vesicles	
phospholipid vesicles	107
liposome binding	108, 109
Transil particles	110–112
fluorosomes	113
SPR biosensor	114, 115
colorimetric assay	116
Artificial membranes	
impregnated membranes	72
PAMPA	117–123
filter IAM	121–123
hexadecane-coated polycarbonate filters (HDM)	124, 125
Other	
surface activity	126
Cell-based assays	
Caco-2	76, 78
Madin-Darby canine kidney	127

diffusion found to behave very similar [72]. This finding was the basis for the development of the PAMPA for rapid prediction of transcellular absorption potential [117–120]. In this system the permeability through a membrane formed by a mixture of lecithin and an inert organic solvent on a hydrophobic filter support is assessed. Whilst not completely predictive for oral absorption in humans, PAMPA shows definite trends in the ability of molecules to permeate membranes by passive diffusion, which may be valuable in screening large compound libraries. This system is commercially available [121], but can easily be set up in-house. Further optimization of the experimental conditions has been investigated, concluding that predictability increases when a pH of 6.5 or 5.5 is used on the donor side [122, 123]. It was also demonstrated that the effect of a cosolvent such as dimethylsulfoxide (DMSO) could have a marked effect depending on the nature, basic/acid, of the compound [123]. Stirring of the donor compartment to limit the contribution of the unstirred water layer appears to be important to get meaningful results. There have been so far no reports in the literature about using PAMPA data in a drug discovery project.

A similar system has been reported based on polycarbonate filters coated with hexadecane, also called hexadecane membranes (HDM) [124, 125]. Thus, this system consists of a 9- to 10-µm hexadecane liquid layer immobilized between two aqueous compartments. Also here it was observed that in this set-up for lipophilic compounds the diffusion through the unstirred water layer becomes the rate-limiting step. To mimic the *in vivo* environment permeability measurements were repeated at different pH values in the range 4–8 and the highest transport value used for correlation with percentage absorbed in human. This gives a sigmoidal dependence, which is better than when taking values measured at a single pH, e.g. 6.8.

2.8.1.1 *In Silico* PAMPA

The experimental P_{app} data have been used to build predictive models. However, since PAMPA is already a model, an *in silico* model based on this is a model of a model. The predictability for *in vivo* permeability or absorption of such *in silico* PAMPA model can be questioned (see Eq. 11), since it is two steps from reality:

$$\text{model} \times \text{model} = \text{random} \qquad (11)$$

2.8.2 IAM, Immobilized Liposome Chromatography (ILC), Micellar Electrokinetic Chromatography (MEKC) and Biopartitioning Micellar Chromatography (BMC)

IAM columns are another means of measuring lipophilic characteristics of drug candidates and other chemicals [99–103]. IAM columns may better mimic membrane interactions than the isotropic octanol–water or other solvent–solvent partitioning system. These chromatographic indices appear to be a significant predictor of passive absorption through the rat intestine [128].

A related alternative is called ILC [104, 105]. Compounds with the same log *P* were shown to have very different degrees of membrane partitioning on ILC depending on the charge of the compound [105].

Another relatively new lipophilicity scale proposed for use in ADME studies is based on MEKC [106]. A further variant is called BMC and uses mobile phases of Brij35 [polyoxyethylene(23)lauryl ether] [129]. Similarly, the retention factors of 16 β-blockers obtained with micellar chromatography with sodium dodecyl sulfate as micelle-forming agent correlates well with permeability coefficients in Caco-2 monolayers and apparent permeability coefficients in rat intestinal segments [130].

Each of these scales produce a lipophilicity index related but not identical to octanol–water partitioning.

2.8.3 Liposome Partitioning

Liposomes, which are lipid bilayer vesicles prepared from mixtures of lipids, also provide a useful tool for studying passive permeability of molecules through lipid. This system has, for example, been used to demonstrate the passive nature of the absorption mechanism of monocarboxylic acids [131]. Liposome partitioning of ionizable drugs can be determined by titration and has been correlated with human absorption [108, 109, 132]. Liposome partitioning is only partly correlated with octanol–water distribution and might contain some additional information.

A further partition system based on the use of liposomes, and commercialized under the name Transil [110, 111], has shown its utility as a lipophilicity measure in PBPK modeling [112]. Fluorescent-labeled liposomes, called fluorosomes, are another means of measuring the rate of penetration of small molecules into membrane bilayers [113, 120]. Similarly, a colorimetric assay amenable to HTS for evaluating membrane interactions and penetration has been presented [116]. The platform comprises vesicles of phospholipids and the chromatic lipid-mimetic polydiacetylene. The polymer undergoes visible concentration-dependent red–blue transformations induced through interactions of the vesicles with the studied molecules.

2.8.4 Biosensors

Liposomes have been attached to a biosensor surface, and the interactions between drugs and the liposomes can be monitored directly using surface plasmon resonance (SPR) technology. SPR is measuring changes in refractive index at the sensor surface caused by changes in mass. Drug–liposome interactions have been measured for 27 drugs and compared to fraction absorbed in humans [114]. A reasonable correlation is obtained, but it is most likely that this method represents just another way of measuring "lipophilicity". The throughput was 100 substances per 24 h, but further progress seems possible. In more recent work using this method it is proposed to use two types of liposomes to separate compounds according to their absorption potential [115].

2.9 Amphiphilicity

The combination of hydrophilic and hydrophobic parts of a molecule defines its amphiphilicity. A program has been described to calculate this property and calibrated against experimental values obtained from surface activity measurements [133]. These values can possibly be used to predict effect on membranes leading to cytotoxicity or phospholipidosis, but may also contain information, not yet unraveled, on permeability. Surface activity measurements have also been used to make estimates of oral absorption [126].

2.10 Drug-like Properties

The various properties described above are important for drugs, in particular for those given orally. The important question arises whether such properties of drugs are different from chemicals used in other ways. This has been subject of a number of investigations [134, 135]. Using neural networks [136, 137] or a decision tree approach [138], a compound can be predicted as being "drug-like" with an error rate of around 20%. A further approach to predict drug-likeness consists of training of the program PASS (prediction of activity spectra for substances) [139], which originally was intended to predict activity profiles and thus is suitable to predict potential side effects.

From an analysis of the key properties of compounds in the World Drug Index the now well accepted "Rule-of-5" has been derived [25, 26]. It was concluded that compounds are most likely to have poor absorption when $MW > 500$, calculated octanol–water partition coefficient $\text{Clog } P > 5$, number of H-bond donors >5 and number of H-bond acceptors >10. Computation of these properties is now available as a simple but efficient ADME screen in commercial software. The "Rule-of-5" should be seen as a qualitative absorption/permeability predictor [43], rather than a quantitative predictor [140]. The "Rule-of-5" is not predictive for bioavailability as sometimes mistakenly is assumed. An important factor for bioavailability in addition to absorption is liver first-pass effect (metabolism). The property distribution in drug-related chemical databases has been studied as another approach to understand "drug-likeness" [141, 142].

Other attempts have been made to try to define good leads. In general lead-like properties are lower/smaller than drug-like properties. Thus, $MW < 350$ and $\text{Clog } P < 3$ should be good starting points for leads [143, 144]. A "Rule-of-3" has been proposed [145] for screening of small fragments, which says the good lead fragments have $MW < 300$, $\text{Clog } P < 3$, H-bond donors and acceptors <3 and rotatable bonds <3.

Similarly, in a study on drugs active as central nervous system (CNS) agents and using neural networks based on Bayesian methods, CNS-active drugs could be distinguished from CNS-inactive ones [145]. A CNS rule-of-thumb says that if the sum of the nitrogen and oxygen (N + O) atoms in a molecule is less than 5 and if the $\text{Clog } P - (N + O) > 0$, then compounds are likely to penetrate to the BBB [146]. Another "rule" is $PSA < 90\,\text{Å}^2$, $MW < 450$ and log D at pH 7.4 of 1–3 [147]. In designing CNS drugs it is important to distinguish BBB penetration and CNS efficacy. The latter is a subtle balance between permeability, effect of BBB transporters, lipophilicity, and free fraction in blood and brain [148].

These aforementioned analyses all point to a critical combination of physicochemical and structural properties [149], which to a large extent can be manipulated by the medicinal chemist. This approach in medicinal chemistry has been called property-based design [2]. Under properties in this context we intend physicochemical as well as PK and toxicokinetic properties. These have been neglected for a long time by most medicinal chemists, who in many cases in the past only had the quest for strongest receptor binding as the ultimate goal. However, this strategy has changed dramatically, and the principles of drug-like compounds are now being used in computational approaches towards the rational design of combinatorial libraries [150] and in decision making on acquisition of outsourced libraries.

2.11
Computation versus Measurement of Physicochemical Properties

2.11.1
QSAR Modeling

Calculation of many different one-, two- and three-dimensional descriptors for building predictive QSAR models for physicochemical (and ADME/toxicity) properties is possible using a range of commercially available software packages, such as ACD, SYBYL, Cerius^2, Molconn-Z, HYBOT, VolSurf, MolSurf, Dragon, MOE, BCUT, etc. Several descriptor sets are based on quantification of three-dimensional molecular surface properties [151, 152] and these have been explored for the prediction of, for example, Caco-2 permeability and oral absorption [16]. It is pointed out here that a number of these "new" descriptors are often strongly correlated to the more traditional physicochemical properties. An aspect largely neglected so far is the concept of molecular-property space looking at the conformational effects on physicochemical properties [153].

Numerous QSAR tools have been developed [152, 154] and used in modeling physicochemical data. These vary from simple linear to more complex nonlinear models, as well as classification models. A popular approach more recently became the construction of consensus or ensemble models ("combinatorial QSAR") combining the predictions of several individual approaches [155]. Or, alternatively, models can be built by running the same approach, such as a neural network of a decision tree, many times and combining the output into a single prediction.

To build robust predictive models good quality training set and sound test set are required. Criteria for a good set include sufficient coverage of chemical space, good distribution between low- and high-end values of the property studied, and a sufficiently large number of compounds. Models can be global (covering many types of chemistry) or local (project-specific). There are many reasons why predictions can fail [156] and medicinal chemists need to be aware of these. There is also a difference between a useful model and a perfect model. The latter does not exist!

In-house physicochemical data collections are growing rapidly through the use of HTS technologies [157]. Therefore, the need for rapidly building and updating is also increasing. Systems for automatic and regular updating of QSAR predictive models have been reported [158] and we expect these to become more widespread. A consequence of regularly updated *in silico* models is that the predicted values will change too. This will require adapted ways of working by the chemists and DMPK scientists in projects using more dynamic data generation and interpretation tools.

2.11.2
In Combo: Using the Best of two Worlds

In modern drug discovery speed and cost control are important in addition to high quality. *In silico* virtual screening for drugability [159] is a good first step in library design and compound acquisition. Once compounds have been made for a targeted project a well-balanced approach using both *in silico* predictions and *in vitro* screening will be a good strategy to guide the programme in a cost-efficient manner. New experimental data can be used to update predictive models regularly so that the ongoing projects can benefit from the latest local and global models available [158, 160].

2.12 Outlook

Physical chemistry plays a key role in the behavior of drugs. Measurement of the key properties has been automated and industrialized to high throughput. The data can and are used to build robust predictive models. These can in turn be used to limit the use of experiments when not strictly needed. This is of course compound saving and more cost-effective. Predictive models are also great tools in virtual screening, prioritization decision making and guiding projects. The rest of this book provides in-depth insight into some of the properties briefly discussed in this introductory chapter.

References

1. Smith, D. A., Jones, B. C., Walker, D. K. Design of drugs involving the concepts and theories of drug metabolism and pharmacokinetics. *Med. Res. Rev.* **1996**, *16*, 243–266.
2. Van de Waterbeemd, H., Smith, D. A., Beaumont, K., Walker, D. K. Property-based design: optimization of drug absorption and pharmacokinetics. *J. Med. Chem.* **2001**, *44*, 1313–1333.
3. Smith, D. A., Van de Waterbeemd, H., Walker, D. K. *Pharmacokinetics and Metabolism in Drug Design*, 2nd edn. (*Methods and Principles in Medicinal Chemistry*), Wiley-VCH, Weinheim, **2006**.
4. Testa, B., Krämer, S. D., Wunderli-Allenspach, H., Folkers, G. (eds.). *Pharmaco-leinetic Profiling in Drug Research*, VHCA, Zurich and Wiley-VCH, Weinheim, **2006**.
5. Testa, B., Van de Waterbeemd, H. (vol. eds.). ADME/Tox Approaches, Vol. 5 in *Comprehensive Medicinal Chemistry*, 2nd edn., Taylor, J. B., Triggle. D. J. (eds.), Elsevier, Oxford, **2007**.
6. Pliska, V., Testa, B., Van de Waterbeemd, H. (eds.). *Lipophilicity in Drug Action and Toxicology*, VCH, Weinheim, **1996**.
7. Testa, B., Van de Waterbeemd, H., Folkers, G., Guy, R. (eds.). *Pharmacokinetic Optimization in Drug Research: Biological, Physicochemical and Computational Strategies*, VHCA, Zurich and Wiley-VCH, Weinheim, **2001**.
8. Kerns, E. H., Di, L. Physicochemical profiling: overview of the screens. *Drug Discov. Today Technol.* **2004**, *1*, 343–348.
9. Van de Waterbeemd, H. Physicochemical approaches to drug absorption. In *Drug Bioavailability (Methods and Principles in Medicinal Chemistry)*, Van de Waterbeemd, H., Lennernäs, H., Artursson, P. (eds.), Wiley-VCH, Weinheim, **2003**, pp. 3–20.
10. Van de Waterbeemd, H. Physicochemistry. In *Pharmacokinetics and Metabolism in Drug Design*, 2nd edn. (*Methods and Principles in Medicinal Chemistry*), Smith, D. A., Van de Waterbeemd, H., Walker, D. K. (eds.), Wiley-VCH, Weinheim, **2006**, pp. 1–18.

11 Van de Waterbeemd, H. Property-based lead optimization. In *Biological and Physicochemical Profiling in Drug Research*, Testa, B., Krämer, S. D., Wunderli-Allenspach, H., Folkers, G. (eds.), VHCA, Zurich and Wiley-VCH, Weinheim, **2006**, pp. 25–45.

12 Kerns, E. H., Di, L. Chemical stability. In *ADME/Tox Approaches*, Testa, B., Van de Waterbeemd, H. (vol. eds.), Vol. *5* in *Comprehensive Medicinal Chemistry*, 2nd edn., Taylor, J. B., Triggle, D. J. (eds.), Elsevier, Oxford, **2007**, pp. 489–507.

13 Kola, I., Landis, J. Can the pharmaceutical industry reduce attrition rates? *Nat. Rev. Drug Discov.* **2004**, *3*, 711–716.

14 Avdeef, A., Tsinman, O. PAMPA – a drug absorption *in vitro* model 13. Chemical selectivity due to membrane hydrogen bonding: *in combo* comparisons of HDM-, DOPC-, and DS-PAMPA models. *Eur. J. Pharm. Sci.* **2006**, *28*, 43–50.

15 Avdeef, A., Artursson, P., Neuhoff, S., Lazorova, L., Gråsjö, J., Tavelin, S. Caco-2 permeability of weakly basic drugs predicted with the double-sink PAMPA pK_a^{flux} method. *Eur. J. Pharm. Sci.* **2005**, *24*, 333–349.

16 Van de Waterbeemd, H. *In silico* models to predict oral absorption. In *ADME/Tox Approaches*, Testa, B., Van de Waterbeemd, H. (vol. eds.), Vol. *5* in *Comprehensive Medicinal Chemistry*, 2nd edn., Taylor, J. B., Triggle, D. J. (eds.), Elsevier, Oxford, **2007**, pp. 669–697.

17 Van de Waterbeemd, H. Which *in vitro* screens guide the prediction of oral absorption and volume of distribution? *Basic Clin. Pharmacol. Toxicol.* **2005**, *96*, 162–166.

18 Lombardo, F., Obach, R. S., Shalaeva, M. Y., Gao, F. Prediction of human volume of distribution values for neutral and basic drugs. 2. Extended data set and leave-class-out statistics. *J. Med. Chem.* **2004**, *47*, 1242–1250.

19 Valko, K., Nunhuck, S., Bevan, C., Abraham, M. H., Reynolds, D. P. Fast gradient HPLC method to determine compounds binding to human serum albumin. Relationships with octanol/water and immobilized artificial membrane lipophilicity. *J. Pharm. Sci.* **2003**, *92*, 2236–2248.

20 Van de Waterbeemd, H., Smith, D. A., Jones, B. C. Lipophilicity in PK design: methyl, ethyl, futile. *J. Comput.-Aided Mol. Des.* **2001**, *15*, 273–286.

21 Avdeef, A., Voloboy, A., Foreman, A. Solubility and dissolution. In *ADME/Tox Approaches*, Testa, B., Van de Waterbeemd, H. (vol. eds.), Vol. *5* in *Comprehensive Medicinal Chemistry*, 2nd edn., Taylor, J. B., Triggle, D. J. (eds.), Elsevier, Oxford, **2007**, pp. 399–423.

22 Bevan, C. D., Lloyd, R. S. A high-throughput screening method for the determination of aqueous drug solubility using laser nephelometry in microtiter plates. *Anal. Chem.* **2000**, *72*, 1781–1787.

23 Avdeef, A. High-throughput measurements of solubility profiles. In *Pharmacokinetic Optimization in Drug Research: Biological, Physicochemical and Computational Strategies*, Testa, B., Van de Waterbeemd, H., Folkers, G., Guy, R. (eds.), VHCA, Zurich and Wiley-VCH, Weinheim, **2001**, pp 305–325.

24 Avdeef, A., Berger, C. M. pH-metric solubility. 3. Dissolution titration template method for solubility determination. *Eur. J. Pharm. Sci.* **2001**, *14*, 281–291.

25 Lipinski, C. A., Lombardo, F., Dominy, B. W., Feeney, P. J. Experimental and computational approaches to estimate solubility and permeability in drug discovery and development settings. *Adv. Drug. Del. Rev.* **1997**, *23*, 3–25.

26 Lipinski, C. Drug-like properties and the causes of poor solubility and poor permeability. *J. Pharmacol. Toxicol. Methods* **2000**, *44*, 235–249.

27 Johnson, K., Swindell, A. Guidance in the setting of drug particle size specifications to minimize variability in absorption. *Pharm. Res.* **1996**, *13*, 1795–1798.

28 Curatolo, W. Physical chemical properties of oral drug candidates in the discovery and exploratory development settings. *Pharm. Sci. Technol. Today* **1998**, *1*, 387–393.

29 Dressman, J. B., Amidon, G. L., Reppas, C., Shah, V. P. Dissolution testing as a prognostic tool for oral drug absorption: immediate release dosage forms. *Pharm. Res.* **1998**, *15*, 11–22.

30 Amidon, G. L., Lennernäs, H., Shah, V. P., Crison, J. R. A. A theoretical basis for a biopharmaceutic drug classification: the correlation of *in vitro* drug product dissolution and *in vivo* bioavailability. *Pharm. Res.* **1995**, *12*, 413–420.

31 Van de Waterbeemd, H. The fundamental variables of the biopharmaceutics classification system (BCS): a commentary. *Eur. J. Pharm. Sci.* **1998**, *7*, 1–3.

32 Huuskonen, J. Estimation of aqueous solubility in drug design. *Comb. Chem. High-Throughput Screen.* **2001**, *4*, 311–316.

33 McFarland, J. W., Avdeef, A., Berger, C. M., Raevsky, O. A. Estimating the water solubilities of crystalline compounds from their chemical structures alone. *J. Chem. Inf. Comput. Sci.* **2001**, *41*, 1355–1359.

34 Livingstone, D. J., Ford, M. G., Huuskonen, J. J., Salt, D. W. Simultaneous prediction of aqueous solubility and octanol/water partition coefficient based on descriptors derived from molecular structure. *J. Comput.-Aided Mol. Des.* **2001**, *15*, 741–752.

35 Bruneau, P. Search for predictive generic model of aqueous solubility using Bayesian neural nets. *J. Chem. Inf. Comput. Sci.* **2001**, *41*, 1605–1616.

36 Liu, R., So, S.-S. Development of quantitative structure–property relationship models for early ADME evaluation in drug discovery. 1. Aqueous solubility. *J. Chem. Inf. Comput. Sci.* **2001**, *4*, 1633–1639.

37 Taskinen, J., Norinder, U. *In silico* prediction of solubility. In *ADME/Tox Approaches*, Testa, B., Van de Waterbeemd, H. (vol. eds.), Vol. *5* in *Comprehensive Medicinal Chemistry*, 2nd edn., Taylor, J. B., Triggle, D. J. (eds.), Elsevier, Oxford, **2007**, pp. 627–648.

38 Dearden, J. C. *In silico* prediction of aqueous solubility. *Exp. Opin. Drug Discov.* **2006**, *1*, 31–52.

39 Bergstrom, C. A. S. Computational models to predict aqueous drug solubility, permeability and intestinal absorption. *Exp. Opin. Drug Metab. Toxicol.* **2005**, *1*, 613–627.

40 Van de Waterbeemd, H. High-throughput and *in silico* techniques in drug metabolism and pharmacokinetics. *Curr. Opin. Drug Dis. Dev.* **2002**, *5*, 33–43.

41 Parrott, N., Paquereau, N., Coassolo, P., Lavé, Th. An evaluation of the utility of physiologically based models of pharmacokinetics in early drug discovery. *J. Pharm. Sci.* **2005**, *94*, 2327–2343.

42 Camenisch, G., Van de Waterbeemd, H., Folkers, G. Review of theoretical passive drug absorption models: historical background, recent development and limitations. *Pharm. Acta. Helv.* **1996**, *71*, 309–327.

43 Pagliara, A., Resist, M., Geinoz, S., Carrupt, P.-A., Testa, B. Evaluation and prediction of drug permeation. *J. Pharm. Pharmacol.* **1999**, *51*, 1339–1357.

44 Caron, G., Gaillard, P., Carrupt, P. A., Testa, B. Lipophilicity behavior of model and medicinal compounds containing a sulfide, sulfoxide, or sulfone moiety. *Helv. Chim. Acta* **1997**, *80*, 449–461.

45 Boisset, M., Botham, R. P., Haegele, K. D., Lenfant, B., Pachot, J. L. Absorption of angiotensin II antagonists in Ussing chambers, Caco-2, perfused jejunum loop and *in vivo*: importance of drug ionization in the *in vitro* prediction of *in vivo* absorption. *Eur. J. Pharm. Sci.* **2000**, *10*, 215–224.

46 Palm, K., Luthman, K., Ros, J., Grasjo, J., Artursson, P. Effect of molecular charge on intestinal epithelial drug transport: pH-dependent transport of cationic drugs. *J. Pharmacol. Exp. Ther.* **1999**, *291*, 435–443.

47 Comer, J. Ionization constants and ionization profiles. In *ADME/Tox Approaches*, Testa, B., Van de Waterbeemd, H. (vol. eds.), Vol. *5* in *Comprehensive Medicinal Chemistry*, 2nd edn., Taylor, J. B., Triggle, D. J. (eds.), Elsevier, Oxford, **2007**, pp. 357–397.

48 Saurina, J., Hernandez-Cassou, S., Tauler, R., Izquierdo-Ridorsa, A., Spectrophotometric determination of pK_a values based on a pH gradient flow-injection system. *Anal. Chim. Acta* **2000**, *408*, 135–143.

49 Jia, Z., Ramstad, T., Zhong, M. Medium-throughput pK_a screening of pharmaceuticals by pressure-assisted capillary electrophoresis. *Electrophoresis* **2001**, *22*, 1112–1118.

50 Comer, J., Tam, K. Lipophilicity profiles: theory and measurement. In *Pharmacokinetic Optimization in Drug Research: Biological, Physicochemical and Computational Strategies*, Testa, B., Van de Waterbeemd, H., Folkers, G., Guy, R. (eds.), VHCA, Zurich and Wiley-VCH, Weinheim, **2001**, pp 275–304.

51 Caron, G., Reymond, F., Carrupt, P. A., Girault, H. H., Testa, B. Combined molecular lipophilicity descriptors and their role in understanding intramolecular effects. *Pharm. Sci. Technol. Today* **1999**, *2*, 327–335.

52 Caron, G., Scherrer, R. A., Ermondi, G. Lipophilicity, polarity and hydrophobicity. In *ADME/Tox Approaches*, Testa, B., Van de Waterbeemd, H. (vol. eds.), Vol. *5* in *Comprehensive Medicinal Chemistry*, 2nd edn., Taylor, J. B., Triggle, D. J. (eds.), Elsevier, Oxford, **2007**, pp. 425–452.

53 Franczkiewicz, R. *In silico* prediction of ionization. In *ADME/Tox Approaches*, Testa, B., Van de Waterbeemd, H. (vol. eds.), Vol. *5* in *Comprehensive Medicinal Chemistry*, 2nd edn., Taylor, J. B., Triggle, D. J. (eds.), Elsevier, Oxford, **2007**, pp. 603–626.

54 Chan, O. H., Stewart, B. H. Physicochemical and drug-delivery considerations for oral drug bioavailability. *Drug Discov. Today* **1996**, *1*, 461–473.

55 Fischer, H., Gottschlich, R., Seelig, A. Blood–brain barrier permeation: Molecular parameters governing passive diffusion. *J. Membr. Biol.* **1998**, *165*, 201–211.

56 Fischer, H. Passive diffusion and active transport through biological membranes – binding of drugs to transmembrane receptors. PhD Thesis, University of Basel, **1998**

57 Van de Waterbeemd, H., Testa, B. The parameterization of lipophilicity and other structural properties in drug design. *Adv. Drug Res.* **1987**, *16*, 85–225.

58 El Tayar, N., Testa, B., Carrupt, P. A. Polar intermolecular interactions encoded in partition coefficients: an indirect estimation of hydrogen-bond parameters of polyfunctional solutes. *J. Phys. Chem.* **1992**, *96*, 1455–1459.

59 Abraham, M. H., Chadha, H. S. Applications of a solvation equation to drug transport properties. In *Lipophilicity in Drug Action and Toxicology*, Pliska, V., Testa, B., Van de Waterbeemd, H. (eds.), VCH, Weinheim, **1996**, pp. 311–337.

60 Young, R. C., Mitchell, R. C., Brown, Th. H., Ganellin, C. R., Griffiths, R., Jones, M., Rana, K. K., Saunders, D., Smith, I. R., Sore, N. E., Wilks, T. J. Development of a new physicochemical model for brain penetration and its application to the design of centrally acting H_2 receptor histamine antagonists. *J. Med. Chem.* **1988**, *31*, 656–671.

61 Von Geldern, T. W., Hoffmann, D. J., Kester, J. A., Nellans, H. N., Dayton, B. D., Calzadilla, S. V., Marsch, K. C., Hernandez, L., Chiou, W., Dixon, D. B., Wu-Wong, J. R., Opgenorth, T. J. Azole endothelin antagonists. 3. Using $\Delta \log P$ as a tool to improve absorption. *J. Med. Chem.* **1996**, *39*, 982–991.

62 Dearden, J. C., Ghafourian, T. Hydrogen bonding parameters for QSAR: comparison of indicator variables, hydrogen bond counts, molecular orbital and other parameters. *J. Chem. Inf. Comput. Sci.* **1999**, *39*, 231–235.

63 Raevsky, O. A., Schaper, K.-J. Quantitative estimation of hydrogen bond contribution to permeability and absorption processes of some chemicals and drugs. *Eur. J. Med. Chem.* **1998**, *33*, 799–807.

64 Raevsky, O. A., Fetisov, V. I., Trepalina, E. P., McFarland, J. W., Schaper, K.-J. Quantitative estimation of drug absorption in humans for passively transported compounds on the basis of their physico-chemical parameters.

Quant. Struct.-Act. Relat. **2000**, *19*, 366–374.

65 Van de Waterbeemd, H., Camenisch, G., Folkers, G., Raevsky, O. A. Estimation of Caco-2 cell permeability using calculated molecular descriptors. *Quant. Struct.-Act. Relat.* **1996**, *15*, 480–490.

66 Van de Waterbeemd, H. Intestinal permeability: prediction from theory. In *Oral Drug Absorption*, Dressman, J. B., Lennernäs, H. (eds.), Dekker, New York, **2000**, pp. 31–49.

67 Österberg, Th., Norinder, U. Prediction of polar surface area and drug transport processes using simple parameters and PLS statistics. *J. Chem. Inf. Comput. Sci.* **2000**, *40*, 1408–1411.

68 Van de Waterbeemd, H., Kansy, M. Hydrogen-bonding capacity and brain penetration. *Chimia* **1992**, *46*, 299–303.

69 Van de Waterbeemd, H., Camenisch, G., Folkers, G., Chrétien, J. R., Raevsky, O. A. Estimation of blood–brain barrier crossing of drugs using molecular size and shape, and H-bonding descriptors. *J. Drug Target.* **1998**, *2*, 151–165.

70 Kelder, J., Grootenhuis, P. D. J., Bayada, D. M., Delbressine, L. P. C., Ploemen, J.-P. Polar molecular surface as a dominating determinant for oral absorption and brain penetration of drugs. *Pharm. Res.* **1999**, *16*, 1514–1519.

71 Clark, D. E. Rapid calculation of polar molecular surface area and its application to the prediction of transport phenomena. 2. Prediction of blood–brain barrier penetration. *J. Pharm. Sci.* **1999**, *88*, 815–821.

72 Camenisch, G., Folkers, G., Van de Waterbeemd, H. Comparison of passive drug transport through Caco-2 cells and artificial membranes. *Int. J. Pharm.* **1997**, *147*, 61–70.

73 Palm, K., Luthman, K., Ungell, A.-L., Strandlund, G., Beigi, F., Lundahl, P., Artursson, P. Evaluation of dynamic polar molecular surface area as predictor of drug absorption: comparison with other computational and experimental predictors. *J. Med. Chem.* **1998**, *41*, 5382–5392.

74 Clark, D. E. Rapid calculation of polar molecular surface area and its application to the prediction of transport phenomena. 1. Prediction of intestinal absorption. *J. Pharm. Sci.* **1999**, *88*, 807–814.

75 Ertl, P., Rohde, B., Selzer, P., Fast calculation of molecular polar surface area as a sum of fragment-based contributions and its application to the prediction of drug transport properties. *J. Med. Chem.* **2000**, *43*, 3714–3717.

76 Stenberg, P., Norinder, U., Luthman, K., Artursson, P. Experimental and computational screening models for the prediction of intestinal drug absorption. *J. Med. Chem.* **2001**, *44*, 1927–1937.

77 Winiwarter, S., Bonham, N. M., Ax, F., Hallberg, A., Lennernäs, H., Karlén, A. Correlation of human jejunal permeability (*in vivo*) of drugs with experimentally and theoretically derived parameters. A multivariate data analysis approach. *J. Med. Chem.* **1998**, *41*, 4939–4949.

78 Artursson, P., Karlsson, J. Correlation between oral drug absorption in humans and apparent drug permeability coefficients in human intestinal epithelial (Caco-2) cells. *Biochem. Biophys. Res. Commun.* **1991**, *175*, 880–885.

79 Van de Waterbeemd, H., Carter, R. E., Grassy, G., Kubinyi, H., Martin, Y. C., Tute, M. S., Willett, P. Glossary of terms used in computational drug design. *Pure Appl. Chem.* **1997**, *69*, 1137–1152 and *Ann. Rep. Med. Chem.* **1998**, *33*, 397–409.

80 Hartmann, T., Schmitt, J. Lipophilicity – beyond octanol/water: a short comparison of modern technologies. *Drug Discov. Today Technol.* **2004**, *1*, 431–439.

81 Leahy, D. E., Morris, J. J., Taylor, P. J., Wait, A. R. Membranes and their models: towards a rational choice of partitioning system. In *QSAR: Rational Approaches to the Design of Bioactive Compounds*, Silipo, C., Vittoria, A. (eds.), Elsevier, Amsterdam, **1991**, pp. 75–82.

82 Steyeart, G., Lisa, G., Gaillard, P., Boss, G., Reymond, F., Girault, H. H., Carrupt, P. A., Testa, B. Intermolecular forces expressed in 1,2-dichloroethane–water partition coefficients. A solvatochromic analysis. *J. Chem. Soc. Faraday Trans.* **1997**, *93*, 401–406.

83 Caron, G., Steyaert, G., Pagliara, A., Reymond, F., Crivori, P., Gaillard, P., Carrupt, P. A., Avdeef, A., Comer, J., Box, K. J., Girault, H. H., Testa, B. Structure–lipophilicity relationships of neutral and protonated β-blockers. Part 1. Intra- and intermolecular effects in isotropic solvent systems. *Helv. Chim. Acta* **1999**, *82*, 1211–1222.

84 Hitzel, L., Watt, A. P., Locker, K. L. An increased throughput method for the determination of partition coefficients. *Pharm. Res.* **2000**, *17*, 1389–1395.

85 Faller, B., Grimm, H. P., Loeuillet-Ritzler, F., Arnold, S., Briand, X. High-throughput lipophilicity measurement with immobilized artificial membranes. *J. Med. Chem.* **2005**, *48*, 2571–2576.

86 Avdeef, A. pH-metric log *P*. II: refinement of partition coefficients and ionization constants of multiprotic substances. *J. Pharm. Sci.* **1993**, *82*, 183–190.

87 Valko, K., Du, C. M., Bevan, C. D., Reynolds, D. P., Abraham, M. H., Rapid-gradient HPLC method for measuring drug interactions with immobilized artificial membrane: comparison with other lipophilicity measures. *J. Pharm. Sci.* **2000**, *89*, 1085–1096.

88 Lombardo, F., Shalaeva, M. Y., Tupper, K. A., Gao, F. ElogD$_{oct}$: a tool for lipophilicity determination in drug discovery. 2. Basic and neutral compounds. *J. Med. Chem.* **2001**, *44*, 2490–2497.

89 Buchwald, P., Bodor, N. Octanol–water partition: searching for predictive models. *Curr. Med. Chem.* **1998**, *5*, 353–380.

90 Carrupt, P. A., Testa, B., Gaillard, P. Computational approaches to lipophilicity: Methods and applications. *Rev. Comp. Chem.* **1997**, *11*, 241–315.

91 Mannhold, R., Van de Waterbeemd, H. Substructure and whole molecule approaches for calculating log *P*. *J. Comput.-Aided Mol. Des.* **2001**, *15*, 337–354.

92 Wildman, S. A., Crippen, G. M. Prediction of physicochemical parameters by atomic contributions. *J. Chem. Inf. Comput. Sci.* **1999**, *39*, 868–873.

93 Spessard, G. O. ACD Labs/logP dB 3.5 and ChemSketch 3.5. *J. Chem. Inf. Comput. Sci.* **1998**, *38*, 1250–1253.

94 Tetko, I., Livingstone, D. J. Rule-based systems to predict lipophilicity. In *ADME/Tox Approaches*, Testa, B., Van de Waterbeemd, H. (vol. eds.), Vol. 5 in *Comprehensive Medicinal Chemistry*, 2nd edn., Taylor, J. B., Triggle, D. J. (eds.), Elsevier, Oxford, **2007**, pp. 649–668.

95 Leo, A. J., Hoekman, D., Calculating log P_{oct} with no missing fragments; the problem of estimating new interaction parameters. *Perspect. Drug Discov. Des.* **2000**, *18*, 19–38.

96 Mannhold, R., Rekker, R. F. The hydrophobic fragmental constant approach for calculating log *P* in octanol/water and aliphatic hydrocarbon/water systems. *Perspect. Drug Discov. Des.* **2000**, *18*, 1–18.

97 Caron, G., Ermondi, G. Calculating virtual log *P* in the alkane/water system ($\log P^{N}_{alk}$) and its derived parameters $\Delta \log P^{N}_{oct-alk}$ and $\log D^{pH}_{alk}$. *J. Med. Chem.* **2005**, *48*, 3269–3279.

98 Bruneau, P., McElroy, N. R. Log $D^{7.4}$ modeling using Bayesian regularised neural networks. Assessment and correction of errors of prediction. *J. Chem. Inf. Model.* **2006**, *46*, 1379–1387.

99 Yang, C. Y., Cai, S. J., Liu, H., Pidgeon, C. Immobilized artificial membranes – screens for drug-membrane interactions. *Adv. Drug Del. Revs.* **1996**, *23*, 229–256.

100 Ong, S., Liu, H., Pidgeon, C. Immobilized-artificial-membrane chromatography: measurements of membrane partition coefficient and predicting drug membrane permeability. *J. Chromatogr. A.* **1996**, *728*, 113–128.

101 Stewart, B. H., Chan, O. H. Use of immobilized artificial membrane chromatography for drug transport applications. *J. Pharm. Sci.* **1998**, *87*, 1471–1478.

102 Ducarne, A., Neuwels, M., Goldstein, S., Massingham, R. IAM retention and blood brain barrier penetration. *Eur. J. Med. Chem.* **1998**, *33*, 215–223.

103 Reichel, A., Begley, D. J. Potential of immobilized artificial membranes for predicting drug penetration across the blood–brain barrier. *Pharm. Res.* **1998**, *15*, 1270–1274.

104 Lundahl, P., Beigi, F. Immobilized liposome chromatography of drugs for model analysis of drug–membrane interactions. *Adv. Drug Deliv. Rev.* **1997**, *23*, 221–227.

105 Norinder, U., Österberg, Th. The applicability of computational chemistry in the evaluation and prediction of drug transport properties. *Perspect. Drug Discov. Des.* **2000**, *19*, 1–18.

106 Trone, M. D., Leonard, M. S., Khaledi, M. G. Congeneric behavior in estimations of octanol–water partition coefficients by micellar electrokinetic chromatography. *Anal. Chem.* **2000**, *72*, 1228–1235.

107 Austin, R. P., Davis, A. M., Manners, C. N. Partitioning of ionising molecules between aqueous buffers and phospholipid vesicles. *J. Pharm. Sci.* **1995**, *84*, 1180–1183.

108 Balon, K., Riebesehl, B. U., Muller, B. W. Drug liposome partitioning as a tool for the prediction of human passive intestinal absorption. *Pharm. Res.* **1999**, *16*, 882–888.

109 Balon, K., Riebesehl, B. U., Muller, B. W. Determination of liposome partitioning of ionizable drugs by titration. *J. Pharm. Sci.* **1999**, *88*, 802–806.

110 Escher, B. I, Schwarzenbach, R. P., Westall, J. C. Evaluation of liposome–water partitioning of organic acids and bases. 2. Comparison of experimental determination methods. *Environ. Sci. Technol.* **2000**, *34*, 3962–3968.

111 Loidl-Stahlhofen, A., Eckert, A., Hartmann, T., Schöttner, M. Solid-supported lipid membranes as a tool for determination of membrane affinity: high-throughput screening of a physicochemical parameter. *J. Pharm. Sci.* **2001**, *90*, 599–606.

112 Willmann, S., Lippert, J., Schmitt, W. From physicochemistry to absorption and distribution: predictive mechanistic modeling and computational tools. *Exp. Opin. Drug Metab. Toxicol.* **2005**, *1*, 159–168.

113 Melchior, D. L. A rapid emprirical method for measuring membrane bilayer entry equilibration of molecules. *J. Pharm. Sci.* **2002**, *91*, 1075–1079.

114 Danelian, E., Karlén, A., Karlsson, R., Winiwarter, S., Hansson, A., Löfås, S., Lennernäs, H., Hämäläinen, D. SPR biosensor studies of the direct interaction between 27 drugs and a liposome surface: correlations with fraction absorbed in humans. *J. Med. Chem.* **2000**, *43*, 2083–2086.

115 Frostell-Karlsson, A., Widegren, H., Green, C. E., Hämäläinen, M. D., Westerlund, L., Karlsson, R., Fenner, K., Van De Waterbeemd, H. Biosensor analysis of the interaction between drug compounds and liposomes of different properties; a two-dimensional characterization tool for estimation of membrane absorption. *J. Pharm. Sci.* **2005**, *94*, 25–37.

116 Katz, M., Ben-Shlush, I., Kolusheva, S., Jelinek, R. Rapid colorimetric screening of drug interaction and penetration through lipid barriers. *Pharm. Res.* **2006**, *23*, 580–588.

117 Kansy, M., Senner, F., Gubernator, K. Physicochemical high throughput screening: parallel artificial membrane permeation assay in the description of passive absorption processes. *J. Med. Chem.* **1998**, *41*, 1007–1010.

118 Kansy, M., Fischer, H., Kratzat, K., Senner, F., Wagner, B., Parrilla, I. High-throughput artificial membrane permeability studies in early lead discovery and development. In *Pharmacokinetic Optimization in Drug Research: Biological, Physicochemical and Computational Strategies*, Testa, B., Van de Waterbeemd, H., Folkers, G., Guy, R. (eds.), VHCA, Zurich and Wiley-VCH, Weinheim, **2001**, pp. 447–464.

119 Kansy, M., Avdeef, A., Fischer, H. Advances in screening for membrane permeability: high resolution PAMPA for medicinal chemists. *Drug Discov. Today Technol.* **2004**, *1*, 349–355.

120 Sugano, H. Artificial membrane technologies to assess transfer and permeation of drugs in drug discovery. In *ADME/Tox Approaches*, Testa, B., Van de Waterbeemd, H. (vol. eds.), Vol. 5 in *Comprehensive Medicinal Chemistry*, 2nd edn., Taylor, J. B., Triggle, D. J. (eds.), Elsevier, Oxford, **2007**, pp. 453–487.

121 Avdeef, A., Strafford, M., Block, E., Balogh, M. P., Chambliss, W., Khan, I. Drug absorption *in vitro* model: filter-immobilized artificial membranes 2. Studies of the permeability properties of lactones in *Piper methysticum* Forst. *Eur. J. Pharm. Sci.* **2001**, *14*, 271–280.

122 Sugano, K., Hamada, H., Machida, M., Ushio, H. High throughput prediction of oral absorption: improvement of the composition of the lipid solution used in parallel artificial membrane permeation assay. *J. Biomol. Screen.* **2001**, *6*, 189–196.

123 Sugano, K., Hamada, H., Machida, M., Ushio, H., Saitoh, K., Terada, K. Optimized conditions of bio-mimetic artificial membrane permeation assay. *Int. J. Pharm.* **2001**, *228*, 181–188.

124 Faller, B., Wohnsland, F. Physicochemical parameters as tools in drug discovery and lead optimization. In *Pharmacokinetic Optimization in Drug Research: Biological, Physicochemical and Computational Strategies*, Testa, B., Van de Waterbeemd, H., Folkers, G., Guy, R. (eds.), VHCA, Zurich and Wiley-VCH, Weinheim, **2001**, pp. 257–274.

125 Wohnsland, F., Faller, B. High-throughput permeability pH profile and high-throughput alkane/water log *P* with artificial membranes. *J. Med. Chem.* **2001**, *44*, 923–930.

126 Suomalainen, P., Johans, C., Soderlund, T., Kinnunen, P. K. Surface activity profiling of drugs applied to the prediction of blood–brain barrier permeability. *J. Med. Chem.* **2004**, *47*, 1783–1788.

127 Irvine, J. D., Takahashi, L., Lockhart, K., Cheong, J., Tolan, J. W., Selick, H. E., Grove, J. R. MDCK (Madin-Darby canine kidney) cells: a tool for membrane permeability screening. *J. Pharm. Sci.* **1999**, *88*, 28–33.

128 Genty, M., Gonzalez, G., Clere, C., Desangle-Gouty, V., Legendre, J.-Y. Determination of the passive absorption through the rat intestine using chromatographic indices and molar volume. *Eur. J. Pharm. Sci.* **2001**, *12*, 223–229.

129 Molero-Monfort, M., Escuder-Gilabert, L., Villanueva-Camanoas, Sagrado, S., Medina-Hernandez, M. J. Biopartitioning micellar chromatography: an *in vitro* technique for predicting human drug absorption. *J. Chromatogr. B* **2001**, *753*, 225–236.

130 Detroyer, A., VanderHeyden, Y., Cardo-Broch, S., Garcia-Alvarez-Coque, M. C., Massart, D. L. Quantitative structure–retention and retention–activity relationships of β-blocking agents by micellar liquid chromatography. *J. Chromatogr. A* **2001**, *912*, 211–221.

131 Takagi, M., Taki, Y., Sakane, T., Nadai, T., Sezaki, H., Oku, N., Yamashita, S. A new interpretation of salicylic acid transport across the lipid bilayer: implications of pH-dependent but not carrier-mediated absorption from the gastrointestinal tract. *J. Pharmacol. Exp. Ther.* **1998**, *285*, 1175–1180.

132 Avdeef, A., Box, K. J., Comer, J. E. A., Hibbert, C., Tam, K. Y. pH-metric log P. 10. Determination of liposomal membrane–water partition coefficients of ionizable drugs. *Pharm. Res.* **1998**, *15*, 209–215.

133 Fischer, H., Kansy, M., Bur, D. CAFCA: a novel tool for the calculation of amphiphilic properties of charged drug molecules. *Chimia* **2000**, *54*, 640–645.

134 Lipinski, C. A. Filtering in drug discovery. *Annu. Rep. Comput. Chem.* **2005**, *1*, 155–168.

135 Leeson, P. D., Davis, A. D., Steele, J. Drug-like properties: guiding principles for design – or chemical prejudice? *Drug Discov. Today Technol.* **2004**, *1*, 189–195.

136 Ajay, Walters, W. P., Murcko, M. A. Can we learn to distinguish between drug-like and nondrug-like molecules? *J. Med. Chem.* **1998**, *41*, 3314–3324.

137 Sadowski, J., Kubinyi, H. A scoring scheme for discriminating between drugs and nondrugs. *J. Med. Chem.* **1998**, *41*, 3325–3329.

138 Wagener, M., Van Geerestein, V. J. Potential drugs and nondrugs: prediction and identification of important structural features. *J. Chem. Inf. Comput. Sci.* **2000**, *40*, 280–292.

139 Anzali, S., Barnickel, G., Cezanne, B., Krug, M., Filimonov, D., Poroikov, V. Discriminating between drugs and nondrugs by prediction of activity spectra for substances (PASS). *J. Med. Chem.* **2001**, *44*, 2432–2437.

140 Stenberg, P., Luthman, K., Ellens, H., Lee, C. P., Smith, Ph. L., Lago, A., Elliott, J. D. Artursson, P. Prediction of the intestinal absorption of endothelin receptor antagonists using three theoretical methods of increasing complexity. *Pharm. Res.* **1999**, *16*, 1520–1526.

141 Ghose, A. K., Viswanadhan, V. N., Wendoloski, J. J. A knowledge-based approach in designing combinatorial or medicinal chemistry libraries for drug discovery. 1. A qualitative and quantitative characterization of known drug databases. *J. Comb. Chem.* **1999**, *1*, 55–68.

142 Oprea, T. L. Property distribution of drug-related chemical databases. *J. Comput.-Aided Mol. Des.* **2000**, *14*, 251–264.

143 Leeson, P. D., Davis, A. D. Time-related differences in the physical property profile of oral drugs. *J. Med. Chem.* **2004**, *47*, 6338–6348.

144 Carr, R. A. E., Congreve, M., Murray, C. W., Rees, D. C. Fragment-based lead discovery: leads by design. *Drug Discov. Today* **2005**, *10*, 987–992.

145 Ajay, Bemis, G. W., Murcko, M. A. Designing libraries with CNS activity. *J. Med. Chem.* **1999**, *42*, 4942–4951.

146 Norinder, U., Haeberlein, M. Computational approaches to the prediction of the blood–brain distribution. *Adv. Drug Deliv. Rev.* **2002**, *54*, 291–313.

147 Van de Waterbeemd, H., Camenisch, G., Folkers, G., Chretien, J. R., Raevsky, O. A. Estimation of blood–brain barrier crossing of drugs using molecular size and shape, and H-bonding descriptors. *J. Drug Target.* **1998**, *6*, 151–165.

148 Summerfield, S. G., Stevens, A. J., Cutler, L., Del Carmen Osuna, M., Hammond, B., Tang, S.-P., Hershey, A., Spalding, D. J., Jeffrey, P. Improving the *in vitro* prediction of *in vivo* central nervous system penetration: Integrating permeability, P-glycoprotein efflux, and free fractions in blood and brain. *J. Pharmacol. Exp. Ther.* **2006**, *316*, 1282–1290.

149 Blake, J. F. Chemoinformatics – predicting the physicochemical properties of drug-like molecules. *Curr. Opin. Biotechnol.* **2000**, *11*, 104–107.

150 Matter, H., Baringhaus, K. H., Naumann, T., Klabunde, T., Pirard, B. Computational approaches towards the rational design of drug-like compound libraries. *Comb. Chem. High-Throughput Screen.* **2001**, *4*, 453–475.

151 Winiwarter, S., Ridderström, M., Ungell, A.-L., Andersson, T. B., Zamora, I. Use of molecular descriptors for ADME predictions. In *ADME/Tox Approaches*, Testa, B., Van de Waterbeemd, H. (vol. eds.), Vol. 5 in *Comprehensive Medicinal Chemistry*, 2nd edn., Taylor, J. B., Triggle, D. J. (eds.), Elsevier, Oxford, **2007**, pp. 531–554.

152 Dudek, A. Z., Arodz, T., Galvez, J. Computational methods in developing quantitative structure–activity relationships (QSAR): a review. *Comb. Chem. High-Throughput Screen.* **2006**, *9*, 213–228.

153 Testa, B., Vistoli, G., Pedretti, A. Musings on ADME predictions and structure–activity relations. *Chem. Biodivers.* **2005**, *2*, 1411–1428.

154 Van de Waterbeemd, H., Rose, S. *Quantitative approaches to structure–activity relationships. In The Practice of Medical Chemistry*, 2nd edn., Wermuth, C. G. (ed.), Academic Press, London, **2003**, pp. 351–369.

155 De Cerqueira Lima, P., Golbraikh, A., Oloff, S., Xiao, Y., Tropsha, A.

Combinatorial QSAR modeling of P-glycoprotein substrates. *J. Chem. Inf. Model.* **2006**, *46*, 1245–1254.

156 Stouch, T. R., Kenyon, J. R., Johnson, S. R., Chen, X. Q., Doweyko, A., Li, Y. *In silico* ADME/tox: why models fail. *J. Comput.-Aided Mol. Des.* **2003**, *17*, 83–92.

157 Saunders, K. Automation and robotics in ADME screening. *Drug Discov. Today Technol.* **2004**, *1*, 373–380.

158 Cartmell, J., Enoch, S., Krstajic, D., Leahy, D. E. Automated QSPR through competitive workflow. *J. Comput.-Aided. Mol. Des.* **2005**, *19*, 821–833.

159 Van de Waterbeemd, H., Gifford, E. ADMET *in silico* modeling: towards *in silico* paradise? *Nat. Rev. Drug Discov.* **2003**, *2*, 192–204.

160 Smith, D. A., Cucurull-Sanchez, L. The adaptive *in combo* strategy. In *ADME/Tox Approaches*, Testa, B., Van de Waterbeemd, H. (vol. eds.), Vol. *5* in *Comprehensive Medicinal Chemistry*, 2nd edn., Taylor, J. B., Triggle, D. J. (eds.), Elsevier, Oxford, **2007**, pp. 957–969.

Part II
Electronic Properties and H-Bonding

3 Drug Ionization and Physicochemical Profiling

Alex Avdeef

Abbreviations

ABL	aqueous boundary layer
ADME	absorption, distribution, metabolism and excretion
BBB	blood–brain barrier
BCS	Biopharmaceutics Classification System
Caco-2	adenocarcinoma cell line derived from human colon
DMSO	dimethylsulfoxide
HEPES	*N*-2-hydroxyethyl-piperazine-*N′*-2-ethanesulfonic acid
GIT	gastrointestinal tract
MDCK	Madin-Darby canine kidney
MES	2-(*N*-morpholino)ethanesulfonic acid
PAMPA	parallel artificial membrane permeation assay
UV	ultraviolet

Symbols

pK_a	ionization constant
$\log P_{oct}$	octanol–water partition coefficient

3.1 Introduction

This chapter considers ionizable drug-like molecules and the effect of such ionization on pharmaceutic properties. Most medicinal substances are ionizable [1]. The biological medium into which these substances distribute embraces a range of pH values. The ionization constant, pK_a, can tell the pharmaceutical scientist to what degree the molecule is charged in solution at a particular pH. This is important to know, since the charge state of the molecule strongly influences its other physicochemical properties.

Molecular Drug Properties. Measurement and Prediction. R. Mannhold (Ed.)

ISBN: 978-3-527-31755-4

The ionization constant plays an important role in absorption, distribution, metabolism and excretion (ADME) of medicinal substances [2]. The effect of the pK_a in oral drug absorption arises from the pH dependence of drug permeability across membrane barriers and the pH dependence of drug solubility in luminal fluid [3–8]. The excretion of drugs can be modulated by pH control. For example, the pH of urine can be altered with oral doses of ammonium chloride or sodium bicarbonate to effect reabsorption of uncharged species (for therapeutic reasons) or to ease excretion of ionized species (which may be toxic). Weak acids may be excreted in alkaline urine and weak bases may be eliminated in acidic urine, possibly a lifesaving principle in the case of drug overdose [9]. The rate of dissolution of an orally delivered solid-dosage form of the drug in acidic gastric fluid can be strongly influenced by the pK_a [10]. The strength and extent of protein binding or of enzymatic transformation of the drug can depend on the pK_a of the drug and/or the ionizable functional group(s) in the binding or catalytic site [11, 12]. In this chapter, the effect of the pK_a in oral absorption (the "A" in ADME) will be stressed. Underlying absorption are physicochemical properties such as dissolution, solubility, permeability and ionization, the four key components in the Biopharmaceutics Classification System (BCS) [3].

3.1.1 Absorption, the Henderson–Hasselbalch Equation and the pH-partition Hypothesis

Nonionized molecules are usually better absorbed in the gastrointestinal tract (GIT) than ionized molecules. In the intestine, water-soluble weak bases are better absorbed from neutral pH regions (e.g. in the distal ileum), and weak acids are better absorbed from mildly acidic regions (e.g. proximal jejunum). This was rationalized by Brodie et al. [13], who introduced the pH-partition hypothesis to explain the influence of pH on the intestinal absorption of ionizable drugs. In the classic absorption experiments using rats, a drug solution at pH 7.4 was injected intravenously and also perfused intestinally using solutions of varied pH. The concentration of the drug in the luminal perfusate was adjusted until there was no net transport across the intestinal wall, so that it was possible to define the blood–lumen barrier ratio:

$$D = \frac{[\mathrm{drug}]_{\mathrm{BLOOD}}}{[\mathrm{drug}]_{\mathrm{LUMEN}}} = \frac{(1+10^{-\mathrm{p}K_a+\mathrm{pH_{BLOOD}}})}{(1+10^{-\mathrm{p}K_a+\mathrm{pH_{LUMEN}}})} \quad (1)$$

The last part of Eq. (1) is derived from the pH dependence of permeability, given a pH gradient between the two sides of the intestinal barrier, based on the well known Henderson–Hasselbalch equation. Direct measurement of *in situ* intestinal perfusion absorption rates confirmed this pH dependence [14].

The pH-partition hypothesis, now widely accepted in pharmaceutical research, suggests that membrane permeability will be highest at the pH where the molecule is least charged; however, this is also the pH where the molecule is *least* soluble. (In Brodie's work, the compounds tested have relatively high water solubility.) At the site of absorption, the amount of the uncharged species and the tendency of the neutral species to cross the phospholipid membrane barrier are both important predictors of absorption. The intrinsic permeability coefficient, P_0, characterizes the membrane transport of the uncharged species. The concentration of the uncharged species, C_0, depends on the dose, the solubility, the pK_a of a molecule and the pH at the site of absorption, often according to the Henderson–Hasselbalch equation [8].

However, for poorly soluble molecules, the classical view of pH dependence of absorption needs to be qualified. Bergström et al. [15] have shown that many sparingly soluble molecules show solubility pH profiles that cannot be simply predicted by the Henderson–Hasselbalch equation. Avdeef et al. [16] have shown that the classical Brodie pH-partition hypothesis can break down when low-solubility and high-permeability drugs are considered, where the pH effect is actually *inverted* (i.e. the pH where the drug is more charged showing *higher* absorptive flux potential than the pH where the drug is more neutral).

3.1.2 "Shift-in-the-pK_a"

The aim of this chapter is to focus on three physicochemical properties where the ionization constant, pK_a, relates to a critical distribution or transport function: (i) octanol–water and liposome–water partitioning, (ii) solubility (ionized species and neutral species aggregation, and salt effects) and (iii) permeability (artificial membrane and cultured cell models). The logarithm of the physicochemical property versus pH ("log-log") plots can indicate both the true pK_a and an *apparent* pK_a, in various circumstances called: the "octanol pK_a" (pK_a^{OCT}), "membrane pK_a" (pK_a^{MEM}), "ionized aggregate pK_a" (p$K_a^{AGG,I}$), "neutral aggregate pK_a" (p$K_a^{AGG,N}$), "Gibbs' pK_a" (pK_a^{GIBBS}) and "flux pK_a" (pK_a^{FLUX}). Such log-log plots are either hyperbolic or sigmoidal in shape, with domains characterized by (0, ±1, ±2, . . .) slopes. At the points where the curves bend (slope at half-integral values), the pH is equal either to the true or the apparent pK_a.

The evaluation of the apparent ionization constants (i) can indicate in partition experiments the extent to which a charged form of the drug partitions into the octanol or liposome bilayer domains, (ii) can indicate in solubility measurements, the presence of aggregates in saturated solutions and whether the aggregates are ionized or neutral and the extent to which salts of drugs form, and (iii) can indicate in permeability measurements, whether the aqueous boundary layer adjacent to the membrane barrier, limits the transport of drugs across artificial phospholipid membranes [parallel artificial membrane permeation assay (PAMPA)] or across monolayers of cultured cells [Caco-2, Madin-Darby canine kidney (MDCK), etc.].

Section 3.2 begins with pK_a definitions and a brief description of the state-of-the-art pK_a measurement methods, stressing the needed accuracy, especially with molecules which possess very low aqueous solubility. In a practical way, the ionization constant is treated as a property of the molecule, usually defined at 25 °C in a nonbuffered medium of 0.15 M potassium (or sodium) chloride aqueous solution. This reference medium defines the "true" pK_a of the molecule [8, 17]. In physicochemical profiling, the environment may contain membranes, cells and proteins, and may not be suitable for completely dissolving the molecule. In this "environmental" medium, the pK_a may *appear* to be different from the "true" value, as graphically indicated in the above-mentioned log-log plots. The *appearance* of a shift-in-the-pK_a and how this shift is used to determine physicochemical properties is the subject of Sections 3.3–3.5. Section 3.3 discusses the octanol–water distribution profile, log D_{oct}, as a function of pH. The sigmoidal curve in the log-log plot indicates both the true and the "octanol pK_a" [18]. Section 3.4 considers solubility pH profiles, which can indicate the true pK_a, and, under certain conditions, the "Gibbs' pK_a" [19]. Furthermore, from the shape of the hyperbolic-sigmoidal curves and the value of the "apparent pK_a," it is possible to recognize the presence and the charge state of aggregates of low-soluble molecules [20]. Section 3.5 deals with permeability profiles, based on cellular models (Caco-2, MDCK, etc.) and artificial membrane models (PAMPA). From the apparent permeability profile, log P_{app} versus pH, the "flux pK_a" can indicate the extent to which membrane transport is limited by the aqueous boundary layer [21]. Another apparent pK_a can indicate permeability of ionic form of the molecule or active transport in the case of cellular permeability [22].

3.2 Accurate Determination of Ionization Constants

The classic (long out-of-print) book by Albert and Serjeant [17] has taught several generations of physical chemists (including the author) how to properly measure ionization constants. The two methods of choice for the measurement of ionization constants are potentiometry [2, 23–31] and ultraviolet (UV) spectrophotometry [32–47]. The UV method is usually more sensitive of the two and thus requires less sample (10–50 versus 200–5000 μM, respectively). However, the potentiometric method can be universally applied, whereas the spectrophotometric method requires not only a measurable UV chromophore, but also one that measurably alters with pH changes. A small but significant number of drug-like substances cannot be characterized by UV. With care, both methods can result in good pK_a reproducibility (±0.02) for water-soluble compounds. For sparingly soluble compounds, the reproducibility may be poorer (±0.1). Many other techniques for determining pK_a have been reported [8], but the above two methods are best suited for pharmaceutical applications.

3.2.1 Definitions – Activity versus Concentration Thermodynamic Scales

The ionization reactions for acids, bases and ampholytes (diprotic) may be represented by the general forms

$$HA \leftrightarrows A^- + H^+ \qquad K_a = [A^-][H^+]/[HA] \tag{2a}$$

$$BH^+ \leftrightarrows B + H^+ \qquad K_a = [B][H^+]/[BH^+] \tag{2b}$$

$$XH_2^+ \leftrightarrows XH + H^+ \qquad K_{a1} = [XH][H^+]/[XH_2^+] \tag{2c}$$

$$XH \leftrightarrows X^- + H^+ \qquad K_{a2} = [X^-][H^+]/[XH] \tag{2d}$$

Listed after the equilibria are the corresponding equilibrium quotients. The law of mass action sets the concentration relations of the reactants and products in a reversible chemical reaction. The negative log (logarithm, base 10) of the quotients in Eqs. (2a)–(2d) produces the familiar Henderson–Hasselbalch equations, where "p" represents the operator "–log":

$$pK_a = pH + \log([HA]/[A^-]) \tag{3a}$$

$$pK_a = pH + \log([BH^+]/[B]) \tag{3b}$$

$$pK_{a1} = pH + \log([XH_2^+]/[XH]) \tag{3c}$$

$$pK_{a2} = pH + \log([XH]/[X^-]) \tag{3d}$$

Equations (3a)–(3d) indicate that when the concentration of the free acid, HA (or conjugate acid, BH^+), equals that of the conjugate base, A^- (or free base, B), the pH has the special name, "pK_a".

All equilibrium constants in the present discussion are based on the *concentration* (not activity) scale. This is a perfectly acceptable thermodynamic scale, provided the ionic strength of the solvent medium is kept fixed at a "reference" level (therefore, sufficiently higher than the concentration of the species assayed). This is known as the "constant ionic medium" thermodynamic state. Most modern results are determined at 25 °C in a 0.15 M KCl solution. If the ionic strength is changed, the ionization constant may be affected. For example, at 25 °C and 0.0 M ionic strength, the pK_a of acetic acid is 4.76, but at ionic strength 0.15 M, the value is 4.55 [24].

The ionic-strength dependence of ionization constants can be predicted by the Debye–Hückel theory [8, 17, 27]. In the older literature, values are reported most often at "zero sample and ionic strength" and are called "thermodynamic" constants. The constants reported at 0.15 M ionic medium are no less thermodynamic. Fortunately, a result determined at 0.15 M KCl background can be transformed to another background salt concentration, provided the ionic strength remains below about 0.3 M [27]. It is sometimes convenient to convert constants to "zero ionic strength" to compare values to those reported in older literature. A general ionic strength correction equation is described in the literature [26, 27].

The effect of temperature on acid or base pK_a values cannot be reliably predicted [2, 17, 23]. For many nitrogenous bases, the pK_a decreases by 0.1–0.3 for every 10 °C rise in temperature. For some carboxylic acids (e.g. acetic, benzoic, salicylic acids), the pK_a remains essentially unchanged between 25 and 37 °C.

3.2.2 Potentiometric Method

In pH-metric titration, precisely known volumes of a standardized strong acid (e.g. HCl) or base (e.g. KOH) are added to a vigorously stirred solution of a protogenic substance, during which pH is continuously measured with a precision combination glass electrode, in a procedure confined to the interval pH 1.5–12.5. The substance (200 μM or higher in concentration) being assayed is dissolved in 2–20 mL of water or in a mixed solvent consisting of water plus an organic water-miscible cosolvent [e.g. simple alcohols, acetonitrile or dimethylsulfoxide (DMSO)]. An inert water-soluble salt (0.15 M KCl) is added to the solution to improve the measurement precision, and to mimic the physiological ionic strength. Usually, the reaction vessel is thermostated at 25 °C and a blanket of argon (*not* helium) bathes the solution surface.

The plot of pH against titrant volume added is called a potentiometric titration curve. The latter curve is usually transformed into a Bjerrum plot [8, 24, 27], for better visual indication of overlapping pK_as or for pK_as below 3 or above 10. The actual values of pK_a are determined by weighted nonlinear regression analysis [25–27].

3.2.3 pH Scales

To establish the *operational* pH scale, the pH electrode can be calibrated with a single aqueous pH 7.00 phosphate buffer, with the ideal Nernst slope assumed. As Eqs. (2a)–(2d) require the "free" hydrogen ion *concentration*, an additional electrode standardization step is necessary. That is where the operational scale is converted to the *concentration* scale p_cH (= –log [H^+]) as described by Avdeef and Bucher [24]:

$$\mathrm{pH} = \alpha + k_s p_c H + j_\mathrm{H}[\mathrm{H}^+] + j_\mathrm{OH} K_\mathrm{w}/[\mathrm{H}^+] \tag{4}$$

where K_w is the ionization constant of water. The four parameters are empirically estimated by a weighted nonlinear least-squares procedure using data from alkalimetric titrations of known concentrations of HCl (from pH 1.7 to 12.3). Typical aqueous values of the adjustable parameters at 25 °C and 0.15 M ionic strength are $\alpha = 0.08 \pm 0.01$, $k_s = 1.001 \pm 0.001$, $j_H = 1.0 \pm 0.2$ and $j_{OH} = -0.6 \pm 0.2$. In cosolvent solutions, these values usually change and can be readily determined. Such a standardization scheme extends the range of accurate pH measurements, and allows pK_as to be assessed as low as 0.6 and as high as 13.0 [8].

3.2.4 Cosolvent Methods

If the compound is virtually insoluble (solubility < 1 μg mL^{-1}), then a potentiometric mixed solvent approach can be tried [2, 28–30]. For example, the pK_a of the antiarrhythmic amiodarone, 9.06 ± 0.14, was estimated from water–methanol mixtures, notwithstanding that the intrinsic solubility of the molecule in water is about 0.006 $\mu g\,mL^{-1}$ [8].

The most explored solvent systems are based on water–alcohol mixtures, acetonitrile–water and DMSO–water [8]. Where possible, methanol is the solvent of choice, because its general effect on pK_as has been studied so extensively. It is thought to be the least "error-prone" of the common solvents [28].

Mixed-solvent solutions of various cosolvent–water proportions are titrated and p_sK_a (the apparent pK_a) is measured in each mixture. The pK_a of acids increases and that of bases decreases with increasing proportion of organic solvent. This depression of ionization is due to decreases in the dielectric constant of the mixed solvent. The estimated aqueous pK_a is deduced by extrapolation of the p_sK_a values to zero cosolvent. Plots of p_sK_a versus weight percent organic solvent, R_w = 0–60 wt%, at times show either a "hockey-stick" shape or a "bow" shape [28]. For $R_w > 60$ wt%, "S"-shaped curves are sometimes observed. For values of $R_w < 60$ wt%, the nonlinearity in p_sK_a plots can be ascribed partly to electrostatic long-range ion–ion interactions [28].

3.2.5
Recent Improvements in the Potentiometric Method Applied to Sparingly Soluble Drugs

In a recent study of very-sparingly soluble drugs, four methods of pK_a determination were compared, using both methanol–water and DMSO–water solutions [48]. Potentiometric methods performed slightly better than UV based methods. In one of the recently modified [48] potentiometric methods [19, 49, 50], it is possible to determine the pK_a *even if there is precipitation* during a portion of the titration, in either aqueous or cosolvent solutions. This is because the improved potentiometric method can determine solubility and ionization constants *simultaneously* in the same titration. It may very well be that at least some of the "hockey-stick" nonlinearity mentioned in the preceding section arises from the presence of some precipitation of the sample, for which no provision had been made in the early studies. This is an interesting conjecture, worthy of further investigation.

Furthermore, pH electrode calibration can be performed *in situ* by the new method [48], *concurrently* with the pK_a determination. This is a substantial improvement in comparison to the traditional procedure of first doing a "blank" titration to determine the four Avdeef–Bucher parameters [24]. The traditional cosolvent methods used with sparingly soluble molecules can be considerably limited in the pH < 4 region when DMSO–water solutions are used. This is no longer a serious problem, and routine "blank" titrations are now rarely needed in the new *in situ* procedure.

3.2.6
Spectrophotometric Measurements

The most effective spectrophotometric procedures for pK_a determination are based on the processing of whole absorption curves over a broad range of wavelengths, with data collected over a suitable range of pH. Most of the approaches are based on mass balance equations incorporating absorbance data (of solutions adjusted to various pH values, with or without buffers) as dependent variables and equilibrium constants as parameters, refined by nonlinear least-squares refinement, using Gauss–Newton, Marquardt or Simplex procedures.

Since the spectroscopic analysis methods are nonlinear, it is necessary to initiate calculations with an approximate pK_a and single-species molar absorptivity profiles. In complicated equilibria, uninformed guessing of pK_as and individual-species molar absorptivity coefficients can be problematic. Elegant mathematical methods have evolved to help this process of supervised calculation. Since not all species in a multiprotic compound possess detectible UV chromophores, and often more than one species have nearly identical molar absorptivity curves, methods have been devised to assess the number of spectrally active components [33]. With mathematically ill-conditioned equations, parameter shift damping procedures are required. Gampp et al. [36] considered principal component analysis and evolving factor analysis methods in identifying the presence and stoichiometries of the absorbing species.

Tam et al. [37–47] developed an impressive generalized method for the determination of ionization constants and molar absorptivity curves of individual species, using diode-array UV spectrophotometry, coupled to an automated pH titrator. Species selection was effected by target factor analysis. Multiprotic compounds with overlapping pK_as have been investigated; binary mixtures of ionizable compounds have been considered; assessment of microconstants have been reported.

3.2.7 Use of Buffers in UV Spectrophotometry

In the UV method, the control of pH is most often done by the use of buffers. Good et al. [51] pointed out the shortcomings of traditional buffers such as phosphate, Tris and borate, due to their reactivity with biological systems, as well as with the analyte. Phosphate buffers are especially problematic with sparingly soluble basic drugs. A number of zwitterionic buffers were then synthesized to overcome the limitations [51]. These included such buffers as MES [2-(*N*-morpho lino)ethanesulfonic acid], HEPES (*N*-2-hydroxyethyl-piperazine-*N*′-2-ethanesulfonic acid), and a series of similar molecules, collectively known as the "Good buffers". In concentrations at or below 0.05 M, minimal interferences are encountered. The monograph by Perrin and Dempsey [52] compiled extensive (and practical) tabulations of buffer properties (another valuable book, long out of print).

Avdeef and Bucher [24] investigated the use of universal buffers in potentiometric titrations. Recently, such a buffer system, formulated with several of the Good components, has been designed specifically for robotic applications, where automated pH control in 96-well microtiter plates is required, with minimal interference to the UV measurement [48]. This universal buffer has a nearly perfectly linear pH response to additions of standard titrant in the pH 3–10 region [8, 48].

3.2.8
pK_a Prediction Methods and Software

Perrin et al. [53] published a monograph on the prediction of pK_a values, compiling a large number of molecular fragment equations, based on linear free energy relationships. Their monograph (unfortunately, long out of print) is often the starting point in commercial pK_a prediction programs. One of the early computer programs, from Advanced Chemistry Development [54] can predict pK_a values with fair precision (±0.4 and sometimes much better). A more recent product comes from Pharma Algorithms [55]. A web-based pK_a prediction program is available from ChemAxon [56]. All three prediction programs are excellent and it is often useful to average the predictions from all three sources, to build a "consensus" value.

The prediction programs do better with established drug molecules than with test compounds from newly synthesized classes. Usually, prediction programs do poorly with molecules that possess internal H-bonds associated with the ionizing groups, although the more actively supported programs seem to be improving in this area. For example, some of the predicted pK_a values in flavonoids, fluoroquinolones and in molecules like doxorubicin can be in error by 3–4 orders of magnitude. The predicted pK_a values are generally quite good for giving the pharmaceutical scientist an idea of the charge state of a molecule at a particular pH. However, for the purpose of using the pK_a values to assess physicochemical properties, as presented later in this chapter, predicted pK_a values are not yet good enough. Accurately measured pK_a values are still required, as will become apparent in Sections 3.3–3.5.

3.2.9
Tabulations of Ionization Constants

The "blue book" compilations [57–59] recommended for experts in the field are probably the most comprehensive sources of ionization constants collected from the literature. On the other hand, the "red books" contain critically selected values [60]. A useful list of 400 pK_a values of pharmaceutically important molecules has been published [23]. Additionally, a more recent compilation of pK_a values of about 250 drug-like molecules may be found in Ref. [8].

3.3
"Octanol" and "Membrane" pK_a in Partition Coefficients Measurement

The octanol–water partition coefficient, P_{oct} (often reported as log P_{oct}), is a particularly useful parameter in quantitative structure–activity relationships, applied to prediction of properties related to drug absorption, distribution, metabolism and excretion [61, 62]. Although the traditional log P_{oct} measurements have been done by the shake-flask method [63, 64], high-performance liquid chromatography-based approaches have become the methods of choice in pharmaceutical research [65–69], with many protocols adapted to 96-well microtiter plate formats [68, 69]. The Dyrssen dual-phase potentiometric log P_{oct} technique [8, 18, 27, 31] in certain applications is singly valuable, especially when the drug-like substance does not have a sensitive UV chromophore, or is partially ionized at physiological pH, since the ionization constant, pK_a, may be needed for log P_{oct} determination (*cf.* Eq. 8 below). In the potentiometric method both pK_a and log P_{oct} are determined in the same assay [8, 27]. The method can be applied to substances with several overlapping pK_as and to substances which undergo ion-pair partitioning.

Many excellent computer programs are available for predicting log P_{oct} from two-dimensional structures. The quality of predictions has risen over the years to the point that routine log P_{oct} measurements are not regularly done at some pharmaceutical companies, but rather, calculated values are used. It is worth noting that log P_{oct} values of newly synthesized classes of drug-like compounds sometimes are still poorly predicted and probably there will be the need for judicious log P_{oct} measurements for years to come.

3.3.1 Definitions

The partition coefficient of an acid or a base in the nonionized species form, P^N_{oct}, is defined by the equilibria and quotients:

$$HA \leftrightarrows HA_{oct} \qquad P^N_{oct} = [HA]_{oct}/[HA] \tag{5a}$$

$$B \leftrightarrows B_{oct} \qquad P^N_{oct} = [B]_{oct}/[B] \tag{5b}$$

where $[HA]_{oct}$ and $[B]_{oct}$ are the concentrations of the neutral species in the organic phase, with reference to the volume of octanol; the unsubscripted concentrations are in aqueous solution, based on aqueous volume. Rigorously, activity units would be used in Eqs. (5). However, the practice in the pharmaceutical sciences is to associate concentration units (mol L^{-1}) with the partition coefficient equations.

The partition coefficient is a thermodynamic constant, defined in terms of a single species (neutral *or charged*) in the dual-phase distribution, and does not itself depend on pH. Since protogenic substances undergo pH-dependent ionization, the actual distribution of different species between the two phases can change with pH. The term describing this pH-dependent distribution is called the octanol–water distribution coefficient (also, the *apparent* partition coefficient) and is defined as the ratio of the concentrations of all of the drug species dissolved in the organic phase divided by the sum of the concentration of all of the drug species dissolved in the aqueous phase, as, for example:

$$D_{oct} = \frac{[X^-]_{oct} + [XH]_{oct} + [XH_2^+]_{oct}}{[X^-] + [XH] + [XH_2^+]} = \frac{[X^-]^*_{oct} + [XH]^*_{oct} + [XH_2^+]^*_{oct}}{[X^-] + [XH] + [XH_2^+]} \cdot \frac{1}{r_{ow}} \tag{6}$$

The asterisked quantity is defined in concentration units of moles of species dissolved in the organic phase per liter of *aqueous* phase solution. The octanol–water volume ratio, $r_{ow} = V_{oct}/V_{water}$, takes into account differences in volumes of the two phases. In pharmaceutical applications, practical values of r_{ow} range from 0.003 to 3 (and sometimes higher).

Consider the partitioning of a lipophilic base, B. The uncharged species, B, will distribute into the organic phase more extensively than the charged species, BH^+. For the partitioning of the latter ion, the ion-pair partition coefficient may be defined as:

$$P^{I}_{oct} = \frac{[BH^+]_{oct}}{[BH^+]} \tag{7}$$

Such ion-pair constants are "conditional", in that they depend on the concentration of the counterion with which the charged drug molecule enters into the octanol phase as an ion-pair. This is due to the low dielectric property of octanol, inducing charge neutrality upon uptake of charged drug molecules. Extraction constants may be used to explicitly include the participation of the counterion [18].

With mass balance consideration, along with Eqs. (2b), (5b) and (7), Eq. (6) may be transformed into:

$$D_{oct} = \frac{P^{I}_{oct} \cdot 10^{\pm(\mathrm{pH}-\mathrm{p}K_a)} + P^{N}_{oct}}{1 + 10^{+(\mathrm{p}K_a-\mathrm{pH})}} \tag{8}$$

with the "+" sign in "±" referring to acids and "–" to bases. Note also that D_{oct} in Eq. (8) is a function of pH and equilibrium constants, and is independent of concentration of drug (provided that the concentration of the counterion is kept constant and significantly greater in concentration than that of the drug). Distribution functions for more complicated cases are described elsewhere [18].

In the traditional shake-flask method, the apparent partition coefficient, log D_{oct}, is measured, usually at pH 7.4 (sometimes at pH 6.5). Different buffers are used to control each pH used in the determinations [70]. Usually, in a comprehensive study, several pH measurements are made, and values of log D_{oct} are plotted against the pH. This plot is often called the "lipophilicity profile". One can determine the true partition coefficients (log P_{oct}) and the ionization constants from the features in such a curve.

3.3.2 Shape of the Log D_{oct}–pH Lipophilicity Profiles

Figure 3.1 shows an example of a monoprotic base, pindolol, lipophilicity profile, as a function of pH. The two limiting slopes in the curve are zero and the slope in between is +1 (–1 for acids). The maximum value of log D_{oct} is equal to $\log P^{N}_{oct}$

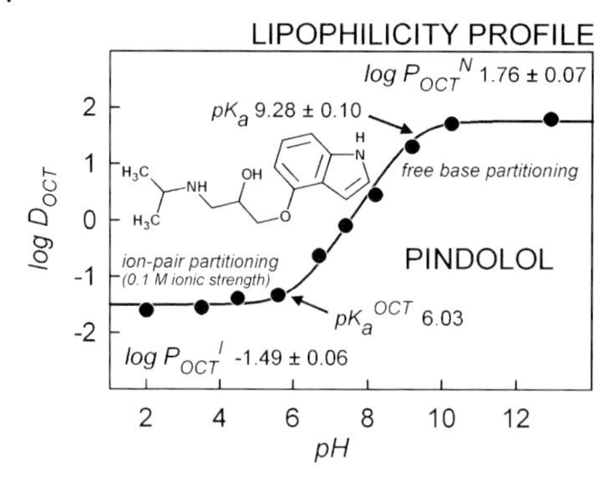

Fig. 3.1 Octanol–water distribution (log D_{oct}) versus pH profile for pindolol, based on data reported by Barbato et al. [70]. The *p*CEL-X computer program (*p*ION) was used to reanalyze the data, with the refined results listed in the figure. The root mean square deviation in the fit was 0.1 log unit.

(and the minimum value is equal to $\log P^{\mathrm{I}}_{\mathrm{oct}}$ in a fully formed sigmoidal plot, as in Fig. 3.1). At the upper bend in the curve, pH is equal to the pK_a of the molecule; at the lower bend in the curve, the pH is equal to the "octanol" pK_a, i.e. pK_a^{OCT}. The "bend" in the lipophilicity curve spans 3.3 pH units, 1.66 pH units to each side of each pK_a, as suggested in Ref. [18]. The "octanol" pK_a is defined by the aqueous pH at the point where the ionized and nonionized drug concentrations are equal *in the octanol phase,* according to the *conditional* equilibrium expression $(\mathrm{HA})_{\mathrm{oct}} \leftrightarrows (\mathrm{K^+A^-})_{\mathrm{oct}} + \mathrm{H^+}$.

The lipophilicity profile of an acid is a mirror image of the shape in Fig. 3.1, i.e. the maximum (neutral-species) partitioning occurs at low pH and the minimum (ion-pair) partitioning at high pH. Other shapes are described in the review by Avdeef [18].

3.3.3
The "*diff* 3–4" Approximation in log D_{oct}–pH Profiles for Monoprotic Molecules

For multi-pH octanol–water distribution measurements in 0.15 M NaCl or KCl solutions, the difference between the true pK_a and pK_a^{OCT} (and between $\log P^{\mathrm{N}}_{\mathrm{oct}}$ and $\log P^{\mathrm{I}}_{\mathrm{oct}}$) is about 3 log units for bases and 4 log units for acids. This approximation appears to hold for simple compounds, where atomic charge is largely localized on an atom. Smaller differences are seen in aromatic systems, where the charge can be substantially delocalized [8]. Table 3.1 shows octanol–water examples of the relative pK_a "shifts" supporting the "*diff* 3–4" approximation. Note that the *diff* in Fig. 3.1 is 3.25 for pindolol; the value is greater than 3, since the ionic strength, 0.1 M, is less than the 0.15 M on which the approximation is based.

Tab. 3.1 Octanol–water and liposome–water partition coefficients[1].

Compound	**pK_a**	***Octanol–water***			***Liposome–water***		
		$\log P^N_{oct}$	pK_a^{OCT}	$\lvert pK_a^{OCT} - pK_a \rvert$	$\log P^N_{MEM}$	pK_a^{MEM}	$\lvert pK_a^{MEM} - pK_a \rvert$
Acids							
ibuprofen	4.45	3.97	8.47	4.02	3.80	6.44	1.99
diclofenac	3.99	4.51	7.82	3.83	4.34	5.80	1.81
phenylvaleric acid	4.59	2.92	8.46	3.87	3.17	6.10	1.51
warfarin	4.90	3.25	8.61	3.71	3.46	6.98	2.08
average shift ± SD				3.9 ± 0.1			1.8 ± 0.3
Bases							
lidocaine	7.96	2.45	4.98	2.98	2.39	6.79	1.17
phenylbutylamine	10.50	2.39	7.66	2.84	3.02[4]	10.59	−0.09
procaine	9.04[2]	2.14	6.09	2.95	2.38	7.42	1.62
propranolol	9.53	3.48	6.83	2.70	3.45	8.69	0.84
tetracaine	8.49[3]	3.51	5.20	3.29	3.23	7.37	1.12
average shift ± SD				3.0 ± 0.2			0.9 ± 0.6

1 Liposome–water partition potentiometric determinations, 25°C, 0.15 M KCl [8, 71]. Liposomes were made of large (phosphatidylcholine) unilamellar vesicles.
2 Second pK_a is 2.29.
3 Second pK_a is 2.39.
4 Predicted from $\log P^N_{oct}$ [8].

The relationship between *diff* log P_{oct} and the pK_a values in monoprotic substances is:

$$\log P^N_{oct} - \log P^I_{oct} = \pm(pK_a^{OCT} - pK_a) \qquad (9)$$

with the "+" sign in "±" applicable to acids and "−" to bases.

3.3.4 Liposome–Water Partitioning and the "*diff* 1–2" Approximation in $\log D_{MEM}$–pH Profiles for Monoprotic Molecules

For multi-pH liposome–water distribution measurements in 0.15 M NaCl or KCl solutions, the difference between the true pK_a and pK_a^{MEM} (and between $\log P^N_{MEM}$ and $\log P^I_{MEM}$) is about 1 log unit for bases and 2 log units for acids. Liposomes formed from phosphatidylcholine have a tendency to stabilize the charged drug more effectively than that in the octanol–water system [71]. Table 3.1 shows liposome–water examples of the relative pK_a "shifts." The average values cited in the examples are close the "*diff* 1–2" approximation noted above.

3.4
"Gibbs" and Other "Apparent" pK_a in Solubility Measurement

3.4.1
Interpretation of Measured Solubility of Ionizable Drug-Like Compounds can be Difficult

The measurement of solubility of drug-like substances is often surprisingly problematic [8, 31]. This is because modern drug discovery programs tend to select active compounds that are in high molecular weight, are lipophilic and are very sparingly soluble in aqueous solution [4]. The dissolution of compounds from solid dosage forms may accompany polymorphic transformations, as active solids transform into thermodynamically more stable forms, often notably less soluble than the original form of the solid. The resultant aqueous solutions are often complicated by the presence of not only multiply charged forms of drug, but also by the presence of aggregates, micelles and complexes (formed with constituents of the dissolution medium) [20, 72–75]. The interpretation of solubility–pH curves can be challenging, and the inference of pH dependence from single pH measurements based on the use of the Henderson–Hasselbalch equation can be unreliable [15, 20]. If not long enough a time is taken for the solubility equilibration, measured results are complicated by time-dependent changes.

3.4.2
Simple Henderson–Hasselbalch Equations

The basic relationships between solubility and pH can be derived for any given equilibrium model. The "model" refers to a set of equilibrium equations and the associated equilibrium quotients. In a saturated solution, three additional equations need to be considered, along with the ionization Eqs. (2a)–(2d), which describe the equilibria between the dissolved acid, base or ampholyte in solutions containing a suspension of the (usually crystalline) solid form of the compounds:

$$\mathrm{HA(s)} \leftrightarrows \mathrm{HA} \quad S_0 = [\mathrm{HA}]/[\mathrm{HA(s)}] = [\mathrm{HA}] \tag{10a}$$

$$\mathrm{B(s)} \leftrightarrows \mathrm{B} \quad S_0 = [\mathrm{B}]/[\mathrm{B(s)}] = [\mathrm{B}] \tag{10b}$$

$$\mathrm{XH(s)} \leftrightarrows \mathrm{XH} \quad S_0 = [\mathrm{XH}]/[\mathrm{XH(s)}] = [\mathrm{XH}] \tag{10c}$$

The concentrations of species in the solid phase, [HA(s)], [B(s)] and [XH(s)], by convention are taken as unity. Hence, the quotients in Eqs. (10) reduce to the concentrations of the neutral species in the saturated solution, each called the intrinsic solubility of the compound, S_0.

In a saturated solution, solubility, S, at a particular pH is defined as the sum of the concentrations of all of the species dissolved in the aqueous solution:

$$S = [A^-] + [HA] \tag{11a}$$

$$S = [BH^+] + [B] \tag{11b}$$

$$S = [XH_2^+] + [XH] + [X^-] \tag{11c}$$

In Eqs. (11), [HA], [B] and [XH] are constant (intrinsic solubility), but the other concentrations are variable. The next step involves conversions of all variables into expressions containing only constants and $[H^+]$ (as the independent variable). Substitution of Eqs. (2) and (10) into (11) produces the desired equations.

$$S = S_0 \cdot (10^{\pm(\mathrm{pH} - \mathrm{p}K_a)} + 1) \tag{12a}$$

$$S = S_0 \cdot (10^{+(\mathrm{pH} - \mathrm{p}K_{a2})} + 10^{-(\mathrm{pH} - \mathrm{p}K_{a1})} + 1) \tag{12b}$$

where in Eq. (12a), the "+" in "±" refers to acids and "–" refers to bases. Equation (12b) is that of an ampholyte. The equations for other more complicated cases are summarized elsewhere [76].

Figure 3.2(a) shows a plot of log S versus pH for naproxen, based on re-analysis (unpublished) of the shake-flask [49, 77] and microtiter plate [20] data reported in the literature. The dashed curves in Fig. 3.2 were calculated with the simple Henderson–Hasselbalch equations. For $\mathrm{pH} \ll \mathrm{p}K_a$, the function reduces to the horizontal line log S=log S_0. For $\mathrm{pH} \gg \mathrm{p}K_a$, log S is a straight line as a function of pH, exhibiting a slope of 1 (and an intercept of log $S_0 - \mathrm{p}K_a$). Where the slope is 0.5, the pH equals to the pK_a.

Figure 3.2(b) shows a plot of log S versus pH for atenolol, based on the shake-flask data reported in the literature [49]. For $\mathrm{pH} \gg \mathrm{p}K_a$, the function again reduces to the horizontal line log S=log S_0. For $\mathrm{pH} \ll \mathrm{p}K_a$, log S is a straight line as a function of pH, exhibiting a slope of –1. Where the slope is 0.5, the $\mathrm{pH} = \mathrm{p}K_a$.

Figure 3.2(c) shows an example of an ampholyte, labetolol. The log S versus pH shake-flask data were taken from the literature [49]. Ampholyte parabolic-shaped curves show features of both an acid and a base profile.

3.4.3
Gibbs' pK_a and the "*sdiff* 3–4" Approximation

Although Fig. 3.2 properly conveys the shapes of solubility–pH curves in saturated solutions of uncharged species, according to the Henderson–Hasselbalch equation, the indefinite ascendancy of the dashed curves in the plots can be misleading. When pH changes elevate the solubility, at some value of pH, the solubility product of the salt will be reached, causing the shape of the solubility–pH curve to level off, as indicated in Fig. 3.2(a) for pH > 8.38.

As a "rule of thumb," in 0.15 M NaCl (or KCl) solutions titrated with NaOH (or KOH), acids start to precipitate as salts above log $(S/S_0) = 4$ and bases above log $(S/S_0) = 3$. This has been called the "*sdiff* 3–4" approximation [49]. With other counterions, such as phosphate, different trends are evident [15].

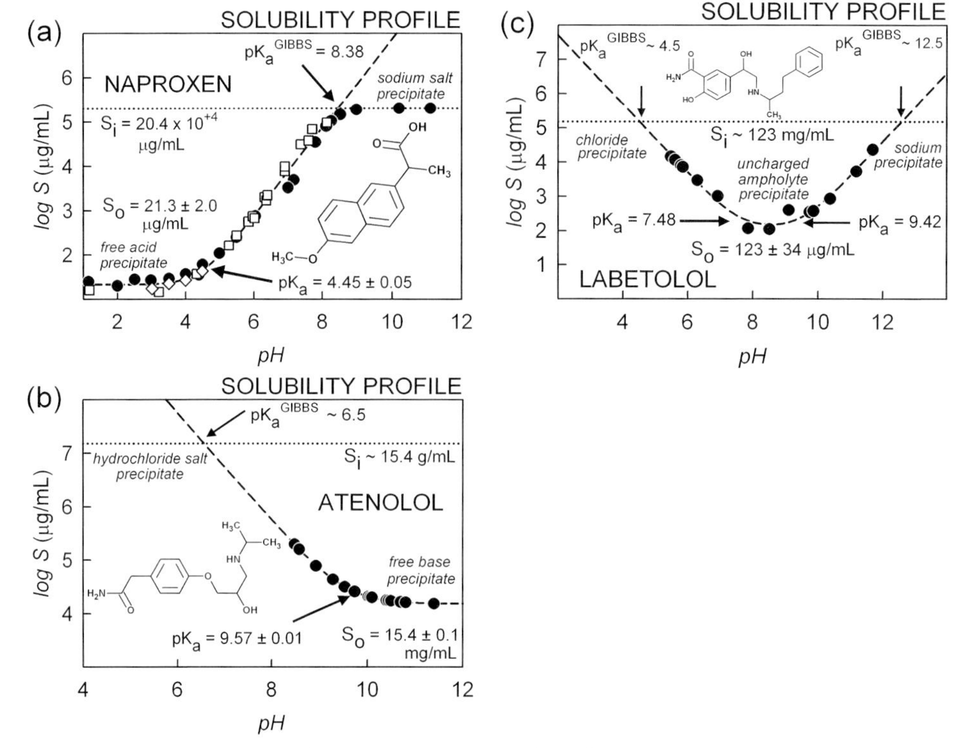

Fig. 3.2 Solubility profiles: log S–pH. The dashed curves, representing uncharged precipitate in equilibrium with solution of the drugs, were calculated by Henderson–Hasselbalch equations. The dotted horizontal lines are estimates of the solubility of the charged form of the drugs, using either actual data (naproxen) or estimates based on the *sdiff* 3–4 approximation (atenolol and labetolol). (a) Naproxen (acid): data reported from three sources: circles representing shake-flask data [49], squares representing shake-flask data [77], diamonds representing high-throughput miniaturized shake-flask (microtiter plate) data [16]. (b) Atenolol (base): data based on shake-flask determination [49]. (c) Labetolol (ampholyte): data based on shake-flask method [49].

Consider the case of the monoprotic acid, HA, which forms the sodium salt (in 0.15 M NaCl) when the solubility product, K_{sp}, is exceeded. In additions to Eqs. (2a) and (10a) above, one needs to add the following equation to treat the case.

$$Na^+A^-(s) \leftrightarrows Na^+ + A^- K_{sp} = [Na^+][A^-]/[Na^+A^-(s)] = [Na^+][A^-] \qquad (13)$$

Effective solubility is still defined by Eq. (11a). However, Eq. (11a) is solved under three limiting conditions with reference to a special pH value. (i) If the solution pH is below the conditions which lead to salt formation, the solubility-pH curve has the shape described by Eq. (12a) (dashed curve in Fig. 3.2a). (ii) If pH is above the characteristic value where salt starts to form (given high enough a sample concentration), Eq. (11a) is solved by taking [A$^-$] to be a constant and [HA] a variable:

$$S = [\mathrm{A}^-] + [\mathrm{A}^-]\cdot\frac{[\mathrm{H}^+]}{K_\mathrm{a}} = S_i \cdot (1 + 10^{-(\mathrm{pH}-\mathrm{p}K_\mathrm{a})}) \tag{14}$$

where S_i (= K_{sp}/[Na$^+$]) refers to the solubility of the conjugate base of the acid, which depends on the value of [Na$^+$] and is hence a conditional constant. Since pH >> pK_a and [Na$^+$] may be assumed to be constant, Eq. (14) reduces to that of a horizontal dotted line in Fig. 3.2(a): log S=log S_i for pH > 8.38. (iii) If the pH is exactly at the special point marking the onset of salt precipitation, the equation describing the solubility–pH relationship may be obtained by recognizing that both terms in Eq. (11a) become constant, so that:

$$S = S_0 + S_i \tag{15}$$

Consider the case of a very concentrated solution of the acid hypothetically titrated from pH well below its pK_a to the point where the solubility product is first exceeded. At first, the saturated solution can only have the unionized molecular species precipitated. When the solubility reaches the solubility product, at a particular elevated pH, salt starts to precipitate, but at the same time there may be remaining free acid precipitate. The simultaneous presence of the solid free acid and its solid conjugate base invokes the Gibbs' phase rule constraint, fixing the pH and the solubility, as long as the two interconverting solids are both present. As the titration progresses, the alkali titrant converts the remaining free acid solid into the solid salt of the conjugate base. During this process, pH is rigorously fixed, in a manner of a "perfect" buffer. This special pH point has been designated the Gibbs' pK_a, i.e. pK_a^{GIBBS} [19]. The equilibrium equation associated with this phenomenon may be stated as:

$$\mathrm{HA(s)} \leftrightarrows \mathrm{A^-(s)} + \mathrm{H^+} K_\mathrm{a}^\mathrm{GIBBS} = [\mathrm{H^+}][\mathrm{A^-(s)}]/[\mathrm{HA(s)}] = [\mathrm{H^+}] \tag{16}$$

This is a conditional constant, depending on the value of the background [Na$^+$].

Since solubility is fixed during the solids interconversion, one may set Eq. (12a) equal to Eq. (14), to get in logarithmic form the expression

$$\log S_0 - \log S_i = \pm(\mathrm{p}K_\mathrm{a}^\mathrm{GIBBS} - \mathrm{p}K_\mathrm{a}) \tag{17}$$

with "+" in "±" for acids and "–" for bases (*cf.* Eq. 17 to 9). Figure 3.2(a) shows the solubility–pH profile for naproxen, where the difference, Eq. (17), is very close to 4 log units, typically found with simple acids in the presence of 0.15 M NaCl. For basic drugs, the difference is approximately 3 log units. These relations have been called the "*sdiff* 3–4" approximation [8].

3.4.4
Aggregation Equations and "Shift-in-the-pK_a" Analysis

When a compound forms a dimer or a higher-order oligomer in aqueous solution, the characteristic solubility–pH profile takes on a shape not predicted by the Henderson–Hasselbalch equation and often indicates an *apparent* pK_a that is different from the true pK_a. Figure 3.3 shows several examples of sparingly-soluble

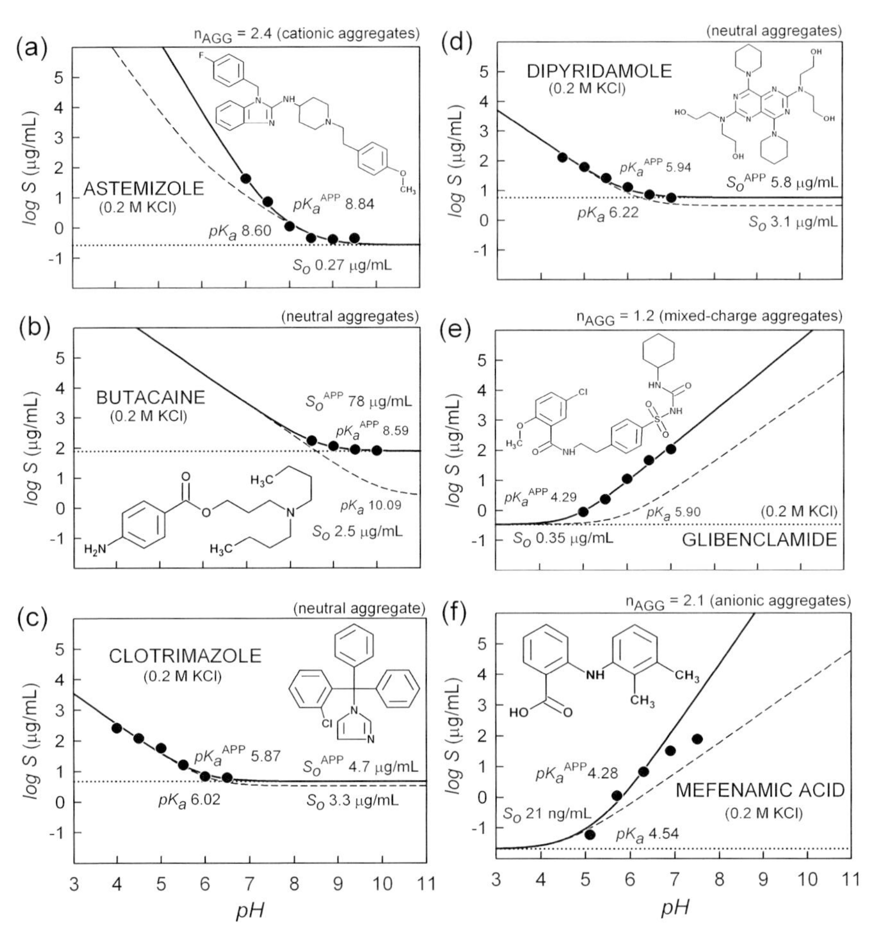

Fig. 3.3 Solubility profiles of sparingly soluble drugs, based on data taken from Avdeef et al. [20]. The solutions consisted of robotically adjusted universal buffers, based on a mixture of Good buffers (see text), and contained 0.2 M KCl. The dashed lines were calculated by the Henderson–Hasselbalch equation and, as can be seen, did not accurately describe the solubility profiles. The solid curves were calculated with equations where aggregation was taken into account [20]. The bases butacaine, clotrimazole and dipyridamole indicated uncharged aggregates. Astemizole indicated cationic aggregates and mefenamic acid appeared to indicate anionic aggregates (with evidence for salt precipitation for pH > 6.5). Glibenclamide consistently behaved as a mixed-charge aggregate.

molecules forming both charged and uncharged aggregates [20]. Bases in Fig. 3.3(b=butacaine, c=clotrimazole and d=dipyridamole) indicate an apparent pK_a that is *lower* than the true value. This has been hypothesized to indicate the formation of *uncharged* aggregates [8, 10, 20, 31]. Conversely, the apparent pK_a is *greater* than the true value in the remaining base, Fig. 3.3(a=astemizole), indicating the formation of *cationic* aggregates [20]. Furthermore, for charged aggregates, the slope in the solubility profile curve (in the diagonal region) is greater than one, and indicates the average degree of aggregation of the charged species [20]. Mefenamic acid (Fig. 3.3f) indicates formation of dimeric anionic aggregates. Glibenclamide (Fig. 3.3c) with a near unit slope, but a pK_a "shift," indicates a mixed-charge aggregate [20].

As the above examples suggest, the pK_a-shift method can be used as a quick alert tool. When a log S versus pH plot is inspected and the true pK_a is known independently, it can be quickly surmised whether aggregates are present and whether these are due to the neutral or the charged form of the drug. Moreover, the intrinsic solubility may be calculated from the magnitude and the direction of the pK_a shift. If an uncharged molecule forms aggregates, weak acids will indicate an apparent pK_a *higher* than the true pK_a and weak bases will indicate an apparent pK_a *lower* than the true pK_a. If the observed shifts are opposite of what is stated above, then the charged (rather than the neutral) species is involved in the aggregation.

Consider the formation of the oligomeric mixed-charge weak acid species, $(AH{\cdot}A)_n^{n-}$. In addition to the required equilibrium Eqs. (2a) and (10a), the additional reaction is needed to describe the model:

$$nA^- + nHA \leftrightarrows (AH{\cdot}A)_n^{n-} \quad K_n^* = \left[(AH{\cdot}A)_n^{n-}\right]/[HA]^n\,[A^-]^n \tag{18}$$

with solubility defined by:

$$S = [A^-] + [HA] + 2n\left[(AH{\cdot}A)_n^{n-}\right] \tag{19}$$

The $[A^-]$ and $\left[(AH{\cdot}A)_n^{n-}\right]$ components in Eq. (19) may be expanded in terms of [HA], pH and the various equilibrium constants [20]:

$$S = S_0 \cdot (1 + 10^{+(\mathrm{pH}-\mathrm{p}K_a)}) + 2n \cdot 10^{+n(\mathrm{pH}-\mathrm{p}K_a)+2n\cdot\log S_0 + \log K_n^*} \tag{20}$$

Two limiting forms of Eq. (20) in log form may be posed as:

$$\log S = \log S_0 \ @\ \mathrm{pH} << \mathrm{p}K_a^{\mathrm{APP}} \tag{21a}$$

$$\log S = \log 2n + \log K_n^* + 2n \log S_0 - n\mathrm{p}K_a + n\mathrm{pH} \ @\ \mathrm{pH} >> \mathrm{p}K_a \tag{21b}$$

Derivations of other aggregation models are described elsewhere [8, 10, 16, 20, 31].

3.5 "Flux" and other "Apparent" pK_a in Permeability Measurement

The Caco-2 cell line is a popular model for characterizing drug permeability [3, 22]. In this chapter, more emphasis is given to a more recently developed permeability model, introduced by Kansy et al. [78], i.e. PAMPA, which is now widely used to assess passive permeability in screening programs [7, 8, 16, 21, 22, 31, 48, 76, 79–82]. For the purposes of this chapter, permeability profiles from both models may be treated in a similar way. Traditionally, pH-dependent cellular permeability coefficients are designated "apparent", expressed as log P_{app}, and PAMPA-based values are designated "effective," log P_e. The two terms have the same meaning, with only minor operational differences in definitions (corrections for filter permeability and mass balance deficiencies).

Permeability–pH profiles, log P_e – pH curves in artificial membrane models (log P_{app} – pH in cellular models), generally have sigmoidal shape, similar to that of log D_{oct} – pH (*cf.* Fig. 3.1). However, one feature is unique to permeability profiles: the upper horizontal part of the sigmoidal curves may be vertically depressed, due to the drug transport resistance arising from the aqueous boundary layer (ABL) adjacent to the two sides of the membrane barrier. Hence, the true membrane contribution to transport may be obscured when water is the rate-limiting resistance to transport. This is especially true if sparingly soluble molecules are considered and if the solutions on either or both sides of the membrane barrier are poorly stirred (often a problem with 96-well microtiter plate formats).

Figure 3.4 shows the log P_e – pH profiles of an acid (warfarin), a base (propranolol) and an ampholyte (morphine). All data were collected by an automated robotic system using 96-well microtiter plates (*p*ION). The PAMPA membrane in Fig. 3.4 was formulated from phospholipids extracted from animal brain to model blood–brain barrier (BBB) permeability [83]. In Fig. 3.4(a and b), vigorous magnetic stirring produced ABL thicknesses matching those expected to be in the GIT environment. In Fig. 3.4(c), the solutions were not stirred and indicated an ABL thickness greater than 2000 μM. In all cases, there is evidence of permeation by both the neutral and the charged forms of the drugs. The observed permeability was ABL-limited above pH 7 for propranolol (Fig. 3.4b) and morphine (Fig. 3.4c). The permeability of warfarin was largely membrane-limited across the pH range studied.

3.5.1 Correcting Permeability for the ABL Effect by the pK_a^{FLUX} Method

In the GIT epithelial environment, the ABL thickness is expected to be 30–100 μM, whereas in unstirred permeability assays, the ABL thickness can be as high as 1000–4000 μM [22, 48, 79]. By taking permeability (stirred or unstirred) data over a range of pH, it is possible to match the effect of the ABL to that expected for the GIT, by applying the pK_a^{FLUX} method [8, 22], briefly described below.

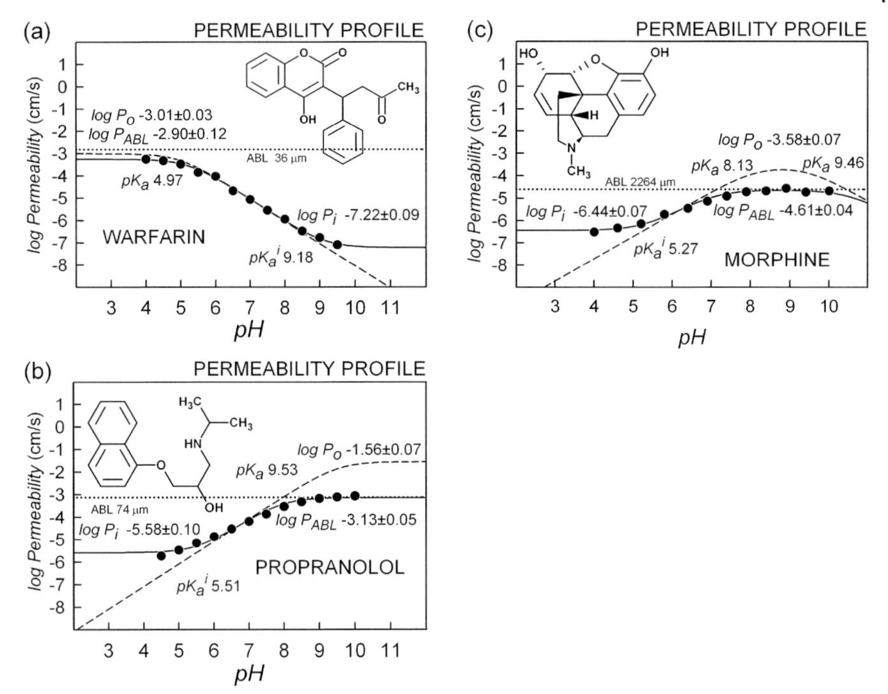

Fig. 3.4 Permeability profiles for (a) warfarin (acid), (b) propranolol (base) and (c) morphine (ampholyte) based on a BBB PAMPA model (*p*ION) composed of animal brain extract of lipids. The data (unpublished) were analyzed with the *p*CEL-X program (*p*ION), with the refined parameters indicated in the three frames. In all three cases, there was evidence for the permeation of charged forms of the drugs. Warfarin showed predominantly membrane-limited transport, whereas propranolol and morphine were ABL-limited in transport for pH > 7. Warfarin and propranolol solutions were vigorously stirred, to match the expected thickness of the ABL (30–100 μm). Morphine solutions were not stirred.

In PAMPA, the effective permeability coefficient, P_e, is related to the membrane and ABL permeability coefficients, P_m and P_{ABL}, respectively, as

$$\frac{1}{P_e} = \frac{1}{P_{ABL}} + \frac{1}{P_m} \tag{22}$$

(For cellular models, a more complicated form of the above equation is needed, to factor in paracellular, facilitated uptake and efflux transport, etc. [22].)

For ionizable molecules, the membrane permeability, P_m (P_c in cellular models), depends on pH of the bulk aqueous solution. The maximum possible P_m is designated P_0, the intrinsic permeability of the uncharged species. For monoprotic weak acids and bases, the relationship between P_m and P_0 may be stated in terms of the fraction of the uncharged species, f_0, as $P_m = P_0 f_0$, i.e.:

$$\frac{1}{P_m} = \frac{10^{\pm(\mathrm{pH}-\mathrm{p}K_a)}+1}{P_0} \tag{23}$$

with "+" used for acids and "–" used for bases. Other cases are described elsewhere [8, 21]. The logarithmic form of the above equation is

$$\log P_m = \log P_0 - \log(10^{\pm(\mathrm{pH}-\mathrm{p}K_a)}+1) \tag{24}$$

which describes a hyperbolic curve (*cf.* Fig. 3.4a and b, dashed curves). In the bend of such curves (where the slope in the curve is one-half), the pH is indicative of the pK_a of the molecule.

Combining Eqs. (22) and (23) leads to

$$\frac{1}{P_e} = \frac{1}{P_{ABL}} + \frac{10^{\pm(\mathrm{pH}-\mathrm{p}K_a)}+1}{P_0} \tag{25}$$

The logarithmic form of Eq. (25) is a hyperbolic curve, just like Eq. (24), with an *apparent* pK_a associated with the pH at half-slope positions (cf. Fig. 3.4, solid-line curves).

For highly permeable molecules it is useful to consider the "flux" ionization constant, pK_a^{FLUX}, which refers to the pH value where the resistance to transport across a permeation barrier is 50% due to the ABL and 50% due to the membrane [21]. The approximate hyperbolic log-log equation (which is accurate when P_0 is at least 10 times greater than P_{ABL})

$$\log P_e \approx \log P_e^{max} - \log(10^{\pm(\mathrm{pH}-\mathrm{p}K_a^{FLUX})}+1) \tag{26}$$

describes the relationship between the effective permeability and the apparent ionization constant [48]. The maximum possible effective (measured) permeability, P_e^{max}, is defined as $\log P_e^{max} = \log P_{ABL} - \log(1 + P_{ABL}/P_0)$. When $P_0 >> P_{ABL}$ (highly permeable molecules), $P_e^{max} \approx P_{ABL}$, indicating water-limited rather than membrane-limited diffusion.

3.5.2
Membrane Rate-Limiting Transport (Hydrophilic Molecules)

If the ABL is vanishingly thin (exceedingly vigorous stirring) or if $P_0 << P_{ABL}$ (common case with hydrophilic molecules), Eq. (26) reduces to Eq. (24), characterized by a horizontal region (indicating intrinsic permeability) and a diagonal region (slope of ±1). Warfarin (Fig. 3.4a) predominantly shows this membrane-limited transport since $P_e < P_{ABL}$ across the entire pH range. For the other two molecules, propranolol and morphine, the transport is membrane-limited only for pH < 7.

The dashed curves in Fig. 3.4 were calculated using Eq. (24), without consideration of the resistance of the ABL, and represent the membrane part of the overall transport resistance. With bases (e.g. Fig. 3.4b), for $pH >> pK_a$, Eq. (24) is that of a (dashed) horizontal line, corresponding to the intrinsic permeability, P_0, and for $pH << pK_a$, Eq. (24) is that of a diagonal line, with a slope of +1, with membrane permeability decreasing with decreasing pH, in accordance with the pH-partition hypothesis. With acids (e.g. Fig. 3.4a), a mirror relationship holds: for $pH << pK_a$, Eq. (24) is that of a horizontal line, and for $pH >> pK_a$, Eq. (24) is that of a diagonal line, with a slope of –1. It is possible to determine the P_0 and pK_a of a molecule from the log P_e – pH data when the transport is primarily membrane-limited.

3.5.3 Water Layer Rate-Limiting Transport (Lipophilic Molecules)

With a substantial ABL thickness (in the absence of stirring) and $P_0 \geq P_{ABL}$ (typical of lipophilic molecules), the transport of molecules is said to be ABL-limited. This is generally the case with highly lipophilic drugs, where the *same* permeability is often measured ($30\text{–}80 \times 10^{-6}\,\mathrm{cm\,s^{-1}}$ in poorly stirred solutions), *regardless of the molecules,* indicating a property of water (P_e^{max}) rather than membrane. The solid-line curves for pH > 7 in Fig. 3.4(b = propranolol and c = morphine) are examples of this. With bases (e.g. Fig. 3.4b), for $pH >> pK_a^{FLUX}$, Eq. (26) is that of a horizontal line, and for $pH << pK_a^{FLUX}$, Eq. (26) is that of a diagonal line, with a slope of +1. With acids (e.g. Fig. 3.4a), for $pH << pK_a^{FLUX}$, Eq. (26) is that of a horizontal line, and for $pH >> pK_a^{FLUX}$, Eq. (26) is that of a diagonal line, with a slope of –1. As long as the ABL contributes resistance to transport, and thus pK_a^{FLUX} is defined (i.e. $P_0 \geq P_{ABL}$):

$$\log P_0 - \log P_{ABL} = \pm(pK_a - pK_a^{FLUX}) \geq 0 \tag{27}$$

If the true pK_a of the molecule is known, then a simple inspection of the plot of log P_e (or P_{app} in the case of cellular assays) versus pH can often reveal the values of both log P_0 and log P_{ABL}.

3.5.4 Ionic-species Transport in PAMPA

In all three frames of Fig. 3.4, there is evidence of ionic-species transport, labeled as log P_i in Fig. 3.4. The pH at the bend in the curves corresponding to the onset of ionic permeability is labeled pK_a^{I} and corresponds to the pH where 50% of the transport is by the neutral species and 50% by the ionic species. This is a conditional constant, but unlike pK_a^{OCT} and pK_a^{GIBBS}, it is dependent mainly on the constituent makeup of the artificial phospholipid mixture. As before:

$$\log P_0 - \log P_i = \pm(pK_a^{I} - pK_a) \tag{28}$$

It is useful to compare Eq. (28) to (9), (17) and (27).

In the BBB-PAMPA lipid formulation illustrated in Fig. 3.4, the *diff* values, defined as the difference $\log P_0 - \log P_i$, range from 2.9 (morphine) to 4.2 (warfarin), somewhat similar to the values observed in the octanol–water system. However, it is premature to propose a "*pdiff* 3–4" approximation, given the limited amount of data reported. With other lipid formulations, larger differences are usually observed. In Double-Sink PAMPA, and especially in hexadecane-PAMPA, transport of ionized drugs has not been reported [84].

3.6 Conclusions

This chapter considered ionizable drug-like molecules. Absorption properties that are influenced by the pK_a were explored. The impact of the pK_a–absorption relationship on key physicochemical profiling underlying absorption (solubility, permeability and ionization) was examined in detail and several simplifying equations were discussed. The various "*diff*" relationships considered in the chapter are systematized in Table 3.2. Table 3.3 summarizes the "apparent pK_a shift" method for detecting aggregates in solubility profiles, when the apparent pK_a value derived from Henderson–Hasselbalch analysis of log S–pH profile does not agree with the

Tab. 3.2 Apparent pK_a*diff* relations (monoprotic substances).

diff Equation	*Approximate value*[1]	*Type*	*Apparent pK_a = aqueous pH when ...*
$\log P_{\text{oct}}^{\text{N}} - \log P_{\text{oct}}^{\text{I}} = \pm(\text{p}K_a^{\text{OCT}} - \text{p}K_a)$	3–4[2]	ion-pair partition (octanol–water)	neutral and charged species equal in concentration in octanol phase
$\log P_{\text{mem}}^{\text{N}} - \log P_{\text{mem}}^{\text{I}} = \pm(\text{p}K_a^{\text{MEM}} - \text{p}K_a)$	1–2[2]	ion-pair partition (liposome–water)	neutral and charged species equal in concentration in lipid bilayer
$\log S_0 - \log S_i = \pm(\text{p}K_a^{\text{GIBBS}} - \text{p}K_a)$	3–4[2]	solubility (salt)	neutral and salt solids coprecipitated
$\log P_0 - \log P_{\text{ABL}} = \pm(\text{p}K_a - \text{p}K_a^{\text{FLUX}})$	>0	permeability (ABL)	50% transport resistance due to ABL and 50% due to membrane
$\log P_0 - \log P_i = \pm(\text{p}K_a^{\text{I}} - \text{p}K_a)$	>3	permeability (ionic transport)	50% transport due to neutral species and 50% due to charged species

1 Approximate values in solutions containing 0.15 M ionic strength. Lower value refers to bases and higher value to acids.
2 See Table 3.1.

Tab. 3.3 Aggregation and apparent pK_a (monoprotic substances)[1].

Compound type	$\Delta = pK_a^{APP} - pK_a$	$\log S_0 =$	*Type of aggregate*
Acid	$\Delta > 0$	$\log S_0^{APP} - \lvert\Delta\rvert$	neutral
Acid	$\Delta < 0$	$\log S_0^{APP}$	anionic
Base	$\Delta > 0$	$\log S_0^{APP}$	cationic
Base	$\Delta < 0$	$\log S_0^{APP} - \lvert\Delta\rvert$	neutral

1 pK_a^{APP} is the apparent pK_a, determined from the application of the Henderson–Hasselbalch equation.

true pK_a value or when the slope in the diagonal portion of the log solubility–pH profile is greater than unity.

References

1 Wells, J. I. *Pharmaceutical Preformulation*, Ellis Horwood, London, **1988**.

2 Benet, L. Z., Goyan, J. E. Potentiometric determination of dissociation constants. *J. Pharm. Sci.* **1967**, *56*, 665–680.

3 Amidon, G. L., Lennernäs, H., Shah, V. P., Crison, J. R. A theoretical basis for a biopharmaceutic drug classification: the correlation of *in vitro* drug product dissolution and *in vivo* bioavailability. *Pharm. Res.* **1995**, *12*, 413–420.

4 Lipinski, C. A. Drug-like properties and the causes of poor solubility and poor permeability. *J. Phamacol. Toxicol. Methods* **2000**, *44*, 235–249.

5 Van de Waterbeemd, H., Smith, D. A., Beaumont, K., Walker, D. K. Property-based design: optimization of drug absorption and pharmacokinetics. *J. Med. Chem.* **2001**, *44*, 1313–1333.

6 Curatolo, W. Physical chemical properties of oral drug candidates in the discovery and exploratory development settings. *Pharm. Sci. Technol. Today* **1998**, *1*, 387–393.

7 Avdeef, A. HT solubility and permeability: MAD-PAMPA analysis. In *Pharmacokinetic Profiling in Drug Research*, Testa, B., Krämer, S. D., Wunderli-Allenspach, H., Folkers, G. (eds.), VHCA, Zurich and Wiley-VCH, Weinheim, **2006**, pp. 221–241.

8 Avdeef, A. *Absorption and Drug Development*, Wiley-Interscience, Hoboken, NJ, **2003**.

9 Ritschel, W. A. In *Perspectives In Clinical Pharmacy*, 1st edn., Fraske, D. E., Whitney, H. A. K., Jr. (eds.), Drug Intelligence Publications, Hamilton, IL, **1972**, pp. 325–367.

10 Avdeef, A., Voloboy, D., Foreman, A. Dissolution–solubility of multiprotic drugs: pH, buffer, salt, dual-solid, and aggregation effects. In *ADME/Tox Approaches*, Van de Waterbeemd, H., Testa, B. (vol. eds.), Vol. 5 in *Comprehensive Medicinal Chemistry*, 2nd edn., Taylor, J. B., Triggle, D. J. (eds.), Elsevier, Oxford, **2007**, pp. 399–423.

11 Tanford, C., Roxby, R. Interpretation of protein titration curves. Application to lysozyme. *Biochemistry* **1972**, *11*, 2192–2198.

12 Fersht, A. *Enzyme Structure and Mechanism*, Freeman, Reading, **1977**, pp. 134–155.

13 Shore, P. A., Brodie, B. B., Hogben, C. A. M. The gastric secretion of drugs: a pH Partition Hypothesis. *J. Pharmacol. Exp. Ther.* **1957**, *119*, 361–369.

14 Schanker, L. S., Tocco, D. J., Brodie, B. B., Hogben, C. A. M. Absorption of drugs from the rat small intestine. *J. Pharmacol. Exp. Ther.* **1958**, *123*, 81–88.

15 Bergström, C. A. S., Luthman, K., Artursson, P. Accuracy of calculated pH-dependent aqueous drug solubility. *Eur. J. Pharm. Sci.* **2004**, *22*, 387–398.

16 Avdeef, A., Kansy, M., Bendels, S., Tsinman, K. Biopharmaceutics classification gradient maps and the pH partition antithesis. Review by Eur. J. Pharm. Sci.

17 Albert, A., Serjeant, E. P. *The Determination of Ionization Constants – A Laboratory Manual*, 3rd edn., Chapman & Hall, New York, **1984**.

18 Avdeef, A. Assessment of distribution–pH profiles. In *Methods and Principles in Medicinal Chemistry, Vol. 4*, Pliska, V., Testa, B., Van de Waterbeemd, H. (eds.), VCH, Weinheim, **1996**, pp. 109–139.

19 Avdeef, A., pH-metric solubility. 1. Solubility-pH profiles from Bjerrum plots. Gibbs buffer and pK_a in the solid state. *Pharm. Pharmacol. Commun.* **1998**, *4*, 165–178.

20 Avdeef, A., Bendels, S., Tsinman, O., Tsinman, K., Kansy, M. Solubility-excipient classification gradient maps. *Pharm. Res.* **2007**, *24*, 530–545.

21 Avdeef, A. High-throughput Measurement of Membrane Permeability. In *Drug Bioavailability (Methods and Principles in Medicinal Chemistry)*, Van de Waterbeemd, H., Lennernäs, H., Artursson, P. (eds.), Wiley-VCH, Weinheim, **2003**, pp. 46–71.

22 Avdeef, A., Artursson, P., Neuhoff, S., Lazarova, L., Gråsjö, J., Tavelin, S. Caco-2 permeability of weakly basic drugs predicted with the double-sink PAMPA pK_a^{FLUX} method. *Eur. J. Pharm. Sci.* **2005**, *24*, 333–349.

23 Newton, D. W., Kluza, R. B. pK_a values of medicinal compounds in pharmacy practice. *Drug Intell. Clin. Pharm.* **1978**, *12*, 548–554.

24 Avdeef, A., Bucher, J. J. Accurate measurements of the concentration of hydrogen ions with a glass electrode: calibrations using the Prideaux and other universal buffer solutions and a computer-controlled automatic titrator. *Anal. Chem.* **1978**, *50*, 2137–2142.

25 Avdeef, A. Weighting scheme for regression analysis using pH data from acid–base titrations. *Anal. Chim. Acta* **1983**, *148*, 237–244.

26 Avdeef, A. STBLTY: methods for construction and refinement of equilibrium models. In *Computational Methods for the Determination of Formation Constants*, Leggett, D. J. (eds.), Plenum, New York, **1985**, pp. 355–473.

27 Avdeef, A. pH-metric log *P*. 2. Refinement of partition coefficients and ionization constants of multiprotic substances. *J. Pharm. Sci.* **1993**, *82*, 183–190.

28 Avdeef, A., Comer, J. E. A., Thomson, S. J. pH-metric log *P*. 3. Glass electrode calibration in methanol–water, applied to pK_a determination of water-insoluble substances. *Anal. Chem.* **1993**, *65*, 42–49.

29 Takács-Novák, K., Box, K. J., Avdeef, A. Potentiometric p*Ka* determination of water-insoluble compounds. Validation study in methanol/water mixtures. *Int. J. Pharm.* **1997**, *151*, 235–248.

30 Avdeef, A., Box, K. J., Comer, J. E. A., Gilges, M., Hadley, M., Hibbert, C., Patterson, W., Tam, K. Y. pH-metric log *P*. 11. p*Ka* determination of water-insoluble drugs in organic solvent–water mixtures. *J. Pharm. Biomed. Anal.* **1999**, *20*, 631–641.

31 Avdeef, A., Testa, B. Physicochemical profiling in drug research: a brief state-of-the-art of experimental techniques. *Cell. Mol. Life Sci.* **2003**, *59*, 1681–1689.

32 Lingane, P. J., Hugus, Z. Z., Jr. Normal equations for the Gaussian least-squares refinement of formation constants with simultaneous adjustment of the spectra of the absorbing species. *Inorg. Chem.* **1970**, *9*, 757–762.

33 Hugus, Z. Z., Jr., El-Awady, A. A. The determination of the number of species present in a system: a new matrix rank treatment of spectrophotometric data. *J. Phys. Chem.* **1971**, *75*, 2954–2957.

34 Maeder, M., Gampp, H. Spectrophotometric data reduction by eigenvector analysis for equilibrium and kinetic studies and a new method of fitting exponentials. *Anal. Chim. Acta* **1980**, *122*, 303–313.

35 Kralj, Z. I., Simeon, V. Estimation of spectra of individual species in a multisolute solution. *Anal. Chim. Acta* **1982**, *129*, 191–198.

36 Gampp, H., Maeder, M., Meyer, C. J., Zuberbühler, A. D. Calculation of equilibrium constants from multiwavelength spectroscopic data. I. Mathematical considerations. *Talanta* **1985**, *32*, 95–101.

37 Allen, R. I., Box, K. J., Comer, J. E. A., Peake, C., Tam, K. Y. Multiwavelength spectrophotometric determination of acid dissociation constants of ionizable drugs. *J. Pharm. Biomed. Anal.* **1998**, *17*, 699–712.

38 Tam, K. Y., Takács-Novák, K. Multiwavelength spectrophotometric determination of acid dissociation constants. Part II. First derivative versus target factor analysis. *Pharm. Res.* **1999**, *16*, 374–381.

39 Mitchell, R. C., Salter, C. J., Tam, K. Y. Multiwavelength spectrophotometric determination of acid dissociation constants. Part III. Resolution of multi-protic ionization systems. *J. Pharm. Biomed. Anal.* **1999**, *20*, 289–295.

40 Tam, K. Y., Hadley, M., Patterson, W. Multiwavelength spectrophotometric determination of acid dissociation constants. Part IV. Water-insoluble pyridine derivatives. *Talanta* **1999**, *49*, 539–546.

41 Box, K. J., Comer, J. E. A., Hosking, P., Tam, K. Y., Trowbridge, L., Hill, A. Rapid physicochemical profiling as an aid to drug candidate selection. In: *High Throughput Screening: The Next Generation*, Dixon, G. K., Major, J. S., Rice, M. J. (eds.), Bios, Oxford, **2000**, pp. 67–74.

42 Tam, K. Y., Takács-Novák, K. Multiwavelength spectroscopic determination of acid dissociation constants: a validation study. *Anal. Chim. Acta* **2001**, *434*, 157–167.

43 Tam, K. Y. Multiwavelength spectrophotometric resolution of the micro-equilibria of a triprotic amphoteric drug: methacycline. *Mikrochim. Acta* **2001**, *136*, 91–97.

44 Takács-Novák, K., Tam, K. Y. Multiwavelength spectrophotometric determination of acid dissociation constants. Part V. Microconstants and tautomeric ratios of diprotic amphoteric drugs. *J. Pharm. Biomed. Anal.* **2000**, *17*, 1171–1182.

45 Tam, K. Y. Multiwavelength spectrophotometric determination of acid dissociation constants. Part VI. Deconvolution of binary mixtures of ionizable compounds. *Anal. Lett.* **2000**, *33*, 145–161.

46 Tam, K. Y., Quéré, L. Multiwavelength spectrophotometric resolution of the micro-equilibria of cetirizine. *Anal. Sci.* **2001**, *17*, 1203–1208.

47 Hendriksen, B. A., Sanchez-Felix, M. V., Tam, K. Y. A new multiwavelength spectrophotometric method for the determination of the molar absorption coefficients of ionizable drugs. *Spectrosc. Lett.* **2002**, *35*, 9–19.

48 Bendels, S., Tsinman, O., Wagner, B., Lipp, D., Parrilla, I., Kansy, M., Avdeef, A. PAMPA-excipient classification gradient maps. *Pharm. Res.* **2006**, *23*, 2525–2535.

49 Avdeef, A., Berger, C. M., Brownell, C. pH-metric solubility. 2. Correlation between the acid–base titration and the saturation shake-flask solubility-pH methods. *Pharm. Res.* **2000**, *17*, 85–89.

50 Avdeef, A., Berger, C. M. pH-metric solubility. 3. Dissolution titration template method for solubility determination. *Eur. J. Pharm. Sci.* **2001**, *14*, 271–280.

51 Good, N. E., Winget, G. D., Winter, W., Connolly, T. N., Izawa, S., Singh, R. M. M. Hydrogen ion buffers for biological research. *Biochemistry* **1966**, *5*, 467–477.

52 Perrin, D. D., Dempsey, B. *Buffers for pH and Metal Ion Control*, Chapman & Hall, London, **1974**.

53 Perrin, D. D., Dempsey, B., Serjeant, E. P. *pK_a Prediction for Organic Acids and Bases*, Chapman & Hall, London, **1981**.

54 Advanced Chemistry Development Inc., Toronto, Canada. ACD/Solubility DB computer program, http://www.acdlabs.com.

55 Pharma Algorithms, Toronto, Canada. Algorithm Builder V1.8 and ADME Boxes

V2. 5 computer programs, http://www.ap-algorithms.com.

56 ChemAxon Ltd, Budapest, Hungary, http://www.chemaxon.com/marvin.

57 Sillén, L. G., Martell, A. E. *Stability Constants of Metal–Ion Complexes (Spec. Publ. No. 17)*, Chemical Society, London, **1964**.

58 Sillén, L. G., Martell, A. E. *Stability Constants of Metal–Ion Complexes (Spec. Publ. No. 25)*, Chemical Society, London, **1971**.

59 Serjeant, E. P., Dempsey, B. *Ionization Constants of Organic Acids in Aqueous Solution*, Pergamon, Oxford, **1979**.

60 Smith, R. M., Martell, A. E. *Critical Stability Constants, Vols. 1–6*, Plenum Press, New York, **1974**.

61 Hansch, C., Leo, A. *Substituted Constants for Correlation Analysis in Chemistry and Biology*, Wiley-Interscience, New York, **1979**.

62 Mannhold, R., Dross, K. P., Rekker, R. F. Drug lipophilicity in QSAR practice: I. A comparison of experimental with calculated approaches. *Quant. Struct.-Act. Relat.* **1990**, *9*, 21–28.

63 Dearden, J. C., Bresnen, G. M. The measurement of partition coefficients. *Quant. Struct.-Act. Relat.* **1988**, *7*, 133–144.

64 Hersey, A., Hill, A. P., Hyde, R. M., Livingstone, D. J. Principles of method selection in partition studies. *Quant. Struct.-Act. Relat.* **1989**, *8*, 288–296.

65 Mirrlees, M. S., Moulton, S. J., Murphy, C. T., Taylor, P. J. Direct measurement of octanol–water coefficients by high-pressure liquid chromatography. *J. Med. Chem.* **1976**, *19*, 615–619.

66 Valkó, K., Bevan, C., Reynolds, D. Chromatographic hydrophobicity index by fast-gradient RP-HPLC: a high-throughput alternative to log *P*/log D. *Anal. Chem.* **1997**, *69*, 2022–2029.

67 Lombardo, F., Shalaeva, M. Y., Tupper, K. A., Gao, F. ElogD$_{oct}$: a tool for lipophilicity determination in drug discovery. 2. Basic and neutral compounds. *J. Med. Chem.* **2001**, *44*, 2490–2497.

68 Hitzel, L., Watt, A. P., Locker, K. L. An increased throughput method for the determination of partition coefficients. *Pharm. Res.* **2000**, *17*, 1389–1395.

69 Comer, J. E. A. High-throughput measurement of log *D* and pK_a. In *Drug Bioavailability (Methods and Principles in Medicinal Chemistry)*, Van de Waterbeemd, H., Lennernäs, H., Artursson, P. (eds.), Wiley-VCH, Weinheim, **2003**, pp. 21–45.

70 Barbato, F., Caliendo, G., Larotonda, M. I., Morrica, P., Silipo, C., Vittoria, A. Relationships between octanol–water partition data, chromatographic indices and their dependence on pH in a set of beta-adrenoceptor blocking agents. *Farmaco* **1990**, *45*, 647–663.

71 Avdeef, A., Box, K. J., Comer, J. E. A., Hibbert, C., Tam, K. Y. pH-metric log *P*. 10. Determination of vesicle membrane–water partition coefficients of ionizable drugs. *Pharm. Res.* **1997**, *15*, 208–214.

72 Streng, W. H., Yu, D. H.-S., Zhu, C. Determination of solution aggregation using solubility, conductivity, calorimetry, and pH measurements. *Int. J. Pharm.* **1996**, *135*, 43–52.

73 Zhu, C., Streng, W. H. Investigation of drug self-association in aqueous solution using calorimetry, conductivity, and osmometry. *Int. J. Pharm.* **1996**, *130*, 159–168.

74 Ritschel, W. A., Alcorn, G. C., Streng, W. H., Zoglio, M. A. Cimetidine–theophylline complex formation. *Methods Find. Exp. Clin. Pharmacol.* **1983**, *5*, 55–58.

75 Glomme, A., März, J., Dressman, J. B. Comparison of a miniaturized shake-flask solubility method with automated potentiometric acid/base titrations and calculated solubilities. *J. Pharm. Sci.* **2005**, *94*, 1–16.

76 Avdeef, A. High-throughput measurements of solubility profiles. In *Pharmacokinetic Optimization in Drug Research (Methods and Principles in Medicinal Chemistry)*, Testa, B., Van de Waterbeemd, H., Folkers, G., Guy, R. (eds.), Wiley-VCH, Weinheim, **2001**, pp. 305–326.

77 Chowhan, Z. T. pH-solubility profiles of organic carboxylic acids and their salts. *J. Pharm. Sci.*, **1978**, *67*, 1257–1260.

78 Kansy, M., Senner, F., Gubernator, K. Physicochemical high throughput screening: parallel artificial membrane permeability assay in the description of passive absorption processes. *J. Med. Chem.* **1998**, *41*, 1007–1010.

79 Avdeef, A., Nielsen, P. E., Tsinman, O. PAMPA – a drug absorption *in vitro* model. 11. Matching the *in vivo* unstirred water layer thickness by individual-well stirring in microtitre plates. *Eur. J. Pharm. Sci.* **2004**, *22*, 365–374.

80 Ruelle, J. A., Tsinman, O., Avdeef, A. Acid–base cosolvent method for determining aqueous permeability of amiodarone, itraconazole, tamoxifen, terfenadine and other very insoluble molecules. *Chem. Pharm. Bull.* **2004**, *52*, 561–565.

81 Kansy, M., Avdeef, A., Fischer, H. Advances in screening for membrane permeability: high-resolution PAMPA for medicinal chemists. *Drug Discov. Today Technol* **2005**, *1*, 349–355.

82 Avdeef, A. The rise of PAMPA. *Expert Opin. Drug Metab. Toxicol.* **2005**, *1*, 325–342.

83 Dagenais, C., Avdeef, A., Tsinman, O., Dudley, A., Beliveau, R. P-glycoprotein deficient mouse *in situ* blood–brain barrier permeability and its prediction using an *in combo* PAMPA model. (under review)

84 Avdeef, A., Tsinman, O. PAMPA – a drug absorption *in vitro* model. 13. Chemical selectivity due to membrane hydrogen bonding: *in combo* comparisons of HDM-, DOPC-, and DS-PAMPA. *Eur. J. Pharm. Sci.* **2006**, *28*, 43–50.

4
Electrotopological State Indices

Ovidiu Ivanciuc

Abbreviations

3D	three-dimensional
AC	accuracy
ADMET	absorption, distribution, metabolism, excretion and toxicity
ALL-QSAR	automated lazy learning quantitative structure–activity relationships
ANN	artificial neural networks
AUROC	area under the receiver operator characteristic
BBB	blood–brain barrier
Caco-2	adenocarcinoma cell line derived from human colon
CART	classification and regression trees
CL_{tot}	total clearance
COX	cyclooxygenase
CYP	cytochrome P450
DA	dopamine antagonists
DF	decision forest
Dual	serotonin-dopamine dual antagonists
ER	estrogen receptor
E-state	electrotopological state
HE-state	hydrogen electrotopological state
HIA	human intestinal absorption
HSA	human serum albumin
IAM	immobilized artificial membrane
KHE	Kier–Hall electronegativity
*k*NN	*k*-nearest neighbors
LDA	linear discriminant analysis
LLNA	local lymph node assay
MARS	multivariate adaptive regression splines
MCC	Matthews correlation coefficient
MLR	multiple linear regression
NPY	neuropeptide Y

Molecular Drug Properties. Measurement and Prediction. R. Mannhold (Ed.)

ISBN: 978-3-527-31755-4

PCR	principal component regression
PFB	percent fraction bound to serum proteins
P-gp	P-glycoprotein
PLS	partial least squares
PM-CSVM	positive majority consensus support vector machines
PNN	probabilistic neural network
PP-CSVM	positive probability consensus support vector machines
QSAR	quantitative structure–activity relationship
QSPkR	quantitative structure–pharmacokinetic relationship
QSPR	quantitative structure–property relationship
RFE	recursive feature elimination
RP	recursive partitioning
RR	ridge regression
SA	serotonin antagonists
SAR	structure–activity relationship
SIMCA	soft independent modeling of class analogy
SVM	support vector machines
TdP	*torsade de pointes*
TMARS	two-step multivariate adaptive regression splines

4.1 Introduction

The electrotopological state (E-state) combines electronic information and molecular topology to describe the chemical structure at the atomic level. The E-state index for an atom is the sum between the intrinsic state of that atom and a perturbation term representing the influence of the remaining atoms in the molecule. The intrinsic state of an atom encodes its electronic information corresponding to a valence state and bonding state, thus providing a measure of the local topology. The influence of the other atoms in the molecule is represented as a perturbation that decreases as the square of the graph distance between atoms. The atomic E-state index measures the electron accessibility as described by the molecular topology alone. The E-state indices and related structural descriptors are used with success in quantitative structure–property relationship (QSPR) and quantitative structure–activity relationship (QSAR) models. The atomic-level structural information encoded into the E-state indices generates a chemical space that can be efficient in measuring the molecular similarity and in screening chemical libraries.

The E-state is based solely on atom connectivity information obtained from the molecular graph, without any input from the molecular geometry or sophisticated quantum calculations. We start this chapter with a brief presentation of the relevant notions of graph theory and continue with the definitions of a couple of important graph matrices. Then the molecular connectivity indices are mentioned because the valence δ index, which defines these indices, is also an important component of the intrinsic state. Next, the family of atomic E-state indices is presented together with several methodologies for their application in QSPR and QSAR. E-state indices may describe individual atoms in sets of molecules that have a common skeleton or the E-state values may be summed for all atoms of a certain type. Other types of E-state indices may be computed for hydrogen atoms, for bonds, as well as for three-dimensional (3D) grid points that are used as descriptors for 3D QSAR. In the last part of the chapter we review recent applications of E-state indices in QSPR and QSAR, with a special emphasis for application in drug design and in modeling biological properties of chemical compounds.

4.2 E-state Indices

Organic molecules may be represented as molecular graphs in which graph vertices correspond to atoms and graph edges represent covalent bonds between atoms [1]. The graph model of the chemical structure describes the chemical bonding pattern of atoms, without reference to the molecular geometry [2]. Structural descriptors derived from the molecular graph, such as fingerprints, structural keys and topological indices, are highly successful in modeling a broad range of physical, chemical or biological properties, thus demonstrating that these properties depend mainly on the bonding relationships between atoms. The molecular graph representation of the chemical structure encodes mainly the connectivity of the atoms and is less suitable for the modeling of those properties that are determined mostly by the molecular geometry, conformation or stereochemistry. Graph descriptors have a clear advantage in screening large chemical libraries, or in modeling various physical, chemical or biological properties of chemical compounds.

4.2.1 Molecular Graph Representation of Chemical Structures

Graphs may be represented in algebraic form as matrices [3–5]. This numerical description of the structure of chemical compounds is essential for the computer manipulation of molecules and for the calculation of various topological indices and graph descriptors [6]. The computation of the E-state indices is based on the adjacency and distance matrices.

The adjacency matrix $\mathbf{A}(G)$ of a molecular graph G with N vertices is the square $N \times N$ symmetric matrix in which $[\mathbf{A}]_{ij}=1$ if vertex v_i is adjacent to vertex v_j and $[\mathbf{A}]_{ij}=0$ otherwise. The adjacency matrix is symmetric, with all elements on the main diagonal equal to zero. The sum of entries over row i or column i in $\mathbf{A}(G)$ is the degree of vertex v_i, δ_i. Usually, the adjacency matrix is based on weighted molecular graphs in which heteroatoms are represented as vertex parameters and multiple bonds are represented as edge parameters [1, 5, 7, 8]. However, these parameters are not considered in computing the δ values for the E-state indices.

In a simple (nonweighted) connected graph, the graph distance d_{ij} between a pair of vertices v_i and v_j is equal to the length of the shortest path connecting the two vertices, i.e. the number of edges on the shortest path. The distance between two adjacent vertices is 1. The distance matrix $\mathbf{D}(G)$ of a simple graph G with N vertices is the square $N \times N$ symmetric matrix in which $[\mathbf{D}]_{ij} = d_{ij}$ [9, 10].

The δ connectivity index (atom degree), that has a central role in computing the E-state, was used in the definition of the Zagreb topological indices [11]. Randić modified the Zagreb index M_2 to obtain the connectivity index χ [12].

4.2.2 The Randić–Kier–Hall Molecular Connectivity Indices

Kier and Hall extended the definition of the δ connectivity index in order to incorporate heteroatoms and multiple bonds in the definition of the connectivity index χ [13–15]. They noticed that the δ connectivity (atom degree) may be expressed as:

$$\delta_i = \sigma_i - h_i \tag{1}$$

where σ is the number of σ electrons and h is the count of hydrogen atoms bonded to atom i. A simple modification of the δ connectivity index can accommodate the presence of heteroatoms and multiple bonds:

$$\delta_i^{\mathrm{v}} = Z_i^{\mathrm{v}} - h_i = \sigma_i + \pi_i + lp_i - h_i \tag{2}$$

where Z_i^{v} is the number of valence electrons of atom i, π_i is the number of electrons in π orbitals and lp_i is the number of electrons in lone pair orbitals. Equation (2), which is valid only for second row atoms, was extended to cover all atoms:

$$\delta_i^{\mathrm{v}} = \frac{Z_i^{\mathrm{v}} - h_i}{Z_i - Z_i^{\mathrm{v}} - 1} \tag{3}$$

where Z_i is the count of all electrons of atom i. The valence δ index encodes the atomic electronic state, because it takes into account the number of valence electrons, the number of core electrons and the number of bonded hydrogens. Pogliani experimented with other similar functions for the δ connectivity index [16–18].

Kier and Hall used the valence δ index from Eq. (3) to define the family of molecular connectivity indices ${}^{m}\chi_t^{\mathrm{v}}$ [13–15]:

$${}^{m}\chi_t^{\mathrm{v}} = \sum_{j=1}^{s} \prod_{i=1}^{n} (\delta_i^{\mathrm{v}})^{-1/2} \tag{4}$$

where s is the number of connected subgraphs of type t with m edges and n is the number of vertices in the subgraph. These connectivity indices represent a weighted sum over all molecular fragments with the same topology in a molecule.

4.2.3 The E-state Index

Kier and Hall noticed that the quantity $(\delta^v - \delta)/n^2$, where n is the principal quantum number and δ^v is computed with Eq. (2), correlates with the Mulliken–Jaffe electronegativities [19, 20]. This correlation suggested an application of the valence delta index to the computation of the electronic state of an atom. The index $(\delta^v - \delta)/n^2$ defines the Kier–Hall electronegativity KHE and it is used also to define the hydrogen E-state (HE-state) index.

The E-state indices are atomic descriptors composed of an intrinsic state value I and a perturbation ΔI that measures the interactions with all other atoms in a molecule. The Kier–Hall electronegativity is the starting point in the definition of the intrinsic state of an atom, which encodes its potential for electronic interactions and its connectivity with adjacent atoms. The intrinsic state of an atom i is [19, 21]:

$$I_i = \frac{(2/n_i)^2 (Z_i^v - h_i) + 1}{\delta_i} = \frac{(2/n_i)^2 \delta_i^v + 1}{\delta_i} \tag{5}$$

where n_i is the principal quantum of atom i, and the valence δ index δ^v is computed with Eq. (2).

The second contribution to the E-state index comes from the interactions between an atom i and all other atoms in the molecular graph. The perturbation on the intrinsic state value I of atom i due to another atom j depends on the difference between the corresponding intrinsic state values, $(I_i - I_j)$, and on the graph distance between atoms i and j. The overall perturbation on the intrinsic state value I of atom i is:

$$\Delta I_i = \sum_{i=1}^{N-1} \sum_{j=i+1}^{N} \frac{I_i - I_j}{(d_{ij} + 1)^2} = \sum_{i=1}^{N-1} \sum_{j=i+1}^{N} \frac{I_i - I_j}{r_{ij}^2} \tag{6}$$

where d_{ij} is the topological distance between atoms i and j, equal to the minimum topological length of the paths connecting the two atoms, i.e. the minimum number of bonds between atoms i and j, and r_{ij} is the number of atoms on the shortest path between atoms i and j, i.e. $r_{ij} = d_{ij} + 1$. Due to the r^2 term, the perturbation term decreases very fast when the topological distance between atoms increases, thus limiting the effect of distant atoms.

Finally, the E-state index S_i of atom i is the sum of the intrinsic state and of the perturbation term:

$$S_i = I_i + \Delta I_i \tag{7}$$

4.2.4
Hydrogen Intrinsic State

The Kier–Hall electronegativity is used to define the HE-state index *HS* [19]:

$$HS_i = KHE_i - KHE(\mathrm{H}) - \sum_{i=1}^{N-1}\sum_{j=i+1}^{N}\frac{KHE_i - KHE_j}{r_{ij}^2} \tag{8}$$

where $KHE(\mathrm{H}) = -0.20$. Kellogg et al. proposed an alternative definition for the HE-state indices, in which the intrinsic state of a hydrogen atom depends on the δ indices of the attached atom [22]:

$$I(\mathrm{H})_i = \frac{(\delta_i^{\mathrm{v}} - \delta_i)^2}{\delta_i} \tag{9}$$

where the hydrogen atom is attached to atom *i*. The HE-state descriptor is:

$$HS_i = I(\mathrm{H})_i + \Delta I_i \tag{10}$$

Thus, in each molecule, there are two sets of E-state values: one for all non-hydrogen atoms and the second for the hydrogen atoms. The *HS* values are zero for atoms without hydrogens.

4.2.5
Bond E-state Indices

Another extension of the E-state descriptors describes the bond parameters derived from intrinsic states. Bond E-state indices are based on intrinsic values that are the geometric mean of the atom intrinsic value [20]:

$$BI_{ij} = (I_i I_j)^{1/2} \tag{11}$$

The bond E-state indices add a perturbation of the *BI* value under the influence of all other bonds in the molecule:

$$BS_{ij} = BI_{ij} + \sum_{e_{kl} \in E(G)} \frac{BI_{ij} - BI_{kl}}{r_{ij,kl}^2} \tag{12}$$

where the graph distance between two bonds is:

$$r_{ij,kl} = \frac{(r_{ik} + r_{il} + r_{jk} + r_{jl})}{4} \tag{13}$$

4.2.6
E-state 3D Field

The E-state and HE-state indices were used by Kellogg et al. to compute field values on a 3D grid superimposed over the molecules [22]. At each grid point w the E-state interaction energy is:

$$E_w = \sum_i S_i f(r_{iw}) \tag{14}$$

where the summation goes over all atoms i in a molecule, S_i is the E-state or HE-state index for atom i, and $f(r_{iw})$ is a distance function between an atom i and the grid point w. Several distance functions f were tested, such as r^{-1}, r^{-2}, r^{-3}, r^{-4} and e^{-r}.

4.2.7
Atom-type E-state Indices

1

The E-state indices are computed for each atom in a molecule and these topological indices are best suited for applications to datasets in which all molecules have a common skeleton. For example, in the case of a series of chemical compounds based on the general structure **1**, the QSAR descriptors may be the E-state indices S_1, S_2, S_3, S_4, S_5 and S_6. Similar descriptors may be computed from HS-state values. However, this approach limits the application of the E-state indices only to series of compounds that have a common skeleton.

A different approach for obtaining E-state descriptors is the classification of the atoms in atom types, followed by the summation of the E-state values for all atoms of a certain type in a molecule. The definition of atom-type E-state groups is based on several rules, i.e. chemical type of the atom, valence state, aromaticity, number of bonded hydrogen atoms and, only in a few cases, the nature of other adjacent atoms. For example, $-CH_3$ groups are denoted with sCH3 and $-CH$ groups in benzene are denoted with aaCH. The sum of atom-type E-state values for $-CH_3$ is denoted with SsCH3, and the sum of atom-type E-state values for an aromatic $-CH$ group is denoted SaaCH.

4.2.8
Other E-state Indices

Important series of E-state and HE-state descriptors are derived from the atomic E-state values, such as maximum group type E-state, minimum group type E-state, maximum group type HE-state, minimum group type HE-state, maximum E-state, minimum E-state, maximum HE-state, minimum HE-state, etc. A list with the definitions of these and many other E-state indices may be found in the Molconn-Z manual (http://www.edusoft-lc.com/molconn). The E-state and the HE-state indices may be used as atomic parameters to generate other topological indices. For example, the intrinsic state and the E-state indices were inserted on the diagonal of the Burden matrix, thus generating an entire family of new descriptors [23].

Voelkel used the formula of the J index [24] to define the E-state topological parameter TI_E [25]:

$$TI_E(G) = \frac{M}{\mu + 1} \sum_{E(G)} (S_i \times S_j)^{-1/2} \quad (15)$$

where M is the number of edges in G, μ is the cyclomatic number of G (the number of cycles in G) and the summation goes over all edges from the edge set $E(G)$. Several QSAR applications of this E-state index will be presented in the next section.

Lin et al. combined atomic electronegativity with molecular graph distances to obtain a new electrotopological descriptor, the molecular electronegativity topological distance vector (METDV) [26]. The nonhydrogen atoms in a molecule are characterized by their relative Pauling electronegativity, i.e. the Pauling electronegativity divided by that of carbon. The METDV descriptors are defined as:

$$METDV_k = \sum_{i<j}^{d_{ij}=k} \frac{RE_i RE_j}{k} \quad (16)$$

where $k = 1, 2, 3, \ldots$, up to the maximum length of the METDV vector, RE_i is the relative Pauling electronegativity for atom i, and d_{ij} is the topological distance between atoms i and atom j. The METDV descriptors were successful in modeling the pIC_{50} for peptide inhibitors of the angiotensin converting enzyme [26].

4.3 Application of E-State Indices in Medicinal Chemistry

In this section we review several recent applications of the E-state indices. Software programs that may be used to compute these topological indices include Molconn-Z (see above), Cerius^2 (http://www.accelrys.com), Dragon (http://www.talete.mi.it) and E-Dragon (http://www.vcclab.org/lab/edragon). Apart from the E-state indices, all these programs compute a large variety of other structural descriptors, which enables an unbiased comparison between different descriptors. E-state indices are selected with a high frequency in the best QSAR models, thus demonstrating their important role in developing predictive models.

4.3.1
Prediction of Aqueous Solubility

Aqueous solubility is selected to demonstrate the E-state application in QSPR studies. Huuskonen et al. modeled the aqueous solubility of 734 diverse organic compounds with multiple linear regression (MLR) and artificial neural network (ANN) approaches [27]. The set of structural descriptors comprised 31 E-state atomic indices, and three indicator variables for pyridine, aliphatic hydrocarbons and aromatic hydrocarbons, respectively. The dataset of 734 chemicals was divided into a training set ($n=675$), a validation set ($n=38$) and a test set ($n=21$). A comparison of the MLR results (training, $r^2=0.94$, $s=0.58$; validation $r^2=0.84$, $s=0.67$; test, $r^2=0.80$, $s=0.87$) and the ANN results (training, $r^2=0.96$, $s=0.51$; validation $r^2=0.85$, $s=0.62$; test, $r^2=0.84$, $s=0.75$) indicates a small improvement for the neural network model with five hidden neurons. These QSPR models may be used for a fast and reliable computation of the aqueous solubility for diverse organic compounds.

4.3.2
QSAR Models

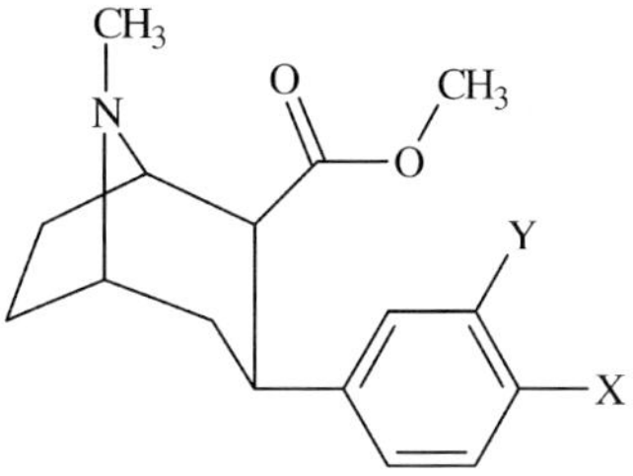

2

Due to its role in cocaine addiction, the dopamine transporter is investigated as a target for cocaine abuse. Maw and Hall modeled the IC_{50} binding affinity for the dopamine transporter of a set of 25 phenyl tropane analogs **2** with QSAR models based on E-state indices [28]. The best QSAR, with $r^2=0.84$ and $q^2=0.77$, has four E-state indices, i.e. the sum of HE-state indices for all nonpolar hydrogen atoms, the sum of HE-state indices for all groups that act as H-bond donors (—CONH, —OCONH, $-NH_2$ and —OH in this dataset), the HE-state index for the substituent X and the atom type E-state index for $-CH_3$ groups.

The HIV-1 protease is responsible for processing the protein precursors to the enzymes (integrase, protease and reverse transcriptase) and the structural proteins of the HIV-1 virus. Maw and Hall found that topological indices provide reliable QSAR models for the IC_{50} data of 32 HIV-1 protease inhibitors [29]. The best QSAR model, with $r^2=0.86$, $s=0.60$ and $q^2=0.79$, was obtained with the shape index ${}^2\kappa_\alpha$, the connectivity index ${}^2\chi^v$, the sum of HE-state indices for all groups that act as

H-bond donors and the sum of HE-state indices for all nonpolar hydrogen atoms in a molecule. These four topological indices highlight the structural features that determine the potency of these inhibitors, i.e. the molecular globularity, the skeletal branching, the H-bond-donating ability and the presence of nonpolar groups. The QSAR model was validated through the prediction of 15 compounds from an external test set, yielding a mean absolute error MAE of 0.82. The QSAR model has a direct structural interpretation that facilitates the design of better HIV-1 protease inhibitors.

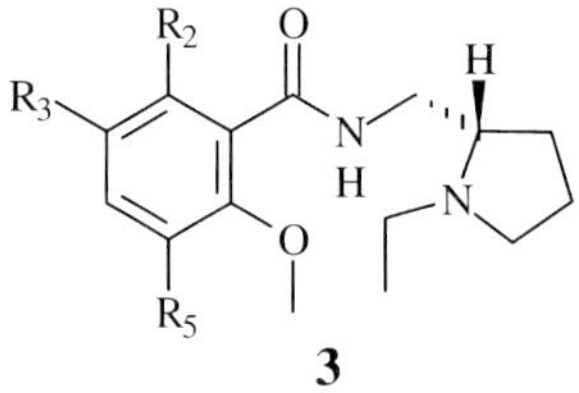

3

Derivatives of (*S*) *N*-[(1-ethyl-2-pyrrolidinyl)methyl]-6-methoxy benzamide **3** are dopamine D_2 receptor antagonists. Samanta et al. obtained the following MLR QSAR for 49 derivatives with the general structure **3** [30]:

$$\begin{aligned} pIC_{50} = {} & 7.180(\pm 0.188) + 0.761(\pm 0.144)\pi\_R_3 - 1.657(\pm 0.351)R\_R_5 \\ & + 0.550(\pm 0.155)R_3\_Et - 0.947(\pm 0.316)I\_NO_2 \\ & + 0.773(\pm 0.320)R_3\_I - 0.111(\pm 0.050)SaaCH \end{aligned} \tag{17}$$

$$n = 49,\ r^2 = 0.801,\ s = 0.406,\ q^2_{LOO} = 0.727,\ F = 28.3$$

where π_R_3 is the hydrophobicity of the substituent R_3, R_R_5 is the resonance effect of the substituent R_5, R_3_Et is an indicator variable for ethyl group in the position R_3, I_NO_2 is an indicator variable for nitro group, R_3_I is an indicator variable for I in the position R_3 and SaaCH is the E-state index corresponding to the aromatic –CH group. This classical QSAR model may suggest chemical transformations that improve the IC_{50} of dopamine receptor antagonists.

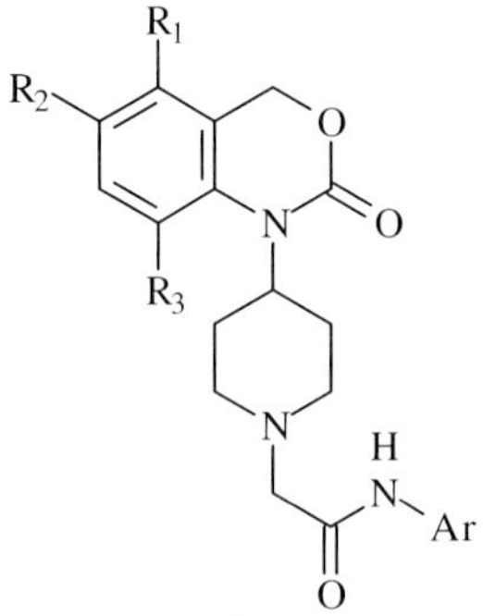

4

The neuropeptide Y (NPY) belongs to a family of peptides that includes peptide YY and pancreatic polypeptide, and it is associated with several diseases such as asthma, immune system disorders, inflammatory diseases, anxiety, depression and diabetes mellitus. NPY is found in the central and peripheral nervous system, and its biological functions are mediated by interactions with five receptor subtypes, i.e. Y1, Y2, Y4, Y5 and Y6. Several studies indicate that the feeding behavior is influenced by interactions between NPY and Y1 and Y5. Deswal and Roy used $Cerius^2$ descriptors and genetic function approximation QSAR to investigate the structural determinants for the inhibition potency of 24 compounds with the general structure **4** for the NPY Y5 receptor [31]. The best QSAR ($r^2=0.720$, $q^2_{LOO}=0.616$, $F=12.2$) was obtained with four indices, i.e. the E-state index for a >N– group SsssN, the molecular connectivity index χ^2, the area of the molecule projected on the XZ plane ShadowXZ, and the AlogP atom type count AtypeC8. The pIC_{50} values predicted for a test set of six compounds have a good correlation with the experimental values, $r^2=0.706$, indicating that the QSAR model is stable and reliable.

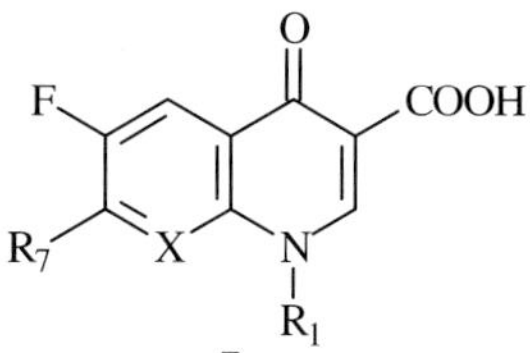

F, O, COOH, R_7, X, N, R_1

5

The steady increase in the frequency of tuberculosis infections resistant to conventional drug therapy highlights the need for new drugs that are efficient against *Mycobacterium tuberculosis* infections. Experimental studies showed that some quinolone derivatives are efficient antibacterials for *M. tuberculosis* as well as other mycobacterial infections, such as those with *M. fortuitum* and *M. smegmatis*. Bagchi *et al.* used a dataset of 68 quinolone derivatives **5** to model their MIC against *M. fortuitum* and *M. smegmatis* with ridge regression (RR), principal component regression (PCR) and partial least squares (PLS) [32]. The QSAR models were developed from a pool of 247 topological indices computed with Polly and Molconn-Z, and included the entire spectrum of E-state indices. The best LOO predictions for *M. fortuitum* MIC were obtained with ridge regression, i.e. $r^2=0.900$ and $q^2=0.796$ for RR, $q^2=0.566$ for PCR, and $q^2=0.792$ for PLS, whereas the best predictions for *M. smegmatis* MIC were obtained with partial least squares, i.e. $r^2=0.967$ and $q^2=0.849$ for RR, $q^2=0.595$ for PCR, and $q^2=0.854$ for PLS. The E-state descriptors used in combination with other topological indices are effective in modeling the MIC of quinolone derivatives against *M. fortuitum* and *M. smegmatis*.

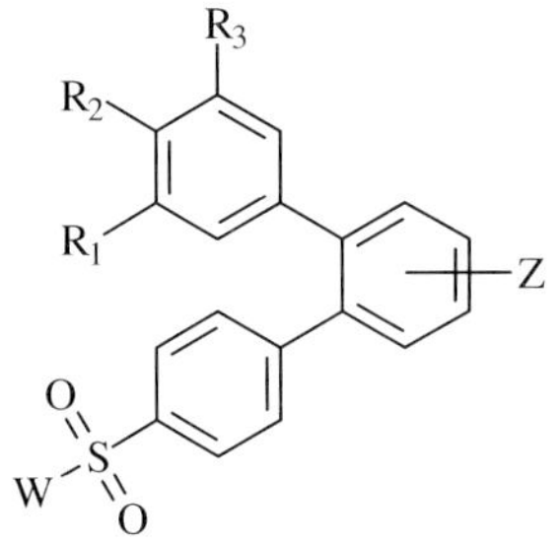

6

The structural features that determine the selectivity for cyclooxygenase (COX)COX-2 versus COX-1 binding affinity to 1-(substituted phenyl)-2-(4-aminosulfonyl/methylsulfonyl)-substituted benzenes **6** was investigated by Chakraborty et al. with QSAR models based on E-state indices and indicator variables [33]. The electrotopological indices represented atomic E-state values for atoms from the common skeleton and sums of E-state indices for groups of atoms. Significant QSAR models were obtained for all three properties investigated, i.e. $r^2 = 0.815$ and $q^2_{\mathrm{LOO}} = 0.675$ for pIC_{50}(COX-1), $r^2 = 0.887$ and $q^2_{\mathrm{LOO}} = 0.842$ for pIC_{50}(COX-2), and $r^2 = 0.746$ and $q^2_{\mathrm{LOO}} = 0.601$ for [pIC_{50}(COX-2) − pIC_{50}(COX-1)].

4.3.3 Absorption, Distribution, Metabolism, Excretion and Toxicity (ADMET)

Depending on its designated target, a drug may be required to have a minimum or a maximum penetration of the blood–brain barrier (BBB). Rose et al. found that a QSAR equation with three topological indices is a reliable model for the blood–brain partitioning of 102 drugs and drug-like compounds [34]. The QSAR model, with $r^2 = 0.66$, $s = 0.45$ and $q^2 = 0.62$, was obtained with the sum of HE-state values for all groups that act as H-bond donors, the sum of HE-state values for all aromatic hydrogens in a molecule and the molecular connectivity difference $\mathrm{d}^2\chi^{\mathrm{v}}$ that measures the molecular branching. The model may offer a structural interpretation of the blood–brain partitioning of a chemical, i.e. molecules that penetrate the BBB have large aromatic groups, few or week H-bond donors, and small branching.

Deconinck et al. investigated the application of Classification And Regression Trees (CART) with boosting for the classification of compounds according to their BBB passage properties [35]. The structural descriptors of 147 chemical compounds were computed with Dragon, and then CART and boosting CART were used to identify the best descriptors. The dataset was divided into a training set of 132 molecules and a test set of 15 molecules, and then a classification model is generated by using 150 classification trees in a boosting approach. The average prediction is computed over 20 such boosting CART models. A significant improvement is obtained for the boosting CART classifiers, that have a percentage of correctly classified molecules of 94.0% compared to 80.6% for a single tree classifier. Although the sum of E-state values is the single electrotopological index selected in the boosting CART models, two other indices are derived from E-state values, i.e. a WHIM index and a total accessibility index weighted by E-state indices.

Membrane transporters, such as P-glycoprotein (P-gp), play an important role in the metabolism and the *in vivo* disposition of drugs. P-gp, which is a member of the ATP-binding cassette superfamily, is a transmembrane efflux pump that can transport various drugs, thus changing their pharmacokinetic properties and leading to multidrug resistance. The structural features that characterize P-gp ligands have been investigated in several computational studies and it was found that E-state indices are important components of SAR models that predict if a chemical is a P-gp ligand.

Gombar et al. used results from *in vitro* monolayer efflux assays to calibrate a linear discriminant analysis (LDA) model that can identify P-gp ligands [36]. The dataset of 95 drugs and drug-like compounds was comprised of 63 P-gp ligands and 32 nonligands. The pool of structural descriptors was mainly composed of E-state and fragment counts indices, supplemented with several constitutional and topological descriptors, such as the number of hydrogen donors and acceptors. The final LDA model contains 27 descriptors, i.e. 13 E-state indices, 12 fragment counts, molar refraction and log P. In calibration, only one nonligand is classified as ligand, whereas in the leave-one-out cross-validation test three nonligands are predicted to be ligands. The 63 ligands are correctly computed both in calibration and prediction. A more rigorous test of the predictive power of the LDA model was performed with a dataset of 58 compounds that were not used to develop the classifier. The LDA classifier was able to predict correctly 33 of the 35 ligands and 17 of the 23 nonligands, with an overall accuracy of 86.2%. In addition to the LDA classifier, the study found that a very simple rule, based on the molecular sum of all atomic E-state values, MolES, may discriminate P-gp ligands and nonligands. Thus, among the 95 compounds, those with MolES > 100 are mainly P-gp ligands (18/19 = 95%), whereas the molecules with MolES < 49 are usually P-gp nonligands (11/13 = 84.6%). The rule was also verified for the test of 58 compounds, with a perfect prediction. The LDA model, based mainly on 13 E-state indices and fragment counts, is a fast filter that may be efficient in screening large chemical libraries.

Among the ADMET properties, human intestinal absorption (HIA) is an important parameter for all drug candidates. Deconinck et al. experimented with CART models for the %HIA of 141 drug-like compounds [37]. The CART algorithm was used to classify chemical compounds in one of the five absorption classes 0–25, 26–50, 51–70, 71–90 and 91–100%. The Dragon package was employed to compute more than 1400 molecular descriptors, comprising constitutional descriptors, E-state indices, topological indices and geometrical descriptors. Among the structural descriptors selected by the variable ranking method, one finds several E-state indices, i.e. molecular E-state variation, mean E-state value, sum of E-state values, and maximal E-state negative variation, as well as several WHIM indices weighted by atomic E-state values.

The %HIA, on a scale between 0 and 100%, for the same dataset was modeled by Deconinck et al. with multivariate adaptive regression splines (MARS) and a derived method two-step MARS (TMARS) [38]. Among other Dragon descriptors, the TMARS model included the TI_E E-state topological parameter [25], and MARS included the maximal E-state negative variation. The average prediction error, which is 15.4% for MARS and 20.03% for TMARS, shows that the MARS model is more robust in modeling %HIA.

Norinder and Österberg used electrotopological indices to obtain PLS models for several drug transport parameters, i.e. Caco-2 cell permeability, HIA, BBB partitioning and immobilized artificial membrane (IAM) chromatography [39]. The PLS models were obtained with the hydrophobicity parameter Clog*P*, the calculated molar refraction and four E-state indices, i.e. the sum of HE-state values for hydrogens bonded to oxygen, nitrogen and sulfur, the sum of HE-state values for hydrogens bonded to other atoms, the sum of E-state values for nitrogen atoms, and the sum of E-state values for oxygen atoms. All QSAR models obtained with E-state indices have a good predictive power, as indicated by the statistical indices: $r^2=0.931$ and $q^2_{LOO}=0.888$ for HIA, $r^2=0.871$ and $q^2_{LOO}=0.815$ for Caco-2, $r^2=0.796$ and $q^2_{LOO}=0.774$ for BBB, and $r^2=0.857$ and $q^2_{LOO}=0.844$ for IAM. QSAR models based on the E-state indices represent a fast screening tool for various drug transport parameters. Furthermore, the E-state indices are computed only from the molecular graph, without the need to determine the 3D structure of the chemical compounds.

Xue et al. investigated the application of recursive feature elimination for the following three classification tests: P-gp substrates (116 substrates and 85 nonsubstrates), human intestinal absorption (131 absorbable compounds and 65 nonabsorbable compounds) and compounds that cause *torsade de pointes* (TdP; 85 TdP-inducing compounds and 276 non-TdP-inducing compounds) [40]. With the exception of TdP compounds, the recursive feature elimination (RFE) increases significantly the prediction power of support vector machines (SVM) classifiers [41] with a Gaussian radial basis function kernel. The accuracy (AC) and Matthews correlation coefficient (MCC) for SVM alone and for SVM plus recursive feature elimination (SVM + RFE) using a L20%O cross-validation test demonstrates the importance of eliminating ineffective descriptors: P-glycoprotein substrates, SVM AC = 68.3% and MCC = 0.37, SVM + RFE AC = 79.4% and MCC = 0.59; human intestinal absorption, SVM AC = 77.0% and MCC = 0.48, SVM + RFE AC = 86.7% and MCC = 0.70; and TdP-inducing compounds, SVM AC = 82.0% and MCC = 0.48, SVM + RFE AC = 83.9% and MCC = 0.56.

Plasma protein binding influences oral bioavailability and pharmacodynamic behavior of drugs. The formation of the drug/plasma protein complex decreases the initial free concentration of the drug, whereas the decomposition of this complex may lead to prolonged presence of the drug in the body. Hall et al. investigated the structural features that determine the binding affinity to human serum albumin (HSA) by modeling the high-performance liquid chromatographic retention index of 94 drugs [42]. The stationary phase was immobilized HSA. The optimum QSPR model, with $r^2=0.77$, $s=0.29$ and $q^2=0.70$, was obtained with six topological indices, i.e. the connectivity index corresponding to five-membered rings ${}^{5}\chi^{v}_{CH}$, the connectivity index corresponding to six-membered rings ${}^{6}\chi^{v}_{CH}$, the sum of E-state values for the —OH group, the sum of E-state values for aromatic carbon atoms, the sum of E-state values for the aliphatic groups $—CH_3$, $—CH_2—$ and >CH—, and the sum of E-state values for halogens. The QSAR model indicates that the binding to HSA is strongly influenced by the structural features encoded by these six descriptors.

In a related study, Hall et al. found that E-state and molecular connectivity indices correlate with the binding of β-lactams (penicillins and cephalosporins) to human serum proteins [43]. The percent fraction bound to serum proteins (PFB) for 74 penicillins was modeled with good statistics, i.e. $r^2=0.80$, $s=12.1$ and $q^2=0.76$. In a subsequent test for 13 penicillins, this QSAR gave good predictions for PFB, with $q^2=0.84$ and a mean absolute error MAE = 12.7. The set of 74 penicillins was then combined with a set of 28 cephalosporins and the dataset of 115 β-lactams gave a good QSAR, with $r^2=0.82$, $s=12.7$ and $q^2=0.78$. These two QSAR models may suggest structural factors that modulate the β-lactams binding to human serum proteins, such as aromatic rings, halogens, methylene groups and >N— atoms.

The total clearance CL_{tot} of a chemical compound is a pharmacokinetic parameter that quantifies the relationship between its rate of transfer and its concentration in blood. The CL_{tot} of a drug characterizes its bioavailability and elimination, and thus may be used to determine its dose and steady-state concentration. Yap et al. obtained CL_{tot} quantitative structure–pharmacokinetic relationships (QSPkR) computed with four machine learning procedures, i.e. general regression neural networks, support vector regression, *k*-nearest neighbors (*k*NN) and PLS [44]. The dataset of 503 compounds was separated into a calibration set ($n=398$) and a validation set ($n=105$). The chemical structure was characterized with topological indices, E-state indices and geometrical descriptors. Based on the statistics obtained for the validation set, the best predictions are obtained with the support vector regression followed by the general regression neural network. Although a large number of geometrical descriptors were tested in the QSPkR models, the results indicate that the most important descriptors are E-state indices and other constitutional and topological descriptors. The only two relevant geometrical descriptors are the 3D Wiener index and the 3D gravitational index. The support vector regression QSPkR model is a fast and reliable computational procedure to identify compounds with poor bioavailability during drug development.

TOPKAT is a system of SAR and QSAR models for the computer-assisted prediction of various toxicity data, such as Ames mutagenicity, rodent carcinogenicity, rat oral LD_{50}, skin sensitization, aerobic biodegradability, eye irritancy, rabbit skin irritancy and rat inhalation toxicity [45]. The skin sensitization QSAR models from TOPKAT are based on experimental data for 335 chemicals, and use E-state indices, shape and symmetry descriptors, and molecular transport indices as structural descriptors. Fedorowicz et al. used logistic regression, TOPKAT and the expert system DEREK to model the skin sensitization potential of chemical compounds [46]. The structural descriptors for the logistic regression were computed with Dragon, Cerius[2] and Molconn-Z. The guinea pig skin sensitization dataset of 105 molecules contains 82 sensitizers and 23 nonsensitizers. The correct classification values for the guinea pig dataset are 73.3% for TOPKAT, 82.9% for DEREK and 87.6% for the logistic regression. The logistic regression classifier for the guinea pig dataset is based on the maximum HE-state index and three autocorrelation descriptors.

Another model system used to determine the skin sensitization potential of chemicals is the murine local lymph node assay (LLNA). Fedorowicz et al. used only DEREK and logistic regression for the LLNA dataset because this SAR model is not implemented in TOPKAT [46]. The LLNA dataset of 178 molecules contains 132 sensitizers and 46 nonsensitizers. The correct classification values for the LLNA dataset are 73.0% for DEREK and 83.2% for the logistic regression. The logistic regression classifier for the LLNA dataset is based on the minimum E-state index and four other structural descriptors. The results obtained indicate that logistic regression is a better classifier for the prediction of skin sensitization potential.

Yap and Chen developed a jury SVM method for the classification of inhibitors and substrates of cytochrome P450 (CYP) 3A4 (241 inhibitors and 368 substrates), 2D6 (180 inhibitors and 198 substrates) and 2C9 (167 inhibitors and 144 substrates) [47]. Structural descriptors computed with Dragon were selected with a genetic algorithm procedure and a L10%O or L20%O SVM cross-validation. Two jury SVM algorithms were applied. The first is the positive majority consensus SVM (PM-CSVM) and the second is the positive probability consensus SVM (PP-CSVM). PM-CSVM classifies a compound based on the vote of the majority of its SVM models, whereas PP-CSVM explicitly computes the probability for a compound being in a certain class. Several tests performed by Yap and Chen showed that at least 81 SVM models are necessary in each ensemble. Both PM-CSVM and PP-CSVM were shown to be superior to a single SVM model (MCC for CYP2D6, MCC=0.742 for single SVM, MCC=0.802 for PM-CSVM and MCC=0.821 for PP-CSVM). As PP-CSVM appears to outperform PM-CSVM, the final classification results were generated with PP-CSVM: MCC=0.899 for CYP3A4, MCC=0.884 for CYP2D6 and MCC=0.872 for CYP2C9.

4.3.4 Mutagenicity and Carcinogenicity

The carcinogenicity and mutagenicity assessment is an important step in determining if novel chemical compounds meet the safety standards for industrial or household use. Early phases of drug development evaluate the genotoxic potential of the chemicals involved in the drug design process with a combination of computational and experimental procedures. The Ames mutagenicity test, using *Salmonella typhimurium* strain TA 100 in the presence of S9 liver homogenate, is a reliable procedure in determining the genotoxic potential of chemicals. Many computational tools for the genotoxicity prediction are based on the experimental results of the Ames mutagenicity test. Classification and regression SAR and QSAR models are valuable tools to prioritize, and reduce the number of compounds that are experimentally tested for their genotoxicity potential. In this section we review several QSAR models for genotoxicity prediction that use E-state indices among other structural descriptors.

Votano et al. developed classification models for the mutagenicity of 3363 diverse compounds tested for their Ames genotoxicity [48]. Three classification models were compared, i.e. ANN, *k*NN and decision forest (DF). All SAR models were developed using the same initial set of 148 topological indices that included E-state indices and molecular connectivity indices. The dataset was split into 2963 training compounds and 400 prediction compounds. All three classifiers gave good predictions, with a slight advantage for the neural network, as indicated by the area under the receiver operator characteristic (AUROC) curve, i.e. AUROC=0.93 for ANN, AUROC=0.92 for *k*NN and AUROC=0.91 for DF. Among the 15 most important structural descriptors selected in these three classifiers, one finds 13 E-state indices. The topological indices from the AN, *k*NN and DF models are related to toxicophores linked to genotoxic responses in *S. typhimurium*.

Quinolone derivatives with antibiotic activity block the bacterial replication by interacting with the bacterial DNA gyrase, thus inhibiting the coiling of bacterial DNA. Quinolines with antiinflamatory effects are used to treat autoimmune diseases, such as rheumatoid arthritis. Laboratory studies have shown that both quinolone and quinoline derivatives are liver carcinogens in rodents and exhibit mutagenicity in the Ames test. The ADAPT system [49] was used by He et al. to develop probabilistic neural network (PNN) classification models for the genotoxic potential of quinolone and quinoline derivatives [50]. The experimental genotoxicity of 85 quinolone derivatives and of 115 quinoline derivatives was determined with the SOS Chromotest – a faster alternative to the Ames test. The SOS Chromotest measures the induction of a *lacZ* reporter gene in response to DNA damage. The quinolone dataset contains 23 genotoxic and 62 nongenotoxic compounds, whereas the quinoline dataset contains 44 genotoxic and 71 nongenotoxic chemicals. An ensemble of nine PNN models was developed for each classification model and the final class attribution (genotoxic/nongenotoxic) was decided by a majority vote of the trained classifiers. Simulated annealing was used to select between three and 10 structural descriptors for each PNN classifier. The ensemble PNN model for quinolone derivatives was able to predict correctly 16 of the 23 genotoxic chemicals and 60 of the 62 nongenotoxic compounds, with an overall accuracy of 89.4%, an overall accuracy for genotoxic class of 69.6% and an overall accuracy for nongenotoxic class of 96.8%. Among the structural descriptors selected in the quinolone PNN model, one finds the minimum atomic E-state value and the through-space distance between minimum and maximum atomic E-state values. The committee PNN model for quinoline derivatives was able to predict correctly 39 of the 44 genotoxic compounds and 67 of the 71 nongenotoxic chemicals, with an overall accuracy of 92.2%, an overall accuracy for genotoxic class of 88.6% and an overall accuracy for nongenotoxic class of 94.4%. The descriptors used in the quinoline PNN model include the average E-state values over all heteroatoms and the sum of E-state values over all heteroatoms. These results show that the structural information carried by the descriptors selected in the ensemble PNN models offer reliable predictions for the genotoxic potential of quinolone and quinoline derivatives.

The TOPKAT system for predicting chemical carcinogens has four rodent models, i.e. male rat, female rat, male mouse and female mouse. These four classifiers are based on structural descriptors computed from atomic and bond E-state indices. Prival tested the TOPKAT system by determining its ability to predict the chronic rodent carcinogenicity of 28 chemical compounds tested by the National Toxicology Program [51]. Although the sample used in this test is small, the classification results suggest that the predictions of the TOPKAT carcinogenicity modules do not agree with the experimental findings of the National Toxicology Program. From the 16 carcinogenic compounds, TOPKAT predicted seven as noncarcinogenic, whereas from the 12 noncarcinogenic compounds, TOPKAT predicted four as carcinogenic.

Snyder et al. collected from the 2000–2002 Physicians' Desk Reference data regarding the Ames mutagenicity test and other genotoxicity tests for 394 drugs and compared them with predictions of three computational systems for genotoxicity evaluation, i.e. MCASE, TOPKAT and DEREK [52]. All three systems have a low sensitivity in predicting the Ames mutagenicity, suggesting that they are not suitable for drug safety evaluations. However, these computational systems incorporate SARs for genotoxicity that may be useful in screening large chemical libraries and in prioritizing the chemicals for mutagenicity tests in the early phases of drug discovery.

4.3.5 Anticancer Compounds

Cisplatin, carboplatin and oxaliplatin are effective treatment options for testicular and ovarian cancers, but their use is hindered by their poor selectivity between malignant and normal cells, as well as by the induction of chemoresistance. Monti et al. investigated the cytotoxicity of 16 *cis*-platinum(II) compounds for the A2780 human ovarian adenocarcinoma cell line and on its cisplatin-resistant subline (A2780Cp8) [53]. The chemical structures computed with the PM3 method implemented in SPARTAN were used as input for the Dragon package that provided a total of 626 structural descriptors, which was filtered to a pool of 197 descriptors retained for the QSAR modeling. The best QSAR model for the pIC_{50} of cisplatin resistant cells A2780Cp8 ($r^2=0.973$, $q^2_{LOO}=0.947$, $q^2_{L50\%O}=0.856$, $s=0.144$, $F=97.9$) has four descriptors, i.e. the E-state topological parameter I_E [25], the centric information index I_B, the Geary index $c(4)p$ and the BCUT descriptor BELe7. The predictive power of the model was evaluated with the leave-one-out and leave-50%-out cross-validation procedures, respectively. This QSAR model has a good predictive power and may be used to design *cis*-platinum(II) derivatives that are effective against cisplatin-resistant cells.

4.3.6
Virtual Screening of Chemical Libraries

QSAR models are very useful tools for the identification of structural features that determine various molecular properties and may even suggest the mechanism of action for biochemical processes. Thus, QSAR models start from structure and correlate descriptors with molecular properties. Once a QSAR model is established, an inverse process becomes possible, i.e. setting a target value for a molecular property and then finding all possible chemical structures that might exhibit that property value, within a certain range of variation. This process in called inverse QSAR and it represents an important step in optimizing the drug-like properties of chemical compounds. Lewis proposed an inverse QSAR strategy that may assist medicinal chemists in deciding how to optimize a library of chemical compounds [54]. The starting point is a dataset of chemical compounds with a molecular property and a corresponding QSAR model. The inverse QSAR strategy involves an iterative application of several steps, i.e. generation of new structures, structure filtering based on synthetic feasibility or undesired properties and QSAR filtering. The first step generates a new chemical library by applying simple chemical transformations to the molecules from the initial dataset. Examples of such transformations are modification of the bond order, adding or removing an atom, adding or removing a fragment, or changing C to N or O. The second step filters molecules that have nonspecific reactivity, such as electrophiles, nucleophiles, acylating agents or redox systems. Synthetic feasibility rules are used to eliminate compounds that are difficult to synthesize or those that are expensive. Finally, QSAR models are used to select candidates for chemical synthesis. The inverse QSAR strategy developed by Lewis was tested for a combinatorial library of 150 inhibitors of human carbonic anhydrase II, that was used to develop a MLR genetic function approximation QSAR, as implemented in Cerius^2. The best QSAR model ($r^2=0.81$, $q^2_{LOO}=0.80$, $F=127$) is based on five structural descriptors, i.e. the molecular flexibility index ϕ, the charge of the most positive atom divided by the total positive charge Jurs-RPCG, the E-state index for sp^2 N atom SdsN, the electrotopological count for aromatic S atoms NaaS and the molecular volume inside the contact surface V_m. This QSAR was used as the starting point for performing automated property optimization.

The E-state indices may define chemical spaces that are relevant in similarity/diversity search in chemical databases. This similarity search is based on atom-type E-state indices computed for the query molecule [55]. Each E-state index is converted to a z score, $z_i=(x_i-\mu_i)/\sigma_i$, where x_i is the ith E-state atomic index, μ_i is its mean and σ_i is its standard deviation in the entire database. The similarity was computed with the Euclidean distance and with the cosine index and the database used was the Pomona MedChem database, which contains 21 000 chemicals. Tests performed for the antiinflamatory drug prednisone and the antimalarial drug mefloquine as query molecules demonstrated that the chemicals space defined by E-state indices is efficient in identifying similar compounds from drug and drug-like databases.

Lazy learning methods represent a class of machine learning algorithms that store the entire training dataset and process it only when it is requested to process a query datapoint. Kumar et al. coupled into a QSAR algorithm the nonlinear reduction of dimensionality with robust regressors. Locally linear embedding was used to reduce the nonlinear dimension of the input space [56]. The compressed dataset is then modeled with lazy learning and support vector regression. Zhang et al. developed a new lazy learning procedure for QSAR, the automated lazy learning quantitative structure–activity relationships (ALL-QSAR) model [57]. A molecular property of a query compound is predicted from a locally weighted linear regression model that first selects a training set of compounds that have a high similarity with the query compound, and then uses the structural descriptors and molecular properties of the training set to make the prediction for the query compound. The ALL-QSAR algorithm was tested for 48 anticonvulsant agents with known ED_{50} values and for 48 antagonists of the dopamine D_1 receptor with known competitive binding affinities K_i. The structural descriptors were computed with Molconn-Z and comprise, among other graph descriptors and topological indices, a wide array of E-state indices. The ALL-QSAR models for the anticonvulsant agents ($r^2=0.90$) and D_1 antagonists ($r^2=0.81$) have higher statistics than those obtained with other models, such as *k*NN, PLS, SVM or comparative molecular field analysis. The anticonvulsant agents ALL-QSAR model was applied for a database screening and it identified several known anticonvulsants that were absent from the training set. The ALL-QSAR is an adaptive model that may be used for online training and virtual screening of chemical libraries.

Estrogen receptors (ERs) are members of nuclear receptor family, and they are essential in cell growth and development in various tissues. Chemical compounds that are ER agonists have been used for prostate cancer treatment, contraception, hormone replacement therapy and osteoporosis prevention. In contrast, the ER agonist activity of some industrial chemicals, pesticides, and environmental pollutants, is known to disrupt the human endocrine functions by mimicking endogenous estrogens. Such chemicals may induce cancers and disrupt the development of the reproductive system. Li et al. used several machine learning procedures to discriminate between ER agonists and nonagonists [58]. The dataset of chemical compounds comprised 243 ER agonists and 463 ER nonagonists, for which more than 1000 structural descriptors were computed. Simple statistical filters were used to reduce the number of descriptors to 199 and then recursive feature elimination was applied to select those molecular indices that discriminate ER agonists from ER nonagonists. Four machine learning procedures were evaluated, i.e. SVM, *k*NN, PNN and C4.5 decision tree. The SVM model gives the best predictions, as shown by the leave-20%-out cross-validation accuracies. Among the 31 descriptors selected by the recursive feature elimination procedure there are two E-state indices, i.e. SsCH3 and SaaCH, whereas the other descriptors are various topological, geometrical and quantum indices. The study suggests that SVM classifiers are robust and reliable models for the prediction of ER agonists and to identify the structural features that distinguish ER agonists from ER nonagonists.

An undesirable side-effect of chemical compounds that exhibit antihistaminic activity is sedation, manifested as a reduced concentration capability. Duart et al. used a dataset of 146 chemicals to develop a combination of linear regression and linear discriminant analysis models to identify compounds with antihistaminic activity and low sedative effect [59]. Starting from a diverse collection of topological indices, it was found that the best classification function to identify antihistaminic compounds contains six electrotopological indices, i.e. SdssC, SaaCH, SdsN, SsssN, SsOH and SdO. Together with other equations based on topological indices, the E-state indices were used to screen the chemical compounds from the Merck Index and the most promising eight candidates were selected for experimental tests. Tests performed with female Wistar rats confirmed the antihistaminic activity of all eight compounds, thus demonstrating the practical value of the graph descriptors in screening large chemical libraries.

Antipsychotic compounds may belong to the class of dopamine antagonists (DA), serotonin antagonists (SA) and serotonin–dopamine dual antagonists (Dual). The design of selective antagonists may benefit from robust classifiers that discriminate between DA, SA and Dual compounds. Kim et al. solved this problem with four machine learning methods, i.e. linear discriminant analysis, soft independent modeling of class analogy (SIMCA), recursive partitioning (RP) and ANN [60]. The SAR dataset of 2772 compounds was collected from the MDDR database, and contains 1135 DA (260 D_2, 263 D_3 and 612 D_4), 1251 SA (517 5-HT_{1A}, 447 5-HT_{2A} and 287 5-HT_{2C}) and 386 Dual. The chemical structures were characterized with constitutional and topological descriptors computed with Cerius^2. Using a training set of 2496 compounds and a prediction set of 276 compounds, it was found that recursive partitioning has the highest prediction rate: 69.6% LDA, 63.4% SIMCA, 74.3 RP and 71.7% ANN. The key descriptors for the RP model are 12 topological indices that include six E-state indices, i.e. SsCH2, SssO, SaasC, SdO, SsssN and SssssC. The classifier that combines topological indices and decision trees may be used for the virtual screening of DA, SA and Dual antagonists.

4.4 Conclusions and Outlook

In this chapter we presented an overview of the E-state, its computation from the molecular graph and its applications in drug design. The E-state encodes at the atomic level information regarding the electronic state and the topological accessibility. The computation of the E-state indices is based exclusively on the molecular topology and it can be done efficiently for very large chemical libraries. Comparative studies that develop QSAR models from a large variety of molecular descriptors show that the E-state indices encode a distinct type of structural information. Due to this advantage, E-state indices are frequently selected in the best QSAR models.

The atomic-level structural information encoded into the E-state generates a chemical space that can be efficient in QSAR modeling and in the virtual screening of chemical libraries. The E-state indices are mature, with proven success in QSAR modeling, and should be considered, together with other descriptors, in SAR and QSAR applications that require a comprehensive exploration of the chemical space.

References

1 Ivanciuc, O. Graph theory in chemistry. In *Handbook of Chemoinformatics*, Gasteiger, J. (eds.), Wiley-VCH, Weinheim, **2003**, Vol. 1, pp. 103–138.

2 Ivanciuc, O., Balaban, A. T. Graph theory in chemistry. In *The Encyclopedia of Computational Chemistry*, Schleyer, P. v. R., Allinger, N. L., Clark, T., Gasteiger, J., Kollman, P. A., Schaefer III, H. F. Schreiner, P. R. (eds.), Wiley, Chichester, **1998**, pp. 1169–1190.

3 Ivanciuc, O., Balaban, A. T. The graph description of chemical structures. In *Topological Indices and Related Descriptors in QSAR and QSPR*, Devillers, J. Balaban, A. T. (eds.), Gordon & Breach, Amsterdam, **1999**, pp. 59–167.

4 Ivanciuc, O., Ivanciuc, T. Matrices and structural descriptors computed from molecular graph distances. In *Topological Indices and Related Descriptors in QSAR and QSPR*, Devillers, J. Balaban, A. T. (eds.), Gordon & Breach, Amsterdam, **1999**, pp. 221–277.

5 Ivanciuc, O., Ivanciuc, T., Balaban, A. T. Vertex- and edge-weighted molecular graphs and derived structural descriptors. In *Topological Indices and Related Descriptors in QSAR and QSPR*, Devillers, J. Balaban, A. T. (eds.), Gordon & Breach, Amsterdam, **1999**, pp. 169–220.

6 Ivanciuc, O. Topological indices. In *Handbook of Chemoinformatics*, Gasteiger, J. (eds.), Wiley-VCH, Weinheim, **2003**, Vol. 3, pp 981–1003.

7 Ivanciuc, O., Ivanciuc, T., Cabrol-Bass, D., Balaban, A. T. Comparison of weighting schemes for molecular graph descriptors: application in quantitative structure–retention relationship models for alkylphenols in gas–liquid chromatography. *J. Chem. Inf. Comput. Sci.* **2000**, *40*, 732–743.

8 Ivanciuc, O. QSAR comparative study of Wiener descriptors for weighted molecular graphs. *J. Chem. Inf. Comput. Sci.* **2000**, *40*, 1412–1422.

9 Harary, F. *Graph Theory*, Addison-Wesley, Reading, MA, **1994**.

10 Buckley, F., Harary, F. *Distance in Graphs*, Addison-Wesley, Reading, MA, **1990**.

11 Gutman, I., Trinajstić, N. Graph theory and molecular orbitals. Total π-electron energy of alternant hydrocarbons. *Chem. Phys. Lett.* **1972**, *17*, 535–538.

12 Randić, M. Characterization of molecular branching. *J. Am. Chem. Soc.* **1975**, *97*, 6609–6615.

13 Kier, L. B., Hall, L. H. *Molecular Connectivity in Chemistry and Drug Research*, Academic Press, New York, **1976**.

14 Kier, L. B., Hall, L. H. *Molecular Connectivity in Structure–Activity Analysis*, Research Studies Press, Letchworth, **1986**.

15 Hall, L. H., Kier, L. B. Molecular connectivity chi indices for database analysis and structure–property modeling. In *Topological Indices and Related Descriptors in QSAR and QSPR*, Devillers, J. Balaban, A. T. (eds.), Gordon & Breach, Amsterdam, **1999**, pp. 307–360.

16 Pogliani, L. Encoding the core electrons with graph concepts. *J. Chem. Inf. Comput. Sci.* **2004**, *44*, 42–49.

17 Pogliani, L. The evolution of the valence delta in molecular connectivity theory. *Internet Electron. J. Mol. Des.* **2006**, *5*, 364–375.

18 Pogliani, L. The hydrogen perturbation in molecular connectivity computations. *J. Comput. Chem.* **2006**, *27*, 868–882.

19 Kier, L. B., Hall, L. H. *Molecular Structure Description. The Electrotopological State*, Academic Press, San Diego, CA, **1999**.

20 Kier, L. B., Hall, L. H. The electrotopological state: Structure modeling for QSAR and database analysis.

In *Topological Indices and Related Descriptors in QSAR and QSPR*, Devillers, J. Balaban, A. T. (eds.), Gordon & Breach, Amsterdam, **1999**, pp. 491–562.

21 Hall, L. H., Mohney, B., Kier, L. B. The electrotopological state: Structure information at the atomic level for molecular graphs. *J. Chem. Inf. Comput. Sci.* **1991**, *31*, 76–82.

22 Kellogg, G. E., Kier, L. B., Gaillard, P., Hall, L. H. E-state fields: applications to 3D QSAR. *J. Comput.-Aided Mol. Des.* **1996**, *10*, 513–520.

23 Ivanciuc, O. Design of topological indices. Part 25. Burden molecular matrices and derived structural descriptors for glycine antagonists QSAR models. *Rev. Roum. Chim.* **2001**, *46*, 1047–1066.

24 Balaban, A. T. Highly discriminating distance-based topological index. *Chem. Phys. Lett.* **1982**, *89*, 399–404.

25 Voelkel, A. Structural descriptors in organic chemistry – new topological parameter based on electrotopological state of graph vertices. *Comput. Chem.* **1994**, *18*, 1–4.

26 Lin, Z., Wu, Y., Quan, X., Zhou, Y., Ni, B., Wan, Y. Use of a novel electrotopological descriptor for the prediction of biological activity of peptide analogues. *Lett. Pept. Sci.* **2002**, *9*, 273–281.

27 Huuskonen, J., Rantanen, J., Livingstone, D. Prediction of aqueous solubility for a diverse set of organic compounds based on atom-type electrotopological state indices. *Eur. J. Med. Chem.* **2000**, *35*, 1081–1088.

28 Maw, H. H., Hall, L. H. E-state modeling of dopamine transporter binding. Validation of the model for a small data set. *J. Chem. Inf. Comput. Sci.* **2000**, *40*, 1270–1275.

29 Maw, H. H., Hall, L. H. E-state modeling of HIV-1 protease inhibitor binding independent of 3D information. *J. Chem. Inf. Comput. Sci.* **2002**, *42*, 290–298.

30 Samanta, S., Debnath, B., Gayen, S., Ghosh, B., Basu, A., Srikanth, K., Jha, T. QSAR modeling on dopamine D_2 receptor binding affinity of 6-methoxy benzamides. *Farmaco* **2005**, *60*, 818–825.

31 Deswal, S., Roy, N. Quantitative structure activity relationship of benzoxazinone derivatives as neuropeptide Y Y5 receptor antagonists. *Eur. J. Med. Chem.* **2006**, *41*, 552–557.

32 Bagchi, M. C., Mills, D., Basak, S. C. Quantitative structure–activity relationship (QSAR) studies of quinolone antibacterials against *M. fortuitum* and *M. smegmatis* using theoretical molecular descriptors. *J. Mol. Model.* **2007**, *13*, 111–120.

33 Chakraborty, S., Sengupta, C., Roy, K. Exploring QSAR with E-state index: Selectivity requirements for COX-2 versus COX-1 binding of terphenyl methyl sulfones and sulfonamides. *Bioorg. Med. Chem. Lett.* **2004**, *14*, 4665–4670.

34 Rose, K., Hall, L. H., Kier, L. B. Modeling blood–brain barrier partitioning using the electrotopological state. *J. Chem. Inf. Comput. Sci.* **2002**, *42*, 651–666.

35 Deconinck, E., Zhang, M. H., Coomans, D., Vander Heyden, Y. Classification tree models for the prediction of blood–brain barrier passage of drugs. *J. Chem Inf. Model.* **2006**, *46*, 1410–1419.

36 Gombar, V. K., Polli, J. W., Humphreys, J. E., Wring, S. A., Serabjit-Singh, C. S. Predicting P-glycoprotein substrates by a quantitative structure–activity relationship model. *J. Pharm. Sci.* **2004**, *93*, 957–968.

37 Deconinck, E., Hancock, T., Coomans, D., Massart, D. L., Vander Heyden, Y. Classification of drugs in absorption classes using the classification and regression trees (CART) methodology. *J. Pharm. Biomed. Anal.* **2005**, *39*, 91–103.

38 Deconinck, E., Xu, Q. S., Put, R., Coomans, D., Massart, D. L., Vander Heyden, Y. Prediction of gastro-intestinal absorption using multivariate adaptive regression splines. *J. Pharm. Biomed. Anal.* **2005**, *39*, 1021–1030.

39 Norinder, U., Österberg, T. Theoretical calculation and prediction of drug transport processes using simple parameters and partial least squares projections to latent structures (PLS) statistics. The use of electrotopological state indices. *J. Pharm. Sci.* **2001**, *90*, 1076–1085.

40 Xue, Y., Li, Z. R., Yap, C. W., Sun, L. Z., Chen, X., Chen, Y. Z. Effect of molecular descriptor feature selection in support vector machine classification of pharmacokinetic and toxicological properties of chemical agents. *J. Chem. Inf. Comput. Sci.* **2004**, *44*, 1630–1638.

41 Ivanciuc, O. Applications of support vector machines in chemistry. In *Reviews in Computational Chemistry*, Lipkowitz, K. B. Cundari, T. R. (eds.), Wiley-VCH, Weinheim, **2007**, Vol. 23, pp 291–400.

42 Hall, L. M., Hall, L. H., Kier, L. B. Modeling drug albumin binding affinity with E-state topological structure representation. *J. Chem. Inf. Comput. Sci.* **2003**, *43*, 2120–2128.

43 Hall, L. M., Hall, L. H., Kier, L. B. QSAR modeling of β-lactam binding to human serum proteins. *J. Comput.-Aided Mol. Des.* **2003**, *17*, 103–118.

44 Yap, C. W., Li, Z. R., Chen, Y. Z. Quantitative structure–pharmacokinetic relationships for drug clearance by using statistical learning methods. *J. Mol. Graph. Model.* **2006**, *24*, 383–395.

45 TOPKAT, Accelrys Inc., San Diego, CA, USA, http://www.accelrys.com.

46 Fedorowicz, A., Singh, H., Soderholm, S., Demchuk, E. Structure–activity models for contact sensitization. *Chem. Res. Toxicol.* **2005**, *18*, 954–969.

47 Yap, C. W., Chen, Y. Z. Prediction of cytochrome P450 3A4, 2D6, and 2C9 inhibitors and substrates by using support vector machines. *J. Chem Inf. Model.* **2005**, *45*, 982–992.

48 Votano, J. R., Parham, M., Hall, L. H., Kier, L. B., Oloff, S., Tropsha, A., Xie, Q., Tong, W. Three new consensus QSAR models for the prediction of Ames genotoxicity. *Mutagenesis* **2004**, *19*, 365–377.

49 Stuper, A. J., Brugger, W. E., Jurs, P. C. *Computer-Assisted Studies of Chemical Structure and Biological Function*, Wiley, New York, **1979**.

50 He, L., Jurs, P. C., Kreatsoulas, C., Custer, L. L., Durham, S. K., Pearl, G. M. Probabilistic neural network multiple classifier system for predicting the genotoxicity of quinolone and quinoline derivatives. *Chem. Res. Toxicol.* **2005**, *18*, 428–440.

51 Prival, M. J. Evaluation of the TOPKAT system for predicting the carcinogenicity of chemicals. *Environ. Mol. Mutagen.* **2001**, *37*, 55–69.

52 Snyder, R. D., Pearl, G. S., Mandakas, G., Choy, W. N., Goodsaid, F., Rosenblum, I. Y. Assessment of the sensitivity of the computational programs DEREK, TOPKAT, and MCASE in the prediction of the genotoxicity of pharmaceutical molecules. *Environ. Mol. Mutagen.* **2004**, *43*, 143–158.

53 Monti, E., Gariboldi, M., Maiocchi, A., Marengo, E., Cassino, C., Gabano, E., Osella, D. Cytotoxicity of *cis*-platinum(II) conjugate models. The effect of chelating arms and leaving groups on cytotoxicity: a quantitative structure–activity relationship approach. *J. Med. Chem.* **2005**, *48*, 857–866.

54 Lewis, R. A. A general method for exploiting QSAR models in lead optimization. *J. Med. Chem.* **2005**, *48*, 1638–1648.

55 Hall, L. H., Kier, L. B. The E-state as the basis for molecular structure space definition and structure similarity. *J. Chem. Inf. Comput. Sci.* **2000**, *40*, 784–791.

56 Kumar, R., Kulkarni, A., Jayaraman, V. K., Kulkarni, B. D. Structure–activity relationships using locally linear embedding assisted by support vector and lazy learning regressors. *Internet Electron. J. Mol. Des.* **2004**, *3*, 118–133.

57 Zhang, S., Golbraikh, A., Oloff, S., Kohn, H., Tropsha, A. A novel automated lazy learning QSAR (ALL-QSAR) approach: method development, applications, and virtual screening of chemical databases using validated ALL-QSAR models. *J. Chem Inf. Model.* **2006**, *46*, 1984–1995.

58 Li, H., Ung, C. Y., Yap, C. W., Xue, Y., Li, Z. R., Chen, Y. Z. Prediction of estrogen receptor agonists and characterization of associated molecular descriptors by statistical learning methods. *J. Mol. Graph. Model.* **2006**, *25*, 313–323.

59 Duart, M. J., García-Domenech, R., Gálvez, J., Alemán, P. A., Martín-Algarra, R. V., Antón-Fos, G. M. Application of a mathematical topological pattern of antihistaminic activity for the selection of new drug candidates and pharmacology assays. *J. Med. Chem.* **2006**, *49*, 3667–3673.

60 Kim, H.-J., Choo, H., Cho, Y. S., Koh, H. Y., No, K. T., Pae, A. N. Classification of dopamine, serotonin, and dual antagonists by decision trees. *Bioorg. Med. Chem.* **2006**, *14*, 2763–2770.

5
Polar Surface Area

Peter Ertl

Abbreviations

BBB	blood–brain barrier
Caco-2	adenocarcinoma cell line derived from human colon
Clog *P*	calculated octanol–water partition coefficient
CNS	central nervous system
DM	dipole moment
FA	fraction absorbed
HCPSA	high-charged polar surface area
HTS	high-throughput screening
MV	molecular volume
MW	molecular weight
PSA	polar surface area
QSAR	quantitative structure–activity relationship
SAP	sum of atom polarities
TPSA	topological polar surface area

Symbols

n_{atoms}	number of nonhydrogen atoms
n_{HBA}	number of H-bond acceptors
n_{HBD}	number of H-bond donors
n_{rotb}	number of rotatable bonds

5.1
Introduction

Polar surface area (PSA) – defined simply as the part of a molecular surface that is polar – is probably, together with the octanol–water partition coefficient, one of the most important parameters used to characterize the transport properties of drugs. PSA has been shown to provide very good correlations with intestinal

Molecular Drug Properties. Measurement and Prediction. R. Mannhold (Ed.)

ISBN: 978-3-527-31755-4

absorption, blood–brain barrier (BBB) penetration and several other drug characteristics. It has also been effectively used to characterize drug-likeness during virtual screening and combinatorial library design. The descriptor seems to encode an optimal combination of H-bonding features, molecular polarity and solubility properties. An additional advantage of PSA is that it can be easily and rapidly calculated as a sum of fragment contributions using only the molecular connectivity of a structure.

Molecular surface properties have been used to describe solvation and partitioning processes for a long time. Amidon et al. [1] studied the correlation of surface properties, expressed in terms of hydrocarbon portions and functional group portions, with the aqueous solubility. Pearlman [2] discussed various applications of molecular surface and volume in quantitative structure–activity relationship (QSAR) studies, and Stanton and Jurs [3] suggested using charged partial surface area descriptors, which combined molecular surface area and atomic charges, for the development of various structure–property models. One of the most useful surface properties has been shown to be PSA, characterizing the polar part of the molecular surface, defined simply as the part of the surface corresponding to oxygens and nitrogens, and including also the hydrogens attached to these atoms (Fig. 5.1). One of the first applications of PSA is a study of Van de Waterbeemd and Kansy [4] to predict BBB penetration. Van de Waterbeemd et al. [5] also used this parameter to predict the Caco-2 permeability of drugs (in these papers the name "polar part of the surface" was still used). The popularity of the PSA descriptor for predicting drug transport properties can mainly be attributed to the pioneering work of the Uppsala University Group [6, 7]. They used the so-called "dynamic PSA" in which the polar surface was calculated as a weighted sum of surfaces generated from a representative set of conformations (see the method section for more details). Clark showed later that static, or single-conformer, PSA also provides very good results in the prediction of intestinal absorption [8] and BBB penetration [9]. Finally, Ertl et al. [10] introduced an extremely rapid method to obtain PSA descriptor simply from the sum of contributions of polar fragments in a molecule without the necessity to generate its three-dimensional (3D) geometry. These and many other studies helped to establish PSA as one of the most

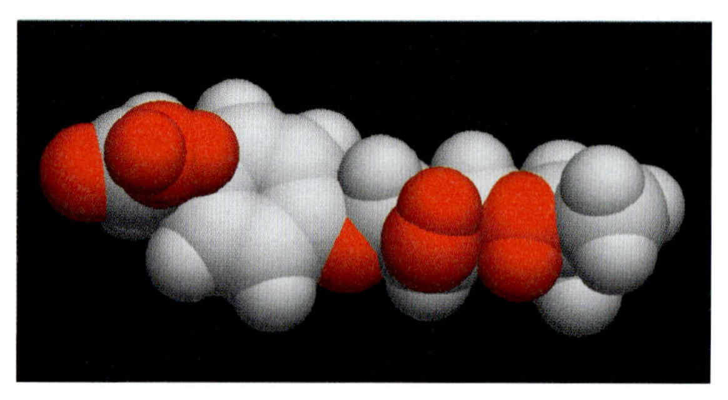

Fig. 5.1 PSA of atenolol.

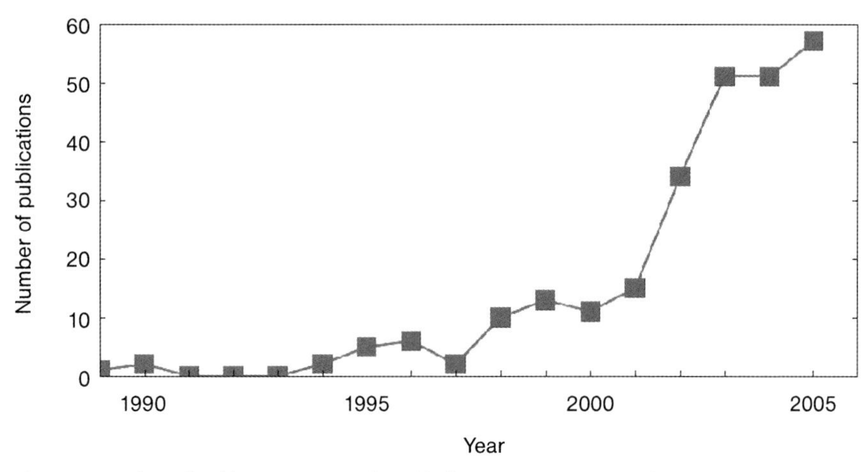

Fig. 5.2 Number of publications using the PSA descriptor in years 1990–2005.

important molecular descriptors used in medicinal chemistry, cheminformatics and QSAR studies.

The recent increase of interest in PSA can also be demonstrated by the number of publications mentioning this descriptor over the past 15 years – as counted by the popular scientific search engine Google Scholar [11]. In Fig. 5.2, one can see that until around 1993 the term "PSA" was practically unused in scientific publications, the first increase is visible in the period 1994 to 2001, while in more recent years a very steep increase in the number of publications using this descriptor has been witnessed.

5.2
Application of PSA for Prediction of Drug Transport Properties

The extreme popularity of PSA descriptors for the prediction of drug absorption [12–14] can be attributed to several reasons. First, PSA is very easy to interpret, with the notion of "molecular polar surface" and its influence on interactions with a molecule's environment similar to a medicinal chemist's own intuition (and probably also a good approximation of physical reality). Second, PSA is easy to calculate. For topological PSA the calculation is particularly easy and fast, requiring only the identification of polar fragments and then a table lookup to find respective fragment contributions. Furthermore, numerous software packages, as well as free resources on the internet, are available for calculating this descriptor. The most important benefit of PSA, however, is that it indeed provides excellent correlations with various drug transport characteristics as documented below. PSA seems to optimally encode those drug properties which play an important role in membrane penetration: molecular polarity, H-bonding features and also solubility.

5.2.1
Intestinal Absorption

For the majority of drugs, the preferred administration route is by oral ingestion which requires good intestinal absorption of drug molecules. Intestinal absorption is usually expressed as fraction absorbed (FA), expressing the percentage of initial dose appearing in a portal vein [15].

PSA has been identified as one of the best parameters for the prediction of intestinal absorption. The dynamic PSA was correlated to *in vitro* intestinal drug transport for a series of β-adrenoreceptor antagonists [6]. The excellent sigmoidal relationship between PSA and FA after oral administration in humans was obtained [7] for a series of structurally diverse drugs that were carefully selected to avoid contributions from factors other than passive permeability (such as metabolism, bad solubility or transport by active mechanism) and covering a broad range of physicochemical properties. Similar sigmoidal relationships can also be obtained for the topological PSA (TPSA) [16] (Fig. 5.3). These results suggest that drugs with a PSA $< 60\,Å^2$ are completely (more than 90%) absorbed, whereas drugs with a PSA $> 140\,Å^2$ are absorbed to less than 10%. This conclusion was later confirmed with the correct classification of a set endothelin receptor antagonists as having either low, intermediate or high permeability [17].

PSA was also shown to play an important role in explaining human *in vivo* jejunum permeability [18]. A model based on PSA and calculated $\log P$ for the prediction of drug absorption [19] was developed for 199 well-absorbed and 35

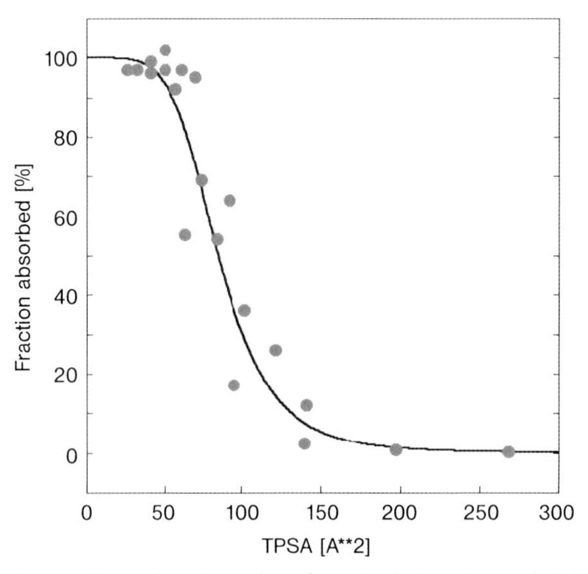

Fig. 5.3 Sigmoidal relationship of intestinal absorption with TPSA for 20 representative drugs.

poorly absorbed compounds. This model allowed the visualization of a "bioavailability area" within an ellipse on a log *P*–PSA plot. Such visualization is quite straightforward for a medicinal chemist to interpret. Winiwarter et al. [20] studied PSA and other molecular descriptors, characterizing H-bond strength, to describe intestinal absorption. The best results were obtained by combining lipophilicity and H-bond donor descriptors. The %PSA descriptor, characterizing the percentage of the molecular surface that was polar, was used to study a series of oral drugs that were launched prior 1983, and also between 1983 and 2002 [21]. Unlike other molecular descriptors, the %PSA ratio remains roughly constant between these two periods, suggesting that it is one of the most important oral drug-like physicochemical properties.

Since experimental determination of intestinal absorption is quite demanding, Caco-2 cell monolayers have been successfully used to model passive drug absorption. Several models for the prediction of Caco-2 permeability using PSA were developed, including those of van de Waterbeemd et al. [5] and Palm et al. [22] who found that relationships between Caco-2 permeability and PSA_d is stronger than with Clog *D*, Krarup et al. [23] who used dynamic PSA calculated for water accessible molecular surface and Bergström et al. [24].

5.2.2
Blood–Brain Barrier Penetration

The BBB is a complex cellular system which protects the central nervous system (CNS) by separating the brain from the systemic blood circulation. Drugs that act on the CNS need to be able to cross the BBB in order to reach their target, while minimal BBB penetration is required for other drugs to prevent CNS side effects. A common measure of BBB penetration is the ratio of drug concentrations in the brain and the blood, which is expressed as log (C_{brain}/C_{blood}).

Van de Waterbeemd and Kansy were probably the first to correlate the PSA of a series of CNS drugs to their membrane transport [4]. They obtained a fair correlation of brain uptake with single-conformer PSA and molecular volume descriptors. Clark [9] derived a good quality model for 55 diverse molecules using single-conformer PSA and calculated log *P*. A very similar equation is also obtained when using fragment based TPSA [16] (Eq. 1) and TPSA in combination with Clog *P* (Eq. 2 and Fig. 5.4).

$$\log \mathrm{BB} = 0.516 - 0.115 \times \mathrm{TPSA} \quad (1)$$

$$n = 55, r^2 = 0.686, r = 0.828, r^2_{cv} = 0.659, F = 115.9, \sigma = 0.42$$

$$\log \mathrm{BB} = 0.070 - 0.014 \times \mathrm{TPSA} + 0.169 \times \mathrm{C}\log P \quad (2)$$

$$n = 55, r^2 = 0.787, r = 0.887, r^2_{cv} = 0.756, F = 95.8, \sigma = 0.35$$

Kelder et al. [25] collected a set of 776 orally administered CNS drugs that are known to be passively transported into the brain and have entered at least phase

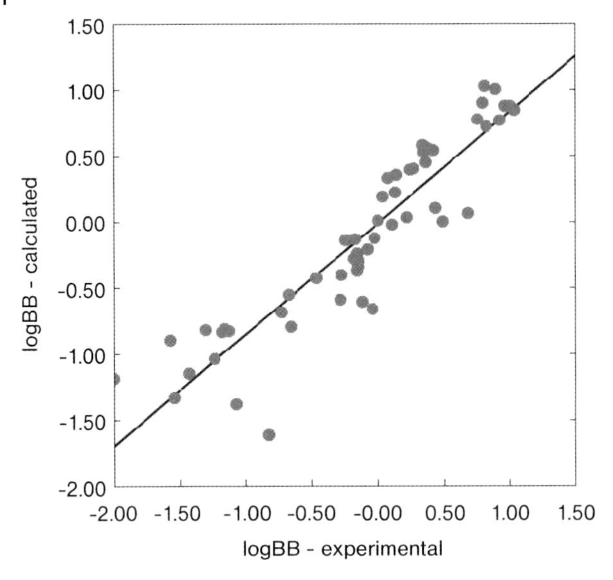

Fig. 5.4 Experimental versus calculated (Eq. 2) log BB for 55 molecules.

II clinical trials. Single-conformer PSA values were calculated for these compounds and the results analyzed as a frequency histogram. This analysis showed that the great majority of orally administered CNS drugs have a PSA of $<70\,Å^2$. A similar analysis of 1590 orally administered non-CNS compounds suggested that the majority of these have a PSA $<120\,Å^2$. Clearly, the BBB provides a significantly tighter constraint on PSA than the intestinal membrane, probably because of the presence of the tight junctions. These results suggest a possible "window of opportunity" for designing non-CNS penetrating, orally absorbed compounds by keeping PSA values between 70 and $120\,Å^2$.

In a recent study involving 150 chemically diverse compounds [26] the following global BBB penetration model was obtained:

$$\log BB = 0.064 - 0.01 \times TPSA + 0.20 \times C\log P \tag{3}$$

$$n = 150, r^2 = 0.69, r^2_{cv} = 0.60$$

The authors noted, however, that the separation of compounds into chemically similar classes considerably improves the construction of predictive BBB penetration models.

Numerous other QSAR models relating BBB penetration to calculated molecular descriptors have also appeared in literature; see for example [27–29]. In each case, PSA was identified as one of the most important parameters determining blood–brain barrier penetration.

5.2.3
Other Drug Characteristics

As already discussed, PSA encodes molecular polarity and H-bonding potential particularly well; therefore, it is not surprising that this descriptor is also useful for the prediction of various other molecular characteristics. PSA has been shown to be one of the most important descriptors for the development of models to predict water solubility of organic molecules [30–32], to explain nonspecific binding [33], critical micelle concentration [34] and to identify promiscuous aggregating inhibitors [35]. Another possible application of PSA is identification of compounds with increased risk of nonspecific toxicity [36]; in this study PSA was used together with other global molecular descriptors. In an interesting study, calculated molecular properties including PSA were used to classify metabolites of *Escherichia coli* [37] to help to understand the metabolome diversity of this organism.

5.3
Application of PSA in Virtual Screening

In the quest for identifying new bioactive molecules, high-throughput screening (HTS) methodologies are routinely used. Many large pharmaceutical companies have set-up whole HTS factories, which are able to screen more than half a million molecules on a particular target. However, even such an enormous screening throughput is not sufficient. The number of small, drug-like compounds available for screening is much larger, not to mention virtual molecules which are in the chemists minds but have not yet been synthesized. Any of these new structures may possess the unique bioactivity and become the next big "blockbuster". One has to find a compromise between spending resources on purchasing/synthesizing and screening samples, and the possibility to cover the largest possible area of a reasonable chemistry space, providing the highest probability of hits. Various cheminformatics methodologies are used in this virtual screening process. These include simple "junk removal" screens discarding molecules containing too reactive or toxic substructures, or molecules with global physicochemical properties outside the ranges generally populated by drugs. Many virtual screening methods are based on the calculation of various scores, which allow the prediction of bioactivity for untested molecules. Such scores are usually obtained by various machine learning methods using a set of known molecules with desired activity as a training set. Finally, the most sophisticated virtual screening technique – virtual docking – is based on the fitting of a putative ligand structure into the respective receptor and selecting molecules with the best fit.

Calculated molecular descriptors are used routinely in the virtual screening process to discard molecules with properties outside the range defined by a set of common drugs, since such outliers would have a high probability of having serious bioavailability problems [38–40]. The most commonly used descriptors used in such screening include calculated log *P*, molecular weight and number of hydrogen donors and acceptors, as suggested by the well known "Rule-of-5" [41]. Of course, PSA itself is also very well suited to distinguishing between bioavailable molecules and those with possible bioavailability problems. As discussed in previous sections, molecules usually require a PSA of below 140–150 Å^2 to show acceptable bioavailability. For drugs acting on the central nervous system, which must therefore pass the BBB, this value should be below 70–80 Å^2. A combination of PSA and the number of rotatable bonds in molecule was shown to correlate well with the oral bioavailability [42]. The authors studied the oral bioavailability in rats on a set of 1100 drug candidates and found that compounds with 10 or fewer rotatable bonds and a PSA less than 140 Å^2 had a high probability of being orally bioavailable in rat.

We have to keep in mind, however, that neither the value of PSA nor actually any other single *in silico* generated descriptor should be used as a "kill criterion" when discarding molecules in virtual screening or selecting structures for follow-up in medicinal chemistry projects. All calculated parameters can only provide hints about the expected properties of a molecule and its bioavailability and should be used together to form a "consensus score" to rank screened molecules. A very nice example of such a "concerted approach" is a study [43] where various 2D and 3D virtual screening techniques were used to identify novel and potent agonists of the melanin-concentrating hormone 1 receptor.

At Novartis, so-called "Bioavailability Radar Plots" [44] are used to visually display the oral absorption potential of molecules. On these plots five important calculated descriptors (log *P*, molecular weight, PSA, number of rotatable bonds and water solubility score [45]) are displayed on the axes of a pentagonal radar plot and compared with predefined property limits (green area) which were determined by the analysis of marketed oral drugs. These plots provide an intuitive tool that displays multiple parameters as a single chart in a straightforward but informative way, providing visual feedback about the molecule's bioavailability potential (Fig. 5.5).

Closely related to the use of PSA in virtual screening is its application in the design of combinatorial libraries with optimal properties. These applications are reviewed further in Refs. [46, 47], for example.

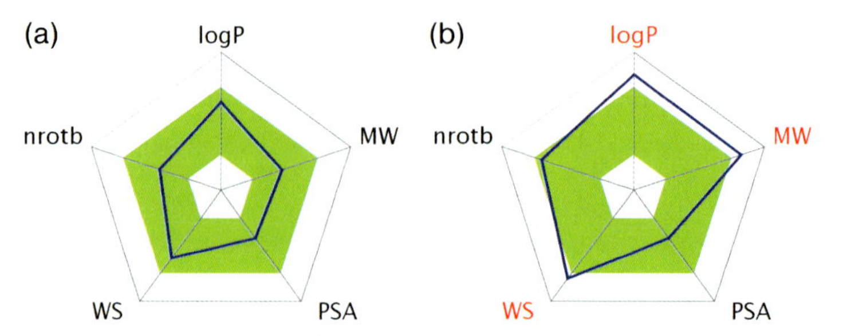

Fig. 5.5 Bioavailability plots for molecules with "good" (a) and "bad" (b) characteristics.

5.4
Calculation of PSA

In its initial application [4, 5], PSA was calculated from a single molecule conformation by summing-up the surface contributions of polar atoms. Per Artursson and coworkers at Uppsala University introduced the so called "dynamic" polar surface (PSA_d) [6] by also taking into account molecular flexibility. The calculation of PSA_d requires a full conformational search for a molecule including geometry optimization to generate a set of low-energy conformers. The PSA is then calculated for all conformers within 2.5 kcal mol^{-1} of the lowest energy conformer. The actual value of PSA_d is obtained by taking the Boltzmann-weighted average of the single-conformer values. Several publications show that PSA_d provides very good correlations with drug transport characteristics. The disadvantage of this approach, however, is that the full conformational search followed up by geometry optimization is computationally expensive. This makes PSA_d unsuitable for processing large datasets and, therefore, for virtual screening applications. This prompted investigations into the applicability of single-conformer, or static, PSA. Clark showed that single-conformer PSA performs very well for prediction of intestinal absorption [8], as well as for BBB penetration [9]. Kelder [25] also reported very good correlations between static and dynamic values (r=0.978) for 45 drugs, with notable differences only observed for cases involving hydrophobic collapse or strong intramolecular interactions.

Several enhancements to the standard 3D PSA approach have been suggested. Hou et al. [48] included contributions to polar surface from only atoms with charges (as calculated by the Gasteiger–Marsili method) above a certain limit. This approach was termed high-charged PSA (HCPSA). Saunders and Platts [49] also considered in PSA calculations the H-bonding strength of particular polar fragments. In this approach, the polar surface belonging to functional groups is multiplied by a scaling factor characterizing their experimentally determined H-bonding strength. Recently, %PSA descriptors, based on the ratio between PSA and total molecular surface area, were also introduced to characterize the properties of oral drugs [21].

The daily cheminformatics business within the pharmaceutical industry requires properties to be calculated for datasets containing millions of molecules, including in-house structures, compound collections from various commercial sample providers or virtual libraries, for example. The rapid calculation of PSA for large numbers of molecules was the main motivation for the development of a method based on fragment contributions. Since the only information required for the calculation is molecular topology (connectivity), this approach is also often referred to as topological PSA-TPSA [10]. The fragment polar surface contributions were obtained by fitting TPSA, calculated as a sum of fragment contributions, with "real" 3D PSA values calculated for a dataset of 34810 drugs from the World Drug Index. The final correlation between TPSA and 3D PSA was excellent, with squared correlation coefficient r^2=0.982 (Fig. 5.6). In the original paper, in addition to contributions of oxygen and nitrogen atoms, also fragments centered on sulfur

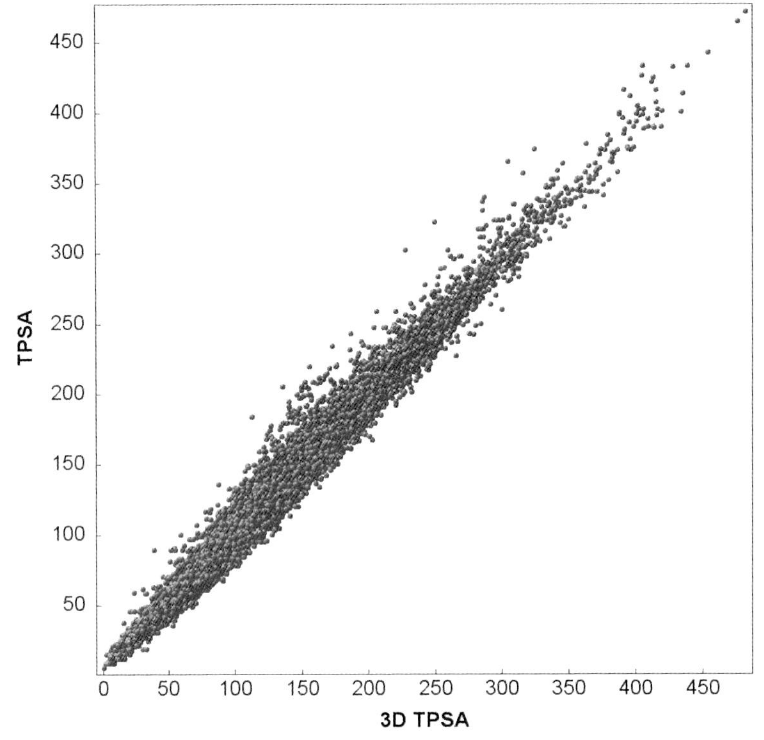

Fig. 5.6 Correlation of 3D PSA and TPSA for 34 810 drug-like molecules ($r^2 = 0.982$).

and phosphorus atoms were examined, although since these fragments did not usually provide an improvement in the correlations [50], the use of only oxygen and nitrogen fragments seems to provide the best choice for PSA calculations.

The most significant differences between TPSA and 3D PSA were observed for large macrocycles containing many polar substituents These substituents are usually buried in the center of the ring and are therefore not accessible to solvent. Fragment-based TPSA provided larger values than 3D PSA in such cases.

The possibility to calculate the PSA descriptor, particularly by the fragment-based approach, is currently available in many commercial and freely available software packages. For instance, the author of this chapter has released an open source C code [51] to calculate PSA. This program requires the Daylight toolkit to process SMILES and generate a molecule object, but the code can be easily modified to work with other chemistry development environments. The program was also later translated to PERL [52] where dependence on the Daylight toolkit was replaced by the PerlMol modules. The method for calculation of TPSA was also implemented in Java in the Chemistry Development Kit [53] and in the JOELib package [54]. Additionally, a free web service to calculate PSA (together with several other useful molecular descriptors) is available on the Internet [55] (Fig. 5.7).

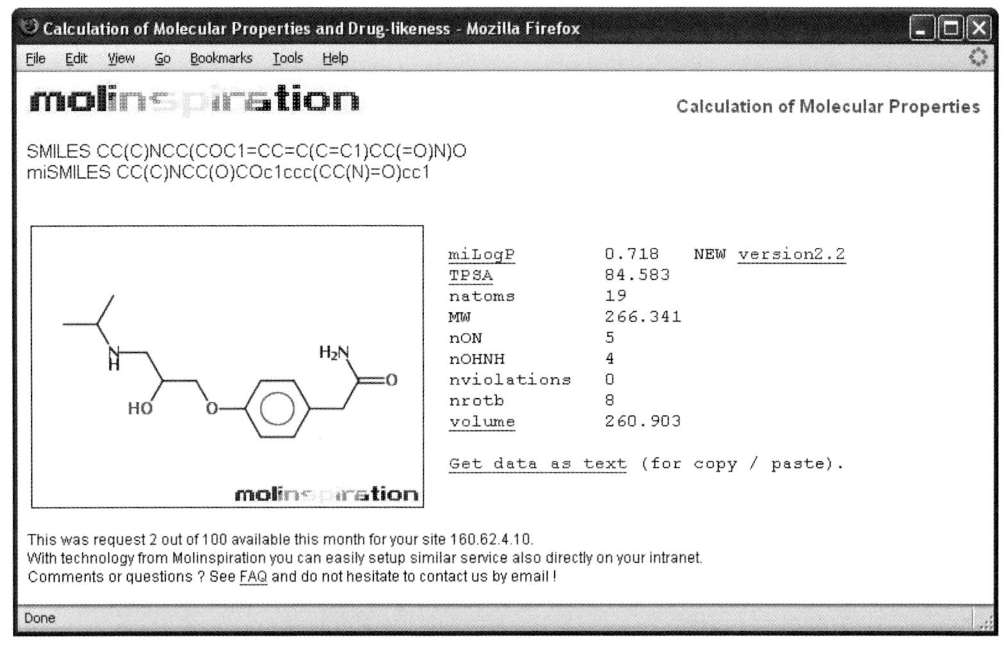

Fig. 5.7 Calculation of PSA and other molecular descriptors on the Internet [55].

5.5 Correlation of PSA with other Molecular Descriptors

Descriptors used to characterize molecules in QSAR studies should be as independent of each other (orthogonal) as possible. When using correlated parameters there is an increased danger of obtaining non-predictive, chance correlation [56]. To examine the correlation between PSA (calculated according to the fragment-based protocol [10]) and other descriptors, we studied a collection of 7010 bioactive molecules from the PubChem database [57]. In addition to PSA, the following parameters were used:

CLOGP calculated octanol–water partition coefficient [58]
n_{HBA} number of H-bond acceptors (any oxygen or nitrogen atom was considered as an "acceptor")
n_{HBD} number of H-bond donors (any –OH or –NH moiety was considered to be a "donor")
MV molecular volume [59]
MW molecular weight
n_{atoms} number of nonhydrogen atoms
n_{rotb} number of rotatable bonds

DM dipole moment calculated by the AM1 semiempirical method [60] for fully optimized molecular structure, starting from the CORINA [61] geometry

SAP sum of atom polarities – sum of absolute values of AM1 charges on nonhydrogen atoms

All of these parameters (with the possible exception of SAP) are frequently used in QSAR studies or as filters in virtual screening. The SAP descriptor was included to check for correlations between PSA and quantum chemically calculated charges.

The correlation matrix for these 10 descriptors is shown in Table 5.1. As expected, PSA shows the highest correlation with the number of H-bond acceptors (r^2=0.924) and number of H-bond donors (r^2=0.736). Thus, in QSAR studies these parameters should not be used together. We recommend the use of PSA because this descriptor provides a more detailed description of H-bonding accessible area than just simple atom counts. Another descriptor with which PSA correlates is SAP (r^2=0.425). When calculating SAP, charges on all nonhydrogen atoms were considered. Probably an even better correlation with PSA would be obtained by considering charges not on all atoms, but only on atoms with charge above (or below) some predefined cut-off value.

To further analyze the relationships within descriptor space we performed a principle component analysis of the whole data matrix. Descriptors have been normalized before the analysis to have a mean of 0 and standard deviation of 1. The first two principal components explain 78% of variance within the data. The resultant loadings, which characterize contributions of the original descriptors to these principal components, are shown on Fig. 5.8. On the plot we can see that PSA, n_{HBD} and n_{HBA} are indeed closely grouped together. Calculated octanol–water partition coefficient CLOGP is located in the opposite corner of the property space. This analysis also demonstrates that CLOGP and PSA are the two parameters with

Tab. 5.1 Cross-correlations (expressed as r^2) between popular molecular descriptors (see text).

	PSA	*CLOGP*	n_{HBA}	n_{HBD}	n_{atoms}	*MW*	*MV*	n_{rotb}	*DM*	*SAP*
PSA	1.000	0.299	0.924	0.736	0.341	0.364	0.258	0.157	0.158	0.425
CLOGP	0.299	1.000	0.221	0.351	0.054	0.041	0.093	0.023	0.034	0.019
n_{HBA}	0.924	0.221	1.000	0.539	0.447	0.466	0.346	0.194	0.188	0.463
n_{HBD}	0.736	0.351	0.539	1.000	0.137	0.150	0.108	0.075	0.055	0.161
n_{atoms}	0.341	0.054	0.447	0.137	1.000	0.952	0.964	0.332	0.093	0.377
MW	0.364	0.041	0.466	0.150	0.951	1.000	0.922	0.336	0.100	0.430
MV	0.258	0.093	0.346	0.108	0.964	0.922	1.000	0.403	0.062	0.340
n_{rotb}	0.157	0.023	0.194	0.075	0.332	0.336	0.403	1.000	0.012	0.195
DM	0.158	0.034	0.188	0.055	0.093	0.100	0.062	0.012	1.000	0.152
SAP	0.425	0.019	0.463	0.161	0.377	0.430	0.340	0.195	0.152	1.000

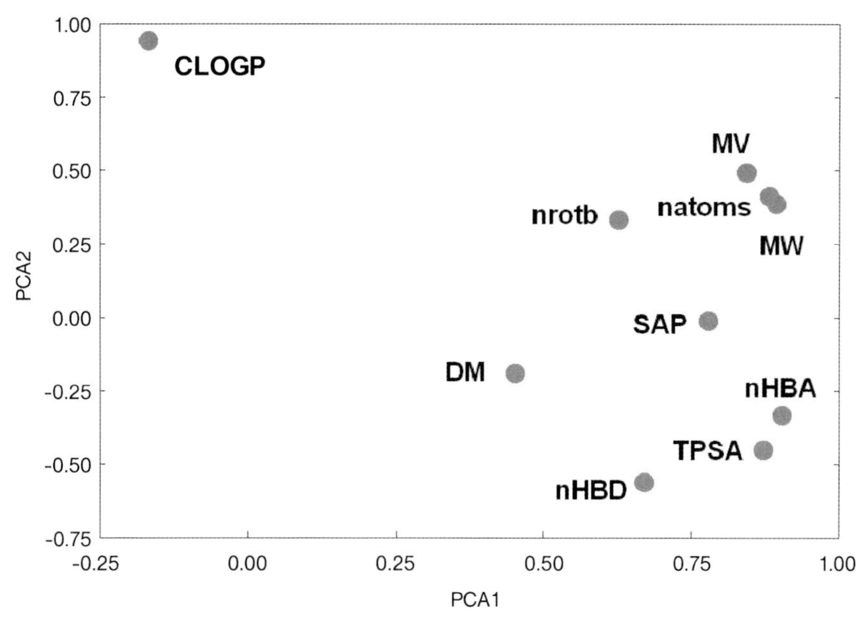

Fig. 5.8 Loadings characterizing 10 molecular descriptors.

the highest information content for characterizing molecular physicochemical properties.

5.6
Conclusions

The diverse examples of PSA applications presented here clearly demonstrate that this descriptor has become a standard tool in the repertoire of medicinal chemists, computational chemists and cheminformatics specialists. Even though PSA can be calculated very easily, it is a very useful parameter for assessing drug-likeness and bioavailability potential of drugs under development, characterizing virtual molecules when selecting molecules for screening, and guiding the design of combinatorial libraries with optimal properties.

References

1 Amidon, G. L., Yalkowski, S. H., Anik, S. T., Valvani, S. C. Solubility of nonelectrolytes in polar solvents V. Estimation of the solubility of aliphatic monofunctional compounds in water using a molecular surface approach. *J. Phys. Chem.* **1975**, *79*, 2239–2246.

2 Pearlman, R. S. Molecular surface areas and volumes and their use in structure–activity relationships. In *Physical Chemical*

Properties of Drugs, Medicinal Research Series, Vol. 10, Yalkowsky, S. H., Sinkula, A. A., Valvani, S. C. (eds.), Marcel Dekker, New York, **1980**, pp. 321–347.
3 Stanton, D. T., Jurs, P. C. Development and use of charged partial surface area structural descriptors in computer-assisted quantitative structure–property relationship studies. *Anal. Chem.* **1990**, *62*, 2323–2329.
4 Van de Waterbeemd, H., Kansy, M. Hydrogen-bonding capacity and brain penetration. *Chimia* **1992**, *46*, 299–303.
5 Van de Waterbeemd, H., Camenisch, G., Folkers, G., Raevsky, O. A. Estimation of Caco-2 cell permeability using calculated molecular descriptors. *Quant. Struct.-Act. Relat.* **1996**, *15*, 480–490.
6 Palm, K., Luthmann, K., Ungell, A.-L., Strandlund, G., Artursson, P. Correlation of drug absorption with molecule surface properties. *J. Pharm. Sci.* **1996**, *85*, 32–39.
7 Palm, K., Stenberg, P., Luthmann, K., Artursson, P. Polar molecular surface properties predict the intestinal absorption of drugs in humans. *Pharm. Res.* **1997**, *14*, 568–571.
8 Clark, D. Rapid calculation of polar molecular surface area and its application to the prediction of transport phenomena. 1. Prediction of intestinal absorption. *J. Pharm. Sci.* **2000**, *88*, 807–814.
9 Clark, D. Rapid calculation of polar molecular surface area and its application to the prediction of transport phenomena. 2. Prediction of blood–brain barrier penetration. *J. Pharm. Sci.* **2000**, *88*, 815–821.
10 Ertl, P., Rohde, B., Selzer, P. Fast calculation of molecular polar surface area as a sum of fragment-based contributions and its application to the prediction of drug transport properties. *J. Med. Chem.* **2000**, *43*, 3714–3717.
11 http://scholar.google.com, we used the search terms "polar surface area" -protein (to exclude publications focused on study of protein surfaces).
12 Stenberg, P., Luthman, K., Artursson, P. Virtual screening of intestinal drug permeability. *J. Control. Rel.* **2000**, *65*, 231–243.
13 Clark, D. E., Grootenhuis, D. J. Predicting passive transport *in silico* – history, hype, hope. *Curr. Top. Med. Chem.* **2003**, *3*, 1193–1203.
14 Artursson, P., Bergström, A. S. Intestinal absorption: the role of polar surface area. In *Drug Bioavailability (Methods and Principles in Midicinal Chemistry)*, Van de Waterbeemd, H., Lennernäs, H., Artursson, P. (eds.), Wiley-VCH, Weinheim, **2003**.
15 Van de Waterbeemd, H., Gifford, E. ADMET *in silico* modelling: towards prediction paradise? *Nat. Rev. Drug. Discov.* **2003**, *2*, 192–204.
16 Ertl, P., Rohde, B., Selzer, P. Calculation of molecular polar surface area as a sum of fragment-based contributions and its application to the prediction of drug transport properties. In *Rational Approaches to Drug Design*, Höltje, H.-D., Sippl, W. (eds.), Prous, Barcelona, **2001**, pp. 451–455.
17 Stenberg, P., Luthman, K., Ellens, H., Lee, C.-P., Smith, P. L., Lago, A., Elliott, J. D., Artursson, P. Prediction of the intestinal absorption of endothelin receptor antagonists using three theoretical methods of increasing complexity. *Pharm. Res.* **1999**, *16*, 1520–1526.
18 Winiwarter, S., Bonham, N. M., Ax, F., Hallberg, A., Lennernäs, H., Karlen, A. Correlation of human jejunal permeability (*in vivo*) of drugs with experimentally and theoretically derived parameters. A multivariate data analysis approach. *J. Med. Chem.* **1998**, *41*, 4939–4949.
19 Egan, W. J., Merz, K. M., Baldwin, J. J. Prediction of drug absorption using multivariate statistics. *J. Med. Chem.* **2000**, *43*, 3867–3877.
20 Winiwarter, S., Ax, F., Lennernäs, H., Hallberg, A., Pettersson, C., Karlen, A. Hydrogen bonding descriptors in the prediction of human *in vivo* intestinal permeability. *J. Mol. Graph. Model.* **2003**, *21*, 273–287
21 Leeson, P. D., Davis, A. M. Time-related differences in the physical property profiles of oral drugs. *J. Med. Chem.* **2004**, *47*, 6338–6348.
22 Palm, K., Luthman, K., Ungell, A.-L., Strandlund, G., Beigi, F., Lundahl, P., Artursson, P. Evaluation of dynamic polar

molecular surface area as predictor of drug absorption: comparison with other computational and experimental predictors. *J. Med. Chem.* **1988**, *41*, 5382–5392.

23 Krarup, L. H., Christensen, I. T., Hovgaard, L., Frokjaer, S. Predicting drug absorption from molecular surface properties based on molecular dynamics simulations. *Pharm. Res.* **1998**, *15*, 972–978.

24 Bergström, C. A. S., Strafford, M., Lazarova, L., Avdeef, A., Luthman, K., Artursson, P. Absorption classification of oral drugs based on molecular surface properties. *J. Med. Chem*, **2003**, *46*, 558–570.

25 Kelder, J., Grootenhuis, P. D. J., Bayada, D. M., Delbressine, L. P. C., Ploeman, J.-P. Polar molecular surface area as a dominating determinant for oral absorption and brain permeation of drugs. *Pharm. Res.*, **1999**, *16*, 1514–1519.

26 Pan, D., Iyer, M., Liu, J., Li, Y., Hopfinger, A. J., Constructing optimum blood brain barrier QSAR models using a combination of 4D-molecular similarity measures and cluster analysis descriptors. *J. Chem. Inf. Model.* **2004**, *44*, 2083–2098.

27 Hou, T. J., Xu, X. J. ADME evaluation in drug discovery. 3. Modeling blood–brain barrier partitioning using simple molecular descriptors. *J. Chem. Inf. Model.* **2003**, *43*, 2137–2152.

28 Mente, S. R., Lombardo, F. A recursive-partitioning model for blood–brain barrier permeation. *J. Comput.-Aid. Mol. Des.*, **2005**, *19*, 465–481.

29 Luco, J. M., Marchevsky, E. QSAR studies on blood–brain barrier permeation. *Curr. Comput.-Aided Drug Des.* **2006**, *2*, 31–55.

30 Liu, R., So, S.-S. Development of quantitative structure–property relationship models for early ADME evaluation in drug discovery. 1. Aqueous solubility. *J. Chem. Inf. Comp. Sci.* **2001**, *41*, 1633–1639.

31 Catana, C., Gao, H., Orrenius, C., Stouten, P. F. Linear and nonlinear methods in modeling the aqueous solubility of organic compounds. *J. Chem. Inf. Model.* **2005**, *45*, 170–176.

32 Bergström, C. A. S. D., Norinder, U., Luthman, K., Artursson, P. Experimental and computational screening models for prediction of aqueous drug solubility. *Pharm. Res.* **2002**, *19*, 182–188.

33 Maurer, T. S., DeBartolo, D. B., Tess, D. A., Scott, D. O. Relationship between exposure and nonspecific binding of thirty-three central nervous system drugs in mice. *Drug Metab. Dispos.* **2005**, *33*, 175–181.

34 Saunders, R. A., Platts, J. A. Correlation and prediction of critical micelle concentration using polar surface area and LFER methods. *J. Phys. Org. Chem.* **2004**, *17*, 431–438.

35 Seidler, J., McGovern, S. L., Doman, T. N., Shoichet, B. K. Identification and prediction of promiscuous aggregating inhibitors among known drugs. *J. Med. Chem.* **2003**, *6*, 4477–4486.

36 Muskal, S. M., Jha, S. K., Kishore, M. P., Tyagi, P. A simple and readily integratable approach to toxicity prediction. *J. Chem. Inf. Comp. Sci.* **2003**, *43*, 1673–1678.

37 Nobeli, I., Ponstingl, H., Krissinel, E. B., Thornton, J. M. A structure–based anatomy of the *E. coli* metabolome. *J. Mol. Biol.* **2003**, *334*, 697–719.

38 Rishton, G. M. Nonleadlikeness and leadlikeness in biochemical screening. *Drug Discov. Today* **2002**, *8*, 86–96.

39 Egan, W. J., Walters, W. P., Murcko, M. A. Guiding molecules towards drug-likeness. *Curr. Opin. Drug Discov. Dev.* **2002**, *5*, 540–549.

40 Pozzan, A. Molecular descriptors and methods for ligand based virtual high throughput screening in drug discovery. *Curr. Pharm. Des.* **2006**, *12*, 2099–2110.

41 Lipinski, C. A., Lombardo, F., Dominy, B. W., Feeney, P. J. Experimental and computational approaches to estimate solubility and permeability in drug discovery and development settings. *Adv. Drug Deliv. Rev.* **1997**, *23*, 3–25.

42 Veber, D., Johnson, S., Cheng, H., Smith, B., Ward, K., Kopple, K. D. Molecular properties that influence the oral bioavailability of drug candidates. *J. Med. Chem.* **2002**, *45*, 2615–2623.

43 Clark, D. E., Higgs, C., Wren, S. P., Dyke, H. J. Wong, M. Norman, D. Lockey, P. M., Roach, A. G. A virtual screening approach to finding novel and potent antagonists at the melanin-concentrating hormone 1 receptor. *J. Med. Chem.* **2004**, *47*, 3962–3971.

44 Ritchie, T. Rapid visualization of bioavailability potential using simple radar, plots, *Current Edge Approaches to Drug Design*, 2004, March, http://www.rscmodelling.org/CEAtoDD/RitchieRadar.ppt.

45 Ertl, P., Mühlbacher, J. Rohde, B., Selzer, P. Web-based cheminformatics and molecular property prediction tools supporting drug design and development at Novartis. *SAR QSAR Environ. Res.* **2003**, *14*, 321–328.

46 Mitchell, T., Showell, G. A. Design strategies for building drug-like chemical libraries. *Curr. Opin. Drug Discov. Dev.* **2001**, *4*, 314–318.

47 Matter, H., Baringhaus, K.-H., Naumann, T., Klabunde, T., Pirard, B. Computational approaches towards the rational design of drug-like compound libraries. *Comb. Chem. High-Throughput Screen.* **2001**, *4*, 453–475.

48 Hou, T. J., Xu, X. J. ADME evaluation in drug discovery. 3. Modeling blood–brain barrier partitioning using simple molecular descriptors. *J. Chem. Inf. Comput. Sci.* **2003**, *43*, 2137–2152.

49 Saunders, R. A., Platts, J. A. Scaled polar surface area descriptors: development and application to three sets of partition coefficients. *New J. Chem.* **2004**, *28*, 166–172.

50 To our knowledge only in a single publication [18] was a better correlation of PSA with drug transport reported after inclusion of S fragments.

51 Ertl, P., http://www.daylight.com/meetings/emug00/Ertl/tpsa.c

52 http://www.perlmol.org/examples/polar_surface_area.

53 The Chemistry Development Kit, http://cdk.sourceforge.net

54 JOELib II, http://joelib.sourceforge.net

55 http://www.molinspiration.com/cgi-bin/properties.

56 Kubinyi, H. *QSAR: Hansch Analysis and Related Approaches*, VCH, Weinheim, **1993**.

57 http://pubchem.ncbi.nlm.nih.gov, molecules with an entry in the "Pharmacological action" field were extracted from the PubChem and standardized before processing.

58 CLOGP 4.0. http://www.biobyte.com.

59 Molinspiration Toolkit 2006.04, http://www.molinspiration.com/services/volume.html.

60 Dewar, M. J. S., Zoebisch, E. G., Healy, E. F., Stewart, J. J. P. Development and use of quantum mechanical molecular models. 76. AM1: a new general purpose quantum mechanical molecular model. *J. Am. Chem. Soc.* **1985**, *107*, 3902–3909.

61 Corina, http://www.mol-net.com.

6
H-bonding Parameterization in Quantitative Structure–Activity Relationships and Drug Design

Oleg Raevsky

Abbreviations

3D	three-dimensional
BCS	Biopharmaceutic Classification System
BP	basal permeability
HBAA	H-bond acceptor–acceptor
HBDD	H-bond donor–donor
HBAD	H-bond acceptor–donor
HMPA	hexamethylphoshoramide
HYBOT	H-bond thermodynamics
INN	international nonproprietary names
MHBP	molecular H-bonding potential
MLR	multivariate linear regression
MOD	mobile order and disorder
NCE	new chemical entities
PSA	polar surface area
QSAR	quantitative structure–activity relationship
SIS	similarity indices of spectra
USAN	US adopted names
WDI	World Drug Index
WEASA	van der Waals enthalpy acceptor surface area
WEDSA	van der Waals enthalpy donor surface area
WFEASA	van der Waals free energy acceptor surface area
WFEDSA	van der Waals free energy donor surface area

Symbols

N_a	number of oxygen and nitrogen atoms in a molecule
N_d	number of hydrogen atoms on oxygen and nitrogen in a molecule
A	H-bond acidity parameter
B	H-bond basicity parameter

Molecular Drug Properties. Measurement and Prediction. R. Mannhold (Ed.)

ISBN: 978-3-527-31755-4

ΣC_a	sum of H-bond acceptor free energy factors
ΣC_d	sum of H-bond donor free energy factors
E_m	optimum H-bond energy
R_m	optimum H-bond length
α	molecular polarizability
$\log P$	partition coefficient
$\log S$	solubility
FA	fraction absorbed

6.1 Introduction

Interactions between H-bond donor and H-bond acceptor molecules result in the formation of many molecular and ionic complexes that are of great importance in chemical and biochemical processes including enzymatic catalysis. H-bond complexes are especially important in biological systems because they play crucial roles in macromolecular structures and in molecular recognition. DNA and proteins are held together in their defined three-dimensional (3D) structures primarily by H-bonds. The double helix of DNA and RNA structures, the peptide and protein secondary structures like α-helices, β-sheets, and β- and γ-loops, and the tertiary structures of proteins are formed by H-bonds (enthalpy contributions) and by hydrophobic contacts (primarily entropy contribution) [1]. In addition, H-bonding also affects membrane transport and the distribution of drugs within biological systems. Accordingly, there have been many attempts to quantify H-bond parameters from the beginning of development of quantitative structure–activity relationship (QSAR) studies and computer-aided drug design. Many different types of H-bond descriptors are used in this field, including indicator variables, numbers of H-bond donors or/and acceptors, atomic and molecular surface area, quantum-chemical descriptors, and thermodynamic and solvatochromatic parameters. Each of these descriptors directly or indirectly describes the complex process of H-bonding on different levels. The hierarchy of its information content is shown in Fig. 6.1 in analogy to the presentation of molecular structure levels in Ref. [2].

Descriptors of the lowest level (indirect H-bond parameters, H-bond indicators, surface H-bond indicators) indicate trends; for example, the more H-bond groups in a molecule the better is its solubility in water, or the greater the polar surface area (PSA) value, the lower is the intestinal absorption in humans. Descriptors at the intermediate level (enthalpy and free energy factors) allow quantitative assessments of H-bond interactions of single pairs of acceptor and donor atoms. Descriptors at the next higher level (distance H-bond potentials, surface thermodynamics parameters) give quantitative descriptions of H-bonding at optimum arrangements of partners. The calculation of thermodynamic parameters, including distance and angle dependencies of potentials as well as the influence of substituents

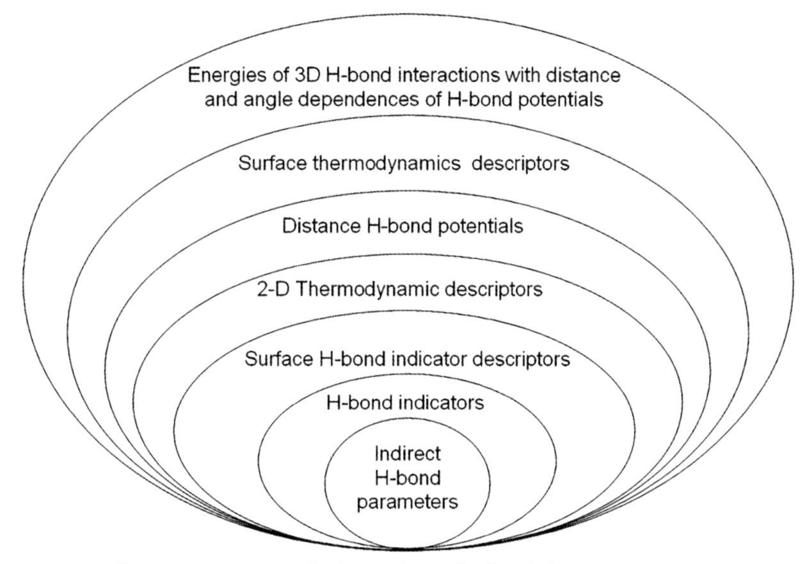

Fig. 6.1 Information content and relationships of H-bond descriptors.

on the H-bond capacity, can be considered as the highest level of quantitative H-bonding description.

This chapter describes and classifies H-bond descriptors, and indicates possible areas of their application in QSAR studies and drug design. Similar analyses were presented in previous articles [3–5].

6.2
Two-dimensional H-bond Descriptors

Two-dimensional H-bond descriptors are included in Table 6.1. Considering information content, they may be classified as indirect descriptors (no direct link with the H-bonding process), H-bond indicators (atoms having potential H-bond capability) and thermodynamic factors (calculated on the basis of experimental thermodynamic data of H-bonding).

6.2.1
Indirect H-bond Descriptors

The first publications in this field appeared in the 1970s. Seiler [6] studied the differences in $\log P$ in the systems octanol–water and cyclohexane–water ($\Delta\log P_{o/w-ch/w}$) to develop some measure of the contribution of H-bonding (I_H). Moriguchi investigated $\log P$ in octanol–water for polar and nonpolar compounds

Tab. 6.1 Two-dimensional H-bond descriptors.

Name or meaning	*Type*	*Symbol*	*Reference*
The difference between octanol–water and cyclohexane–water log *P* values	indirect	I_H	6
The difference log *P* octanol–water for polar and nonpolar chemicals with the same molecular weight	indirect	E_W	7
The atomic charge on the hydrogen atom	indirect	Q_H	4, 8, 9
The energy of the lowest unoccupied molecular orbital	indirect	ELUMO	4, 8, 9
The electron donor superdelocalizability	indirect	D_E	4, 8–11
The self-atom polarizability	indirect	P_E	4, 8–11
The charge on the most negatively charge atom	indirect	Q_{MH}	4, 8, 9
The energy of the highest occupied molecular orbital	indirect	E_{HOMO}	4, 8, 9
Surface electrostatic potential maxima	indirect	$V_{s,max}$	12
Electrostatic potential minima	indirect	V_{min}	12
sum of donors (OHs + NHs), sum of acceptors (Ns + Os)	indicator	N_a, N_d	17
Infrared or nuclear magnetic resonance spectral shifts	thermodynamic	$\Delta\nu_{OH}$, δ_H	20, 21
Parameter of acidity	thermodynamic	*A*	23
Parameter of basicity	thermodynamic	*B*	23
H-bond acceptor enthalpy and free energy factors	thermodynamic	E_a, C_a	27
H-bond donor enthalpy and free energy factors	thermodynamic	E_d, C_d	27
Sum of absolute values of free energy H-bond factors	thermodynamic	ΣC_{ad}	30

with the same molecular volume [7]. The descriptors they found were calculated from experimental values of properties that contain, in hidden form, an H-bond component together with other factors. The separation of these factors is a special problem.

Other indirect H-bond descriptors are generated from quantum-chemical calculations. The atomic charge on hydrogen (Q_H), the energy of the lowest unoccupied molecular orbital (E_{LUMO}), electron donor superdelocalizability (D_E), atomic polarizability (P_E) and surface electrostatic potential maxima ($V_{s,max}$) were used to characterize H-bond donor ability [4, 8–12]. The charge on the most negative atom (Q_{MH}), the energy of the highest occupied molecular orbital (E_{HOMO}) and the electrostatic potential minima (V_{min}) were used to characterize H-bond acceptor ability [4, 8, 9, 11]. Correlations between such theoretical H-bond descriptors and experimental properties of H-bond complexes were not satisfactory. For example, Gancia et al. [11] indicate that none of the calculated properties showed a high correlation to the H-bond equilibrium constants when considering a dataset of 124 compounds (including 31 OH and 27 NH proton donors, and 35 oxygen and 31 nitrogen proton acceptors). Only for separate subsets were reasonable correlations found (correlation coefficients between 0.77 and 0.92). Thus, these theoretical parameters cannot be considered as practical H-bond descriptors despite their rather wide and sometimes successful use in QSAR (see, e.g. Refs. [13, 14]).

6.2.2 Indicator Variables

Another approach is to use indicator variables where a value of 1 is given if an H-bond can be formed or 0 if it cannot. This was first proposed by Fujita et al. [15]. Charton and Charton [16] indicated the types and numbers of H-bonds that a molecule is capable of forming. The simplest such indicator is the number of oxygen and nitrogen atoms (N_a), and the number of hydrogen atoms on oxygen and nitrogen (N_d) in a molecule. These descriptors, which are included in the software of Accelrys, ACD and others, became popular in QSAR and drug design especially after introduction of the "Rule-of-5" [17]. However, such indicator variables have a very small information content and do not reflect the influence of substituents near H-bond acceptor or donor atoms. N_a and N_d are too crude to quantitatively characterize H-bond contributions, and the "Rule-of-5" should be seen as a qualitative absorption/permeability predictor [18].

6.2.3 Two-dimensional Thermodynamics Descriptors

In H-bond complexes, it is possible to consider spectral shifts of appropriate bands in infrared spectra ($\Delta\nu_{OH}$) or proton shifts in nuclear magnetic resonance spectra (δ_H) as quantitative H-bond parameters [19, 20]. However, the most reliable way to describe H-bonding quantitatively is to consider the thermodynamics of H-bond complexation. For any process, Eq. (1) describes the thermodynamic relationships among the following properties: ΔH (the change in enthalpy), ΔG (the change in free energy), ΔS (the change in entropy), the binding constant (K) and the absolute temperature (in K). R is the universal gas constant:

$$\Delta G = -RT \ln K = \Delta H - T\Delta S \quad (1)$$

Binding constants of H-bond complexes are normally used to create H-bond scales [21–23], e.g. in accordance with Ref. [22]:

$$\log K = 7.354\alpha_2^H\beta_2^H \quad (2)$$

where α_2^H and β_2^H are parameters of acidity and basicity scales.

A multiplicative approach to describe the enthalpy and free energy of H-bonding was developed by Raevsky et al. [3, 24–27]. From a large number of their own experimental values and data from the literature, these authors established a database of thermodynamic parameters of H-bond complexes in different solvents. They selected 936 systems to construct a unified scale of H-bond donor and acceptor factors [26]. Those systems satisfied the following criteria:

- The complex stoichiometry had to be 1:1.
- The reaction was carried out in the nonpolar, aprotic solvent carbon tetrachloride.
- Both ΔG and ΔH had to be measured for each reaction.
- Both ΔG and ΔH were estimated by direct experimental procedures. Data obtained using any type of estimation of those values, including spectroscopic absorption shifts and intensities or some other such parameters, were excluded from consideration.

Phenol and hexamethylphoshoramide (HMPA) were selected, respectively, as the standard H-bond donor and H-bond acceptor with their values fixed on free energy and enthalpy H-bond scales: for phenol, –2.50 for the H-bond donor enthalpy factor (E_d) and also for the H-bond donor free energy factor (C_d); for HMPA, 2.50 for the H-bond acceptor enthalpy factor (E_a) and 4.00 for the H-bond acceptor free energy factor (C_a).

The best fit equations for all the calculated factors were:

$$\Delta H = 4.96(\text{kJ mol}^{-1})E_a E_d \tag{3}$$

$$\Delta G = 2.43(\text{kJ mol}^{-1})C_a C_d \tag{4}$$

Correlations of calculated and experimental enthalpy, and calculated and experimental free energy values for the indicated 936 systems gave the following results:

$$\Delta H_{calc} = -0.27(\pm 0.45) + 1.00(\pm 0.02)\Delta H_{exp} \tag{5}$$

$$n = 936, r^2 = 0.91, s = 2.70, F = 9553$$

$$\Delta G_{calc} = -0.07(\pm 0.12) + 1.00(\pm 0.01)\Delta G_{exp} \tag{6}$$

$$n = 936, r^2 = 0.97, s = 1.11, F = 28556$$

where n is the number of compounds, r is the correlation coefficient, s is the standard deviation and F is the Fisher criterion.

Thus, the statistical criteria for Eqs. (5) and (6) show that the approach provides a reasonably reliable method to calculate H-bond enthalpy and free energy.

From 163 calculated H-bond donor and 195 calculated H-bond acceptor factors, one can get enthalpy and free energy values for 31 785 reactions using Eqs. (3) and (4). Later, the number of H-bond factor values was significantly increased. A special program for calculating factor values was created and included in the HYBOT (Hydrogen Bond Thermodynamics) program [28, 29]. The current version, HYBOT-2006, has about 20 000 values of H-bond acceptor factors and about 5000 values of H-bond donor factors of diverse compounds. It allows the calculation of the H-bond strength of any existing or any conceivable new compound with sufficient accuracy.

Table 6.2 demonstrates large intervals of enthalpy and free energy H-bond donor and acceptor factor values as result of a different nature of H-bonding atoms as well as an influence of substituents at those atoms on its H-bond capability.

The satisfactory correlations between Kamlet–Abraham's acidity parameters and H-bond donor free energy factors, and between Kamlet–Abraham's basicity parameters and H-bond acceptor free energy factors for many sets of compounds [26] deserve particular mentioning:

$$C_d = 0.10(\pm 0.08) - 4.47(\pm 0.18)\alpha_2^H \quad (7)$$

$$N = 63, r^2 = 0.98, s = 0.18, F = 2469$$

However, the HYBOT approach utilizes significantly larger data sets of H-bond factor values and is useful not only for calculating binding constants (free energies), but also for calculating the enthalpy of H-bonding. This is especially important in considering the H-bonding of compounds containing donor and acceptor atoms that can simultaneously form different H-bonds with essential entropy term contributions.

The situation differs essentially for correlations between thermodynamic desriptors and H-bond indicators. Graphical comparison of indicator N_a and

Tab. 6.2 Enthalpy and free energy H-bond donor and acceptor factor values for a few chemicals [28].

NN	*Chemicals*	E_a	E_d	C_a	C_d
1	CH_3OH	1.53	−2.34	1.76	−1.63
2	C_4H_9OH	1.50	−1.44	1.60	−1.32
3	CF_3CH_2OH	1.49	−1.51	1.34	−2.43
4	$C(CF_3)_3OH$	1.49	−1.53	1.53	−3.50
5	C_6H_5OH	0.97	−2.47	0.97	−2.49
6	$3\text{-}NO_2C_6H_4OH$	0.97	−2.85	0.69	−3.45
7	$4\text{-}NO_2C_6H_4OH$	0.97	−2.97	0.63	−3.57
8	CH_3COOH	1.37	−3.01	1.37	−2.58
9	$CH_2ClCOOH$	1.14	−3.01	1.20	−3.50
10	CCl_3COOH	1.14	−3.01	0.80	−4.75
11	$CH_3C(O)CH_3$	1.60	–	1.95	–
12	$CH_3C(O)C_6H_5$	1.37	–	1.75	–
13	$CH_3C(O)N(CH_3)_2$	1.86	–	2.88	–
14	$C_2H_5\text{–}O\text{–}C_2H_5$	1.49	–	1.54	–
15	$C_6H_5\text{–}O\text{–}CH_3$	1.00	–	0.90	–
16	$[(CH_3)_2N]_3PO$	2.52	–	4.00	–

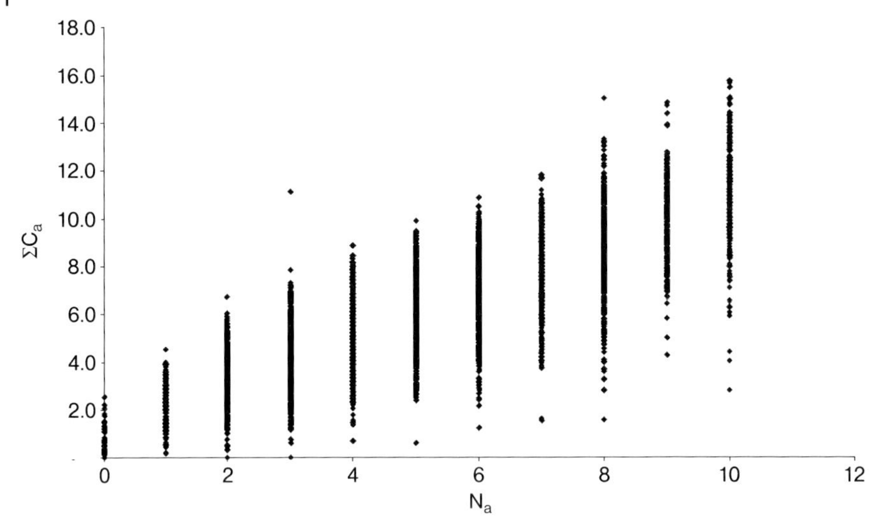

Fig. 6.2 Graphical comparison of indicator and thermodynamics H-bond factors.

thermodynamic ΣC_a H-bond acceptor descriptors for 7826 drugs and "drug-like" chemicals with $N_a \leq 10$ is given in Fig. 6.2, indicating that, for any fixed N_a value, the differences among the corresponding H-bond acceptor factor values can be very large. For example, in a set of compounds with $N_a = 4$ the sums of H-bond acceptor factors range from $\Sigma C_a = 2.30$ for metrazole to 8.17 for pirolazamide. Whereas N_a values are the same for any acceptor atom in a molecule, ΣC_a considers the type of acceptor and donor atoms as well as substituents near those atoms. Thus, despite acceptable inter-relation ($r^2 = 0.71$), these descriptors differ essentially in their information content about H-bonding.

6.3 Three-dimensional H-bond Descriptors

Short description of 3D H-bonding parameters and descriptors is included in Table 6.3.

6.3.1 Surface H-bond Descriptors

PSA is also used as H-bond descriptor to predict various properties of chemicals and drugs. PSA is defined as that part of a molecular surface that arises from oxygen and nitrogen atoms, and also the hydrogens attached to them. Applications of PSA as a QSAR descriptor in correlations with permeability and absorption were carried out first by Van de Waterbeemd et al. [30] and Palm et al. [31]. Clark [32–34] developed this further. Chapter 5 in this book is completely devoted to

Tab. 6.3 Three-dimensional H-bonding parameters and descriptors.

Name or meaning	*Symbol and formula*
Polar surface area [30, 35]	PSA, PSA_d, TPSA
Van der Waals acceptor surface area which is proportional to E_a[38, 39]	WEASA = $\Sigma(1 \ldots n)k_aE_a$ n = number of acceptor atoms, $k_a = 1/5\frac{1/3}{S_O}$ and S_O = sphere surface with a radius of 1.36 Å (O sp^3)
Van der Waals acceptor surface area which is proportional to C_a[38, 39]	WFEASA = $\Sigma(1 \ldots n)k_aC_a$ n = number of acceptor atoms, $k_a = 1/5\frac{1/3}{S_O}$ and S_O = sphere surface with a radius of 1.36 Å (O sp^3)
Van der Waals donor surface area which is proportional to E_d[38, 39]	WEDSA = $\Sigma(1 \ldots n)k_dE_d^n$ n = number of donor atoms, $k_d = 1/5\frac{1/3}{S_H}$ and S_H = sphere surface with a radius of 1.08 Å (H atom)
Van der Waals donor surface area which is proportional to C_d[38, 39]	WFEDSA = $\Sigma(1 \ldots n)k_dC_d^n$ n = number of donor atoms, $k_d = 1/5\frac{1/3}{S_H}$ and S_H = sphere surface with a radius of 1.08 Å (H atom)
Optimum H-bond enthalpy for three types of complexes [41]	E_m(OHO), E_m(OHN), E_m(NHN)
3D H-bond distance descriptors [45]	HBAA,HBDD,HBAD
SIS of H-bond interactions [45]	SIS+ +,SIS− −,SIS+ −
H-bonding potential [47]	MHBP
Surface area around a molecule where optimum enthalpy of interactions of acceptor atoms with H-bond donor probe is realized [38, 39]	$OEASA_{probe} = \Sigma(1 \ldots n)k_{a(Hd)}E_aE_{d(probe)}$ $E_{d(probe)}$ = enthalpy factor of the probe H-bond donor and $k_a = 1/20\frac{1/3}{S_{rm}}$. S_{rm} is the surface area of sphere with a radius of r_m = 2.45 Å for the strongest H-bonding
Surface area around a molecule where optimum free energy of interactions of acceptor atoms with H-bond donor probe is realized [38, 39]	$OFEASA_{probe} = \Sigma(1 \ldots n)k_{a(Hd)}C_aC_{d(probe)}$ $C_{d(probe)}$ = free energy factor of the probe H-bond donor and $k_a = 1/20\frac{1/3}{S_{rm}}$. S_{rm} is the surface area of sphere with a radius of r_m = 2.45 Å for the strongest H-bonding
Surface area around a molecule where optimum enthalpy of interactions of donor atoms with H-bond acceptor probe is realized [38, 39]	$OEDSA_{probe} = \Sigma(1 \ldots n)k_{a(Hd)}E_aE_{a(probe)}$ $E_{a(probe)}$ = enthalpy factor of the probe H-bond donor
Surface area around a molecule where optimum free energy of interactions of donor atoms with H-bond acceptor probe is realized [38, 39]	$OFEDSA_{probe} = \Sigma(1 \ldots n)k_{a(Hd)}C_aC_{a(probe)}$ $C_{a(probe)}$ = free energy factor of the probe H-bond acceptor
Sum of enthalpy values (kcal m^{-1} Å^{-2}) of interactions between the acceptor atoms in a molecule and donor probe on OEASA [38, 39]	$SIEA_{probe} = \oint Hd(s)$
Sum of enthalpy values (kcal m^{-1} Å^{-2}) of interactions between the donor atoms in a molecule and an acceptor probe on OEDSA [38, 39]	$SIED_{probe} = \oint Hd(s)$

PSA. Hence, we only mention here that the definition of PSA is similar to that of N_a and N_d. Thus, PSA possesses the same disadvantages when compared to thermodynamic H-bond descriptors. In fact, it has been estimated that there is a strong linear relationship between the calculated static PSA and the calculated dynamic polar surface [32, 33]. A further possible simplification using only the number of H-bond forming atoms and PLS statistics was proposed [35]. Excellent correlations of PSA with the number of H-bond donors and acceptors were published [35, 36].

Obviously, PSA, like N_a and N_d, is not a perfect descriptor for transport properties, at least in the framework of the initial definition. Clark [32] first proposed to refine PSA by considering the strength of H-bonds. Then, Van de Waterbeemd stated that "a further refinement in the PSA approach is expected to come from taking into account the strength of the H-bonds, which in principle already is the basis of the HYBOT approach" [37]. This idea was realized in [38, 39], where four new 3D HYBOT surface descriptors related to PSA were proposed including WEASA (van der Waals enthalpy acceptor surface area), WFEASA (Van der Waals free energy acceptor surface area), WEDSA (van der Waals enthalpy donor surface area) and WFEDSA (Van der Waals free energy donor surface area); see also Table 6.3.

6.3.2
SYBYL H-bond Parameters

The Tripos force field considers H-bonds as nondirectional and electrostatic in nature [40]. To accommodate this, calculations in which H-bonds are expected to be important should include partial charges and the electrostatic contributions. H-bond energies are included in the evaluation of the force field by scaling the van der Waals interactions between nitrogen, oxygen and fluorine and hydrogens bonded to nitrogen, oxygen or fluorine. For an alternative treatment, the next formula is applied:

$$E_{\text{H-bonds}} = \sum [C_{ij} / R_{ij}^{12} - D_{ij} / R_{ij}^{10}] \quad (8)$$

where C_{ij} is a coefficient depicting repulsive hydrogen atom–hydrogen acceptor interactions, D_{ij} is a coefficient depicting attractive hydrogen atom–hydrogen acceptor interactions, and R_{ij} is a distance between atoms i and j (Å)

6.3.3
Distance H-bond Potentials

Since the beginning of the 1980s, two different approaches to quantify the H-bond contribution to properties at the 2D and 3D levels developed independently. The carefully parameterized methodology of HYBOT allows one to take into account the influence of substituents on H-bond acceptor and donor strengths. Modern procedures based on X-ray data of ligand–macromolecular complexes consider the distance and angle dependences of H-bonding. Goodford et al. exploited this approach in developing the method GRID [41–43]. This method is parameterized by fitting experimental X-ray data from crystalline complexes, and is designed to calculate the interactions of a probe (a small molecule such as water or ammonia) and a macromolecular system.

The GRID energy is usually computed pairwise between the probe at its grid point and each extended atom of the target, one by one. First versions of the programme used only three energy components for each pairwise energy (E):

$$E = \sum E_{\mathrm{lj}} + \sum E_{\mathrm{el}} + \sum E_{\mathrm{hb}} \tag{9}$$

where ΣE_{lj} is Lennard–Jones, ΣE_{el} is electrostatic and ΣE_{hb} is H-bonding terms, and:

$$E_{\mathrm{hb}} = E_r \times E_t \times E_p \tag{10}$$

E_r, E_t and E_p are functions of r, t and p, respectively (r is the distance between the probe group and an atom in the target, t is the angle made by the H-bond at the target and p is the angle at the probe).

The 8-6 function was adopted, and found to give the most satisfactory results and the closest agreement with experimental observation [42]. This was given by following equations:

$$E_{\mathrm{r}} = C / r^8 - D / r^6 \tag{11}$$

where $C = -3E_{\mathrm{m}}r^8$ (kcal Å^8 mol^{-1}), $D = -4E_{\mathrm{m}}r_{\mathrm{m}}^6$(kcal Å^8 mol^{-1}), r is the separation of the acceptor atom and the donor heavy atom in angstroms, E_{m} is the optimum H-bond energy in kcal mol^{-1} for the particular H-bonding atoms considered and r_{m} is the optimum H-bond length in angstroms for the particular H-bonding atoms considered.

There are three fixed H-bond potentials (E_{m}) in the GRID framework: –4.00 kcal mol^{-1} as an optimum H-bond energy for O—H···O complexes, –2.8 kcal mol^{-1} for O—H···N and N—H···O complexes, and –2.00 kcal mol^{-1} for N—H···N complexes, and three fixed optimum H-bond lengths: 2.8 Å for O···O, 3.0 Å for N···O and 3.2 Å for N···N heavy atoms [42]. Nevertheless, the authors of the approach noted that "Further work will be required, but it is already clear that H-bonds involving ether oxygens may be weaker than those to other oxygen atoms". In a recent publication Goodford already indicated that the H-bond term is calculated on the basis of the same equation as for the Lennard–Jones term [($E_{ij} = (Ad^i – Bd^j)F$, where $i = -12$ and $j = -6$)] "but the constants A and B now have values which depend on the chemical nature of the interacting atoms, and the function F depends on their hybridization and the relative positions of the interacting atoms and their bonded neighbors" [44].

In 1987 Raevsky proposed to describe 3D structure by means of the spectra of interatomic interactions [24]. In this approach each pair of atoms in a molecule gives a line in the spectrum for any type of interaction. A line's position corresponds to the distance between the two atoms while its intensity corresponds to the product of physicochemical parameters associated with those atoms. Atomic vibrations transform lines into bands; thus spectra of interatomic interactions are superpositions of all such bands. The computer program MOLTRA (MOLecular TRansfom Analysis) [45] calculates a set of such spectra including interactions of H-bond acceptors between each other (its intensity corresponds to a product of H-bond acceptor factors calculated by HYBOT), interactions of H-bond donors between each other (its intensity corresponds to a product of H-bond donor factors) and interactions of H-bond acceptors with H-bond donors (its intensity corresponds to a product of H-bond acceptor and donor factors). An example of such spectra for H-bond donors in a set of porphyrins is shown in Fig. 6.3. In principle, each point of such spectra can be used as H-bond acceptor–acceptor (HBAA), H-bond donor–donor (HBDD) and H-bond acceptor–donor (HBAD) distance descriptor. Other valuable 3D H-bond descriptors can be estimated by quantitatively comparing the same type of spectra for all compounds in any training set. Any spectral region and all possible distances may be considered and similarity indices of spectra (SIS+ +, SIS– –, SIS+ –) may be used to construct QSAR for any property or activity [46].

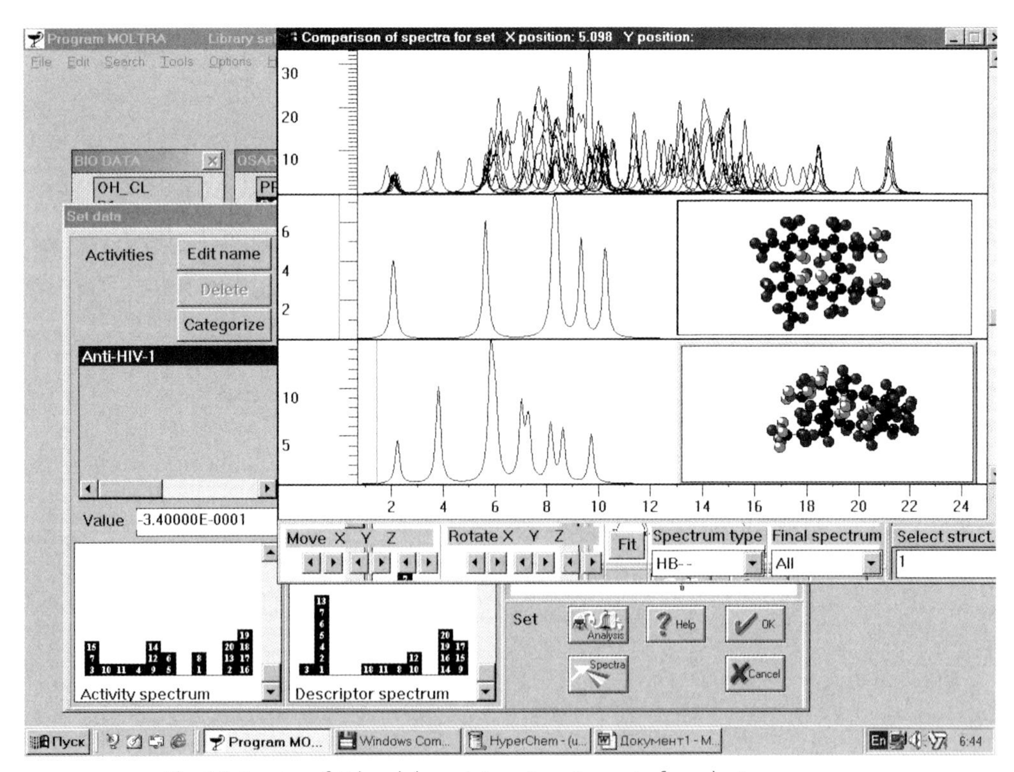

Fig. 6.3 Spectra of H-bond donor interactions in a set of porphyrins.

An attempt to consider the influence of substituents on H-bond potentials was described later in [47]. Here the molecular H-bonding potential (MHBP) is calculated as follows:

$$\mathrm{MHBP_k} = \sum_{i=1}^{N} f_i f_{ct}(D_{ik}) f(U) \tag{12}$$

where k indicates a given point in space, i is a given fragment, N is the total number of fragments in the molecule, f_i is the α or β free energy solvatochromatic parameters of H-bonding of atom i, $f(U)$ is the angular function, f_{ct} is the distance function, D_{ik} is the distance between fragments i and point k. However, the application of this approach is limited by the use of free energy solvatochromatic H-bonding parameters (instead of enthalpy!) and, as in GRID, a fixed parameter of distance function for optimum H-bond potentials.

The development of H-bond potentials on the basis of enthalpy data and distance functions for optimal energies is described in Ref. [48]. The HYBOT program package contains a large database on thermodynamic parameters of H-bond complexes of small organic molecules, including 5984 of the type O—H···O (the range of enthalpy values for this type of complex is 0.9–15.9 kcal mol^{-1}), 3039 of the type O—H···N (0.1–19.1 kcal mol^{-1}), 1016 of the type N—H···O (0.5–10.5 kcal mol^{-1}) and 305 of the type N—H···N (0.5–11.5 kcal mol^{-1}). One can suppose that an optimum arrangement of partners in H-bonding can be realized in cases where small molecules do not have any bulky substituents. Thus, one can presume that the above-mentioned enthalpy interval values correspond to the optimum enthalpy of H-bonding (E_{mi}). Those values depend on the types of functional groups participating in H-bonding and the nature of their substituents.

The wide intervals of enthalpy values for each of those types of H-bonding complexes allow to infer a dependency between energies and distances even in the case of an optimum arrangement of the atoms participating in the H-bond. Hence, to create a realistic platform to quantitatively describe H-bonding, one must recognize that optimum H-bond energies (E_{mi}) depend on optimum H-bond distances (r_{mi}) for the different types of H-bonds found in various complexes. Raevsky and Skvortsov [48] selected specific complexes to assess the relationship between the lengths of H-bonds and their energies. Optimum H-bonding conditions occur when there is a linear arrangement of the heavy donor atom, hydrogen and acceptor atom, and of the acceptor nucleus, electron pair and hydrogen atom. Thus, 58 such "ideal" H-bonding complexes (in which the angles of the donor heavy atom, hydrogen and acceptor atoms were in the range of 173–187°, and angles of the acceptor nucleus, the lone pair of electron and the hydrogen atom were in the range of 170–190°) were selected from the Cambridge Structural Database. There were 13 such O—H···O complexes, 19 O—H···N or N—H···O complexes and 26 N—H···N complexes among the 58 "ideal" representatives. The information about a part of those "ideal" systems is included in Table 6.4.

It is obvious that there are enough large intervals of distances between heavy atoms (D_{ad}) in "ideal" complexes even for the same type of H-bonding (OH···O,

Tab. 6.4 X-ray data, H-bond parameters and optimum energies and distances for few "ideal" complexes [48].

Complex[1]	D_{ad}[2]	D_{ah}[3]	$t_{(H)}$[4]	$p_{(LP)}$[4]	E_a	E_d	*Acceptor*	*Donor*	$E_a \times E_d$	E_m	r_m
DLASPA02	2.542	1.508	179.42	177.97	1.50	−2.63	O	O	−3.93	−5.19	2.59
SIGBEP	2.667	1.719	176.7	175.13	1.24	−2.63	O	O	−3.26	−4.30	2.66
KAPVUS	2.688	2.024	175.55	177.73	1.50	−2.11	O	O	−3.16	−4.17	2.67
CAMALH	2.819	1.814	179.4	173.44	1.50	−1.23	O	O	−1.84	−2.43	2.844
WAZLIS	2.934	1.826	178.42	173.51	1.44	−0.88	O	O	−1.27	−1.67	2.953
BEQVUO	2.62	1.782	175.47	175.78	2.07	−2.74	N	O	−5.67	−7.49	2.578
NEDDOP	2.802	1.984	175.11	177.37	1.99	−2.11	N	O	−4.19	−5.54	2.776
DEPGAG	2.897	1.951	178.25	176.93	1.78	−1.91	O	N	−3.40	−4.49	2.884
ICRERD10	2.946	2.049	179.11	177.42	1.78	−1.83	O	N	−3.27	−4.31	2.903
JICWIB10	3.045	2.146	177.05	175.54	2.06	−1.29	O	N	−2.66	−3.51	2.995
ZUFDEJ	2.745	1.902	178.31	173.48	2.17	−1.87	N	N	−4.06	−5.36	2.741
BARIMZ10	2.871	1.759	175.71	177.46	2.14	−1.83	N	N	−3.92	−5.17	2.772
RIWJIQ	2.896	2.084	173.98	176.16	1.99	−1.83	N	N	−3.66	−4.83	2.828
REMTUY	3.087	2.232	176.69	170.98	1.99	−1.29	N	N	−2.57	−3.39	3.073
GIFZAW01	3.059	2.008	178.57	175.94	1.99	−1.29	N	N	−2.57	−3.39	3.073

1 Complex code in accordance with Ref. [49].
2 D_{ad} is distance between donor heavy and acceptor atoms.
3 D_{ah} is distance between acceptor atom and hydrogen.
4 $t_{(H)}$ and $p_{(LP)}$ are angles in accordance with Ref. [42].

NH···O and NH···O). Because there are significant differences between covalent and van der Waals radii of oxygen and nitrogen atoms, the relationships between optimum energy and optimum distances were sought separately for the three above-mentioned subsets [48]. Those relationships were not expected to be linear because energy values approach zero as distance increases. In that study a sigmoid function was used:

$$E_{mi} = k_1 / (1 + 10^{k_2 + k_3 r} mi) \tag{13}$$

Values of k_1 are fixed and limited by the maximum values for the enthalpies. Reasonably good correlations of E_{mi} with r_{mi} were estimated.

Rearranging Eq. (13) and simplifying the constants, they obtained the following equation was obtained:

$$r_{mi} = k_4 \log[(k_1 - E_{mi})/(E_{mi})] + k_5 \tag{14}$$

Using Eq. (14) it is now possible to calculate r_{mi} from E_{mi} values (see the results of such calculation for 15 "ideal" complexes in Table 6.4). Thus, for each specific pair of atoms participating in an H-bond, the H-bonding potential can be calculated on the basis of Eqs. (3), (11) and (14). Examples of such potentials for O—H···O complexes are presented in Fig. 6.4.

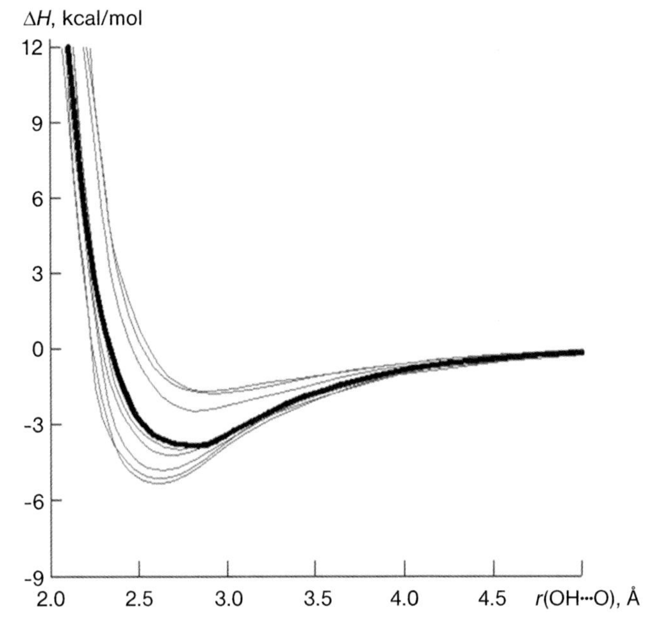

Fig. 6.4 H-bond potentials for OH · · · O complexes. The bold curve corresponds to Goodford's fixed potential for all OH · · · O complexes.

Hence, a scheme to calculate the H-bonding energy of atoms in a 3D arrangement can be carried out by the following steps:

- Use HYBOT's acceptor and donor enthalpy factors to estimate the optimum H-bonding energy via Eq. (3),
- Estimate the optimum H-bonding distance corresponding to the calculated energy optimum via Eq. (14),
- Use Eq. (11) or other functions to reflect distance and angle deviations from "ideal" arrangement of the atoms.

The above given calculation scheme for energies of H-bond complexes was realized in the program 3D HYBOT [50]. This program also calculates 10 3D descriptors including already mentioned descriptors WEASA, WFEASA, WEDSA and WFEDSA, and also additional six descriptors: $OEASA_{probe}$, $OFEASA_{probe}$, $OEDSA_{probe}$, $OFEDSA_{probe}$, $SIEA_{probe}$ and $SIED_{probe}$, which are defined in Table 6.3. The last ones quantitatively characterize surface areas around a molecule where the optimum enthalpy (free energy) of interactions of donor (acceptor) atoms with H-bond acceptor (donor) probe is realized. The HYBOT surface descriptors gave greatly improved descriptions of H-bond capabilities of compounds compared to PSA and ensured better statistical parameters in correlating human intestinal absorption with 154 passively transported drugs [39].

Concluding this section on H-bond potentials, the important role of surrounding water molecules in ligand–protein complex formation deserves mentioning. Binding of the ligand to its specific site will be favored if the energy of H-bonds in the complex and the entropy gain in realizing some bound water molecules are more favorable than the free energy contribution of the H-bonds between the binding partners, in their free state, and these water molecules [1]. Due to its vital importance for any area of human activity, there were many attempts to study structure of liquid water. In spite of the apparent simplicity of the water molecule, "liquid water is one of the most mysterious substances in our world" [51]. Up to the present time experimental data on liquid and solid water were interpreted on the basis of a water model with *four strong hydrogen bridge bonds*, two of them being formed by two lone pair electrons of oxygen and two coordinated hydrogen atoms of two water molecules in the first hydration shell. However, recently Wernet et al. studied the first hydration shell of a water molecule in liquid water by probing its electronic structure using X-ray absorption spectroscopy and X-ray Raman scattering [52]. They concluded that liquid "water consists mainly of structures with only *two strong H-bonds*, one donating and one accepting". This result "nonetheless implies that most molecules are arranged in strongly H-bond chains or rings embedded *in a disordered cluster network connected mainly by weak H-bonds*".

6.4
Application of H-bond Descriptors in QSAR Studies and Drug Design

H-bonding is an important, but not the sole, interatomic interaction. Thus, total energy is usually calculated as the sum of steric, electrostatic, H-bonding and other components of interatomic interactions. A similar situation holds with QSAR studies of any property (activity) where H-bond parameters are used in combination with other descriptors. For example, five molecular descriptors are applied in the solvation equation of Kamlet–Taft–Abraham: excess of molecular refraction (R_2), which models dispersion force interactions arising from the polarizability of π- and n-electrons; the solute polarity/polarizability (π_2^H) due to solute–solvent interactions between bond dipoles and induced dipoles; overall or summation H-bond acidity ($\Sigma\alpha_2^H$); overall or summation H-bond basicity ($\Sigma\beta_2^H$); and McGowan volume (V_x) [53]:

$$\log \mathrm{SP} = c + rR_2 + s\pi_2^H + a\sum \alpha_2^H + b\sum \beta_2^H + vV_x \tag{15}$$

In the framework of the mobile order and disorder (MOD) theory five components contribute most to the Gibbs free energy of partitioning of a solute in a biphasic system of two essentially immiscible solvents [23]:

$$\log P = \Delta B + \Delta D + \Delta F + \Delta \mathrm{O} + \Delta \mathrm{OH} \tag{16}$$

where the entropy of missing term, ΔB, informs about differences between the two phases in the entropy of the solute/solvent exchange; the hydrophobic effect term, ΔF, accounts for differences in the propensities of the solvent phases to squeeze the solute out of the solution; two H-bond interaction terms, ΔO and ΔOH, express differences in the strengths of the H-bonds that bind the solute and solvent molecules in each phase; and the term ΔD is similar to the two previous ones, but accounts for nonspecific forces only.

The thermodynamic approach followed by Raevsky considers the property P to be based on contributions from three main intermolecular interactions: steric, electrostatic and H-bonding [38]:

$$P = f\left(\alpha, \sum q, \sum C\right) \tag{17}$$

where α is molecular polarizability (a volume-related term), Σq is a sum of partial atomic charges (an electrostatics-related term) in a molecule and ΣC comprises free energy H-bonding factors.

6.4.1 Solubility and Partitioning of Chemicals in Water–Solvent–Gas Systems

The solubility of chemicals, drugs or pollutants in water (S_w), in octanol (S_o), their saturation concentration in air (C_{air}), as well as their partitioning in the corresponding two-phase systems [octanol–water ($P_{o/w} = C_o/C_w$), air–water ($P_{air/w} = C_{air}/C_w$) and air–octanol ($P_{air/o} = C_{air}/C_o$)] are important physicochemical parameters in medicinal chemistry and in environmental research. The following correlations of those properties with HYBOT descriptors have been published recently [54–58]:

$$\log P_{o/w} = 0.267(\pm 0.008)\alpha - 1.00(\pm 0.02)\sum C_a \tag{18}$$

$$n = 2850,\ r^2 = 0.94,\ s = 0.23$$

$$\log P_{air/w} = 0.032(\pm 0.008)\alpha - 1.63(\pm 0.07)\sum C_a + 1.04(\pm 0.07)\sum C_d \tag{19}$$

$$n = 322,\ r^2 = 0.91,\ s = 0.65$$

$$\log P_{air/o} = -0.258(\pm 0.017)\alpha - 0.43(\pm 0.16)\sum C_a + 0.73(\pm 0.19)\sum C_d \tag{20}$$

$$n = 98,\ r^2 = 0.86,\ s = 0.61$$

$$\log S_w = 0.53(\pm 0.11) - 0.275(\pm 0.008)\alpha + 0.90(\pm 0.05)\sum C_a - 0.33(\pm 0.06)\sum C_d \tag{21}$$

$$n = 569,\ r^2 = 0.89,\ s = 0.49$$

$$\log S_o = 1.06(\pm 0.15) - 0.063(\pm 0.006)\alpha + 0.03(\pm 0.03)\sum C_a - 0.14(\pm 0.04)\sum C_d \tag{22}$$

$$n = 23,\ r^2 = 0.92,\ s = 0.16$$

$$\log C_{\text{air}} = 0.36(\pm 0.25) - 0.257(\pm 0.014)\alpha - 0.33(\pm 0.11)\sum C_{\text{a}} + 0.78(\pm 0.04)\sum C_{\text{d}} \quad (23)$$

$n = 90, r^2 = 0.90, s = 0.40$

Thus, there is a significant influence of H-bond capability on partitioning and solubility of chemicals. From Eqs. 18–23, we can conclude that partitioning in the system octanol–water is the result of the competing volume-related term (expressed by α) and the molecular H-bond acceptor ability (expressed by HYBOT's ΣC_{a}). Partitioning in the air–water system almost completely depends on the H-bond acceptor and donor abilities of chemicals. H-bond ability contributes negatively to partitioning in the air–octanol system. H-bond strength strongly increases the concentration in water and decreases the concentration in air.

Partitioning in the octanol–water system was characterized by the solvation equation [59]:

$$\log P_{\text{o/w}} = 0.088 + 0.562E - 1.054S - 0.032A - 3.460B + 3.814V \quad (24)$$

$n = 813, r^2 = 995, s = 0.12, F = 23161.6$

Statistical criteria of Eq. (24) are too good; the standard deviation, which was created on the basis of different measurements by various authors, is much less than even the experimental error of determination. This could be due to mutual intercorrelation of descriptors "leading to over-optimistic statistics" [18]. Another reason may be the lack of diversity in the training set. The application of the solvation equation to data extracted from the MEDchem97 database gave much more modest results: n = 8844, r^2 = 0.83, root mean square error = 0.674, F = 8416 [60].

An amended solvation energy relationship was used for correlation of solubility of compounds in water [61]:

$$\log S_{\text{w}} = 0.518 - 1.004R_2 + 0.771\pi_2^{\text{H}} + 2.168\sum \alpha_2^{\text{H}} + 4.238\sum \beta_2^{\text{H}} - 3.262\sum \alpha_2^{\text{H}} \sum \beta_2^{\text{H}} - 3.967\nu V_x \quad (25)$$

$n = 659, r^2 = 92, s = 0.56, F = 1256$

The excellent correlation between calculated and experimental $\log P$ values was obtained by vast investigations of the partitioning of simple chemicals in different mutually immiscible two-phase liquid systems by means of universal model based on the MOD theory [23]:

$$\log P_{\text{calc}} = 0.993(\pm 0.005) \log P_{\text{exp}} \quad (26)$$

$n = 2207, r^2 = 0.94, s = 0.55$

The author of this approach indicated that partitioning incorporates two major factors, namely, a "bulk or volume" component ($\Delta B + \Delta F$) favoring lipophilicity and a "solvation" component ($\Delta D + \Delta O + \Delta OH$) opposing lipophilicity. From this point of view the MOD model, the "solvation equation" and the HYBOT approach do not differ fundamentally.

6.4.2 Permeability and Absorption in Humans

Calculated molecular descriptors including H-bond parameters were used for QSAR studies on different types of permeability. For example, the new H-bond descriptor ΣC_{ad}, characterizing the total H-bond ability of a compound, was successfully applied to model Caco-2 cell permeability of 17 drugs [30]. A similar study on human jejunal *in vivo* permeability of 22 structurally diverse compounds is described in Ref. [62]. An excellent one-parameter correlation of human red cell basal permeability (BP) was obtained using the H-bond donor strength [63]:

$$\log \mathrm{BP} = -0.70(\pm 0.64) + 1.08(\pm 0.16)\sum C_d \quad (27)$$

$$n = 10, r^2 = 0.97, r_{cv}^2 = 0.95, s = 0.43$$

A quantitative description of human skin permeability (k_p) based on H-bond donor and acceptor factors was obtained with 22 alcohols and steroids [63]:

$$\log k_p = -4.88(\pm 0.48) + 0.23(\pm 0.17)\sum C_a - 0.31(\pm 0.10)\sum C_d \quad (28)$$

$$n = 22, r^2 = 0.83, r_{cv}^2 = 0.79, s = 0.50$$

QSAR studies of the pH-dependent partitioning of acidic and basic drugs into liposomes [64] yielded following equations:

$$\log P_{ui} = 3.22 + 0.076\alpha - 0.216\sum C_{aui} - 0.91\sum q_{ui}^{+} \quad (29)$$

$$\log P_{(-)} = 1.48 + 0.076\alpha - 1.32\sum q_{(-)}^{+} + 0.97 Ip \quad (30)$$

$$\log P_{(+)} = 2.39 + 0.076\alpha - 0.092\sum C_{a(+)} - 0.93\sum q_{(+)}^{+} \quad (31)$$

$$n = 54, R = 0.924, s = 0.36, Q = 0.889$$

For 31 passively transported drugs, excellent sigmoidal relationships were found between human intestinal absorption and their H-bond acceptor and donor factors [65]:

$$\mathrm{FA} = 1/(1 + 10^{-[5.02 - 0.31\Sigma Cad]}) \quad (32)$$

$$n = 31, r^2 = 0.89, s = 0.12$$

$$\mathrm{FA} = 1/(1 + 10^{-[5.05 - 0.36\Sigma Ca + 0.26\Sigma Cd]}) \quad (33)$$

$$n = 31, r^2 = 0.95, s = 0.09$$

A volume-related term (expressed by polarizability) and electrostatics (expressed by partial atomic charge) made minor contributions to intestinal absorption in humans. Lipophilicity, expressed by $\log P$ or $\log D$ values, shows no correlation with the human absorption data. Recently, similar results were obtained for 154 passively transported drugs on the basis of surface thermodynamics descriptors [39]:

$$\mathrm{FA} = 1/(1+10^{-[2.2-0.016(\mathrm{OFEASA}+\mathrm{OFEDSA})]}) \quad (34)$$

$$n = 154, r^2 = 0.83, s = 0.15$$

The application of the "solvation equation" for the effective rate of absorption ($\log \mathrm{GI}\ k_{\mathrm{eff}}$) gave rather modest results [66]:

$$\log \mathrm{GI}k_{\mathrm{eff}} = 0.544 - 0.025E + 0.141S - 0.409A - 0.513B + 0.204V \quad (35)$$

$$n = 127, r^2 = 0.79, s = 0.29, F = 0.84$$

The influence of physicochemical properties, including lipophilicity, H-bonding capacity and molecular size and shape descriptors on brain uptake has been investigated using a selection of 45 known CNS-active and 80 CNS-inactive drugs [67]. A combination of a H-bonding and a molecular size descriptor, i.e. the major components of lipophilicity and permeability, avoiding knowledge of distribution coefficients, is proposed to estimate the BBB penetration potential of new drug candidates.

The permeability of "drug-like" chemicals and real drugs was comprehensively studied by Lipinski [68]. Four databases which contain the information about many thousands of chemicals including compounds from the Derwent World Drug Index (WDI), International Non-Proprietary Names (INN), US Adopted Names (USAN), New Chemical Entities (NCE) and New Drugs were used to compare ADME properties. The graphical comparison (Fig. 6.5a [68]) of ΣC_{ad} with the fraction of chemicals within a given database with (ΣC_{a} + ΣC_{d}) up to a certain value of ΣC_{ad} permitted to conclude that "the Raevsky sum as a global measure of the influence of H-bond donor and acceptor groups on permeability shows the very large difference and hence likely poorer permeability between the newer drugs and the older drugs". Figure 6.5(b, prepared by the author of this chapter) shows for passively transported drugs a plot of the fraction absorbed in humans (FA) against their ΣC_{ad} values. The comparison of the two diagrams convincingly confirms the possibility to use thermodynamic H-bond acceptor and donor factors for permeability estimation. It is obvious from the right diagram that drugs having $\Sigma C_{\mathrm{ad}} \leq 15.00$ are well absorbed compounds. Thus, there are about 70% such compounds among INN/USAN drugs, 65% among WDI, 60% among NCE and only 40% among New Drugs.

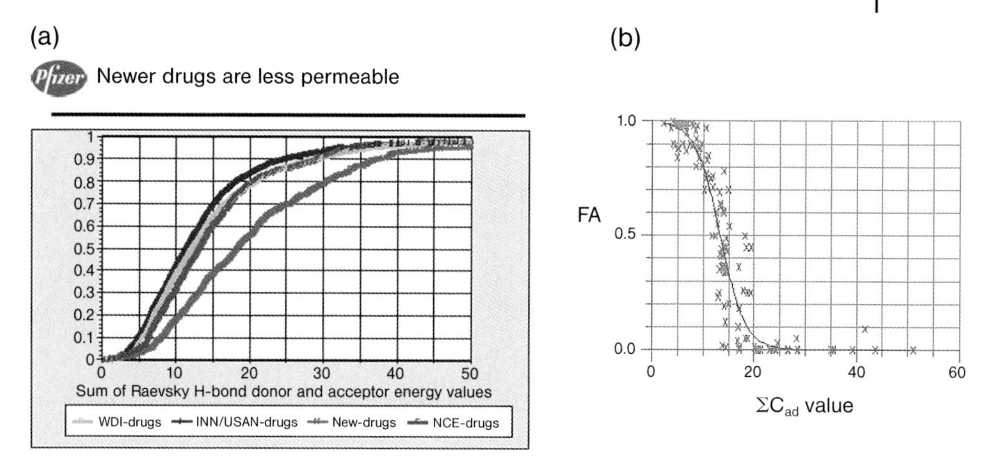

Fig. 6.5 (a) Plot of the fraction of compounds within a database with "$\Sigma C_{ad} \leq$ values shown on the *x*-axis" against ΣC_{ad} (e.g. 40% of "New Drugs" have $\Sigma C_{ad} \leq 15$) [68]. (b) Fraction of a compound absorbed (FA) in humans plotted against the compounds' ΣC_{ad} value.

6.4.3
Classification of Pharmacokinetic Properties in Computer-aided Selection of Useful Compounds

The Biopharmaceutic Classification System (BCS) [69] separates drugs into four different classes depending on their solubility and permeability: class 1 (high solubility, high permeability), class 2 (low solubility, high permeability), class 3 (high solubility, low permeability) and class 4 (low solubility, low permeability). The rate-limiting step to drug absorption will vary according to the class to which the drug belongs. For a class 2 drug, the limiting step is dissolution; permeability plays only a minor role. On the other hand, for a class 3 drug, permeability is rate-limiting while dissolution has very little influence on absorption [70].

Of course, class 4 is not valuable in medicinal chemistry. Such compounds have to be excluded from drug discovery processes as early as possible. At present, there are computer "alert" programs based on the "Rule-of-5" or similar approaches that are used in preliminary screening to select and exclude compounds of class 4 [71]. Van de Waterbeemd indicated in 1998 that the four BCS classes of drugs can be determined solely by considering physicochemical descriptors such as molecular weight and PSA [72]. However, as mentioned in this chapter, those descriptors are too crude for the quantitative description of molecular size and H-bonding ability.

For this purpose, we decided to apply molecular polarizability and H-bond factors, as calculated by HYBOT, to published water solubility and oral absorption data of 254 drugs. Among them, 156 had a solubility $>50\,\mu g\,mL^{-1}$ and FAs >0.50; they were assigned to class 1. The 74 drugs with a solubility $<50\,\mu g\,mL^{-1}$ and FAs >0.50 were grouped in class 2. The 17 drugs with a solubility $>50\,\mu g\,mL^{-1}$ but FAs

Tab. 6.5 Biopharmaceutics classification for 254 Drugs on the basis of HYBOT descriptors.

		Biopharmaceutical class			
		1 (S > 50 μg mL⁻¹; FA > 0.50)	*2 (S < 50 μg mL⁻¹; FA > 0.50)*	*3 (S > 50 μg mL⁻¹; FA < 0.50)*	*4 (S < 50 μg mL⁻¹; FA < 0.50)*
Drugs number		156	74	17	7
Percent		61.4	29.1	6.7	2.8
Correct recognition	solubility	115 (73.7%)	61 (83.4%)	16 (94.1%)	5 (74.1%)
	FA	135 (86.5%)	69 (93.2%)	13 (76.5%)	3 (42.9%)

<0.5 were placed in class 3, while the 7 drugs with a solubility $<50\,\mu g\ mL^{-1}$ and FAs <0.50 were put in class 4.

For a sufficient discrimination of highly soluble from poorly soluble drugs, a combination of two descriptors was needed: α and ΣC_a. Out of the 173 drugs in class 1 and class 3 (i.e. compounds with $S \geq 50\,\mu g\,mL^{-1}$), 131 drugs (about 76%) fulfilled the condition of $0.25\,\alpha - \Sigma C_a \leq 2.0$. In contrast, of the 81 drugs in classes 2 and 4 (i.e. those with $S < 50\,\mu g\,mL^{-1}$), 66 (about 82%) fulfilled the condition of $0.25\alpha - \Sigma C_a > 2.0$.

The most important parameter in classifying absorption is ΣC_{ad}. Of the 230 drugs in classes 1 and 2 (those having FA ≥ 0.50), 204 (about 89%) have sums of H-bond acceptor and donor factors of less than 15.00. On the basis of the three descriptors considered, the probability of correct classification is about 75%. Biopharmaceutical classification based on HYBOT descriptors is presented in Table 6.5. In our previous publications, for solubility and for absorption, we indicated that there were few significant deviations between the calculated values and those observed as estimated by QSAR models. For example, the presence of C(O)NH and SO_2NH groups in drugs significantly decreases its solubility. Drugs containing a $PhCHNH_3^+$ group in β-lactams can be actively transported and so move from class 3 to class 1 or from class 4 to class 2. It is possible that a biopharmaceutical classification of drugs could be made more correctly by combining physicochemical descriptors with empirical rules related to structural fragments.

An analysis of the solubility and absorption rates for the 254 drugs considered here shows that the 25 compounds fulfilling the condition of $0.25\alpha - \Sigma C_a > 5.0$ and $\Sigma C_{ad} > 20.0$ have solubility of only a few micrograms per milliliter, and are absorbed at the level of only a few percents. Such properties are too poor for drug development, so these parameters can be useful as an "alert" in computer-aided compound selection.

6.4.4 Chemical Interactions with Biological Targets

Beyond successful modeling of important physicochemical drug properties, there exist several examples in the literature that document the prime impact of H-bond descriptors for pharmacodynamic drug properties. These include, for example, the affinities (K_i) to muscarinic receptors [27], the inhibition of phosphorylation of polyGAT by α-substituted benzilidenemalononitrile-5-*S*-aryltyrphostins [46], as well as the inhibition of dihydrofolate reductase by 4,6-diamino-1,2-dehydro-2,2-dimethyl-1-(*p*-phenyl)-*S*-triazines [46].

6.4.5
Aquatic Toxicity

Usually aquatic toxicity of chemicals with general narcosis mechanism of action is described by the octanol/water partition coefficient [73]. However, $\log P_{o/w}$ is a composite descriptor which has components of molecular volume and H-bond acceptor terms. Raevsky and Dearden [74] therefore used molecular polarizability (as a volume-related term) and the H-bond acceptor factor instead of $\log P_{o/w}$ to model aquatic toxicity ($\log LC_{50}$) to the guppy for 90 chemicals with general narcosis mechanisms. This excellent correlation has statistical criteria better than that obtained for the same data using $\log P_{o/w}$:

$$\log LC_{50} = 5.14(\pm 0.12) - 0.259(\pm 0.008)\alpha + 0.79(\pm 0.03)\sum C_a \quad (36)$$

$$n = 90, r^2 = 0.95, s = 0.32$$

6.5
Conclusions

It was indicated in a remarkable publication of Kubinyi in 2001 that "despite of all attempts to arrive at a better understanding of the role of water and of H-bonds in biological systems and of all the individual enthalpy and entropy terms that are involved in disolvation, H-bond formation, and hydrophobic interactions we are far from a satisfactory situation" [1]. One of the main reasons for this situation is the ubiquitous application of indirect and/or indicator parameters of H-bonding processes in QSAR studies and drug design up to recent times. Studies based on direct thermodynamic parameters of H-bonding and exact 3D structures of H-bonding complexes have essentially improved our understanding of complex processes of solvation and specific intermolecular interactions. These studies consider the structure of liquid water, new X-ray data for specific H-bonding complexes, quantitative estimation of contribution of H-bond acceptor and donor factors and volume-related terms in chemicals solvation processes, partitioning in water–solvent–air systems, a refinement in the PSA approach, improvement of GRID potentials, and calculation schemes of optimum H-bonding potential values for any concrete H-bonding atoms in any complexes which consider the nature of interacting atoms and the influence of substituents. These developments ensure real quantitative description of H-bonding and the successful application of direct H-bonding descriptors in QSAR and drug design.

Acknowledgments

This investigation was supported by the International Science & Technology Center (project 888) and the Russian Ministry of Education & Sciences (contract 02.434.11.1014). The author is very grateful to Drs. J. W. McFarland (Reckon, USA), H. Van de Waterbeemd (AstraZeneca, UK) and K.-J. Schaper (Borstel Research Center, Germany) for collaboration and very useful discussions, and Professor Dr. R. Mannhold for valuable editorial remarks and advice.

References

1 Kubinyi, H. Hydrogen bonding, the last mystery in drug design? In *Pharmacokinetic Optimization in Drug Research*, Testa, B., Van de Waterbeemd, H., Folkers, G., Guy, R. (eds.), Wiley-VCH, Weinheim and VHCA, Zurich, **2001**, pp. 513–524.

2 Testa, B., Kier, L. B. The concept of molecular structure in structure–activity relationship studies and drug design. *Med. Res. Rev.* **1991**, *11*, 35–48.

3 Raevsky, O. A. Quantification of non-covalent interactions on the basis of the thermodynamic hydrogen bond parameters. *J. Phys. Org. Chem.* **1997**, *10*, 405–413.

4 Dearden, J. C., Ghafourian, T. Hydrogen bonding parameters for QSAR: comparison of indicator variables, hydrogen bond counts, molecular orbital and other parameters. *Chem. Inf. Comput. Sci.* **1999**, *39*, 231–235.

5 Winiwarter, S., Ax, F., Lennernäs, H., Hallberg, A., Pettersson, C., Karlen, A. Hydrogen bonding descriptors in the prediction of human *in vivo* intestinal permeability. *J. Mol. Graph. Model.* **2003**, *21*, 273–287.

6 Seiler, P. Interconversion of lipophilicities from hydrocarbon/water into the octanol/water system. *Eur. J. Med. Chem.* **1974**, *9*, 473–479.

7 Moriguchi, I. Quantitative structure–activity studies I. Parameters relating to hydrophobicity. *Chem. Pharm. Bull.* **1975**, *23*, 247–257.

8 Wilson, L. Y., Famini, G. R. Using theoretical descriptors in quantitative structure–activity relationships: some toxicological indices. *J. Med. Chem.* **1991**, *34*, 1668–1674.

9 Dearden, J. C., Ghafourian, T. Investigation of calculated hydrogen bond parameters for QSAR. In *Computational Tools and Biological Applications*, Sanz, F., Giraldo, J., Manaut, F. (eds.), Prous, Barcelona, **1995**, pp. 117–119.

10 Dearden, J. C., Cronin, M. T. D., Wee, D. Prediction of hydrogen bond donor ability using new quantum chemical parameters. *J. Pharm. Pharmacol.* **1997**, *49* (*Suppl. 4*), 110.

11 Gancia, E., Montana, J. G., Manallack, D. T. Theoretical hydrogen bonding parameters for drug design. *J. Mol. Graph. Model.* **2001**, *18*, 349–362.

12 Murray, J. S., Politzer, P. Relationships between solute hydrogen-bond acidity/basicity and the calculated electrostatic potential. *J. Chem. Res. (S)* **1992**, 110–114.

13 Bakken, G. A., Jurs, P. C. Prediction of hydroxyl radical rate constants from molecular structure. *Chem. Inf. Comput. Sci.* **1999**, *39*, 1064–1075.

14 Katritzky, A. R., Tamm, T., Wang, Y., Sild, S., Karelson, M. QSPR treatment of solvent scales. *Chem. Inf. Comput. Sci.* **1999**, *39*, 684–691.

15 Fujita, T., Nishioka, T., Nakajima, M. Hydrogen-bonding parameter and its significance in quantitative structure–activity studies. *J. Med. Chem.* **1977**, *20*, 1071–1081.

16 Charton, M., Charton, B. I. The structural dependence of amino acid hydrophobicity parameters. *J. Theor. Biol.* **1982**, *99*, 629–644.

17 Lipinski, A. A., Lombardo, F., Deminy, B. W., Feeney, P. J. Experimental and computational approaches to estimate solubility and permeability in drug discovery and development settings. *Adv. Drug Deliv. Rev.* **1997**, *23*, 3–25.

18 Van de Waterbeemd, H., Smith, D. A., Beaumont, K., Walker, D. K. Property-based design: optimization of drug absorption and pharmacokinetics. *J. Med. Chem.* **2001**, *44*, 1313–1333.

19 Taft, R. W., Shuely, W. J., Doherty, R. H., Kamlet, M. J. Linear solvation relationships. 39. A multiple parameter equation for β values of (H-bond acceptor basicities) of the XYZP=O compounds. *J. Org. Chem.* **1988**, *53*, 1737–1741.

20 Yokono, S., Shieh, D. D., Goto, H, Arakawa, K. Hydrogen bonding and anesthetic potency. *J. Med. Chem.* **1982**, *25*, 873–876.

21 Abraham, M. H., Duce, P. P., Prior, D. V., Barratt, D. G., Morris, J. J., Taylor, P. J. Hydrogen bonding. Part 9. Solute proton donor and proton acceptor scales for use in drug design. *J. Chem. Soc. Perkin Trans.* **1989**, *2*, 1355–1375.

22 Abraham, M. H., Grellier, P. I., Prior, D. V., Taft, R. W., Morris, J. J., Taylor, P. J., Laurence, C., Berthelor, M., Doherty, R. M., Kamlet, M. G., About, J. I., Straidi, K., Gutheneuf, G. General treatment of hydrogen bond complexation constants in tetrachloromethane. *J. Am. Chem. Soc.* **1988**, *110*, 8634–8536.

23 Ruelle, P. Universal model based on the mobile order and disorder theory for predicting lipophilicity and partition coefficients in all mutually immiscible two-phase liquid systems. *Chem. Inf. Comput. Sci.* **2000**, *40*, 681–700.

24 Raevsky, O. A. QSAR description of molecular structure. In *QSAR in Drug Design and Toxicology*, Hadzi, D., Jerman-Blazic, B. (eds.), Elsevier, Amsterdam, **1987**, pp. 31–36.

25 Raevsky, O. A., Grigor'ev, V. Ju., Solov'ev, V. P., Kireev, D. B., Sapegin, A. M., Zefirov, N. S. Drug design H-bonding scale. In *QSAR: Rational Approaches in the Design of Bioactive Compounds*, Silipo, C., Vittoria, A. (eds.), Elsevier, Amsterdam, **1991**, pp. 135–138.

26 Raevsky, O. A., Grigor'ev, V. Ju., Kireev, D., Zefirov, N. S. Complete Thermodynamic description of H-bonding in the framework of multiplicative approach. *Quant. Struct.-Act. Relat.* **1992**, *11*, 49–64.

27 Raevsky, O. A. Hydrogen bond strength estimation by means of HYBOT. In *Computer-Assisted Lead Finding and Optimization*, Van de Waterbeemd, H., Testa, B., Folkers, G. (eds.), Wiley-VCH, Weinheim, **1997**, pp. 367–378.

28 Raevsky, O. A., Grigor'ev, V. Ju., Trepalin, S. V. HYBOT (Hydrogen Bond Thermodynamics) program package. Registration by Russian State Patent Agency N 990090 of 26 February **1999**.

29 Raevsky, O. A., Skvortsov, V. S., Grigor'ev, V. Ju., Trepalin, S. V. HYBOT In UNIX/LINUX. Registration by Russian State Patent Agency N 2002610496 of 5 February **2002**.

30 Van de Waterbeemd, H., Camenisch, G., Folkers, G., Raevsky, O. Estimation of Caco-2 cell permeability using calculated molecular descriptors. *Quant. Struct.-Act. Relat.* **1996**, *15*, 480–490.

31 Palm, K., Stenberg, P., Luthmann, K., Artursson, P. Polar molecular surface properties predict the intestinal absorption of drugs in human. *Pharm. Res.* **1997**, *14*, 568–571.

32 Clark, D. E. Rapid calculation of polar molecular surface area and its application in the prediction of transport phenomena. 1. Prediction of intestinal absorption. *J. Pharm. Sci.* **1999**, *88*, 807–814.

33 Clark, D. E. Rapid calculation of polar molecular surface area and its application in the prediction of transport phenomena. 2. Prediction of blood–brain barrier penetration. *J. Pharm. Sci.* **1999**, *88*, 815–821.

34 Clark, D. E., Pickett, S. D. Computational methods for prediction of "drug-likeness". *Drug Discov. Today* **2000**, *5*, 49–58.

35 Osterberg, T., Norinder, U. Prediction of polar surface area and drug transport processes using simple parameters and PLS statistics, *J. Chem. Inf. Comput. Sci.* **2000**, *40*, 1408–1414.

36 Norinder, U., Haeberlein, M. Calculated molecular properties and multivariate statistical analysis in absorption prediction.

In *Drug Bioavailability, Estimation of Solubility, Permeability, Absorption and Bioavailability (Methods and Principles in Medicinal Chemistry)*, Van der Waterbeemd, H., Lennernäs, H., Artursson, P. (eds.), Wiley-VCH, Weinheim, **2003**, pp. 358–405.

37 Van de Waterbeemd, H. Physico-chemical approaches to drug absorption. In *Drug Bioavailability Estimation of Solubility, Permeability, Absorption and Bioavailability (Methods and Principles in Medicinal Chemistry)*, Van der Waterbeemd, H., Lennernäs, H., Artursson, P. (eds.), Wiley-VCH, Weinheim, **2003**, pp. 1–20.

38 Raevsky, O. A. Physicochemical descriptors in property-based drug design. *Minirev. Med. Chem.* **2004**, *4*, 1041–1052.

39 Raevsky, O. A., Skvortsov, V. S. Quantifying hydrogen bonding in QSAR and molecular modeling. *SAR QSAR Environ. Res.* **2005**, *12*, 1–14.

40 Sybyl 6.7, Tripos Inc., St Louis, MO, USA, http://www.tripos.com.

41 Goodford, P. J. A computational procedure for determining energetically favorable binding sites on biologically important macromolecules. *J. Med. Chem.* **1985**, *28*, 849–857.

42 Boobbyer, D. N. A., Goodford, P. J., McWhinnie, P. M., Wade, R. C. New hydrogen bond potentials for use in determining energetically favorable binding sites in molecules of known structure. *J. Med. Chem.* **1989**, *32*, 1083–1094.

43 Wade, R. C., Clark, K., Goodford, P. J. Further development of hydrogen bond functions for use in determining energetically favorable binding sites on molecules of known structure. 1. Ligand probe groups with the ability to form two hydrogen bonds. *J. Med. Chem.* **1993**, *36*, 140–146.

44 Goodford, P. J. The basic principles of GRID. In *Molecular Interaction Fields: Application in Drug Discovery and ADME Prediction (Methods and Principles in Medicinal Chemistry)*, Cruciani, G., Mannhold, R. Kubinyi, H., Folkers, G. (eds.), Wiley-VCH, Weinheim, **2005**, pp. 1–25.

45 Raevsky, O. A., Trepalin, S. V., Rasdolsky, A. N. The program package MOLTRA (MOLecular Transform Analysis). Registration by Russian State Patent Agency N 990092 of 26 February **1999**.

46 Raevsky, O. A., Schaper, K.-J., Van de Waterbeemd, H., McFarland, J. Hydrogen bond contribution to properties and activities of chemicals and drugs. In *Molecular Modeling and Prediction of Bioactivity*, Gundertofte, K., Jørgensen, F. S. (eds.), Kluwer/Plenum, New York, **2000**, pp. 221–227.

47 Caron, G., Rey, S., Ermondi, G., Crivori, P., Guillard, P., Carrupt, P.-A., Testa, B. Molecular hydrogen-bonding potentials (MHBPs) in structure–permeation relations. In *Pharmacokinetic Optimization in Drug Research; Biological, Physicochemical, and Computational Strategies*, Testa, B., Van de Waterbeemd, H., Folkers, G., Guy, R. (eds.), Wiley-VCH, Weinheim, **2001**, pp. 513–524.

48 Raevsky, O. A., Skvortsov, V. S. 3D hydrogen bond thermodynamics (HYBOT) potentials in molecular modeling. *J. Comput.-Aided Mol. Des.* **2002**, *16*, 1–10.

49 Cambridge Structural Database, version 5.29, **2000**.

50 Raevsky, O. A., Skvortsov, V. S., Grigor'ev, V. Ju., Trepalin, S. V. 3D HYBOT. Registration by Russian State Patent Agency N 2004612207 of 27 September **2004**.

51 Zubavicus, Y., Grunze, M. New insights into the structure of water with ultrafast probes. *Science* **2004**, *304*, 974–976.

52 Wernet, Ph., Nordlund, D., Bergmann, U., Cavalleri, M., Odelius, M., Ogasawara, H., Naslund, L. A., Hirsch, T. K., Ojamae, L., Glatzel, P., Pettersson, L. G. M., Nilsson, A. The structure of the first coordination shell in liquid water. *Science* **2004**, *304*, 995–999.

53 Abraham, M. H., Chadha, H. S. Application of a Solvation Equation to Drug Transport Properties In *Lipophilicity in Drug Action and Toxicology*, Pliska, V., Testa, B., Van de Waterbeemd, H. (eds.), VCH, Weinheim, **1996**, pp. 311–337.

54 Raevsky, O. A., Schaper, K.-J. Analysis of water solubility data on the basis of HYBOT descriptors. Part 1. Partitioning of volatile chemicals in the water–gas phase system. *QSAR Comb. Sci.* **2003**, *22*, 926–942.

55 Raevsky, O. A., Grigor'ev, V. Ju., Raevskaja, O. E., Schaper, K.-J. Physicochemical properties/descriptors governing the solubility and partitioning in water–solvent–gas systems. Part 1. Partitioning between octanol and air. *SAR QSAR Environ. Res.* **2006**, *17*, 285–297.

56 Schaper, K.-J., Kunz, B., Raevsky, O. A. Analysis of water solubility data on the basis of HYBOT descriptors. Part 2. Solubility of liquid chemicals and drugs. *QSAR Comb. Sci.* **2003**, *22*, 943–958.

57 Raevsky, O. A., Schaper, K.-J. Physicochemical descriptors governing the solubility and partitioning of chemicals in water–solvent–gas systems. In *Abstracts of 12th International Workshop on Quantitative Structure–Activity Relationships in Environmental Toxicology*, Lyon, France, **2006**, p. 23.

58 Raevskaja, O. E., Raevsky, O. A., Schaper, K.-J. QSPRs for vapour pressure and concentration of chemicals above saturated aqueous solutions. In *Abstracts of 12th International Workshop on Quantitative Structure–Activity Relationships in Environmental Toxicology*, Lyon, France, **2006**, p. 91.

59 Abraham, M. H., Chadha, H. S., Whiting, G. S., Michell, R. C. Hydrogen bonding. 32. An analysis of water–octanol and water–alkane partitioning and logp parameter of Seiler. *J. Pharm. Sci.* **1994**, *83*, 1085–1100.

60 Platts, J. A., Abraham, M. H., Butina, D., Hersey, A. Estimation of molecular linear free energy relationship descriptors by group contribution approach. 2. Prediction of partition coefficient. *J. Chem. Inf. Comput. Sci.* **2000**, *40*, 71–80.

61 Abraham, N., Lee, J. The correlation and prediction of solubility of compounds in water using an amended solvation energy relationship, *J. Pharm. Sci.* **1999**, *88*, 868–880.

62 Winiwarter, S., Bonham, N. M., Ax, F., Hallberg, A., Lennernäs, H., Karlen, A. Correlation of human jejunal permeability (*in vivo*) of drugs with experimentally and theoretically derived parameters. A multivariate data analysis approach. *J. Med. Chem.* **1998**, *41*, 4939–4949.

63 Raevsky, O. A., Schaper, K.-J. Quantitative estimation of hydrogen bond contribution to permeability and absorption processes of some chemicals and drugs. *Eur. J. Med. Chem.* **1998**, *33*, 799–807.

64 Schaper, K.-J., Zhang, H., Raevsky, O. A. pH-dependent partitioning of acidic and basic drugs into liposomes – a quantitative structure–activity relationship. *Quant. Struct.-Act. Relat.* **2001**, *20*, 46–54.

65 Raevsky, O. A., Fetisov, V. I., Trepalina, E. P., McFarland, J. W., Schaper, K.-J. Quantitative estimation of drug absorption in human for passively transported compounds on the basis of their physico-chemical parameters. *Quant. Struct.-Act. Relat.* **2000**, *19*, 366–374.

66 Zhao, Y. H., Le, J., Abraham, M. H., Hersey, A., Eddershaw, P. J., Luscombe, C. N., Boutina, D., Beck, G., Sherborn, B., Cooper, J., Platts, J. A. Evaluation of human intestinal absorption data and subsequent derivation of a quantitative structure–activity relationship (QSAR) with the Abraham descriptors. *J. Pharm. Sci.* **2001**, *90*, 749–784.

67 Van de Waterbeemd, H., Camenisch, G., Folkers, G., Chretien, J. R., Raevsky, O. A. Estimation of blood–brain barrier crossing of drugs using molecular size and shape, and H-bonding descriptors. *J. Drug Targeting*, **1998**, *6*, 151–165.

68 Lipinski, C. A. Drugs structure and properties, past and present. Can we design drugs with beautiful properties? http://www.iainm.demon.co.uk/spring99/lipins_n.pdf.

69 Amidon, G. L., Lennernäs, H., Shan, V. P., Crison, J. R. A. A theoretical basis for a pharmaceutic drug classification: correlation of *in vitro* drug product dissolution and *in vivo* bioavailability. *Pharm. Res.* **1995**, *12*, 413–420.

70 Avdeev, A. Physicochemical profiling (solubility, permeation and charge state). *Curr. Top. Med. Chem.* **2001**, *1*, 277–351.

71 ACD/LogP-Rule of 5, http://www.acdlabs.co.uk.

72 Van de Waterbeemd, H. The fundamental variables of the biopharmaceutics classification system (BCS): a commentary. *Eur. J. Pharm. Sci.* **1998**, *7*, 1–3.

73 Lipnick, R. Narcosis: fundamental and baseline toxicity mechanism for nonelectrolyte organic chemicals. In *Practical Application of Quantitative Structure–Activity Relationships (QSAR) in Environmental Chemistry and Toxicology*, Karcher, W., Devillers, J. (eds.), Kluwer, Dordrecht, **1990**, pp. 129–144.

74 Raevsky, O. A., Dearden, J. C. Creation of predictive models of aquatic toxicity of environmental pollutants with different mechanisms of action on the basis of molecular similarity and HYBOT descriptors. *SAR QSAR Environ. Res.* **2004**, *15*, 433–448.

Part III
Conformations

7
Three-dimensional Structure Generation

Jens Sadowski

Abbreviations

2D, 3D	two-, three-dimensional
CCR	close contact ratio
CSD	Cambridge Structural Database
NMR	nuclear magnetic resonance
PDB	Protein Data Bank
QSAR	quantitative structure–activity relationships
RMS	root mean square

7.1
Introduction

Many biological, physical and chemical properties are clearly functions of the three-dimensional (3D) structure of a molecule. Thus, the understanding of receptor–ligand interactions, molecular properties or chemical reactivity requires not only information on how atoms are connected in a molecule (connection table), but also on their 3D structure.

Since the early days of organic chemistry, the tetrahedral nature of tetravalent carbon has been known along with such consequences as chirality or the ability to rotate the plane of polarized light. X-ray crystallography has helped to a deeper insight into the 3D structure of molecules and can be used to even determine the absolute configuration of chiral compounds. In addition, several systems to represent stereochemistry and other 3D features in two dimensions on paper have been proposed over the years such as the Fischer projection or the representation of chiral centers with wedge-like bonds indicating whether they point above of below the paper plane. Even qualitative conformational representations as illustrated in the drawings of the chair and the boat conformations of cyclohexane in Fig. 7.1 are commonly used. However, despite being useful to highlight the 3D nature of some structural aspects, these representation schemes are quantitative

Molecular Drug Properties. Measurement and Prediction. R. Mannhold (Ed.)

ISBN: 978-3-527-31755-4

Fig. 7.1 Symbolic drawings of cyclohexane chair and boat conformations in two dimensions using different line thickness and wedge symbols.

and do not allow for a more detailed analysis of 3D properties of molecules and they cannot reflect conformational flexibility appropriately.

Mechanical 3D molecular models have been used by chemists since the end of the 19th century. In particular, Andre Dreiding's stainless steel models have become rather popular since the 1950s. Some of Dreiding's findings illustrate the experiences of working with mechanical models: "The degree of elasticity of the construction material causes that the Bayer strain of ring systems becomes sensible with the fingers. When transforming the cyclohexane chair form into the flexible form or vice versa, one has first to overcome a certain strain after which the atoms 'snap' themselves into the other form. The Pitzer strain however is not directly visible; it can only be estimated by measurement of the distances of non-bonded atoms". Even the true mechanical nature of the models is highlighted by another experimental result: "The chair–boat transformation was executed 150 00 times by a machine on a number of cyclohexane models until one unit broke. The angles of the non-broken units were not deformed".

The increased interest in 3D aspects of organic chemistry and quantitative structure–activity relationship (QSAR) studies has caused an increasing need for a much broader access to 3D molecular structures from experiment or calculation.

Experimental sources of 3D structure information are X-ray crystallography, microwave spectroscopy, electron diffraction and nuclear magnetic resonance (NMR) spectroscopy. The largest source of experimentally determined molecular structures is the Cambridge Crystallographic Database (CSD) [1], which contains at present about 400 000 X-ray structures of small molecules. In addition, the Brookhaven Protein Data Bank (PDB) [2] contains about 40 000 structures of proteins and other biological macromolecules, including several thousands of drug-sized molecules in their biologically active conformations bound into their receptors. For several reasons, the experimental sources of 3D structures are not sufficient and there is a real need for computer-generated models:

- The number of compounds whose 3D structure has been determined (about 400 000) is small when compared to the number of known compounds (more than 25 million).
- Computational techniques in organic chemistry such as for drug design, structure elucidation or synthesis planning quite often investigate enormous numbers of hypothetical molecules, which are not yet

synthesized or even not stable, as in the case of transition states of chemical reactions.
- Theoretical methods such as quantum mechanics or molecular mechanics can produce 3D molecular models of high quality and predict a number of molecular properties with high precision. Unfortunately, these techniques also require at least some reasonable 3D geometry of the molecule as starting point.
- Very often, it is unknown which conformation of a flexible molecule is needed. For example, in drug design, we hunt often for the so-called bioactive conformation, which is the molecule in its receptor-bound state. In this case, any other experimental structure of the isolated molecule – in vacuum, in solution or in crystal – can be the wrong choice.

The missing link between the constitution of a molecule and its 3D structure in computational chemistry is a technique capable of automatically generating 3D models starting from the connectivity information of a given molecule. Due to its basic role, 3D structure generation is one of the fundamental problems in computational chemistry. As a consequence, in recent years a number of automatic 3D model builders and conformer generators have become available. For two comprehensive reviews, see Refs. [3, 4].

In the following, we will discuss two-dimensional (2D)-to-3D conversion in this context. However, it should be emphasized that we do so only for the sake of brevity. In reality, none of the conversion programs utilizes information of a 2D image of a chemical structure. Only the information on the atoms of a molecule and how they are connected is used (i.e. the starting information is the constitution of the molecule). One could even refer to linear structure representations such as SMILES as one-dimensional. However this is not true since SMILES allows for branches and ring closure which makes its information content essentially 2D. Thus, all structure representations which lack 3D atomic coordinates will in the following simply be referred to as 2D.

Most molecules of organic, biochemical or pharmaceutical interest can adopt more than one conformation. Although this ability to adopt multiple conformations has some implications for generating a single low-energy 3D structure which we will cover here, we will not embark in general into the field of conformation analysis, which is instead covered in the next chapter as a topic of its own. However, having said this, even many conformation analysis approaches need at least one reasonable 3D structure to start with.

The consecutive levels of 3D information are illustrated in Fig. 7.2. The pure connectivity information is usually referred to as 2D. If stereo information is available, it can be referred to as "2.5D" since the stereo descriptors add some 3D information. From this, a single 3D structure is obtained from the program CORINA [5] and, subsequently, a multi-conformer ensemble from the program OMEGA [6] (bottom left). In this chapter, we will refer only to the step from 2D (2.5D) to 3D.

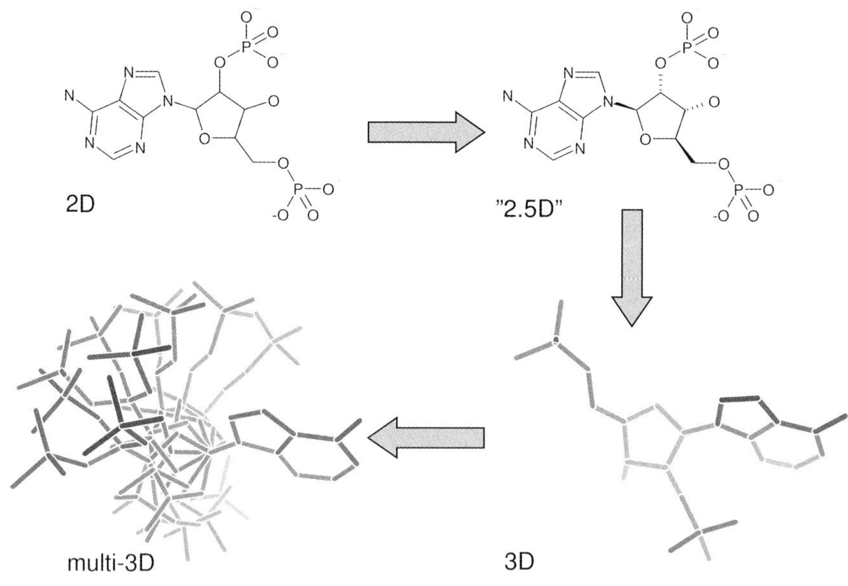

Fig. 7.2 Consecutive levels of 3D information for adenosine diphosphate.

7.2
Problem Description

7.2.1
Computational Requirements

The main area of automatic structure generation is the 2D-to-3D conversion of large sets of drug-like organic compounds. These sets often contain millions of structures, imposing some restrictions on the development of 3D structure generators. Database developers at Molecular Design Ltd formulated the following criteria for a 2D-to-3D conversion program [7]:

- *Robustness.* The program should run for a long time and many molecules before failing, and it should indicate the actions taken on failure rather than simply crash.
- *Large files.* The program should be able to handle large numbers of structures contained in a single file to minimize the number of conversion jobs.
- *Variety of chemical types.* The program should be able to handle a wide variety of chemical types.
- *Stereochemistry.* The stereochemical information contained in the input data must be handled correctly.

- *Rapid and automated.* The large size of the databases to be processed requires the conversion program to run in batch mode and to work with acceptable speed.
- *High-quality models.* The generated models should be of sufficiently high quality without any further energy minimization and should represent at least one low-energy conformation. It should have internal diagnostics to validate the models generated.

7.2.2 General Problems

Each approach to automatic generation of 3D molecular models has to solve a number of general problems. The strategy for building a molecular model can be compared with the use of a mechanical molecular model building kit or its modern replacement – interactive 3D molecule editors. Monocentric fragments that represent different hybridization states and provide the corresponding bond angles are connected using joints with a length corresponding to the required bond lengths.

A basic assumption in this process of 3D structure generation is an allowed transfer of bond lengths and bond angles from one molecular environment to another (i.e. the usage of standard values for bond lengths and bond angles). However, this assumption requires us to distinguish between a sufficiently large number of different atom types, hybridization states and bond types with appropriate bond lengths and bond angles. Usually, the deviations from these standard values are rather small.

A totally different situation is encountered for dihedral or torsional angles, which describe the twisting of a fragment of four atoms connected by a sequence of bonds. As the steric energy may have multiple minima around a rotatable bond with similar energy content, this leads to more than one possibility for constructing a 3D model for such molecules, or in other terms, to multiple conformations.

In acyclic molecules or substructures, the preferred torsional angles are those which simultaneously minimize torsional strain and the steric interactions between nonbonded atoms. The relatively large flexibility of such systems gives rise to multiple solutions (conformations) for the process of structure generation, which have quite similar energy. Account of this flexibility has to be taken and geometrically unacceptable situations such as the overlap of atoms ("clashes") must strictly be avoided. With increasing numbers of possible conformations, it becomes less and less likely that the generated 3D structure corresponds to the experimentally determined geometry, which often is just one of many possible low-energy conformations.

In cyclic structures, ring closure has to be taken into account as an additional geometrical constraint of the 3D structure generation process. Ring closure dramatically reduces the degrees of freedom as expressed in a reduction in the number of possible conformations compared to those in acyclic systems. In particular, the endocyclic torsion angles are mutually dependent. Due to this fact, many of the available programs for 3D structure generation use explicit information about possible single-ring conformations. These so-called ring templates fulfill implicitly the ring closure condition. They can be stored as explicit 3D coordinates or simply as lists of torsion angles. Additional levels of sophistication are reached when the rings have exocyclic substituents or when they are assembled in fused or bridged ring systems. Another challenge arises with increasing ring size. Large rings are apart from the requirement to ring closure nearly as flexible as acyclic systems.

A simple example illustrates the different conformational problems encountered in chain portions and ring fragments of a molecule. When searching all conformations of *n*-hexane by systematically permuting all torsion angles in 60° steps, the theoretical number of conformations is 3^4 (methyl torsions omitted). Only 12 of these 81 conformations are valid. When instead searching conformations of cyclohexane in the same manner, only one conformation is found which fulfills the ring closure condition – the cyclohexane chair conformation. Note that other known cyclohexane conformations such as boat and twist-boat cannot be constructed from torsion angles on a 60° grid.

Due to the specific complications when predicting the geometry of ring systems, many of the approaches to 3D structure generation dedicate most of the program intelligence to this part. Most often, the molecule under consideration is fragmented into acyclic and cyclic portions at the very beginning of the 3D generation process. The fragments are then handled separately and reassembled at the end of the whole process.

7.2.3
What 3D Structures Do You Need?

As already outlined above, often there are many different relevant 3D structures of one and the same molecule, like the receptor-bound conformation of bioactive molecules as well as their conformations in the solvent, crystal or gas phase. Many applications in computational drug design, structure elucidation or prediction of transition states in synthetic chemistry depend on them. In addition to this, there are as many different ways to 3D structures. As an illustration, consider the conformations of biotin shown in Fig. 7.3. There, the conformation of biotin in complex with streptavidin is compared to a small-molecule X-ray structure of biotin alone, to a single 3D structure obtained from CORINA, to the global minimum from a conformational search with OMEGA and to the AM1 optimized global energy minimum. In addition, there might be other, different conformation one would obtain from gas-phase experiments or in solvent. All are relevant and, most often, all are different. Thus, a single 3D structure can never serve all purposes. Consider it rather as a starting point for further investigations based on conformation analysis or experiment.

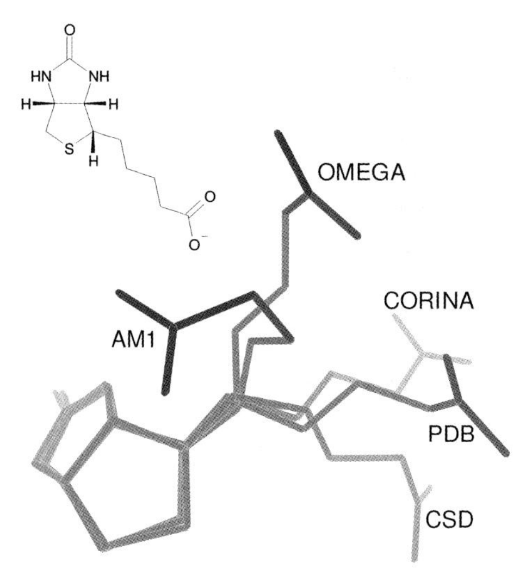

Fig. 7.3 Biotin in the conformation bound to streptavidin (PDB 1NQM) compared to a small-molecule X-ray structure from CSD (BIOTIN10), a single 3D structure generated by CORINA, the lowest energy conformation from a conformation search with OMEGA and the AM1 optimized global minimum.

7.3 Concepts

7.3.1 Classification of Strategies

Here, an attempt to classify different strategies to generate 3D molecular models is undertaken with the aim to specify the remit of methods which will be covered under the term "automatic 3D structure generators". The focus will be on methods designed for small, drug-like molecules. The prediction of the geometry of polymers, in particular of biopolymers, is a task of its own and not even attempted by the approaches discussed here.

Manual methods. In the early beginning of thinking in three dimensions in organic chemistry, 3D molecular models were built by hand, using standard bond length and bond angle units from mechanical molecular model building kits. This technique, still useful today, found its modern expression in the well-known interactive 3D structure editing options incorporated into nearly all graphical modeling programs. The user may construct a 3D molecular model interactively by positioning atoms and bonds on a 3D graphics interface or by connecting predefined fragments. All these approaches clearly do not fulfill the requirement of being automatic and will therefore not be covered here.

Data-based and rule-based methods. Most automatic approaches for 3D generation are based on the knowledge of chemist about geometric and energetic rules and principles for constructing 3D molecular models. This knowledge was originally gained from experimental data and through theoretical investigations. It is built into 2D-to-3D conversion programs in the form of data tables (e.g. standard bond lengths) and rules (e.g. prefer equatorial over axial conformations for mono-substituted cyclohexane).

Fragment-based methods. At the far end of data-based methods are approaches that use data in the form of fragment geometries. Three-dimensional data about geometries of typical multi-atom fragments of molecules are used to build complete 3D structures. The fragments used are often of high quality and obtained either from crystal structures or theoretical calculations. The most common use of fragment data is templates for ring conformations.

Conformation analysis methods. In many cases in the process of building a 3D structure from scratch, decisions have to be made between multiple alternatives with similar energy. A typical example is an sp^3–sp^3 torsion angle with similar energies for the alternatives of +60°, –60° and 180°. In many cases, rules are used to decide (e.g. stretch an open chain portion as much as possible to avoid clashes). Sometimes, the best result cannot be determined without a conformation analysis (e.g. complex ring systems with exocyclic substituents). Despite conformation analysis being a topic of its own covered in the next chapter, many automatic 3D structure generators have to fall back in certain situations to a limited conformation search in order so solve a specific problem and to come up with a reasonable solution.

Numerical methods. Computer-intensive numerical methods like quantum mechanics, molecular mechanics, or distance geometry [8] do not normally fall into the scope of automatic model builders. However, some model builders have built-in fast geometry optimization procedures or make use of distance geometry in order to generate fragment conformations.

Clearly, there is no sharp border between all of the concepts discussed above. Most model builders try to use at least some of them in an efficient mixture in order to achieve the best compromise between computation times and quality.

7.3.2 Standard Values

Much of the knowledge derived from experimental structures and from theory can be systematically expressed as explicit data about certain geometric details. Typical examples are standard bond lengths and bond angles which are stored in tables. Since both bond lengths and bond angles have only one global energy minimum, it is possible to store preferred values for typical bond types and angle types. Most often, these values are derived from experimental structures in the CSD [1] or from textbook knowledge. Programs differ in the level of detail for these constants. For example, a bond between two sp^3 carbons can include details about the chemical context on different levels. The following statistics were obtained from the CSD for

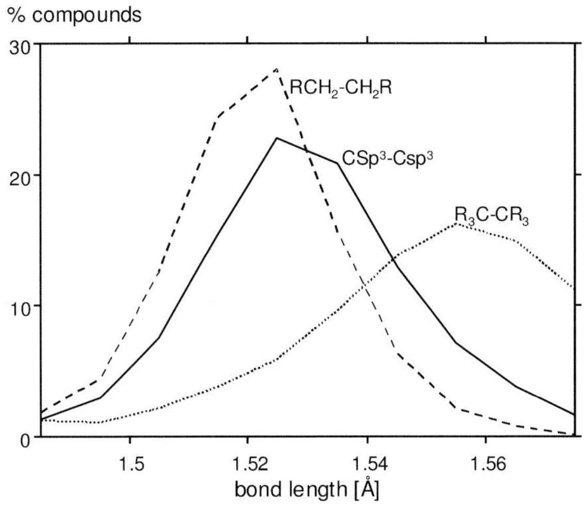

Fig. 7.4 Distributions of bond lengths found in the CSD for all single Csp^3–Csp^3 bonds (closed line), for the methylene analogs RCH_2–CH_2R (dashed line) and for single bonds between quaternary carbons R_3C–CR_3 (dotted line).

Csp^3–Csp^3 single bonds (Fig. 7.4). The first histogram (closed line) refers to the unspecific fragment with any substituents. It has a maximum between 1.52 and 1.53 Å. The second histogram (dashed line) refers to the fragment with two hydrogens attached to each carbon. It is a bit more distinct with its maximum slightly shifted towards shorter bond lengths. Clearly, the increased context information gives a sharper maximum and allows for a more precise guess for an appropriate bond length. In the third case, a histogram was obtained for single bonds between two quaternary sp^3 carbons (dotted line). In this case, the increased steric hindrance causes a shift towards longer bond lengths. However, the differences between the histograms are rather small and in most cases the rough guess of 1.53 obtained from the most general distribution (closed line) is good enough.

Bond angles are another typical example for tabulated geometric parameters. The most general way to treat them is to use standard values for basic atomic geometries as 180° for sp, 120° for sp^2 and 109.47° for sp^3. As in the case of the bond lengths, different substitution patterns around the central atom can cause shifts of the bond angles away from these ideal values. Figure 7.5 shows bond angle distributions found in the CSD for general tetrahedral carbons, their methylene analogs R–CH_2–R and quaternary carbons R–CR_2–R without any hydrogen attached. The most general histogram for all sp^3 carbons is almost symmetrically distributed around 111° (closed line). Asymmetric substitution with two heavy atoms and two hydrogens causes angle widening between the heavy atoms towards 113° (dashed line), whereas quaternary substitution with four heavy atoms attached to the central carbon forces the angle distribution back towards values around 109°. Again, a more detailed bond angle type can help to assign more accurate values but in most cases the ideal value of 109.5° would be good enough for sp^3 carbons.

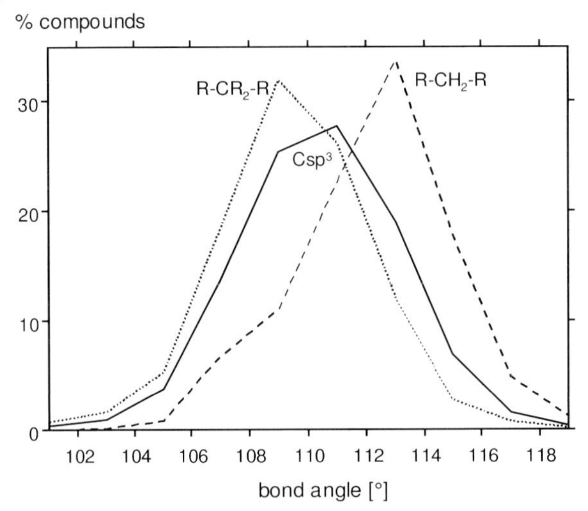

Fig. 7.5 Distributions of bond angles found in the CSD for all sp^3 carbons (closed line), for the methylene analogs R–CH_2–R (dashed line), and for quaternary carbons with four nonhydrogen substituents R–CR_2–R (dotted line).

The last example for commonly used standard values in 3D structure generation is torsion angles. Torsion angles can have multiple local energy minima distributed over 360°. This behavior can be reflected by storing explicit angle values for torsion angles corresponding to local energy minima. In addition, the torsion angles can be augmented by information about the energy level they correspond to. Again, an example based on statistics from the CSD will illustrate this. Figure 7.6 shows the distribution of torsion angles obtained for *ortho*-substituted phenol ethers. The histogram shows a strong preference for a planar configuration around zero° with the ether substituent opposite to the *ortho* substituent. There is another weakly populated maximum around 90°. In this case, a 3D generator could use this knowledge by using preferentially 0° for torsions of this type with an alternative value of 90° with lower preference. In addition, using inverse Boltzmann statistics, an energy equation can be derived from the distribution as $E(\tau) = -A \ln f(\tau)$, where $E(\tau)$ is the energy value corresponding to torsion angle τ, A is an adjustable factor and $f(\tau)$ is the frequency of torsion angle τ. This is done by both the conformation analysis program MIMUMBA [9] and the 3D structure generator CORINA [5]. The closed line graph of the derived energy in Fig. 7.6 illustrates this.

7.3.3
Fragments

Another way to use data about known geometric details is to use multi-atom fragments with explicit 3D coordinates in the 3D generation process. One obvious example is ring templates. By using complete ring geometries as building blocks,

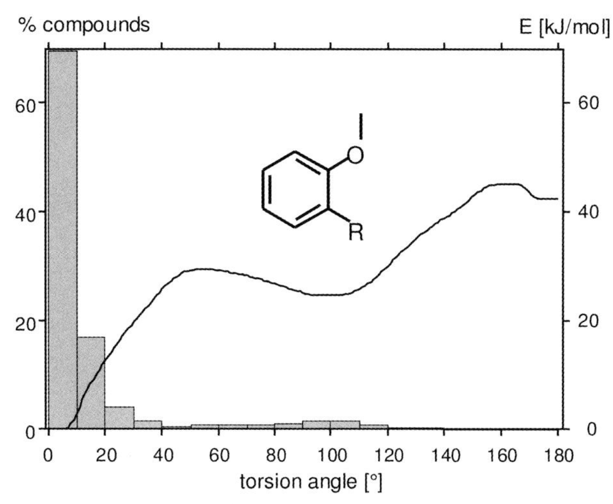

Fig. 7.6 Torsion angle distribution for *ortho*-substituted phenol ethers (bars) and the derived potential energy (closed line).

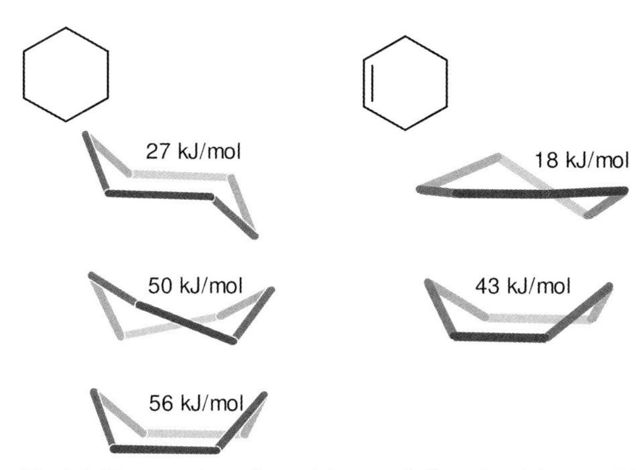

Fig. 7.7 Ring templates for cyclohexane (left) and cyclohexene (right) together with their associated internal energies.

the program does not have to care any longer about ring closure since the template ring geometries will be closed already. By using multiple geometries for individual ring types, this concept can be used even for ring conformation analysis. Figure 7.7 shows templates for cyclohexane and cyclohexene as used by the program CORINA [5] along with the internal energy values used by the program to rank them. Note that the cyclohexane boat conformation very likely will not be interesting as a reasonable low-energy geometry due to its high energy content. Still, the boat template might be interesting for constructing larger multi-ring structures with geometric restrictions.

The ring templates can be further used to construct larger, multicyclic systems as illustrated in Fig. 7.8. For norbornane, two fitting conformations of cyclopentane in the envelope conformation can be joined in order to construct the complete 3D structure of norbornane. In this case, this is the only low-energy conformation known for norbornane due to its rigidity.

In other cases like in atropine (Fig. 7.9), it is beneficial to analyze multiple ring conformations in order to make a good decision for one low-energy conformer. In this case, a cyclopentane envelope is combined with a cyclohexane chair and a cyclohexane boat, respectively, in order to form geometries for the bicyclic ring system of atropine. The internal energies of the templates and the resulting complete ring system are given below. Note that the total energies are not just the sums of the fragment energies. They are corrected by terms for the exocyclic substituents. We come to rules for exocyclic substituents in more detail when discussing rules for constructing 3D models.

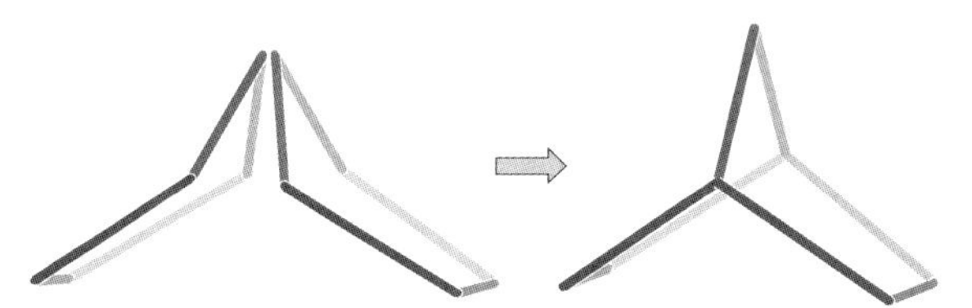

Fig. 7.8 Joining two cyclopentane envelopes to the norbornane 3D structure.

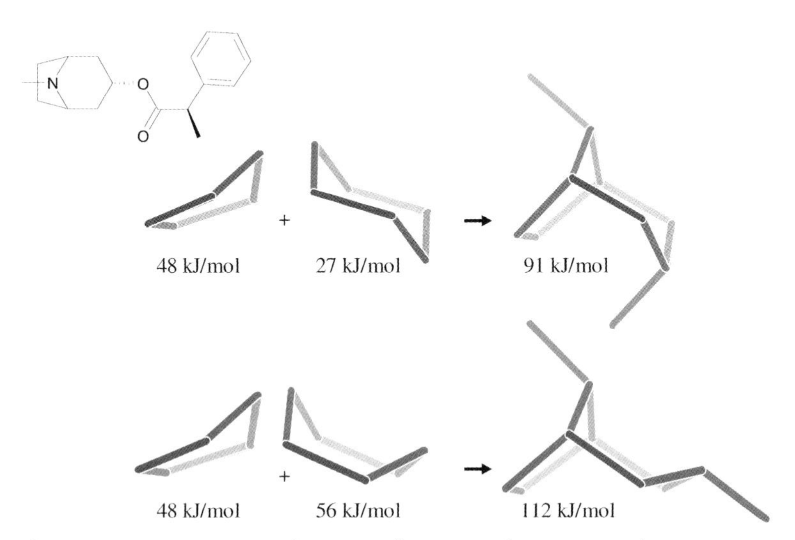

Fig. 7.9 Constructing ring conformations for atropine from ring templates.

7.3.4
Rules

In addition to value tables and fragment data, a certain part of the knowledge about 3D structures can be expressed in more general rules for solving specific problems in the 3D generation process. Here, some typical examples will be illustrated.

Equatorial/axial ring substituents. A particular aspect of ring conformation is the configuration of exocyclic substituents. The most commonly known example is the equatorial versus axial placement of substituents of saturated, nonplanar rings, e.g. cyclohexane. Methyl-cyclohexane can adopt two different chair conformations which place the methyl substituent either in the plane of the ring (equatorial – to the left) or perpendicular to it (axial – to the right) as shown in Fig. 7.10(a). The spectroscopically measured energy difference between the two conformations is 7.1 kJ mol^{-1} (for a good reference for experimental conformational energies, see Ref. [10]). This particular energy contribution is known as the so-called Pitzer strain. The energy difference is mainly caused by the extra steric strain between the axial methyl group and the axial hydrogens in the 3-position as indicated by the dotted line. The main parameter for this steric interaction is size. Figure 7.10(b) shows *t*-butyl-cyclohexane. The larger size of the *t*-butyl substituent causes a significantly higher Pitzer strain of 23 kJ mol^{-1}. Programs for 3D generation normally have rules implemented which prefer the largest substituent in equatorial position.

In addition to this so-called 1,3-interaction between an exocyclic substituent and the hydrogens in 3-position, steric interactions can cause extra strain also in 1,2-disubstituted rings as the *trans*-1,2-dimethyl-cyclohexane in Fig. 7.10(c). The

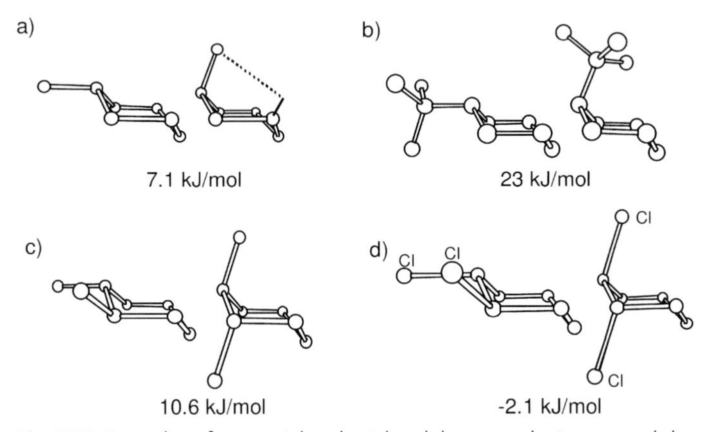

Fig. 7.10 Examples of equatorial and axial cyclohexane substituents and the corresponding energy differences. (a) Methyl-cyclohexane, (b) *t*-butyl-cyclohexane, (c) *trans*-1,2-dimethyl-cyclohexane and (d) *trans*-1,2-dichloro-cyclohexane.

energy difference between the eq,eq and the ax,ax conformations is 10.6 kJ mol^{-1} – much closer than expected to the mono-substituted analog in Fig. 7.10(a) despite the ax,ax form (right-hand side) having twice the number of axial substituents. The reason is that the eq,eq conformation (left-hand side) also causes extra strain by the two methyl groups coming into close contact. This extra contribution is roughly the equivalent of half an axial methyl group. This equatorial–equatorial effect can even invert the order of conformations as shown in Fig. 7.10(e). In the case of *trans*-1,2-dichloro-cyclohexane, the 1,2-diequatorial interaction (left) becomes so strong that the diaxial conformation (right) is energetically preferred by –2.1 kJ mol^{-}. A program for 3D structure generation should also have rules to decide in cases like this correctly.

Note that the 1,2-diequatorial substituted examples in Fig 7.10(c and d) are individual stereoisomers. The corresponding *cis*-species (Fig. 7.11b) is not another conformation, but another stereo isomer. The experimentally by calorimetry determined energy difference between the isomers is 6.5 kJ mol^{-1}.

Trivalent nitrogen. The handling of trivalent nitrogen can potentially be tricky under certain circumstances. One typical case is trivalent nitrogen attached to a π-system, e.g. an aromatic ring. An example is aniline as shown in Fig. 7.12. To a certain extend, the free electron pair of the nitrogen atom is conjugated to the aromatic system with a preferred flat geometry. On the other hand, the conjugation is not very strong which keeps the C–N bond rotatable. If the torsion angle of the C–N bond is turned away from the planar conformation of 0°, the conjugation becomes even weaker and the nitrogen changes its configuration from a flat geometry to a pryramidal configuration. The truth is that nitrogen is flexible and can rather freely change between the two extremes. However, often there is a statistical preference for one of them.

Evidence for this behavior can be found in experimental structures. Analyzing the CSD for the configuration of nitrogen in anilines with two hydrogens at the nitrogen atom, the distribution shown in Fig. 7.13 is found. The so-called out-of-plane angle of the nitrogen center varies between 0° and 60°. The global maximum

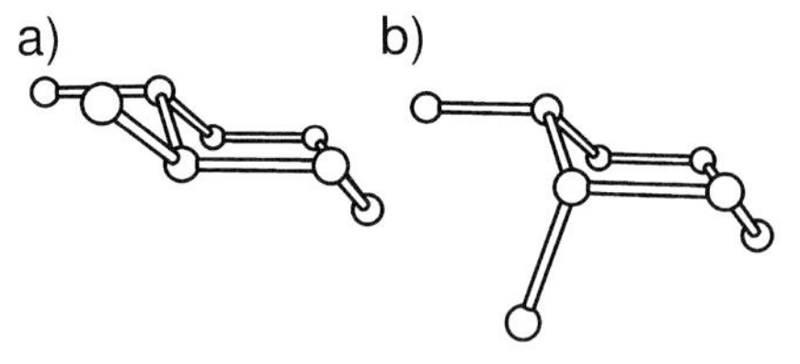

Fig. 7.11 1,2-dimethyl cyclohexane stereoisomers: (a) *trans* and (b) *cis*.

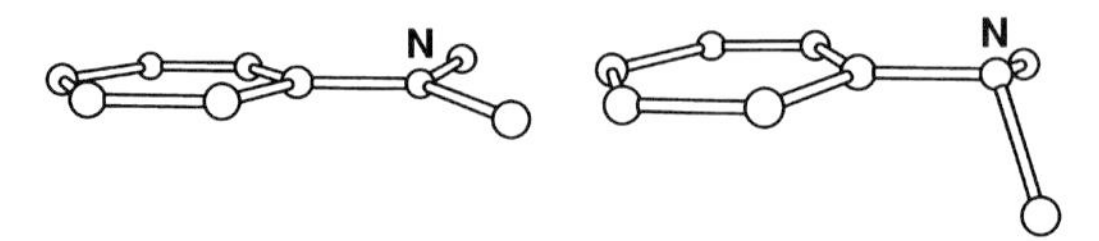

Fig. 7.12 Planar and pyramidal configuration of aniline nitrogen.

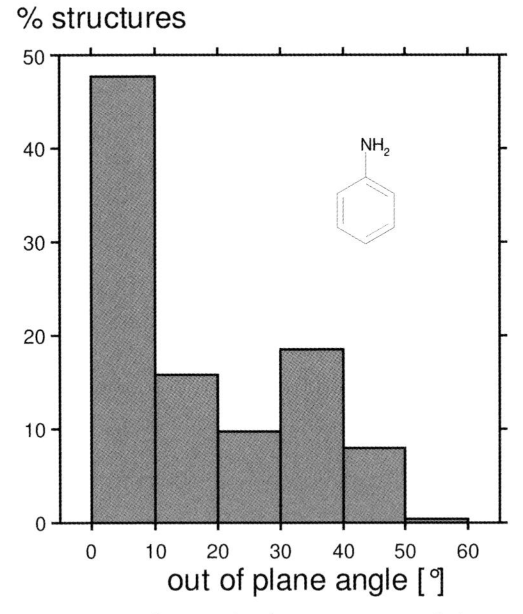

Fig. 7.13 Distribution of aniline nitrogen out-of-plane angles in the CSD.

is around 0° (completely flat). There is another local maximum between 30° and 40° which corresponds to a pyramidal configuration. Nitrogen is flexible! The crux for 3D structure generation is that a program normally generates just one of the possible configurations. Thus, appropriate rules for handling this situation are needed. In the case of a nitrogen with two hydrogen atoms, the majority of cases seem to be flat. Thus, most structure generators will prefer to generate just this configuration.

The situation becomes more complex in the case of nonhydrogen substituents attached to the nitrogen or *ortho*-substituents at the aromatic ring. Both potentially drive the nitrogen out of its preferred planar, conjugated conformation. The program CORINA [5] has a set of rules implemented based on a careful analysis of the CSD as illustrated in Fig. 7.14. Similar rules can be used in related cases, e.g., sulfone amides.

Flexible nitrogen. Another nitrogen-related problem can be addressed by rules as well. Pyramidal nitrogen can normally freely change its configuration between the two pseudo stereo isomers. This conformational exchange has an energy barrier in the order of magnitude of a torsion angle and can be spectroscopically observed, and is sometimes even called pseudo-rotation. For a single 3D structure this does normally not matter, but it should be observed when generating multiple conformations. However, there are cases were a program has to take into account this extra degree of freedom, e.g. when the nitrogen is in a context with other, fixed centers. Figure 7.15 shows two conformations of 1,3-dimethyl-piperidine. A rule

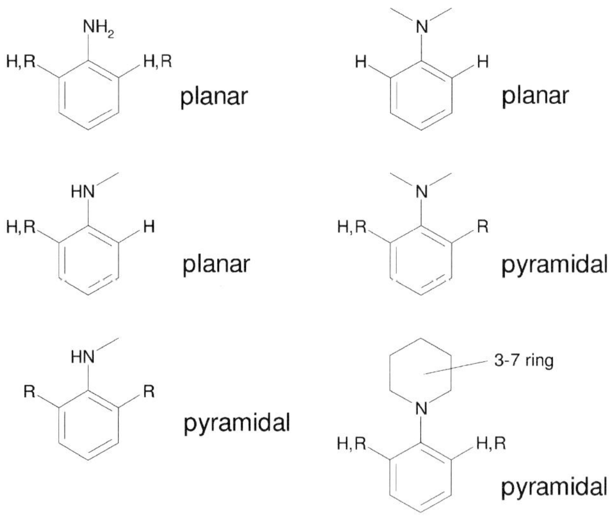

Fig. 7.14 Aniline rules implemented in CORINA.

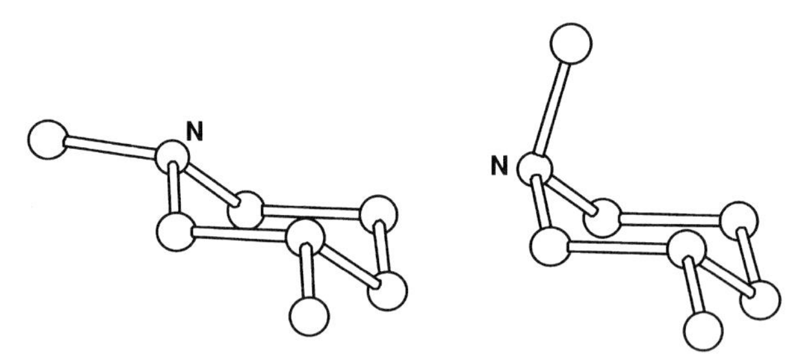

Fig. 7.15 Two conformations of 1,3-dimethyl-piperidine.

implemented in the 3D structure generation program should normally prefer the di-equatorial form on the left-hand side.

Restricted stereo centers. Taking care of chiral centers is an important part of the 3D structure generation process – any given stereo information has to be taken into account and the result has to be the correct stereoisomer. However, in cases with unspecified stereo centers, it is normally good enough to choose just one arbitrary isomer since they do not differ in macroscopic properties as conformational energy. This is different in cases where a chiral center is locked into a ring. One typical case is stereo centers in rings with exocyclic neighbors. Rules have to ensure that in cases with unspecified stereochemistry the correct form with the lowest energy is produced. This is similar to the flexible amine example above. Another important case is bridgehead atoms in cage-like bridged ring systems. A simple example is norbornane Fig. 7.16. Only the stereoisomer with both hydrogens at the junction atoms in exo position is valid. The theoretical endo,exo form is geometrically impossible. A structure generator should have rules in place

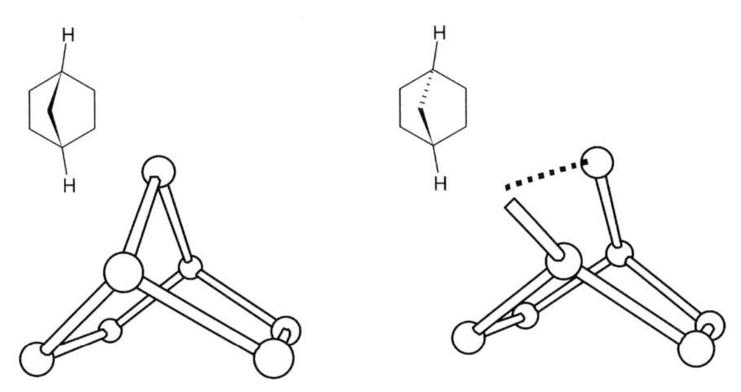

Fig. 7.16 Norbornane stereoisomers and 3D structures. Only the exo,exo form (left) can exist.

which in cases with unspecified chirality automatically produce the correct exo,exo form. However, if the input structure accidentally comes with the incorrect form specified, it is impossible to generate a valid 3D structure.

7.3.5 Quality Control

High quality is one of the criteria defined in the requirements section above. Since the program should run automatically in batch mode, we mean by quality control an internal check of the 3D structures produced by the structure generator itself. In general, the abilities of a fast, automatic structure builder to assess the quality of its models are rather limited since, for example, an exhaustive conformation analysis and energy optimization is impossible in most cases. However, there are a limited number of simple quality checks to avoid trivially distorted structures:

- Comparison of bond lengths and angles in the generated structures with their default values.
- Check of the planarity of functional groups which are by definition planar, e.g., aniline nitrogen, double bonds or amides.
- Close intramolecular contacts (clashes). A rough measure has been proposed and implemented in the program CONCORD [4, 11] – the close contact ratio (CCR). The CCR of a 3D structure is defined as the ratio of the smallest nonbonded distance to the smallest acceptable value for this distance. Normally, structures with CCR > 0.8 are acceptable. Some programs as CORINA [5] or CONCORD [4, 11] have fallback procedures for attempting to relax close contacts in structures with unacceptably low CCR.

If the above sketched quality checks flag for an unacceptably distorted or unrealistic geometry and the program cannot remedy this, it should be good practice not to send the questionable structure to the output file.

7.3.6
Comparison of 3D Structures

Often, one needs to compare different 3D structures or conformations of a molecule. That is done internally by the 3D structure generation program to weed out too similar conformations of fragments. Another aspect is the need of the computational chemist to compare different generated or experimental structures. A well-established measure is the so-called root mean square (RMS) value of all atom–atom distances between two 3D structures. The RMS value needed here is a minimum value achieved by superimposing the two 3D structures optimally. Before calculating the RMS, the sum of interatomic distances is minimized by optimizing the superimposition in 3D.

The RMS value is only a rough measure for the similarity of two conformations. It is summarized over all atoms under consideration. That means that a local drastic deviation between two conformations can be hidden in an overall good fit. Reversely, a deviation in one part of the molecule can hide a perfect fit of another part. However, the RMS is very useful for obtaining a quick, robust measure of conformational similarity. A few recommendations can help to a better understanding:

- Exclude hydrogen atoms since their position is in many cases fixed by the heavy atoms they are attached to. This reduces the amount of noise in the RMS value significantly.
- Make sure that symmetry is taken into account when producing the atom mapping between the two conformations. Consider the example given in Fig. 7.17. The molecule – phenyl-cyclohexane – has two independent symmetry axes through the two rings as indicated by dotted lines. This

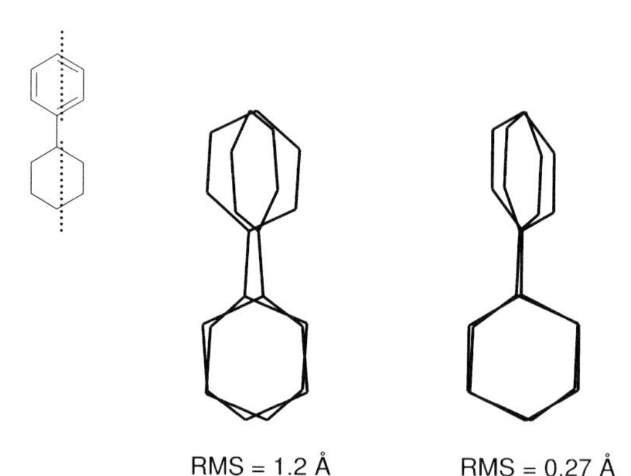

Fig. 7.17 RMS fits of two similar phenyl-cyclohexane conformations. The structural diagram indicates the two independent symmetry axes (dotted lines). The left-hand side fit was obtained with an incorrect atom mapping. The right-hand side fit is obtained from the correct atom mapping.

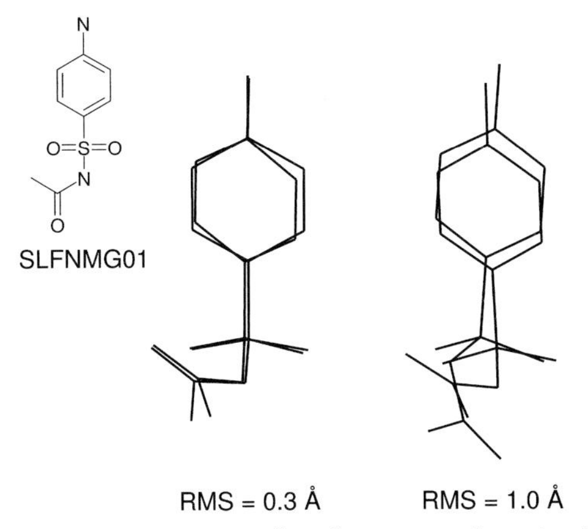

Fig. 7.18 Two computed conformations of a molecule compared to its known crystal structure.

symmetry leads to two sets of topologically equivalent atoms and subsequently to four different ways to map the atoms of two conformations onto each other. The atom mapping used for the fit on the left-hand side causes a suboptimal fit and subsequently misleads to a high RMS value. The correct atom mapping of the symmetrically related atoms leads instead to the optimal low RMS as illustrated on the right.

- The following rules of thumb for interpreting RMS values are given: less than 1.0 Å is similar and less than 0.3 Å is nearly identical. This is illustrated in Fig. 7.18. The crystal structure of a small molecule from the CSD is compared to two conformations generated by the program OMEGA [6]. On the left-hand side, the superimposition of one conformation with the X-ray structure results in an RMS of 0.3 Å. The two conformations are basically the same. For the other conformation on the right-hand side an RMS of 1.0 Å is obtained since the amide moiety is rotated into a different direction. See the next chapter for a more detailed introduction of the use of the RMS value when sampling conformations.

7.4 Practical Aspects

7.4.1 Brief Overview and Evaluation of Available Software

The reliability of scientific work based on computer-generated 3D structures requires a careful evaluation of available 3D generators to find the program best suited for this purpose. The history of concepts and software in this field has been exhaustively reviewed in the literature [3, 4]. In addition, an evaluation study of seven publicly available structure generators using 639 X-ray structures has been published some years ago [12]. In recent years, mainly two programs have been widely used – CORINA [5] and CONCORD [11]. Both have been developed for mainly one purpose – fast and automatic generation of high-quality, single conformation 3D structures of drug-sized molecules. Thus, focus in this section will be exclusively on these two programs. Here, an updated evaluation study based on 25 017 X-ray structures will be presented.

Dataset. The dataset was selected from the CSD [1] under the following restrictions: error-free organic compounds, fully resolved, with the connection table completely assigned and with an *R* factor less than or equal to 5%. After export, all purely inorganic compounds without any carbon atoms, all compounds outside a molecular weight range between 100 and 750, compounds with more than six rotatable bonds, and compounds with rings larger than nine atoms were removed. In cases with multiple species in the unit cell, all fragments but the largest one were removed (i.e. counter ions, solvents, etc.). Finally, all duplicate compounds were removed from the dataset. These criteria should reduce the dataset to reasonably small and moderately flexible compounds, resulting in 25 017 structures. After calculating stereo parity values for stereo centers and converting into the MDL SDFile format [13], this dataset was used for the present evaluation study.

Criteria. It was chosen to base this study mainly on the ability of the programs to reproduce experimental structures or features thereof. Despite the fact that even an experimental structure normally just shows one of potentially many reasonable low-energy conformations of a molecule, and that CONCORD and CORINA also just create one single conformation, coverage of X-ray structures will highlight significant tendencies in the ability of the software to generate high-quality structures. The evaluation procedure is inspired by the computational requirements defined in Section 7.1.1. For both programs, a set of quality criteria was determined: the conversion rate, the number of program crashes, the number of stereo errors, the average computation time per molecule, the percentage of reproduced X-ray geometries, the percentage of reproduced ring geometries, the percentage of reproduced chain geometries and the percentage of structures with too close intramolecular contacts (clashes). An X-ray geometry is considered to be reasonably well reproduced if the RMS deviation of the heavy atom positions is less than 0.3 Å. In cases of ring atoms, only flexible rings were considered and the percentages are based on the number of compounds with flexible rings. This excludes trivial rigid cases as phenyl rings from this criterion. A chain geometry is taken to be well reproduced if the RMS deviation of the torsion angles at rotatable bonds is less than 15°. A 3D structure is considered to be free of close contacts if the CCR (the ratio of the smallest nonbonded distance to the smallest acceptable value for this distance) is greater than 0.8.

Programs. CONCORD version 6.1.0 and CORINA version 3.4 were used for this study. CONCORD was run with the following options:

```
%logfile nobrief
%logdef concord.log
%max_atoms 1000
%max_rotors 1000
%max_ring_len 1000
%relax bumps
%relax_mode ccr
%optimize none
%hbond off
%sybyl off
%mdl no2d
%mdldef out.sdf
%connectivity mdl
%status
%input in.sdf
%exit
```

CORINA was run with the "`-d r2d`" option in order to remove structures without generated 3D coordinates from the output.

Results. Table 7.1 summarizes the results of the evaluation study obtained for CONCORD and CORINA. None of the programs crashed or produced any stereo errors. CORINA had a conversion rate close to 100%, whereas CONCORD converted only 91%. However, CONCORD was faster than CORINA with an average

Tab. 7.1 Comparison of CONCORD and CORINA using 25 017 X-ray structures.

	CONCORD	*CORINA*
Conversion rate (%)	90.8	99.9
Program crashes	–	–
Stereo errors	–	–
CPU time (s/molecule)[1]	0.004	0.014
RMS < 0.3 <Å (%)[2]	20	29
RMS^{rings} < 0.3 Å (%)[3]	71	78
RMS^{TA} < 15° (%)[4]	32	43
CCR > 0.8 (%)[5]	96	98

1 On a 2.8 GHz Pentium running Red Hat Linux 9.
2 Percentage of structures with an RMS deviation of the nonhydrogen atoms less than 0.3 Å.
3 Percentage of structures with an RMS deviation of the atoms in flexible rings less than 0.3 Å.
4 Percentage of structures with an RMS deviation of the torsion angles in open-chain portions of less than 15°.
5 Percentage of structures with a close contact ratio of greater than 0.8.

conversion time of 0.004 s per compound compared to 0.014 s per compound for CORINA. Looking at the structure-related quality criteria, the percentages of reproduced X-ray geometries in all criteria are in favor of CORINA. In summary, both programs perform a robust, fast and reasonably good 3D conversion. CONCORD is about 3.5 times faster than CORINA, whereas CORINA has a significantly higher conversion rate of structures with a better reproduction of the experimental geometries on average.

7.4.2 Practical Recommendations

In this section a few practical recommendations are given. Most of them are generally applicable to all programs, some are specific for CORINA – the program the author of this chapter is best familiar with.

File formats. Most programs for 3D conversion accept a number of different input and output formats. The most common ones are MDL SDFile [13], SMILES [14], SYBYL MOL2 [15]and PDB [2]. SMILES and SDFile are the only formats for encoding chemical structures completely with all information on atomic number, bond types, chiral centers and formal charges. Both SMILES and SDFile are most recommended as input formats for 3D structure generation. SMILES cannot store 3D information and is thus not applicable as output format. MOL2 is based on detailed atom types which encode a lot more of the chemical nature of the atoms. This feature is at the same time the strongest limitation of the MOL2 format since there are many chemical features which are not mapped by the available atom types and thus cannot be expressed correctly. PDB is the least suitable format for small molecules since it lacks information on formal charges, chirality and bond types. Thus, it is normally not supported as input format and should not be used as output format for storing 3D structures. Despite being less useful for encoding general molecules, MOL2 and PDB are popular formats in many application programs, and sometimes the only supported input formats. The following recommendations are given for using input and output formats: Use preferentially SMILES of SDFile for input and MOL2 only if you must. Use preferentially SDFile for output and MOL2 or PDB only if you must.

Stereo input. Both SMILES and SDFile use explicit local atom and double bond stereo descriptors. In addition, chirality can be calculated from 3D coordinates given in SDFiles or MOL2. SDFile supports in addition the opportunity to express tetrahedral stereocenters by using wedge bond symbols pointing above or below the plane of 2D structure diagrams. However, drawing 2D structures and assigning wedge bond descriptors opens for a few common pit-falls which lead to ambiguous chirality. Figure 7.19(a) shows a few examples of what is not recommended. Avoid in particular drawing substituents of tetrahedral atoms with 90 or 180° angles between the bond vectors or centers where all substituents point into the same 180° half circle. The same principles can be applied to chiral double bonds. The recommended way to draw a tetrahedral stereocenter unambiguously is to distribute three substituents symmetrically separated by 120° around the

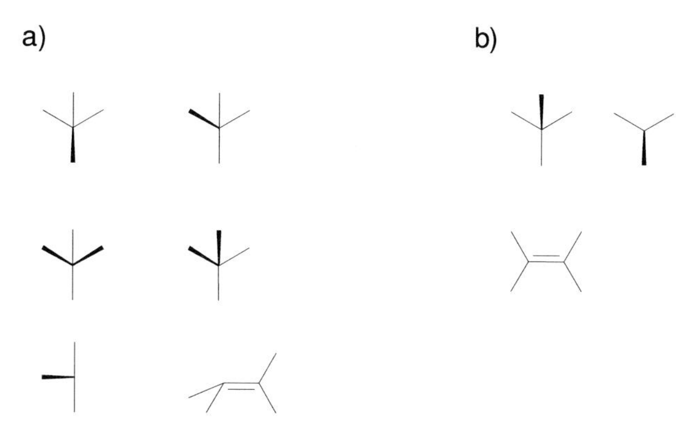

Fig. 7.19 (a) Ambiguous 2D drawings of stereo centers. (b) Recommended encoding of tetrahedral atoms and double bonds.

center and add the fourth substituent with the wedge bond between two of them. In drawings with suppressed hydrogen at the stereocenter, start with the same 120° separation between three substituents and add the wedge symbol to one of them. In case of chiral double bonds, make sure that substituents on the same side of the double bond are placed on opposite sides of the axis along the double bond. Figure 7.19(b) shows examples for correctly drawn tetrahedral atoms and double bonds.

Nitrogen configuration. As discussed above in the section about concepts in 3D structure generation, trivalent nitrogen can adopt several configurations. Three-dimensional generators have to settle for one of them usually by using a set of rules. In cases where this is not sufficient, a few workarounds are available. In case of conjugated trivalent nitrogen, e.g., in aniline, it is possible to force the desired configuration – planar or pyramidal – manually and individually for all atoms by changing the input file. One can use MOL2 as input format and choose an appropriate nitrogen atom type – N.pl3 for planar or N.3 for pyramidal. Alternatively, CORINA accepts an additional feature of SMILES to express atomic hybridization – [N·2] for planar sp^2 and [N·3] for pyramidal sp^3. Thus, a SMILES c1ccccc1[NH2·3] would force the nitrogen in aniline into pyramidal configuration. This is of course not applicable in general when automatically converting large numbers of structures.

Canonical 3D structures. Due to the requirement to generate one single 3D structure rather than multiple low-energy conformations, generator programs have often to choose between a number of possible local conformational details as individual torsion angles. In many cases, these choices are arbitrary between equally reasonable alternatives, e.g. with equal energy. Commonly, in such cases the first alternative of a number of equal possibilities is taken. This causes in turn a dependence of the final geometry from the order of atoms in the molecule and can in consequence lead to different 3D structures depending on the atom numbering. This is normally not a problem since the generated structures are assumed to be equally reasonable. For cases where this is irritating or when generation of exactly the same geometry for different atom numberings is essential, CORINA offers an option for ensuring that a canonical 3D structure independent of the atom order is always obtained. Note, however, that this still will not necessarily lead to similar 3D structures for similar molecules which, for example, share a common fragment. Since a 3D structure generator processes one molecule at a time, this is out of the scope of pure 3D structure generation.

General recommendations. Some additional general recommendations are given here. Before starting the 3D generation process for large databases of molecules, consider removing small fragments, e.g. salts or solvents, to generate stereo information for unspecified centers, to specify the protonation of polar groups and to define the desired tautomers. For some applications which depend on hydrogen information, e.g., docking or pharmacophore searching, consider generating multiple isomers in order to capture all relevant cases. CORINA supports some of these operations implicitly by options for removing small fragments, for charge neutralization and for exhaustive stereoisomer generation. Most programs add missing hydrogen atoms internally during the 3D generation process. Make sure that these hydrogens are also added to the output file since they contain valuable additional 3D information.

7.5 Conclusions

Automatic 3D structure generation has been discussed as a fundamental operation in computational chemistry. It has become a standard procedure in molecular modeling and appropriate software has been available for many years. Several of the most common concepts as well as their strengths and limitations have been shown in detail. An evaluation study of the two most commonly used programs – CONCORD and CORINA – has shown their general applicability for robust, fast and automatic 3D structure generation. Within the limitation of single conformation generation, reasonable rates of reproducing experimental geometries and other quality criteria are reached. For many applications, the obtained 3D structures are good enough for use without any further optimization. In addition, the generated structures can be used for more advanced applications including multiconformer generation as discussed in the next chapter.

References

1 Allen, F. H., Davies, J. E., Galloy, J. J., Johnson, O., Kennard, O., Macrae, C. F., Mitchell, J. F., Smith, J. M., Watson, D. G. The development of versions 3 and 4 of the Cambridge Structural Database system. *J. Chem. Inf. Comput. Sci.* **1991**, *31*, 187–204.

2 Bernstein, F., Koetzle, T. F., Williams, G. J. B., Meyer, E. F., Jr., Brice, M. D., Rodgers, J. R., Kennard, O.,

Schimanouchi, T., Tasumi, M. J. The Protein Data Bank: a computer-based archival file for macromolecular structures. *J. Mol. Biol.* **1977**, *112*, 535–542.

3 Sadowski, J., Gasteiger, J. From atoms and bonds to three-dimensional atomic coordinates: automatic model builders. *Chem. Rev.* **1993**, *7*, 2567–2581.

4 Pearlman, R. S. 3D molecular structures: generation and use in 3D searching. In *3D QSAR in Drug Design*, Kubinyi, H. (ed.), ESCOM, Leiden, **1993**, pp. 41–79.

5 a) Gasteiger, J., Rudolph, C., Sadowski, J. Automatic generation of 3D atomic coordinates for organic molecules. *Tetrahedron Comput. Methodol.* **1990**, *3*, 537–547. b) CORINA, Molecular Networks GmbH, Erlangen, Germany, http://www.mol-net.de.

6 OMEGA, OpenEye Scientific Software, Santa Fe, NM, USA, http://www.eyesopen.com.

7 Henry, D. R., McHale, P. J., Christie, B. D., Hillman, D. Building 3D structural databases: experiences with MDDr-3D and FCD-3D. *Tetrahedron Comput. Methodol.* **1990**, *3*, 531–536.

8 Crippen, G. M., Havel, T. F. *Distance Geometry and Molecular Conformation.* Research Studies Press (Wiley), New York, **1988**.

9 Klebe, G., Mietzner, T., Weber, F. Methodological developments and strategies for a fast flexible superposition of drug-size molecules. *J. Comput.-Aided Mol. Des.* **1999**, *13*, 35–49.

10 Dale, J. *Stereochemistry and Conformational Analysis*, VCH, New York, **1978**.

11 a) Pearlman, R.S. Rapid Generation of high quality approximate 3D Molecular structures. *Chem. Des. Autom. News* **1987**, *2*, 1/5–1/6. b) CONCORD, Tripos Inc., St Louis, MO, USA, http://www.tripos.com.

12 Sadowski, J., Gasteiger, J., Klebe, G. Comparison of automatic three-dimensional model builders using 639 X-ray structures. *J. Chem. Inf. Comput. Sci.* **1994**, *34*, 1000–1008.

13 Dalby, A., Nourse, J. G., Hounshell, W. D., Gushurst, A. K. I., Grier, D. L., Leland, B. A., Laufer, J. Description of several chemical structure file formats used by computer programs developed at Molecular Design Limited. *J. Chem. Inf. Comput. Sci.* **1992**, *32*, 244–255.

14 Weininger, D. SMILES, a chemical language and information system. 1. introduction to methodology and encoding rules. *J. Chem. Inf. Comput. Sci.* **1988**, *28*, 31–36.

15 SYBYL, Tripos Inc., St Louis, MO, USA, http://www.tripos.com.

8
Exploiting Ligand Conformations in Drug Design

Jonas Boström and Andrew Grant

Abbreviations

2D, 3D	two-, three-dimensional
GB/SA	Generalized Born/surface area
HTS	high-throughput screening
MMFF	Merck molecular force field
PDB	Protein Data Bank
PTP1B	protein tyrosine phosphatase-1B
QSAR	quantitative structure–activity relationship
RMS	root mean square
SMARTS	SMiles ARbitrary Target Specification
SMILES	Simplified Molecular Input Line Entry System

8.1
Introduction

Molecules of the simplicity of ethane or the complexity of proteins and DNA adopt different conformations. In the case of ethane this gives rise to the notion of a staggered and eclipsed bond, whereas proteins form an array of complex structural elements and DNA – the famous double helix. The understanding of the conformational properties of small molecules is an important factor in computational approaches contributing to drug discovery.

This chapter summarizes the computational methodologies used for conformational analysis. Specifically, Section 8.1 gives a theoretical outline of the problem and presents details of various implementations of computer codes to perform conformational analysis. Section 8.2 describes calculations illustrative of the current accuracy in generating the conformation of a ligand when bound to proteins (the bioactive conformer) by comparisons to crystallographically observed data. Finally, Section 8.3 concludes by presenting some practical

Molecular Drug Properties. Measurement and Prediction. R. Mannhold (Ed.)

ISBN: 978-3-527-31755-4

applications of using knowledge of molecular conformation in actual drug discovery projects.

8.1.1 Molecular Geometry and Energy Minimizations

The geometry of a molecule determines many of its physical and chemical properties. There are two distinct approaches to the calculation of molecular geometry of molecules, i.e. quantum chemical and molecular mechanics (or force field) methods. These methods are distinguished by the degree in which as models they rely on parameters. *Ab initio* quantum methods invoke approximations of differing levels of sophistication to solve the Schrödinger equation. These methods are generally free from parameters related to conformational energy, but their computational complexity limits their range of applicability. They can be an invaluable tool to obtain accurate information, say about a specific torsional barrier [1]. Molecular mechanics substitutes the quantum description of molecules with classical potential energy functions. These models use simple ideas from physics, e.g. describing the stretching of a chemical bond as a harmonic oscillator. A typical force field expression is given by:

$$E_{\text{tot}} = E_{\text{stretch}} + E_{\text{bond}} + E_{\text{vdw}} + E_{\text{torsion}} + E_{\text{elec}} + E_{\text{other}} \tag{1}$$

in which various terms arise from the stretching of bonds and angles, rotations about bonds, van der Waals and electrostatic interactions between all pairs of atoms, and other terms, e.g. those describing solvation effects. A detailed description of the individual terms in Eq. (1) is given elsewhere [2]. In general, these terms contain parameters that depend on the way atoms are classified. Force fields differ in the precise mathematical representations of the terms in Eq. (1) and in schemes for classifying different atom types. Hence, it is not advisable to attempt to transfer parameters between different force fields. The advantage of force field calculations is that they ideally contain a small number of parameters that can be transferred to a wide range of molecules. The approximations invoked by molecular mechanics are such that calculations are several orders of magnitude faster than *ab initio* quantum methods. It is this performance that facilitates many of the practical applications of conformational analysis to drug discovery described later in this chapter. A comparison of force fields commonly in use is given by the work of Gundertofte et al. [3]. The extent to which force fields are parameterized is a differentiating feature. Recent work proposes reducing the empirical reliance on parameterization and improving the physical description of intermolecular interactions, to address the limitations in the accuracy of current force fields [4]. This work models electron density by Gaussian functions from which accurate energetic contributions are obtained. Currently this approach is applied to the accurate computation of components of the intermolecular energy. However, it is possible that some of the underlying ideas will be adopted in order to obtain better accuracy in the calculation of conformational energetics, ultimately bridging the accuracy gap that exists between quantum and force field methods.

The minimization of the conformational energy given in Eq. (1) as a function of the position of the atoms is central to conformational analysis. Typically the nature of the energy function produces many local minima, referred to as conformers. In order to identify the most energetically favorable conformer it is necessary to combine a global search methodology (for further details, see Section 8.1.2) with local energy minimization. There are many approaches to local energy minimization, which generally differ in the way they use gradient information. The most accurate methods use high-order derivatives of the energy function in Eq. (1). However, the most reasonable compromises between accuracy and computational efficiency tend to be methods that make explicit use of the gradient of the conformational energy, while using various schemes to estimate higher order derivatives. Such techniques are reviewed by Burkert and Allinger [2]. There are two contrasting strategies for carrying out local energy minimization. Perhaps the simplest approach is to allow all of the Cartesian coordinates of atoms to adjust independently during local energy minimization. This approach is straightforward to implement, particularly in consideration of the computation of second- and higher-order gradients. However, it is computationally less efficient than only allowing changes to torsion angles during energy minimization (ensuring bond lengths and angles are constrained). This second strategy is a little harder to implement in the case higher-order derivates are necessary and on occasions can impede the performance of the search for the global minima. However, the significant reduction in the number of degrees of freedom being optimized results in much improved performance.

8.1.2 Conformational Analysis Techniques

Most drug-like molecules adopt a number of conformations through rotations about bonds and/or inversions about atomic centers, giving the molecules a number of different three-dimensional (3D) shapes. To obtain different energy minimized structures using a force field, a conformational search technique must be combined with the local geometry optimization described in the previous section. Many such methods have been formulated, and they can be broadly classified as either systematic or stochastic algorithms.

Systematic searches exhaustively sample conformational space by sequentially incrementing the torsional angles of all of the rotatable bonds in a given molecule. This conceptually simple approach is straightforward to implement, but scales exponentially with respect to the number of rotatable bonds. To control the exponential increase in the number of potential conformers obtained, systematic searches are usually combined with tree-based search techniques taken from computer science. Even the best implementations of systematic searches become impractical beyond several rotatable bonds (typically greater than 10). Stochastic searches are based on probabilistic theories and are better suited to calculations on very flexible molecules. Unlike systematic searches, no attempt is made to enumerate all possible conformations, but rather statistical sampling is used in order to efficiently search conformational space. Typical implementations include Monte Carlo Metropolis sampling, simulated annealing and genetic algorithms. For an extensive review of methods for searching the conformational space of molecules the reader is referred to the excellent work by Leach [5].

8.1.2.1 The Relevance of the Input Structure

Conformational analysis programs require an initial 3D structure, from which conformer ensembles are calculated. However the simplest representation of molecules is generally concerned with describing how atoms are connected to each other and by what type of bonds. This notion of a molecule as a two-dimensional (2D) graph with nodes as atoms and edges as bonds is powerfully exploited by notations such as SMILES (Simplified Molecular Input Line Entry System) [6], SYBYL line notation [7] and InChI [8], which provide compact yet detailed representations of molecules. Such approaches do not provide any information about the 3D arrangement of atoms in molecules. The generation of 3D structures given a 2D graph is a subtle and complex problem, which is reviewed in Chapter 7 of this volume. Ideally both stochastic and systematic approaches to conformational searches should be independent of the initial 3D structure. Nonetheless, practical problems can result in calculated conformational ensembles being significantly influenced by the choice of starting 3D structure. For example, stochastic searches will follow different trajectories in conformational space depending on the starting point of the search. Certain trajectories will encounter energy barriers that "trap" the search in local minima, effectively curtailing the search, and hence potentially producing different conformational ensembles, depending on the choice of the initial conformer. Systematic searches in torsion space can be affected by the choice of bond lengths and particularly bond angles of the initial 3D structure. A poor choice of bond angles that do not accurately reflect local environments for a given 3D structure can introduce erroneous energy barriers arising from steric hindrance. This has been shown to subsequently prevent conformational analysis programs from generating the bioactive conformation [9]. In general, the ideal initial structure is that from experiment (X-ray crystallography), although if not available geometry optimization of the initial structure can prove effective in accounting for the role of local bonding environments on bond-angles.

8.1.3 Software

Early examples of conformational analysis programs with a specialization suitable for rational drug design were WIZARD [10] and MULTIC [11]. The current programs most relevant to drug design are Catalyst [12], ConFlex [13], Confort [14], Flo99 [15], ICM [16], MacroModel [17], MOE [18], OMEGA [19], SYBYL [20] and Tinker [21].

There is a wide variation in computational performance and accuracy of these programs. They differ in their implementation of those details previously described, such as force fields, search algorithms and local optimization methods. Other fundamental differences are the treatment of solvation (see Section 8.2.1.4) and whether the selection of a set of conformers belonging to an ensemble is designed based on conformational energy relative to the global minimum (Section 8.2.1) or aims to be diverse in shape with less attention on conformational energy cutoffs (see Section 8.2.2). To avoid the computational overhead of energy minimization some programs only assign discrete values to torsions. These values are typically based on known experimental distributions of torsions and have, for example, been implemented using SMARTS (SMiles ARbitrary Target Specification) substructure patterns [22].

8.2 Generating Relevant Conformational Ensembles

Critical to computational approaches for supporting drug design projects is the elucidation of bioactive conformations, i.e. the conformation adopted by ligands when bound to a biological target. Given a few ligands known to bind to a certain biological target, determining the details of the bioactive conformation can guide the molecular design of novel compounds. Alternatively for computational screening of large multiconformer databases to identify biological active compounds, it is necessary that sufficient bioactive conformations are generated by the conformational analysis procedure. The need to calculate conformers for a large number of molecules imposes the constraints that the generation is appropriately fast and that conformational ensembles can be stored as efficiently as possible.

The aim of this section is to describe the major issues related to using conformational analysis tools with the goal of maximizing the probability of generating bioactive conformations. We will focus on the two programs that in our experience have proved the most useful for conformational analysis in applications that contribute to drug discovery projects. These are MacroModel [17], which we consider particularly useful for detailed analysis, and OMEGA [19], which is useful both for detailed analysis and for the calculations on large numbers of molecules.

8.2.1 Conformational Energy Cutoffs

A standard approach to conformational analysis is to apply conformational energy cutoffs [23–25]. This pragmatic approach reduces the number of conformations in a calculated ensemble, while at the same time removing energetically unrealistic conformations. In the next section we attempt to obtain optimal values for energy cutoffs guided by analyzing 36 bioactive conformations obtained from crystallographically determined ligand–protein complexes, taken from the Protein Data Bank (PDB) [26].

8.2.1.1 Thermodynamics of Ligand Binding

The free energy required to transform the lowest energy conformation of a ligand in solution to the bound (bioactive) conformation is commonly referred to as the conformational energy penalty. For any chemical equilibrium an increase in the Gibbs free energy of the system of 1.4 kcal mol^{-1} (at 300 °C) decreases the equilibrium constant by a factor of 10. Obviously in terms of protein–ligand binding, conformational penalties reduce the binding affinity in numerically the same fashion. Accordingly, the conformational energy component of ligand binding may significantly influence the affinity of the ligand. One view of protein–ligand binding is that the ligand exchanges its solvation environment from water to that provided by the protein-binding site. The smaller the change in ligand conformation as part of this process the more likely it is to be thermodynamically favorable.

8.2.1.2 Methods and Computational Procedure

To ensure calculation accuracy only very high-quality ligand–protein complexes were considered. The full criteria for selecting the 36 ligands are defined as follows:

- The X-ray structure resolution must be high (≤2.0 Å).
- The "B" factors of the ligands must be low (preferably <30).
- The ligands must not include rotatable bonds that cannot be detected by protein crystallography, e.g. hydroxyl torsions.
- The ligands should not include unusual moieties, for which there are no relevant force field parameters.
- The ligands must be reasonably small, flexible and drug-like.

The ReLiBase+ program [27] was used to select the molecules. The molecular structures are shown in Fig. 8.1, which also displays the PDB entry code.

The conformational energy penalties were calculated as follows. First, the global minimum for each ligand was obtained from an exhaustive conformational analysis using the Mixed Torsional/Low-Mode search [28], implemented in MacroModel version 9.0. This Monte Carlo-based method uses eigenvectors of the Hessian (positional second derivative matrix) with the lowest eigenvalues, to determine search directions in conformational space. These eigenvectors correspond to the low-frequency normal modes and their effectiveness in search algorithms designed to work in Cartesian coordinates is 2-fold: they filter out search directions which effect mainly changes to bond lengths and angles, and, more importantly they identify search directions associated with torsional degrees of freedom including subtle correlated changes between different bond torsions. As such this method is particularly well suited to the difficult task of exploring ring conformations (see Chapter 7). The Merck molecular force fields (MMFFs) [29] and the aqueous (pH 7) ionization states of acids and bases were used. It is important to include the effects of water (see Section 8.2.1.4), therefore the Generalized Born/surface area (GB/SA) solvent model [30] was included.

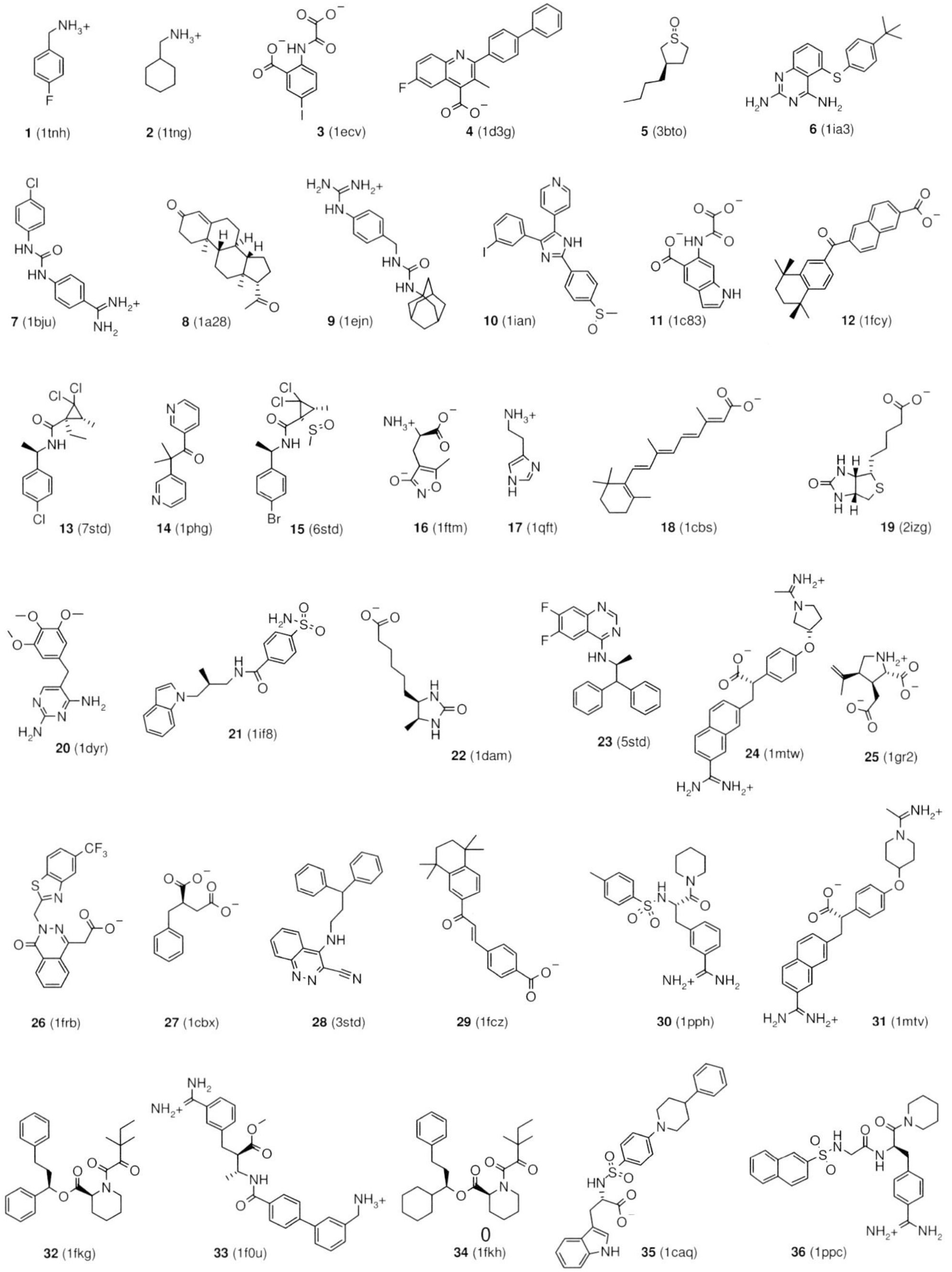

1 (1tnh) 2 (1tng) 3 (1ecv) 4 (1d3g) 5 (3bto) 6 (1ia3)

7 (1bju) 8 (1a28) 9 (1ejn) 10 (1ian) 11 (1c83) 12 (1fcy)

13 (7std) 14 (1phg) 15 (6std) 16 (1ftm) 17 (1qft) 18 (1cbs) 19 (2izg)

20 (1dyr) 21 (1if8) 22 (1dam) 23 (5std) 24 (1mtw) 25 (1gr2)

26 (1frb) 27 (1cbx) 28 (3std) 29 (1fcz) 30 (1pph) 31 (1mtv)

32 (1fkg) 33 (1f0u) 34 (1fkh) 35 (1caq) 36 (1ppc)

Fig. 8.1 Ligand structures and their corresponding PDB codes.

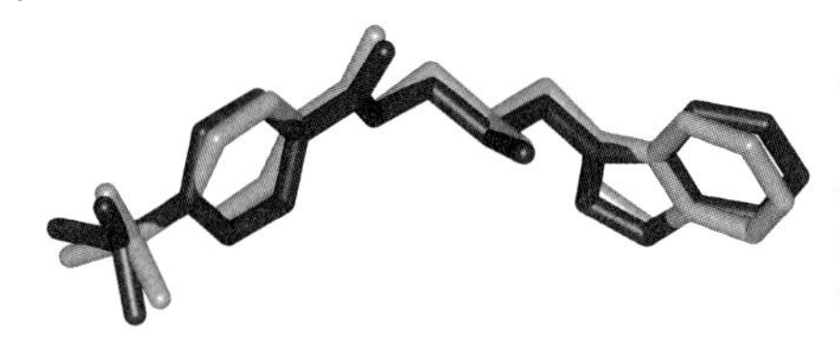

Fig. 8.2 A least-squares superimposition of the unmodified X-ray structure of the protein-bound ligand **21** (dark grey) and the corresponding constrained optimized structure (grey) using flat-bottomed Cartesian constraints with a half-width of 0.8 Å. The RMS value is 0.43 Å. Hydrogens are removed for clarity.

Second, for each ligand the experimentally determined structure was energy minimized also using the solvent model. This procedure is merely to make bond lengths and bond angles consistent with the MMFFs force field, ensuring that deviations in theses terms do not make spurious contributions to the conformational energy. To ensure the integrity of the experimental conformation an additional term is included in the force field, which essentially constrains the maximum positional displacement of any individual atoms. This term is known as a "flat-bottomed" Cartesian constraint, because it permits small atom displacements without introducing any energy penalty. However, beyond a certain displacement the energy penalty increases rapidly. For this study the half-width of the potential was set to 0.8 Å and the force constant was set to 100 kcal mol^{-1} $Å^{-1}$. This resulted in an average root mean square (RMS) deviation between the optimized structure and the X-ray structure of 0.37 Å. An example of a superimposition with an RMS deviation of 0.43 Å is given in Fig. 8.2. Finally the conformational energy penalty was calculated from the difference in the global energy minimum and the conformational energy after constrained optimization.

8.2.1.3 Calculated Conformational Energy Cutoff Values

The results of calculating the conformational energy penalties are displayed in Fig. 8.3. The mean conformational energy penalty is 1.5 kcal mol^{-1} and 94% of the ligands have a value less than 4 kcal mol^{-1}. The two ligands (**35** and **36**) with higher values were calculated to be within an energy threshold of 6 kcal mol^{-1}. It can be seen in Fig. 8.3 that the energy penalty increases with the number of rotatable bonds. The 17 compounds with one to four rotatable bonds all, save one, show conformational energy penalties below 2 kcal mol^{-1}. In addition, ligands **1–4** were bound in their (MMFF–GB/SA) global minimum conformation. The 13 compounds with five to eight rotors were all found within an energy-threshold of 4 kcal mol^{-1}, whereas the six compounds with eight or more rotors all show energy penalties within 6 kcal mol^{-1}. This suggests that energy cutoffs can be applied albeit with different threshold values depending on the number of rotatable bonds.

8.2.1.4 Importance of Using Solvation Models

Molecules with polar atoms or charged groups can form intramolecular interactions such as internal H-bonds or cation–π interactions. As a consequence energy minimization *in vacuo* often results in "electrostatically collapsed" conformations due to the dominance of these intramolecular interactions. Several compounds in

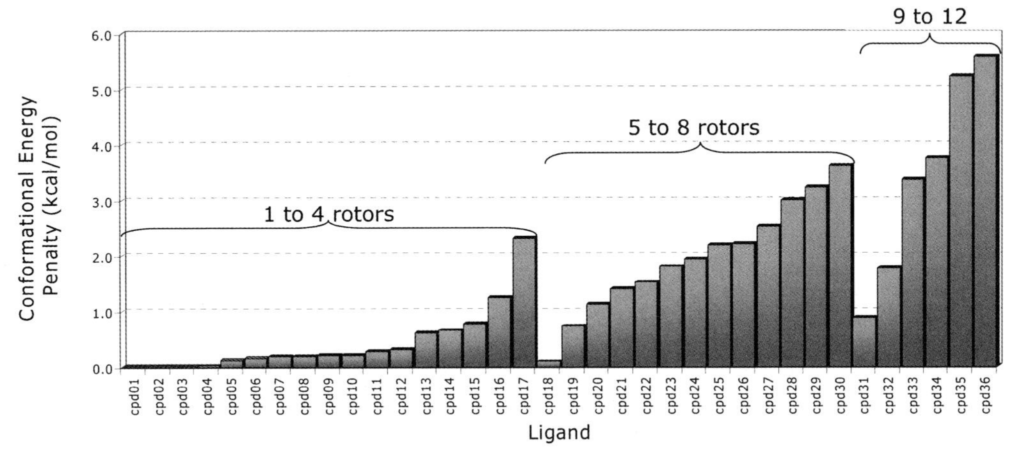

Fig. 8.3 Calculated conformational energy penalties of the protein-bound ligands **1–36**. The energy penalty increases with the number of rotors. Ligands with one to four rotors display values less than 2 kcal mol^{-1}, ligands with five to eight rotors display values less than 4 kcal mol^{-1} and ligands with eight to 11 rotors display values less than 6 kcal mol^{-1}.

the dataset of 36 ligands contain potential intramolecular H-bonds (NH—N, NH—O or NH—Cl). In these cases, a collapsed conformation is more energetically favorable than more extended conformations often observed experimentally. One approach avoiding this is to leave out the electrostatic component of the force field. Obviously this is computationally efficient and relatively effective in not generating conformations dominated by electrostatic intramolecular interactions. However, omitting electrostatic interactions overemphasizes the contribution of favorable van der Waals intramolecular interactions. This in turn can result in "hydrophobically collapsed" conformations in which there is an exaggerated tendency say for bulky nonpolar groups or aromatic rings to interact. A more physically correct treatment is to account for the role of water, particularly its property of screening intramolecular electrostatic interactions. This is a challenging task and not all conformational analysis programs implement solvation models. The most computationally tractable models for solvation treat water as a structureless continuum. The electrostatic component of the solvation energy is calculated by considering the continuum to be a linearly polarizable dielectric. Nonelectrostatic effects are modeled using simple relationships between free energy and surface area, generally derived from the solubility data available for hydrocarbons. For example, MacroModel uses a solvation model that combines a GB model to describe the effects of dielectric polarization with a simple surface-area term to account for nonpolar solvation effects [28]. The overall effect of introducing a solvent term to the force field is that the conformational energy difference between the extended and collapsed forms is reduced. This effect is illustrated in Fig. 8.4 in which the global minimum of compound **17** is calculated to contain an internal H-bond both in gas phase and in solution. The energy difference between the

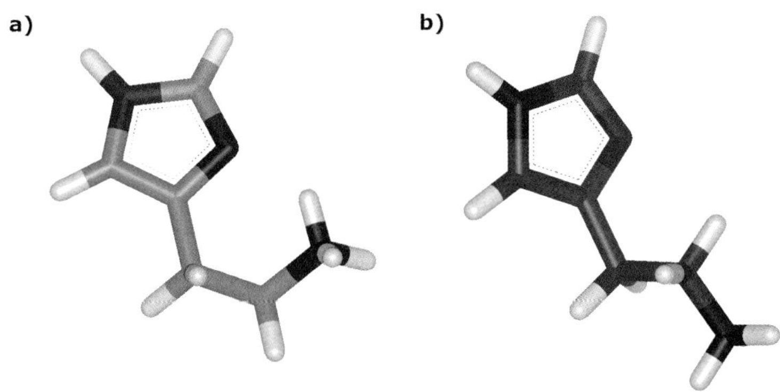

Fig. 8.4 The global energy minima of compound **17** in gas phase and in solution are virtually identical. The energy difference between lowest energy conformation (a) and the bioactive conformation (b) in gas phase is 13.4 kcal mol^{-1}. Although a folded conformation is still the global minimum, the extra competition between solute–solvent interactions and intramolecular H-bonds reduces the energy difference to 2.3 kcal mol^{-1} when using the GB/SA hydration model.

lowest energy conformation and the bioactive conformation (Fig. 8.4) in gas phase is 13.4 kcal mol^{-1}. In solvent, modeled using GB/SA, the corresponding energy difference is only 2.3 kcal mol^{-1}.

8.2.2
Diverse or Low-Energy Conformational Ensembles?

The clearest objective of conformational analysis is to identify the energetically most favored conformer, the global minimum. However, as discussed in Section 8.2.1.1, this is not necessarily the conformation most relevant for ligand binding. Hence, another goal of conformational analysis is to calculate an ensemble of conformers, one of which has a high probability of being similar to the bioactive conformation. There are two distinct strategies for defining which conformers to select for an ensemble. One approach is to keep a very diverse set of conformers, which as best as possible represent the conformational search space. The perceived advantage is that the diversity in conformers offers the best possible chance that a retained conformer is sufficiently like the bioactive conformation. However, the drawback is that often too many of the conformers generated are of no relevance to the bioactive conformation, and are energetically unfeasible. In addition, even the conformation most similar to the bioactive conformation can be too different to be identified as such when using techniques such as pharmacophore searching and 3D shape matching (see Section 8.3.2). An alternative approach is to retain a set of low-energy conformers that are within a certain conformational energy threshold of the global minimum identified by the search procedure. A suitable value for the energy threshold can be inferred from studies such as those described in Section 8.2.1. In general it is unlikely that the bioactive conformation will correspond exactly to that a molecule adopts in solution. Hence there is no advantage in minimizing conformers that are within the conformational energy threshold; indeed this can hinder the identification of the bioactive conformation [22, 25, 35].

The most common measure used to assess differences in conformations is the RMS difference. This is obtained analytically by a least-squares minimization to find the optimal rotational alignment [31, 32]. The extent to which conformers are discarded based on how similar their RMS values are is referred to as duplicate removal. Most conformational analysis programs define this as a parameter that can be adjusted. For example, large values of the duplicate removal parameter can be used to generate diverse conformational ensembles. In the present subsection we demonstrate the influence of the duplicate removal parameter in retrieving bioactive conformations. This is done by comparing the experimental structures of ligands extracted from ligand–protein complexes found in the Brookhaven database [26], with calculated conformational ensembles, computed using different values of the duplicate removal parameter.

8.2.2.1 **Methods and Computational Procedure**

The OMEGA version 1.8 program [19] was used for this experiment. OMEGA couples a systematic search in torsion space with rules for certain torsions. This approach is computationally efficient, and ideal for this large-scale exercise.

OMEGA is controlled by a configuration file where parameters that affect the generated conformational ensembles are defined. For example, "`-ewindow`" defines an upper bound in the conformational energy (relative to the global minimum) and is used to discard high-energy conformations; "`-maxconfs`" is the maximum number of conformations generated for each input structure; and "`-rms`" specifies the RMS value below which two conformations are considered to be the same. In this experiment the "`-ewindow`" and "`-maxconfs`" parameters were kept constant. The "`-ewindow`" parameter was set to 6 kcal mol^{-1} and the "`-maxconfs`" parameter was set to 1000 allowed conformations. These parameter values were established to be appropriate in a previous study using D-optimal design [9]. The effects of the duplicate removal parameter were investigated using the following values: "`-rms`": 0.25, 0.50, 0.75, 1.00, 1.25, 1.50, 1.75, 2.00, 2.25, 2.50, 2.75 and 3.00 Å.

The data set of 36 ligands described in the previous section was used. All conformations in a given ensemble were superimposed on the corresponding unmodified X-ray structure by a least-squares superimposition procedure. Only nonhydrogen atoms were matched. An in-house tool was used for this purpose [33]. A ligand was considered to be reproduced if a conformation of 0.5 Å or less from the unmodified X-ray structure was present in the calculated ensemble. The imposition of the stringent criterion is chosen to reflect our interest in accurately obtaining bioactive conformations, for the purpose of molecular design, pharmacophore elucidation and shape comparison techniques. For example, we have previously shown that a RMS > 0.5 Å, makes the identification of molecules similar to the bioactive conformation, less likely to be successful [9].

8.2.2.2 Reproducing Bioactive Conformations Using Different Duplicate Removal Values

Figure 8.5 shows for different values of the duplicate removal parameter – the number of ligands for which the OMEGA conformation reproduces the bioactive conformation. It can be seen that the value of the duplicate removal parameter has a significant effect on the ability to reproduce the experimental bioactive conformation. Low values of the "-rms" parameter reproduce greater than 65% of the bioactive conformations, whereas for high values the number converges to approximately 30%.

Overall this seems to indicate that attempting to generate diverse conformations by imposing a large value of the duplicate removal parameter is of little utility if the goal is to generate bioactive conformations. Thus, these results cast doubt on the reliability of storing in multiconformer databases "diverse" ensembles for computational screening methods such as shape-matching, rigid-body docking or pharmacophore searching. The success of these methods to identify biologically active compounds depends on a conformation resembling the bioactive conformation being present in the database. Other programs include additional features for removing redundant conformations, like Catalyst's poling algorithm [34] and

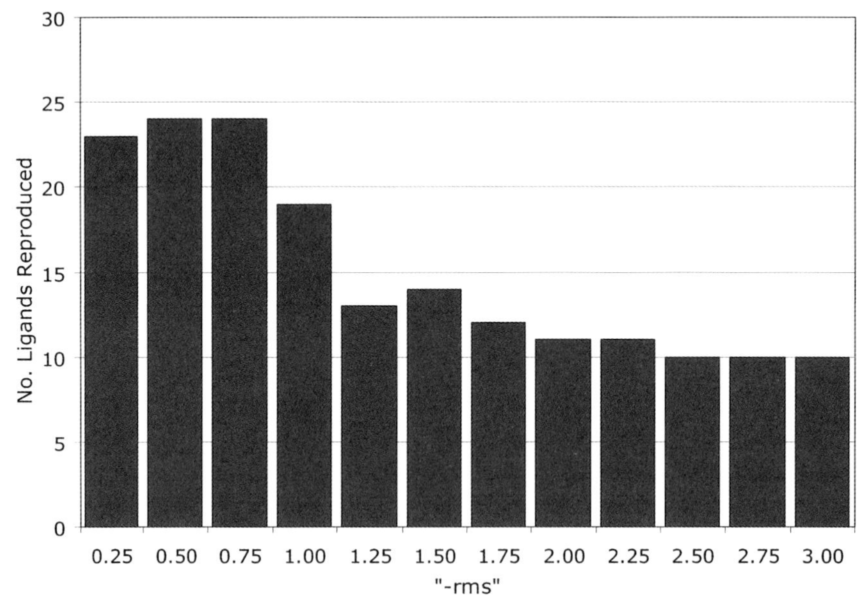

Fig. 8.5 A graph showing how the number of ligands reproduced varies as a function of the duplicate removal parameter "-rms".

Confort's [14] refinement procedure. Previous studies show that these functions do not increase the likelihood of retrieving the bioactive conformation; on the contrary, they have a negative impact [35].

It is also clear from Fig. 8.5 that not all bioactive conformations are found to an accuracy of 0.5 Å. For the eight ligands in the set having eight or more rotatable bonds, none of the bioactive conformations were found. By using a less strict definition of when a bioactive conformation is considered to be reproduced (RMS < 1.0 Å) 31 of the 36 ligands are found in the OMEGA generated ensembles.

8.2.3
Combinatorial Explosion in Conformational Analysis

Conformational analysis is a combinatorial problem. The number of conformations for a molecule with n rotatable bonds is given by:

$$N_{\text{conformations}} = (360/m)^n \quad (2)$$

where m is the torsion angle increment in degrees. For example, given an angle increment of 60°, a molecule with five rotatable bonds has 7776 conformations, whereas for a molecule with six rotatable bonds, using a torsion angle increment of 30° results in 2985984 conformations. That is, the number of conformations increases fast with the number of rotatable bonds and with decreasing angle increment – the so-called combinatorial explosion.

It is, thus, clear that the efficiency of a conformational search method is an important factor to consider. The efficiency depends on how cleverly the method can reduce the number of conformations that are either stored or energy minimized. One of the most common approaches is to exclude conformations prior to any energy minimization by, for example, checking for unphysical close nonbonded contacts. Other alternatives are to apply conformational energy cutoffs (see Section 8.2.1) and to remove "redundant" conformations based on RMS comparisons (see Section 8.2.2). The notion that big pharmaceutical companies at present prefer large collections of molecules to use for both experimental and computational approaches to screening provide a practical imperative for a high performing conformational search algorithm. To illustrate the scale of the problem we have examined the medicinal chemistry part of the GVK BIO compound database [36]. This commercially available database consists of approximately 650000 compounds with chemical and biological information extracted from medicinal chemistry journals. In what follows this database can be considered as a model for a pharmaceutical collection of drug-like molecules. The distribution of the number of rotatable bonds of these compounds is shown in Fig. 8.6. It can be seen that most of these drug-like compounds have three to six rotatable bonds.

To illustrate the number of conformers generated for a typical pharmaceutical like database, OMEGA calculations were carried out for the 450000 compounds

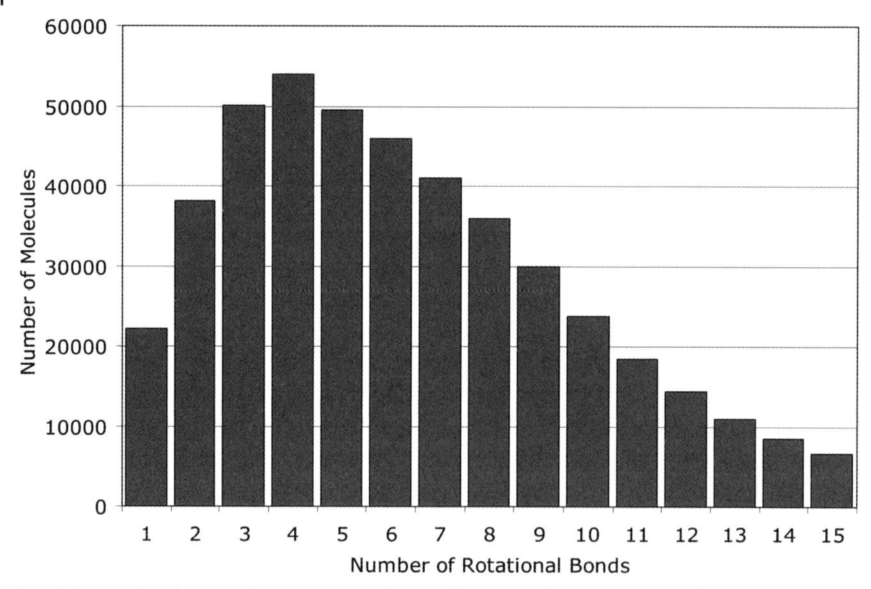

Fig. 8.6 The distribution of the number of rotatable bonds for the compounds in the GVK BIO database.

in GVK BIO with less than 16 rotatable bonds. The computer processor unit requirement for a calculation of this size using OMEGA on a 100 processor (PI 3.0 GHz) Linux cluster is less than 1 h. The average number of conformations as a function of the number of rotatable bonds in a molecule is shown in Fig. 8.7, which shows that the average number of conformations for molecules with three and six rotatable bonds is 24 and 360, respectively. For molecules with eight or more rotatable bonds there are many low-energy conformations, although the number retained is relatively small compared to that expected from a complete systematic search (based on using Eq. 2). The chance of having the bioactive conformation present in such large ensembles has been observed to be low [25, 35]. This is because the approximations made to curtail the combinatorial explosion become less effective as the number of rotatable bonds increases. Current research aimed at improving conformational search algorithms aim to address this issue. It should be noted that these results refer to parameter settings appropriate for OMEGA calculations. However, they can serve as a guide when selecting the number of maximum conformations for other programs as well.

8.2.3.1 Representing a Conformational Ensemble by a Single Conformation

Certain computational methodologies such as some approaches to quantitative structure–activity relationship (QSAR) studies use 3D ligand structures [37, 38]. These methods generally assume that a bioactive conformation has been established for a set of molecules and that these conformers can be aligned in a manner that reflects the relative orientation they would adopt in a binding site. It is thus

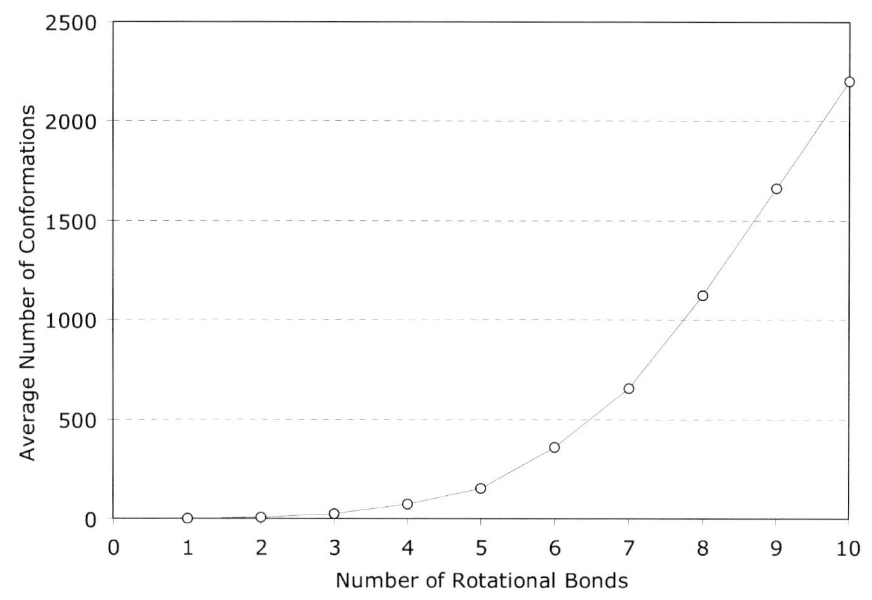

Fig. 8.7 The average number of conformations generated as a function of the number of rotatable bonds for compounds in the GVK Bio database.

obvious that any deviation from an experimentally determined X-ray ligand structure has the potential to introduce some kind of bias into such modeling approaches. In the worse case, this can lead to QSAR models or conclusions being based on erroneous conformations (i.e. not the bioactive conformations) or incorrect molecular alignments.

Some methods claim that their models are insensitive to conformational sampling, and therefore do not require the molecules to be aligned. For example, the ALMOND program is specifically developed for generating and handling alignment independent descriptors [39]. For these types of methods we suggest that it is more appropriate to use rule-based conformer generation, rather than global minimum conformations, when calculating 3D descriptors. The advantage is that the former generates conformers in a consistent manner. CONCORD [40] and CORINA [41] are two frequently used rule-based methods that normally produce just a single conformation. It should be noted that neither developing group claims that their goal, with their 2D-to-3D converter, is to reproduce bioactive conformations. Rather, their aim is to generate one reasonable low-energy conformation (see Chapter 7).

On the basis of the calculations reported in the current study we conclude that conformational energy penalties are in general small, less than 4 kcal mol^{-1}. The inference is that a larger conformational penalty is an impediment to high-affinity binding. Thus, the use of strict energy cutoffs is important, to reduce the number of energetically inaccessible conformations, so long as the force field is a good description of the true energy surface. Energy penalties tend to increase with the number of rotatable bonds. It is important to use the aqueous solution conformational ensembles for these types of calculations. *In vacuo* calculations can be used when only one polar/charged group is present in a ligand. In addition, we suggest that in order to increase the probability of having the bioactive conformation in a calculated ensemble, it is preferable to collect low-energy conformations, as opposed to sampling diverse conformations. Obviously, the more low-energy conformations generated the greater the chance of success in generating the bioactive conformation.

8.3 Using Conformational Effects in Drug Design

This last section presents some practical applications of conformational analysis in drug discovery projects. To illustrate the importance of conformational energy calculations we draw mainly on examples taken from our experience and chosen to reflect different scales of problems that can be addressed.

8.3.1 Conformational Restriction

Ligand binding to a biological target is generally associated with an entropic penalty arising from the loss of conformational degrees of freedom of a molecule. To increase binding affinity a well-known molecular design strategy in medicinal chemistry is to introduce conformational restrictions into molecules, in order to limit the degrees of freedom an unbound molecule can potentially lose. Another reason such "conformational locks" are introduced into molecules is to optimally position certain functional groups so as to mimic structural motifs observed in known inhibitors. Conformational locks can be introduced into molecules by adding substituents to a molecule to create constraints that favor a particular conformation, or by introducing ring closures. The influence on conformation results from intra-molecular interactions (e.g. internal H-bonds) or steric hindrance.

A recent successful example of using conformational locks, leading to high-affinity protein tyrosine phosphatase-1B (PTP1B) inhibitors, was reported by Black et al. [1]. The crystallographically determined X-ray structure of the unsubstituted compound **37** (PDB code: 2bge) revealed that the two rings are orthogonal to each other in the bound conformation, see Fig. 8.8. The energy difference between the orthogonal conformation and the corresponding coplanar conformation (0°) was determined using high-level *ab initio* calculations. The planar form was found to be 4.1 kcal mol^{-1} (MP2/6-31+G*) lower in energy. The introduction of substituents in the *ortho* position was predicted to energetically favor the orthogonal conformation, reversing the orthogonal/coplanar ring relationship and therefore making the bound orthogonal conformation readily accessible. The *ortho*-substituted com-

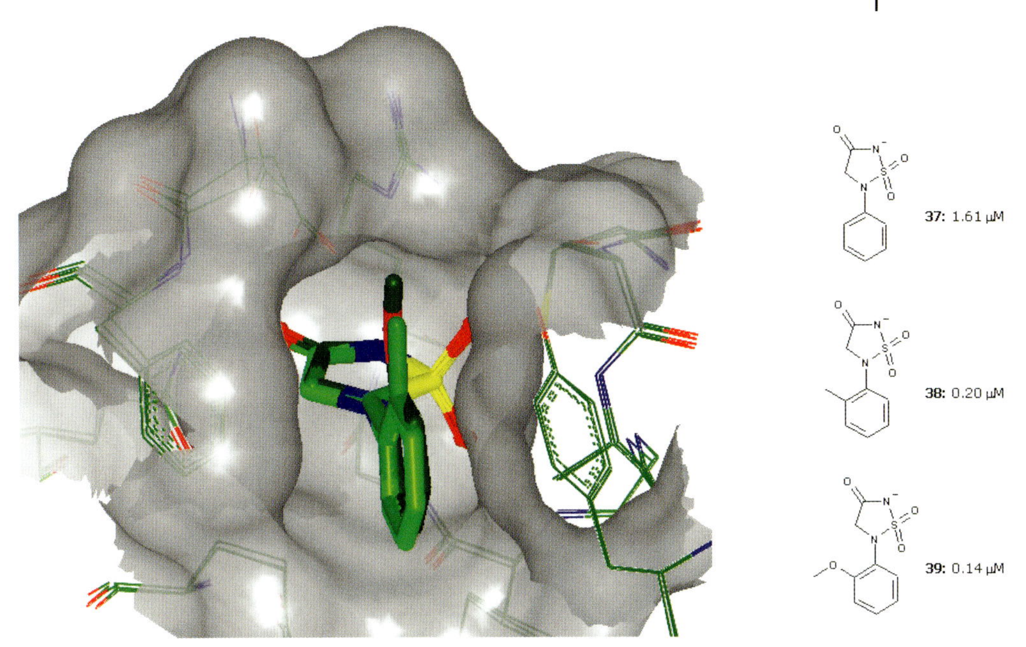

Fig. 8.8 The structures of compounds **37** (carbons colored light green), **38** (carbons colored green) and **39** (carbons colored dark green) bound to PTP1B. The two rings are orthogonal to each other. This is energetically unfavorable for ligand **37**, but favorable for the *ortho*-substituted ligands **38** and **39**. This is reflected in their corresponding IC_{50} values: 1.61 µM for **37**, 0.20 µM for **38** and 0.14 µM for **39** [1].

pounds **38** and **39** were synthesized, and showed significant improvement in inhibiting PTP1B activity, as compared to their unsubstituted counterpart (**37**). The 10-fold increase in affinity can be attributed to the reduced conformational energy penalty of binding. The structures of **37**, **38** and **39** are shown in Fig. 8.8.

Ikegashira et al. reported another recent example of the successful exploitation of conformational locks. They describe the discovery of a novel class of hepatitis C virus NS5B RNA polymerase inhibitors [42]. By designing and synthesizing conformationally constrained analogs of **40** (see Fig. 8.9), they obtained a series of novel compounds with significantly improved potency. Compound **41** was, for example, shown to be 7-fold more potent, see Fig. 8.9.

The rationale for the discovery arose by investigating the effect of various substituents at the phenyl *ortho* position to the indole ring. This showed that a fluorine atom was the most effective substituent for enzyme inhibition [43] and it was established that the potency was influenced by the steric effect of the *ortho* substituent. It was therefore assumed that the fluorine made the dihedral angle between the indole and the central phenyl ring optimal for ligand binding. The angle was calculated to be 47°. It was subsequently hypothesized that fixing this dihedral by a ring closure would increase the potency, by minimizing the entropic loss of the ligand on binding. Various ring systems were designed, synthesized

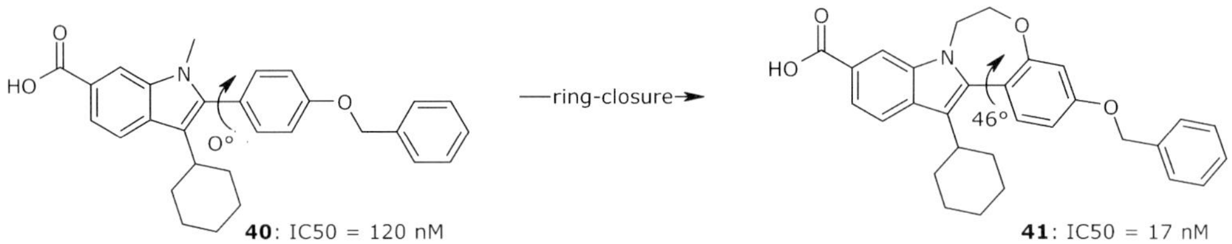

—ring-closure→

40: IC50 = 120 nM

41: IC50 = 17 nM

Fig. 8.9 By fixing a torsional angle to the preferred value (46°) by a ring-closure analog of **40**, a novel series of compounds with significantly better potency was obtained, here illustrated by compound **41**, which is 7 times more potent than **40**.

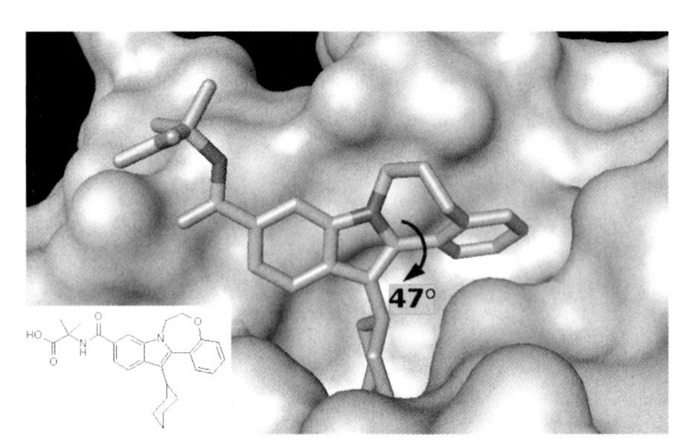

Fig. 8.10 The X-ray crystal structure of a tetracyclic compound bound to hepatitis C virus NS5B544 polymerase. The observed dihedral angle between the indole ring and phenyl ring is 47°, which is in agreement with the predictions.

and tested with the goal of producing a dihedral angle at the ring closure similar to that observed for the compound with fluorine in the *ortho* position. The potency of these compounds was found to correlate with the measured dihedral angle at the ring closure. Compounds with seven-membered ring systems were found to be the most potent and conformational analysis showed that the dihedral angle for these compounds was approximately 50°, which is close in value to the compound prior to introducing the ring closure. The more planar rings (five- or six-membered) and the more twisted rings (eight-membered) were found to be less potent. The actual angle for the bioactive conformation was obtained by solving an analogous compound (see Fig. 8.10) by X-ray crystallography (PDB code: 2dxs). The observed dihedral angle between the indole ring and phenyl ring was found to be 47°, which was in excellent agreement with the angle predicted by calculation.

8.3.2
Shape-Based Scaffold Hopping

Modern assay technologies in drug design make feasible the screening of upward of 1 million compounds. The throughput of such assays enables the rapid identification of molecules that are potential starting points for drug development. Computational screening techniques mirror the goals of high-throughput screening (HTS) experiments. Motivations for such calculations are numerous. For example, they serve as an alternative in projects for which a HTS assay cannot be established or because resource is not available to carry out the experimental screen. Experimental error inherent to HTS can prevent the identification of ligands that only exhibit a weak biological activity, but nonetheless could be useful starting points in a drug discovery project. Computational screening in this case can be used as a useful complement to HTS. There are many techniques for the computational screening of compounds, some of which have a crucial dependence on conformational analysis. An interesting example has been the identification of small-molecule inhibitors of the ZipA–FtsZ protein–protein interaction – a proposed antibacterial target [44].

Experimental screening established that compound **42** shown in Fig. 8.11 disrupts ZipA–FtsZ protein–protein interaction. However, previous studies suggested potential issues with toxicity associated with this class of compounds. Additionally such amine-substituted pyridyl-pyrimidines are heavily patented in the context of kinase inhibition. Both of these factors limit the scope of the subsequent lead optimization process, to transform this compound into a viable drug. Knowledge that compound **42** was a micromolar inhibitor of ZipA–FtsZ was exploited by searching for molecules that were similar in shape.

Crucial to this endeavor was the generation of a multiconformational database of the set of compounds available for experimental screening. This is dependent on the computational performance of the conformational analysis program OMEGA, which readily enables libraries of several million compounds to be generated. The outcome of searching for molecules similar in shape to compound **42** was a new set of micromolar inhibitors. Although weaker than those identified by HTS, the molecules were smaller with scaffolds that did not cause cytotoxicity and were free from patent issues [44]. A subsequent X-ray crystal structure of one of these molecules (**43**) verified the predictions of the molecular conformation produced by OMEGA, as well as the orientation in the ZipA binding site predicted using shape-matching techniques, see Fig. 8.12.

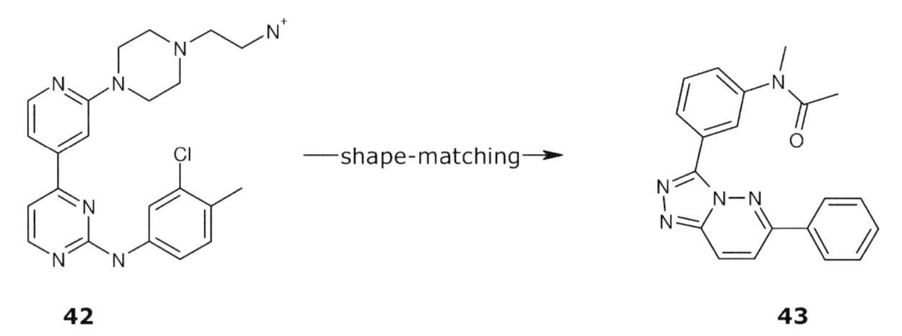

Fig. 8.11 The molecular structures of **42** and **43**. Compound **43** was identified by searching a multiconformational database for molecules that were similar in shape to **42**.

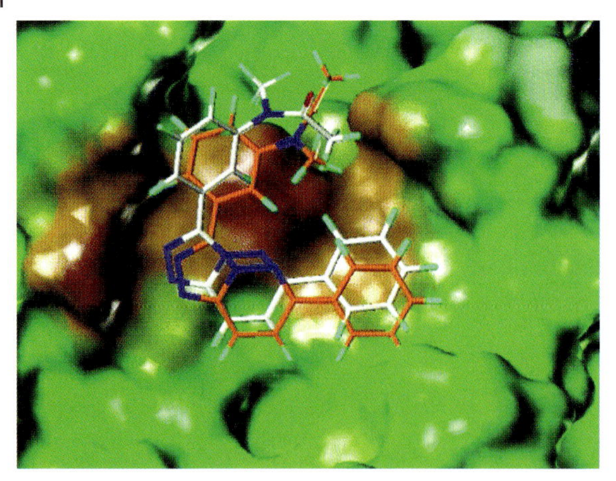

Fig. 8.12 Experimental validation of the predicted conformation and the actual binding mode of compound **43**. The X-ray structure is overlaid with its calculated counterpart (colored in orange).

This approach of combining shape-matching and conformational analysis proved a useful complement to HTS. Some of the compounds identified by the computational screen were not detected in the original experimental screen. This was because their relative weak activity was difficult to separate from the noise of the assay. Nonetheless, these compounds had different scaffolds (i.e. were "lead-hops") compared to the previously known inhibitor. The key contribution from conformational analysis was that the newly discovered inhibitors were not found by the corresponding searches based on 2D methods.

8.4 Conclusions

Drug discovery requires the analysis of increasing amounts of data. Computational chemistry plays a role in organizing and interpreting such data with the goal of making predictions. More specifically, molecular design involves coupling predictions about how modifications to molecular structure are manifested in terms of changes to experimental properties. The dominant methods of establishing similarities between molecules and their relationships to measured data are based on the local connectivity of atoms, and as such reflect how molecules have been historically drawn in medicinal chemistry. A better physically founded approach to molecular modeling is to consider the 3D shape of molecules. However, it has proved difficult to fully realize all of the potential advantages of 3D-based techniques, in part due to the difficulty of obtaining good descriptions of the conformations of molecules. A particular challenge is in determining the conformation of a molecule when it is bound to a given biological target, the so-called bioactive conformation. This chapter has outlined some of the progress to date in conformational analysis, applied to modeling in drug discovery. Despite the numerous approximations involved in both modeling the energetics of conformers and the methodologies involved in searching conformational space, it has been shown that the insights from conformational analysis can contribute to drug discovery projects.

References

1 Black, E., Breed, J., Breeze, A. L., Embrey, K., Garcia, R., Gero, T. W., Godfrey, L., Kenny, P. W., Morley, A. D., Minshull, C. A., Pannifer, A. D., Read, J., Rees, A., Russell, D. J., Toader, D., Tucker, J. Structure-based design of protein tyrosine phospotase-1B inhibitors. *Bioorg. Med. Chem. Lett.* **2005**, *15*, 2503–2507.

2 Burkert, U., Allinger, N. L. *Molecular Mechanics, ACS Monograph 177*, American Chemical Society, Washington, **1982**, pp. 1–72.

3 Gundertofte, K., Liljefors, T., Norrby, P.-O., Pettersson, I. A comparison of conformational energies calculated by several molecular mechanics methods. *J. Comp. Chem.* **1996**, *17*, 429–449.

4 Piquemal, J., Cisneros, G. A., Reinhardt, P., Gresh, N., Darden, T. Towards a force field based on density fitting. *J. Chem. Phys.* **2006**, *124*, 104101–104113.

5 Leach, A. Methods for searching the conformational space of molecules. In *Reviews in Computational Chemistry*, Lipkowitz, K. B., Boyd, D. B. (eds.), VCH, New York, **1991**, pp. 1–47.

6 Weininger, D. SMILES, a chemical language and information system. 1. Introduction to methodology and encoding rules. *J. Chem. Inf. Comput. Sci.* **1988**, *28*, 31–36.

7 Ash, S., Cline, M. A., Homer, R. W. Hurst, T. Smith, G. B. SYBYL line notation (SLN): a versatile language for chemical structure representation. *J. Chem. Inf. Comput. Sci.* **1997**, *37*, 71 –79.

8 The IUPAC International Chemical Identifier (InChI™) is a nonproprietary identifier for chemical substances that can be used in printed and electronic data sources thus enabling easier linking of diverse data compilations. It was developed under IUPAC Project 2000-025-1-800 during the period 2000–2004, http://www.iupac.org/inchi.

9 Boström, J., Greenwood, J. R. Gottfries, J. Assessing the performance of OMEGA with respect to retrieving bioactive conformations. *J. Mol. Graph. Model.* **2003**, *21*, 449–462.

10 Dolata, D. P., Carter, R. E. WIZARD: applications of expert system techniques to conformational analysis. 1. The basic algorithms exemplified on simple hydrocarbons. *J. Chem. Inf. Comp. Sci.* **1987**, *27*, 36–47.

11 Lipton, M., Still, W. C. The multiple minimum problem in molecular modeling. Tree searching internal coordinate conformational space. *J. Comp. Chem.* **1988**, *9*, 343–355.

12 a) Sprague, P. W. Automated chemical hypothesis generation and database searching with catalyst. *Perspect. Drug Discov. Des.* **1995**, *3*, 1–20. b) Sprague, P. W., Hoffman, R. Catalyst pharmacophore models and their utility as queries for searching 3D databases. In *Computer-Assisted Lead Finding and Optimization – Current Tools for Medicinal Chemistry*, Van de Waterbeemd, H., Testa, B., Folkers, G. (eds.),VHCA, Basel, **1990**, pp. 230–240.c)http://www.accelrys.com/products/catalyst.

13 Conflex, San Diego, CA, USA, http://www.conflex.us.

14 Confort, Tripos Inc., St Louis, MO, USA, http://www.tripos.com.

15 a) McMartin, C. Bohacek, R. QXP: powerful, rapid computer algorithms for structure-based drug design. *J. Comput.-Aided. Mol. Des.* **1997**, *11*, 333–344. b) McMartin, C., Bohacek, R. Flexible matching of test ligands to a 3D pharmacophore using a molecular

superposition force field: comparison of predicted and experimental conformations of inhibitors of three enzymes. *J. Comput.-Aided. Mol. Des.* **1995**, *9*, 237–250.

16 ICM, Molsoft L.L.C., La Jolla, CA, USA, http://www.molsoft.com.

17 Mohamadi, F., Richards, N. G. J., Guida, W. C., Liskamp, R., Lipton, M., Caufield, C., Chang, G., Hendrikson, T., Still, W. C. MacroModel – an integrated software system for modeling organic and bioorganic molecules using molecular mechanics. *J. Comput. Chem.* **1990**, *11*, 440–467. http://www.schrodinger.com.

18 MOE (Molecular Operating Environment), Chemical Computing Group Inc., Montreal, Quebec, Canada, http://www.chemcomp.com.

19 OMEGA, OpenEye Scientific Software, Santa Fe, NM, USA, http://www.eyesopen.com.

20 SYBYL molecular modeling software, Tripos Inc., St Louis, MO, USA, http://www.tripos.com.

21 Tinker, Software Tools for Molecular Design, http://dasher.wustl.edu/tinker.

22 Sadowski, J., Boström, J. MIMUMBA revisited: torsion angle rules for conformer generation derived from X-ray structures. *J. Chem. Inf. Model.* **2006**, *46*, 2305–2309.

23 Nicklaus, M. C., Wang, S., Driscoll, J. S., Milne, G. W. Conformational changes of small molecules binding to proteins. *Bioorg. Med. Chem.* **1995**, *3*, 411–428.

24 Boström, J., Norrby, P.-O., Liljefors, T. Conformational energy penalties of protein-bound ligands. *J. Comput.-Aided. Mol. Des.* **1998**, *12*, 383–396.

25 Perola, E., Charifson, P. S. Conformational analysis of drug-like molecules bound to proteins: an extensive study of ligand reorganization upon binding. *J. Med. Chem.* **2004**, *47*, 2499–2510.

26 Bernstein, F., Koetzle, T. F., Williams, G. J. B., Meyer, E. F., Jr., Brice, M. D., Rodgers, J. R., Kennard, O., Schimanouchi, T., Tasumi, M. J. The Protein Data Bank: a computer-based archival file for macromolecular structures. *J. Mol. Biol.* **1977**, *112*, 535–542.

27 Hendlich, M., Bergner, A., Gunther, J., Klebe, G. Design and development of a database for comprehensive analysis of protein–ligand interactions. *J. Mol. Biol.* **2003**, *326*, 607–620.

28 a) Kolossváry, I., Guida, W. C. Low mode search. An efficient, automated computational method for conformational analysis: application to cyclic and acyclic alkanes and cyclic peptides. *J. Am. Chem. Soc.* **1996**, *118*, 5011–5019. b) Kolossváry, I., Guida W. C. Low-mode conformational search elucidated: Application to C39H80 and flexible docking of 9-deazaguanine inhibitors into PNP. *J. Comput. Chem.* **1999**, *20*, 1671–1684.

29 a) Halgren, T. A. Merck molecular force field I–V. *J. Comp. Chem.* **1996**, *17*, 490–641. b) Halgren, T. A. Merck molecular force field VI–VII. *J. Comp. Chem.* **1999**, *20*, 720–748.

30 Still, W. C., Tempczyk, A., Hawley, R. C., Hendrickson, T. Semianalytical treatment of solvation for molecular mechanics and dynamics. *J. Am. Chem. Soc.* **1990**, *112*, 6127–6129.

31 a) Kabsch, W. A solution for the best rotation to relate two sets of vectors. *Acta Crystallogr.* **1976**, *A32*, 922–923. b) Kabsch, W. A discussion of the solution for the best rotation to relate two sets of vectors. *Acta Crystallogr.* **1978**, *A34*, 827–828.

32 Theobald, D. L. Rapid calculation of RMSDs using a quaternion-based characteristic polynomial. *Acta Crystallogr.* **2005**, *A61*, 478–480.

33 Personal communication, Jens Sadowski, AstraZeneca R & D, Mölndal, Sweden.

34 Smellie, A., Teig, S. L., Towbin, P. Poling: promoting conformational variation. *J. Comp. Chem.* **1995**, *16*, 171–187.

35 Boström, J. Reproducing the conformations of protein-bound ligands: a critical evaluation of several popular conformational searching tools. *J. Comput.-Aided Mol. Des.* **2001**, *15*, 1137–1152.

36 GVK Biosciences, Kundanbagh, Begumpet, Hyderabad, India, http://www.gvkbio.com.

37 Cramer, III, R. D., Patterson, D. E., Bunce, J. D. Comparative molecular field analysis (CoMFA). 1. Effect of shape on binding of steroids to carrier proteins. *J. Am. Chem. Soc.* **1988**, *110*, 5959–5967.

38 Klebe, G., Abraham, U., Mietzner, T. Molecular similarity in a comparative analysis (CoMSIA) of drug molecules to correlate and predict their biological activity. *J. Med. Chem.* **1994**, *37*, 4130–4146.

39 Pastor, M., Cruciani, G., McLay, I., Pickett, S., Clementi, S. GRid-INdependent descriptors (GRIND): a novel class of alignment-independent three-dimensional molecular descriptors. *J. Med. Chem.* **2000**, *43*, 3233–3243.

40 CONCORD, Tripos Inc., St Louis, MO, USA, http://www.tripos.com.

41 a) Gasteiger, J., Rudolph, C., Sadowski J. Automatic generation of 3D-atomic coordinates for organic molecules. *Tetrahedron Comput. Methodol.* **1990**, *3*, 537–547. b) CORINA, Molecular Networks GmbH, Erlangen, Germany, http://www.mol-net.de.

42 Ikegashira, K., Oka, T., Hirashima, S., Noji, S., Yamanaka, H., Hara, Y., Adachi, T., Tsuruha, J. I., Doi, S., Hase, Y., Noguchi, T., Ando, I., Ogura, N., Ikeda, S., Hashimoto, H. Discovery of conformationally constrained tetracyclic compounds as potent hepatitis C virus NS5B RNA polymerase inhibitors. *J. Med. Chem.* **2006**, *49*, 6950–6953.

43 Ishidaa, T., Suzuki, T., Hirashimaa, S., Mizutani, K., Yoshida, A., Ando, I., Ikeda, S., Adachi, T., Hashimoto, H. Benzimidazole inhibitors of hepatitis C virus NS5B polymerase: identification of 2-[(4-diarylmethoxy)phenyl]-benzimidazole. *Bioorg. Med. Chem. Lett.* **2006**, *16*, 1859–1863.

44 Rush, T. S., Grant, J. A., Mosyak, L., Nicholls, A. A shape-based 3-D scaffold hopping method and its application to a bacterial protein–protein interaction. *J. Med. Chem.* **2005**, *48*, 1489–1495.

9
Conformational Analysis of Drugs by Nuclear Magnetic Resonance Spectroscopy

Burkhard Luy, Andreas Frank, and Horst Kessler

Abbreviations

1D, 2D, 3D	one-, two-, three-dimensional
CCR	cross-correlated relaxation
DG	distance geometry calculations
DMSO	dimethylsulfoxide
E.COSY	exclusive correlation spectroscopy
etCCR	exchange-transferred CCR
etNOE	exchange-transferred NOE
etNOESY	exchange-transferred NOESY
etPCS	exchange-transferred PCS
etRDC	exchange-transferred RDC
fMD	free MD
HMQC	heteronuclear multiple-quantum correlation
HSQC	heteronuclear single-quantum correlation
MD	molecular dynamics
NMR	nuclear magnetic resonance
NOE/ROE	nuclear Overhauser effect
NOESY/ROESY	nuclear Overhauser enhancement spectroscopy
PDMS	poly(dimethylsiloxane)
PCS	pseudo-contact shift
PRE	paramagnetic relaxation enhancement
RCSA	residual chemical shift anisotropy
RDC	residual dipolar coupling
rMD	restrained MD
RQC	residual quadrupolar coupling
SA	simulated annealing
STD	saturation transfer difference spectroscopy
TOCSY	total correlation spectroscopy

Molecular Drug Properties. Measurement and Prediction. R. Mannhold (Ed.)

ISBN: 978-3-527-31755-4

9.1 Introduction

Nuclear magnetic resonance (NMR) spectroscopy is, next to X-ray diffraction, the most important method to elucidate molecular structures of small molecules up to large biomacromolecules. It is used as a routine method in every chemical laboratory and it is not the aim of this article to give a comprehensive review about NMR in structural analysis. We will concentrate here on liquid-state applications with respect to drugs or drug-like molecules to emphasize techniques for conformational analysis including recent developments in the field.

The power of NMR results from the detection of every atom (except oxygen, for which no suitable isotope exists) in a molecule in its specific environment. Each of the magnetically active isotopes is characterized by its specific resonance frequency at a given magnetic field, determined by its gyromagnetic constant γ. In addition, chemically different nuclei of the same isotope have different resonance frequencies. They are characterized by their so-called chemical shift. The mutual interactions of these nuclei, which occur through chemical bonds or through space, can be used to elucidate the molecular structure in solution. Furthermore, NMR not only yields the information of the connectivity of all atoms within the molecule, the so-called constitution, as well as the spatial arrangements of atoms or groups, the configuration and conformation, but also information about the internal dynamics within the molecules and the interaction to other molecules, such as surrounding solvent molecules or biomacromolecules. Drugs and their interactions can therefore be studied in solution under conditions which are close to the realistic biological conditions.

NMR-based structural studies always begin with the elucidation of the constitution. This is almost identical with the assignment of NMR signals to atoms or groups in the molecule of interest. Whereas assignment of as many as possible signals is required for the elucidation of the three-dimensional (3D) structure, a full assignment of all NMR signals is not necessary for establishing the constitution, e.g. a phenyl ring is easily identified by the characteristic proton chemical shift range of about 7–9 ppm and an assignment of the proton signals within the aromatic ring is not required. However, when distances between protons are used in the conformational analysis, it is important to assign their specific constitutional positions. The main source for establishing the constitution (assignment) are scalar coupling constants (also called J-couplings), which are transmitted via chemical bonds and strongly depend on the kind and number of bonds between the magnetically active nuclei (^{1}H, ^{13}C, ^{15}N, ^{19}F, ^{31}P, etc.). It depends on the questions asked and the amount of material available, if isotope labeling is needed for all or selected nonhydrogen atoms, the so-called heteronuclei like ^{13}C and ^{15}N, which are usually in low natural abundance. This also determines the technique which is used for the assignment. A variety of coupling constants can be used for establishing the connectivities (direct couplings: $^1J_{CH}$, $^1J_{NH}$, $^1J_{CC}$, $^1J_{NC}$; geminal couplings $^2J_{HH}$, $^2J_{NH}$, $^2J_{CH}$ and vicinal couplings $^3J_{HH}$ or long-range couplings $^nJ_{HH}$ with $n>3$, or $^nJ_{NH}$, $^nJ_{CH}$ with $n>1$). Even if the constitution is already known, the partial or complete assignment of the NMR signals to the nuclei in distinct positions in the molecules is required for further analysis.

Elucidation of the stereostructure – configuration and conformation – is the next step in structural analysis. Three main parameters are used to elucidate the stereochemistry. Scalar coupling constants (mainly vicinal couplings) provide information about dihedral bond angles within a structure. Another way to obtain this information is the use of cross-correlated relaxation (CCR), but this is rarely used for drug or drug-like molecules.

The second, but probably the most important parameter is the nuclear Overhauser effect (NOE), which yields direct spatial distances between hydrogen atoms. Sometimes NOE is also used for other nuclei. NOE is caused by relaxation between magnetic dipoles after introducing a nonequilibrium population of a specific nucleus. This phenomenon is observed by determining the change of the population of neighboring protons induced from the nonequilibrium population of a proton. As the build-up of the NOE is inversely proportional to the sixth power of the distances between the nuclei, it can be used to determine intramolecular distances by following the kinetics of the build-up. It should be noted that one-dimensional (1D) NOE-difference measurements do not show a defined distance dependence, since the NOE measured in a selectively excited spectrum via difference spectroscopy is a mixture of the kinetic NOE and the steady-state NOE, which is independent of the distance between the nuclei. Hence, NOE difference measurements are only useful for qualitative interpretation [1]. Accurate distance measurements can be obtained by transient NOE or two-dimensional (2D) NOESY experiments. The sign of the NOE is either negative or positive, where a negative NOE is a decrease of signal intensity when a neighboring nucleus is saturated. The sign of the NOE depends on the size of the molecules, the magnetic field strength, the solvent and the temperature. For molecules of medium size, the NOE can be close to the sign change and the detection of the NOE in the rotating frame (ROE) is recommended. Instead of using ROE for distance determination, it is also possible to change the temperature, e.g. in the measurement of the drug cyclosporin A in $CDCl_3$ at lower temperatures large negative NOEs were observed in NOESY spectra [2].

More recently, residual dipolar couplings (RDCs) and other anisotropic NMR parameters are used for structural analysis (see Sections 9.2.2 and 9.2.3). To measure such anisotropic parameters, the solute must be (partially) aligned. The recent development of suitable alignment media makes the approach applicable to a wide range of drug and drug-like molecules. Since anisotropic parameters are proportional to the averaged orientation of the molecule relative to the static magnetic field as an external reference, long-range correlations within a molecule can be established which allow the determination of the structure with a so far unknown precision. Since the more general applicability of anisotropic NMR parameters was established only very recently, we give a quite detailed description of state-of-the-art techniques in this field.

Highly interesting for most pharmaceutical applications is the conformation of a ligand bound to its receptor. In Section 9.3, exchange-transferred experiments will be reviewed that make it possible to determine the conformation of the ligand in complex with the receptor by detecting the free ligand. This class of methods is not limited by the molecular weight of the receptor and can even be applied to membrane proteins, as long as the ligand binding affinity is in the intermediate range ($10^{-6} < K_D < 10^{-3}$ M).

If all nuclei are assigned and the spectral parameters for the conformational analysis are extracted, a conformation is calculated – usually by distance geometry (DG) or restrained molecular dynamics calculations (rMD). A test for the quality of the conformation, obtained using the experimental restraints, is its stability in a free MD run, i.e. an MD without experimental restraints. In this case, explicit solvents have to be used in the MD calculation. An indication of more than one conformation in fast equilibrium can be found if only parts of the final structure are in agreement with experimental data [3]. Relaxation data and heteronuclear NOEs can also be used to elucidate internal dynamics, but this is beyond the scope of this article.

It is important to note that most molecules are not *rigid* but may prefer a distrinct structure and the conformation of a molecule strongly depends on its specific environment. Hence, the crystal structure of a drug does not have to correspond to the receptor bound conformation. Also, a conformation in solution depends on the nature of the solvent and measuring conditions, and may change when the molecule is bound to the receptor [4]. In addition, different receptors or receptor subtypes can bind the same drug in different conformations. It is a general assumption and observation, but by far not a strict condition, that the conformation in aqueous solution is similar to the bound conformation and is a better representation of the "bioactive conformation" than an X-ray structure of the isolated molecule in the crystalline state.

To give an example: The conformation of cyclosporin A in organic solvents [2, 5] and in the crystal [2] strongly differs from the conformation of cyclosporin A when it is bound to its receptor cyclophilin [6]. When cyclosporin A is dissolved in more polar solvents, several conformations are observed in the NMR spectra [7, 8], but the solubility in water is too low to allow a structure determination in this solvent. Another example is the cyclic pentapeptide Cilengitide [cyclo (Arg-Gly-Asp-D-Phe-[Nme]-Val)] [9], which is being developed as an antiangiogenic drug by Merck (Darmstadt, Germany), where the conformation in water resembles closely the conformation bound to the head group of the $\alpha_v\beta_3$ integrin [10], but the conformation of Cilengitide is drastically changed in the crystal [11] by forming stacked nanotube-like associates.

Summing up, small molecules are more sensitive to surface effects to influence their conformation than large biopolymers. Hence, we are convinced that conformational analyses in solution without explicit treatment of the solvent are artificial and have to be taken with greatest caution.

The size restrictions of this chapter does not allow us to treat all methods and applications which are important in drug design and discovery in more detail. Although the basics for their understanding are described, screening technologies, for example, have been extensively reviewed recently [12–15] and are almost completely neglected here.

9.2 NMR Parameters for Conformational Analysis

The oldest and most widely used structural restraints in NMR spectroscopy are distance restraints derived from NOE experiments [1]. Transient NOE, 2D NOESY and ROESY spectra provide valuable information for interatomic distances up to 5 Å that will be discussed in the following.

Instead of measuring only the time-dependent dipolar interaction via NOE, it is also possible to determine dipolar couplings directly if the solute molecule is partially aligned in so-called alignment media. The most important resulting anisotropic parameters are RDCs, but residual quadrupolar couplings (RQCs), residual chemical shift anisotropy (RCSA) and pseudo-contact shifts (PCSs) can also be used for structure determination if applicable.

Angular restraints are another important source of structural information. Several empirical relationships between scalar couplings and dihedral angles have been found during the last decades. The most important one is certainly the Karplus relation for ^{3}J-couplings. Another, relaxation-based angular restraint is the so-called CCR between two dipolar vectors or between a dipolar vector and a CSA tensor.

9.2.1 NOE/ROE

Nuclear spins can be considered as dipoles that interact with each other via dipolar couplings. While this interaction leads to strongly broadened lines in solid-state NMR spectroscopy, it is averaged out in isotropic solution due to the fast tumbling of the solute molecules. In liquid-state NMR spectroscopy, the dipolar interaction can only be observed indirectly by relaxation processes, where they represent the main source of longitudinal and transverse relaxation.

Relaxation can best be understood by looking at a simple heat model. If we put a hot piece of metal in porridge, the heat of the metal will dissipate into the highly viscous bath (it *relaxes*) until the equilibrium state is reached in which the temperature of the metal and the porridge is identical. Shortly after we put the metal into the porridge, only the porridge molecules very close to the metal will be heated or, in other words, the temperature of molecules depends on their distance to the metal. If we measure the time course of the temperature at a specific distance to the metal, we can derive the effective heat exchange rate at this distance. When the distance dependence of the heat exchange rate is known, we even can back-calculate the distance.

In terms of NMR spectroscopy, the hot piece of metal corresponds to a single excited spin, the porridge is the surrounding spins and the equilibrium state is the Boltzmann equilibrium. The effective heat exchange rate in this case is called cross-relaxation rate and describes the magnetization transfer between spins. For an isolated pair of ^{1}H nuclei with their magnetization oriented along the static magnetic field and considering only dipolar interactions, it is given by:

$$\sigma^{\mathrm{NOE}} = \frac{\hbar^2 \mu_0^2 \gamma^4}{40\pi^2} \cdot \frac{\tau_c}{r^6} \left\{ -1 + \frac{6}{1 + 4\omega_0^2 \tau_c^2} \right\}$$

with $\hbar$ for Planck's constant divided by 2π, μ_0 the permeability of vacuum, γ the ^{1}H gyromagnetic ratio, r the distance between the two relaxing nuclei, ω_0 the Larmor frequency of the protons and τ_c the characteristic correlation time for the tumbling of the molecule.

The decisive part for the cross-relaxation rate is the r^{-6} dependence of σ^{NOE} with respect to the internuclear distance r. Since only τ_c is variable for a given spectrometer frequency and can be considered constant for a rigid molecule under defined conditions, measurement of the internuclear relaxation rate directly provides distance information within the molecule of interest. The cross-relaxation rate can be measured, for example, from cross-peak intensities in 2D NOESY spectra. In NOESY experiments the cross-relaxation induces a build-up of cross-peak intensities as calculated in Fig. 9.1(A) for a typical drug-like molecule with a correlation time of 0.1 ns for several internuclear distances at a spectrometer frequency of 600 MHz (which corresponds to the ^{1}H Larmor frequency ω_0). Integration of cross-peaks at a certain mixing time τ_m, e.g. at $\tau_m = 300$ ms, where the build-up curve still increases approximately linear, directly provides the r^{-6}-encoded distance information (linear two-spin approximation). With the calibration of experimental cross-peak intensities by a known interproton distance, e.g. $r_{HH} \approx 1.76$ Å within a CH_2 group or $r_{HH} \approx 2.49$ Å for vicinal aromatic protons, distance restraints for structure calculations are readily available via the relationship $\sigma^{\mathrm{NOE}} \times r^6 = \text{constant}$.

For a reliable extraction of distances, it is important that dipolar relaxation is strongly dominating other relaxation processes. Hence, it is important to avoid paramagnetic ions or molecules such as transition metals or (paramagnetic) oxygen. Especially solution of small molecules therefore have to be carefully degased.

The method outlined so far is generally applicable for obtaining distance information, but several technical limitations have to be considered when choosing the experimental setup and interpreting the cross-peak integrals. The main limitation for medium-sized drugs is the dependence of σ^{NOE} on ω_0 and τ_c (Fig. 9.1B). For molecules for which $\omega_0\tau_c \approx 1.12$, the cross-relaxation rate σ^{NOE} is close to zero and magnetization transfer between nuclei cannot be observed. On a 600-MHz spectrometer at room temperature this condition is fulfilled for globular molecules with a molecular weight of roughly 500 g mol^{-1} for dimethylsulfoxide (DMSO) and

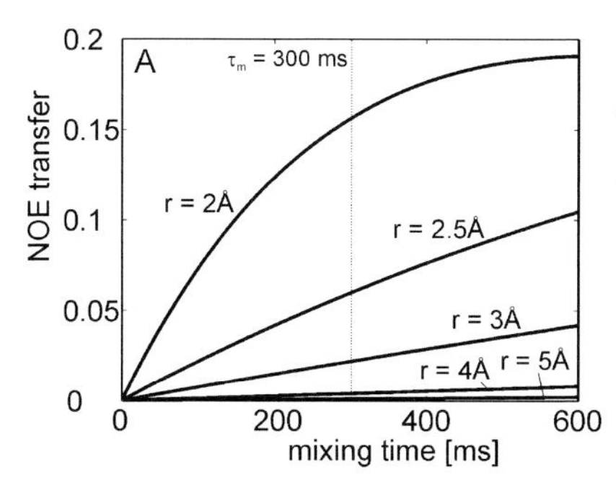

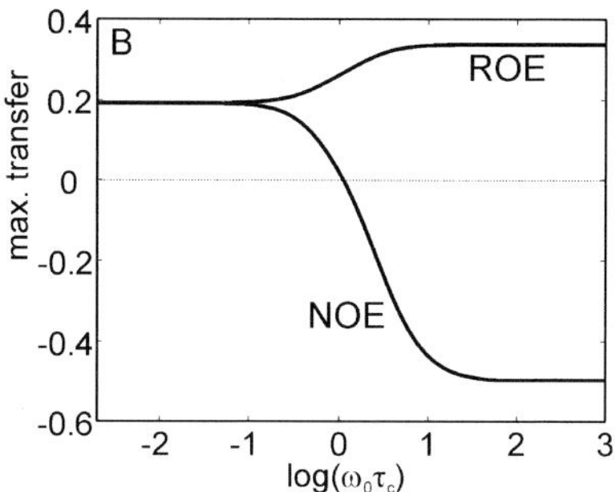

Fig. 9.1 The internuclear transfer of magnetization via NOE cross-relaxation in an isolated spin-pair. (A) Build-up curves for the cross-peak intensity in a 2D NOESY experiment for various internuclear distances *r*. The dashed line indicates a typical mixing time τ_m=300 ms used for drug-like molecules. For the simulation a correlation time τ_c=0.1 ns is assumed for two protons at ω_0=600 MHz. (B) Maximum transfer efficiency for an isolated proton spin pair calculated using only dipolar relaxation processes. Note the sign change for the NOE cross-relaxation at $\omega_0\tau_c$=1.12.

2000 g mol^{-1} for $CDCl_3$ as the solvent. A more general estimate of τ_c can be based on Stokes' law for spherical molecules:

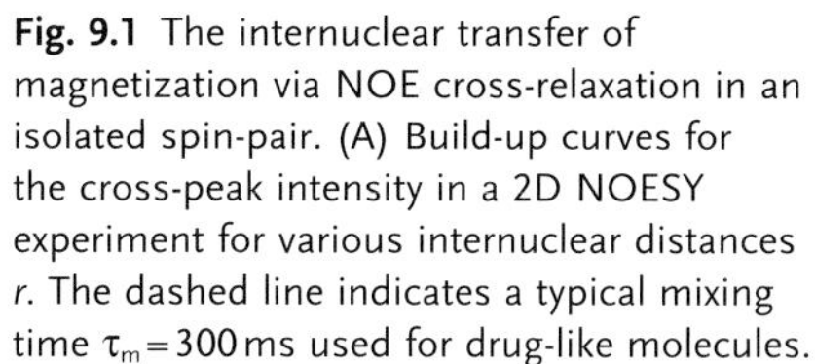

$$\tau_c = \frac{4\pi\eta r_m^3}{3k_B T}$$

with the viscosity of the solvent η, the effective hydrodynamic radius r_m including a potential solvent shell, Boltzmann's constant k_B and the temperature T. This is very roughly $10^{-12} \times$ MW (in Da) at room temperature in an low-viscosity organic solvent. For the relatively viscous and widely used solvent DMSO at room temperature, the majority of drug molecules falls in the τ_c range for weak NOE cross-relaxation and the extraction of distance information out of NOESY spectra is limited if not impossible. In these cases a viable alternative is the rotating frame NOE or ROE, which is based on cross-relaxation in the transverse plane perpendicular to the static magnetic field. The cross-relaxation rate for ROE is given by:

$$\sigma^{\mathrm{ROE}} = \frac{\hbar^2\mu_0^2\gamma^4}{40\pi^2} \cdot \frac{\tau_c}{r^6}\left\{2 + \frac{3}{1+\omega_0^2\tau_c^2}\right\}$$

with the previously described meaning of the variables. The same r^{-6} dependence applies and, as shown in Fig. 9.1(B), the magnetization transfer in the ROE case is always positive. For most drug and drug-like molecules the so-called ROESY experiment therefore is the preferred alternative for obtaining distance information.

The accuracy of obtained cross-relaxation rates and therefore the derived distance restraints is influenced by many practical aspects. A critical point is the choice of the right pulse sequence. The classical difference-NOE experiment with soft CW irradiation on a selected spin does *not* lead to conclusive distance restraints. Selective irradiation requires a continuous irradiation for some time (e.g. 0.2 s) [16]. In the heat model, this case can be compared with the introduction of an active heater into the porridge bath that will retain a certain temperature. The heater will continuously heat the surroundings and there is certainly a distance dependence of the temperature of the porridge with respect to the heater, but since we do not know this distribution *a priori* we cannot determine an exact heat exchange rate. Translated into the difference NOE experiment this means that we have a distance dependence of the strength of the NOE, but it does not obey the r^{-6} dependence with respect to the internuclear distance r and cannot accurately be used in structure calculations.

It is important to note that many structures in the literature are described which are based on such inconclusive data and the results in these cases are highly questionable. Pulse sequences that lead to more conclusive data are the 1D transient-NOE experiment with the selective inversion of a specific spin [1] or, preferably, 2D NOESY [17] and ROESY [18] experiments. The basic pulse sequence schemes for the three experiments are shown in Fig. 9.2. Extensions to the sequences are manifold and will not be discussed here. The central element in all sequences is the mixing time τ_m, which corresponds to the mixing time shown in Fig. 9.1(A). Depending on the correlation time τ_c of a molecule, the mixing time should be chosen short enough to prevent errors from non-linear terms in the NOE/ROE build-up curves and long enough to provide sufficient cross-peak intensities. For small to medium-sized drug-like molecules, mixing times of 1 s to 100 ms are feasible with typical standard values of 200 ms for ROESY and 300 ms

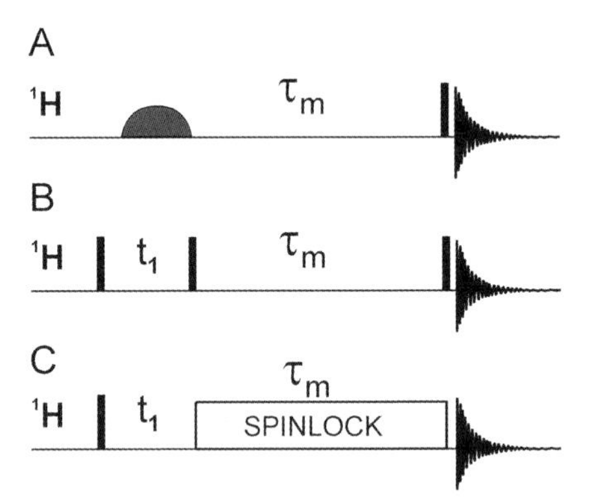

Fig. 9.2 Schematic representation of the three basic experiments useful for the determination of σ^{NOE}: (A) transient NOE experiment, (B) 2D NOESY and (C) 2D ROESY. The gray-filled half-circle represents a frequency-selective inversion pulse which inverts the spin to which the cross-relaxation rate shall be measured. All filled bars are 90° pulses, phases and t_1-time incrementation for sign-sensitive measurement in the indirect dimension can be applied in the usual States, TPPI or States-TPPI manner. For the spin-lock sequence, see references given in the text.

for NOESY experiments. Distances are most reliably determined if a series of NOESY/ROESY experiments with different mixing times are recorded and used for fitting the build-up curves for every cross-peak individually, but measurements for a single, carefully chosen mixing time also give good results.

Positive cross-relaxation as shown in Fig. 9.1 will lead to an increase in signal intensity of a transient-NOE experiment. In 2D NOESY and ROESY spectra, instead, it will lead to cross-peaks with opposite sign relative to the *diagonal peaks*. Negative cross-relaxation, as observed for NOESY spectra for large molecules with long correlation times, yields cross and diagonal peaks of identical sign. Potentially misleading cross-peaks in NOESY and ROESY spectra as a result of chemical exchange, spin diffusion or total correlation spectroscopy (TOCSY) transfer (*vide infra*) all have the same sign as the diagonal peaks. However such cross-peaks can be distinguished in 2D experiments with positive cross-relaxation (small molecules). Sometimes contributions to NOESY/ROESY cross-peaks resulting from such sources only lead to lowered intensities resulting in indistinguishable errors in the distance determination.

In all experiments, so-called zero quantum coherences can evolve via scalar couplings and obscured dispersive antiphase cross-peaks might result. Such coherences relax for longer mixing times τ_m, but can also be suppressed, e.g. by a recently reported zero quantum suppression scheme [19] that generally improves the appearance of NOESY and ROESY spectra. Fortunately, dispersive antiphase signals have an overall integral intensity of zero and do not cause errors in the quantification of NOEs.

Particular care has to be taken when implementing ROESY experiments. The spin-lock, which holds the spins along a defined axis perpendicular to the static magnetic field, can be realized in many different ways and is still an active field of research [18, 20]. In most spin-lock sequences the conditions for undesired TOCSY transfer are partially fulfilled and especially cross-peaks close to the diagonal or antidiagonal might not be accurately interpretable. Since in most cases the effectiveness of the spin-lock also depends on the chemical shift offset, an offset-dependent correction has to be applied to the measured cross-peak intensities [20].

In addition to potential experimental errors, a number of systematic errors have to be considered when interpreting data. As can be seen in Fig. 9.1(A), the NOE/ROE build-up curve is not perfectly linear even in the case of two isolated spins. It is affected by the auto-relaxation rate and higher-order transfers that relax magnetization forth and back to the original spin (Fig. 9.3A). Especially at long mixing times and strong cross-peaks for short interproton distances (as for the $r = 2$ Å case in Fig. 9.1A), cross-relaxation rates will be underestimated. On the other hand, magnetization in larger spin systems can travel from one spin to its neighboring spin and further to a third spin. This effect, which is called spin diffusion, will lead to an overestimation of cross-relaxation rates for distant spins when a "short-cut" exists via intermediate spins (Fig. 9.3B). Spin diffusion is least efficient at short mixing times which therefore must be considered more reliable for extracting distance restraints.

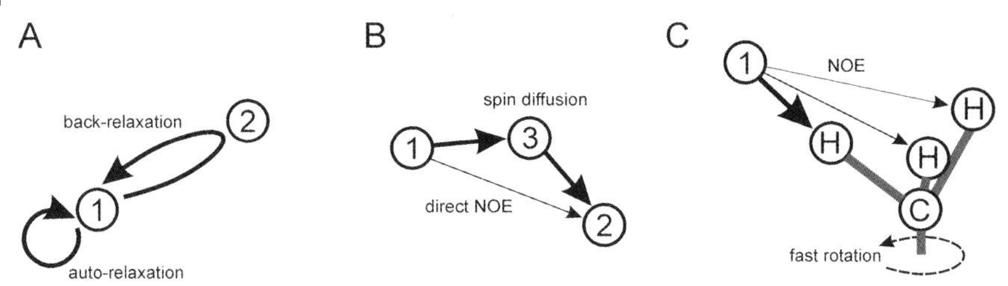

Fig. 9.3 Special NOE transfer pathways that can lead to a wrong interpretation of σ^{NOE} rates. (A) Spin 1 experiences auto-relaxation and back-relaxation that is transferred to spin 2 and relaxed back to 1. (B) A "shortcut" for NOE relaxation via spin 3 close to spins 1 and 2 can lead to a significant contribution to the cross-peak intensity for the direct NOE between spins 1 and 2. (C) For fast rotating methyl groups the intensity of NOE transfer from spin 1 is mainly determined by the closest distance to spin 1. This is generally the case for conformationally averaged distances.

The typical timescale of an NMR experiment is in the millisecond range. All NMR parameters, including $\sigma^{NOE/ROE}$, are averaged over this timescale. If a molecule undergoes significant conformational changes, the r^{-6} dependence of cross-relaxation rates will in turn lead to an overestimation of conformations with shorter interproton distances. One example is the fast rotation of methyl groups, where the measured cross-relaxation rate to another proton is mainly determined by the closest distance (Fig. 9.3C). In such cases, the restraint limits for the upper distances must be increased accordingly in structure calculations.

Localized motion can also lead to local variations in correlation times. Folded peptides with unfolded C- or N-terminal residues, for example, will have varying correlation times for the rigid and flexible parts of the molecule, resulting in different cross-relaxation rates. Such effects can usually be distinguished by the linewidths and intensities of the corresponding diagonal signals, since the auto-relaxation rates also depend on the correlation time.

It should be noted that the internal calibration from one or more cross-peaks between protons of known distances underlies the same experimental and systematic errors. In a high-resolution structure determination the initial calibration therefore should be used only as a first estimate. After first structure calculations, distances can be back-calculated and a larger number of distances can be used for recalibrating the restraints. Best results are obtained, if the complete relaxation matrix is derived and used for NOESY/ROESY cross-peak intensity fitting as for example performed by the program CORCEMA [21].

Considering all potential experimental and systematic errors of NOE/ROE cross-peak intensities, it is remarkable how robust the derived distance restraints still are. The reason lies in the r^{-6} dependence of the cross-relaxation rate: even if a cross-peak intensity is determined wrongly by a factor 2, the resulting distance restraint is only affected by the factor $\sqrt[6]{2} \approx 1.12$, which usually lies within the error range of distance restraints used in structure calculations. It should be further noted that the quality of a resulting structure is not so much determined by the

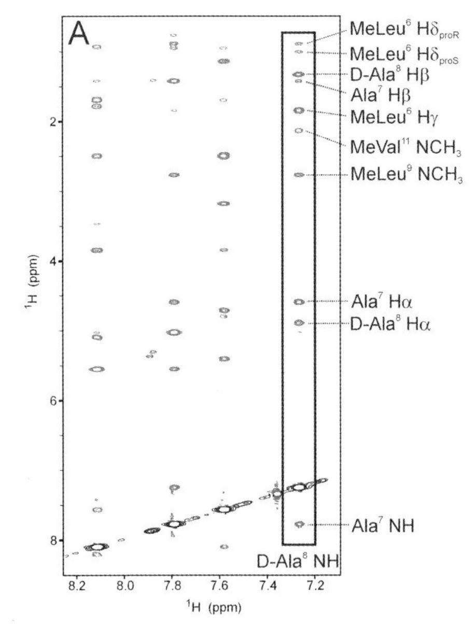

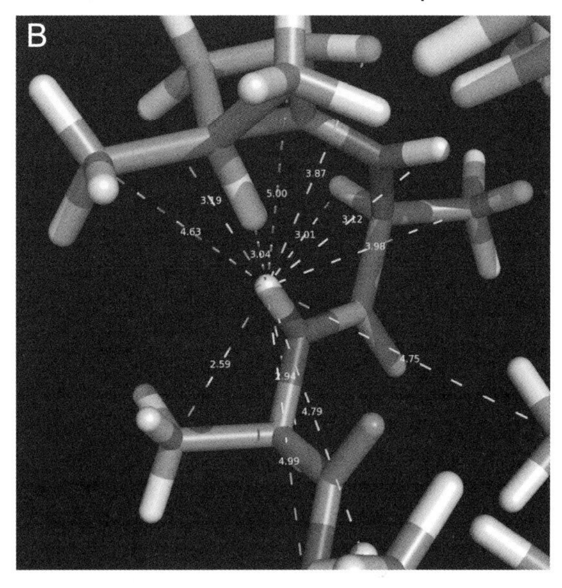

Fig. 9.4 The fixation of the Ala^8 H^N proton in cyclosporin A via ROE-derived distance restraints as an illustrative example for NMR structure determination. (A) The amide region of the 2D ROESY spectrum of cyclosporin A at room temperature with the cross-peaks to Ala^8 H^N indicated by a box. (B) Distances to Ala^8 H^N in the final structure indicated by dashed yellow lines as a result of the derived distance restraints. The distance to the methyl carbon is given as a representative for methyl groups.

error of a single distance restraint, but by the number of distance restraints per atom.

An illustrative example for a structure calculation based on ROE-derived distance restraints is shown in Fig. 9.4 for cyclosporin A dissolved in $CDCl_3$ [2, 22]. After assignment and integration of cross-peaks, signal intensities have been offset-corrected, calibrated and translated into distance restraints. A simulated annealing (SA)-based structural model of an average structure that fulfils all derived distance ranges has been calculated using the program XPLOR-NIH. Although all distance restraints can be considered to have a certain error, the large number of restraints per atom adds up to a highly reliable structure.

9.2.2 Residual Dipolar Couplings (RDCs)

Although RDCs were reported as early as 1963 by Englert and Saupe [23], their measurement and applicability for structural investigations has been limited to very small molecules with typically less than 10 protons until a few years ago. RDCs of a solute molecule can only be obtained if it is at least partially aligned with respect to the magnetic field in a so-called alignment medium. For conventional liquid crystalline alignment media, spectra get very complex with very large, overlapping multiplet patterns. This changed with the introduction of sufficiently weak alignment media that made RDCs a very useful parameter in structure determination. In biomolecular NMR spectroscopy, the measurement of RDCs is a standard technique [24, 25] and few structures are reported today without their use as structural restraints. For water-insoluble molecules arbitrarily scalable alignment media have only recently been introduced. Impressive first results demonstrate the potential of RDCs in the field of drugs and drug-like molecules.

9.2.2.1 Dipolar Interaction

In a simplistic picture, spins can be looked at as magnets with an inherent rotation at the Larmor frequency. Although spins are not oriented directly along the static magnetic field B_0, the integration over time of the fast rotating magnets yields a resulting magnetic moment parallel or antiparallel to B_0 (Fig. 9.5A and B). The magnetic moment of a spin results in the same magnetic field as a classical magnet with the typical r^{-3} dependence of the magnetic field with respect to the distance to the magnet and the $3\cos^2\theta - 1$ dependence with respect to the angle θ relative to the axis of the magnetic moment. Since the magnetic moment of the spin is oriented along the static magnetic field B_0, the angle θ is identical to the angle Θ with respect to B_0.

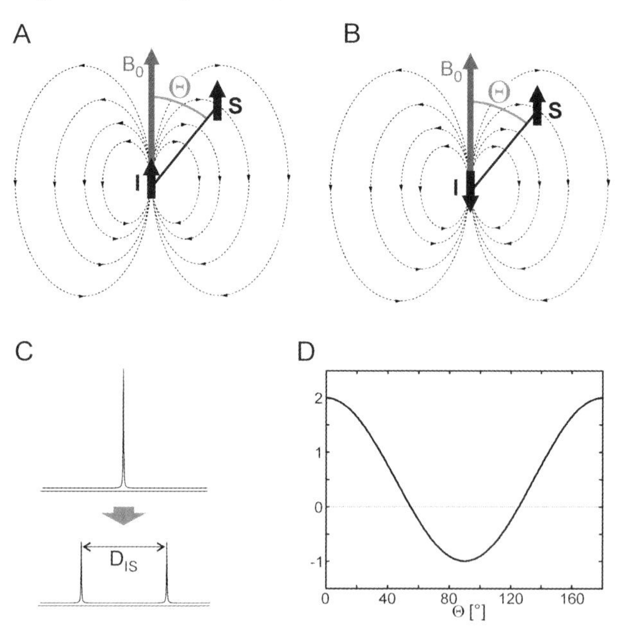

Fig. 9.5 Illustration of the dipolar interaction. (A and B) The magnetic field induced by spin I adds up to the static magnetic field B_0 and leads to a shift of the resonance frequency of the close-by spin S. Since spins parallel and antiparallel to the magnetic field are equally distributed, this leads to a splitting with the dipolar coupling D_{IS} (C). The $3\cos^2\Theta - 1$ angular dependence is shown in (D).

A second spin close in space feels the magnetic field contribution of the magnetic moment of the first spin and therefore resonates at a slightly different frequency as compared to an isolated spin. Since spins parallel and antiparallel to B_0 are about equally populated, the field contributions of neighboring spins lead to a split signal with the distance in Hertz between the doublet components being the so-called dipolar coupling (Fig. 9.5C). The dipolar coupling D_{IS} between two spins I and S is described by:

$$D_{IS} = 2 \cdot \frac{\hbar \gamma_I \gamma_S \mu_0}{16\pi^2 r_{IS}^3} (3\cos^2\Theta - 1)$$

with constants as defined in Section 9.2.1, the angle Θ as the angle between the internuclear vector and B_0, the gyromagnetic ratios γ_I and γ_S of the two spins and their distance r_{IS}.

Dipolar couplings are typically on the order of several thousand Hertz. In solid-state powder spectra, these couplings lead to broad lines devoid of chemical shift resolution. In liquid-state high-resolution spectra, however, dipolar couplings are averaged to zero with a significant loss of valuable structural information. Although distance restraints from dipolar contributions to relaxation processes as, for example, NOE (Section 9.2.1) still provides the most widely used structural NMR parameter, even more information can be obtained if dipolar couplings are detected directly. To be able to measure the desired additional structural information, an intermediate state between solid and liquid has to be reached. With the help of alignment media, solute molecules are only oriented for a time average of about 0.05% which, for example, reduces a 46 kHz dipolar coupling between a carbon and its directly attached proton to a RDC of only 23 Hz. This coupling adds or subtracts to the direct $^1J_{CH}$-coupling in the order of 130–160 Hz. Hence, a coupling of this size can accurately be measured and does not significantly decrease resolution.

9.2.2.2 **Alignment Media**

Three different ways of introducing partial alignment are known today: (i) alignment in a liquid crystalline phase, typically a lyotropic mesophase, (ii) alignment in a stretched gel, and (iii) orientation via paramagnetic ions. Liquid crystals have been the first alignment media introduced and a large number of different systems is known (see, e.g. Ref. [26]). They orient spontaneously in a magnetic field and the weak interaction with the solute produces the desired partial orientation. Liquid crystalline phases, however, have a first-order phase transition and therefore a limitation to a minimum alignment. For most systems, this limit already results in strongly broadened lines and their application is only possible with a special experimental set-up using sample spinning with variable angles [27, 28]. Only few lyotropic mesophases, mostly aqueous systems, result in sufficiently weak orientation to even produce useful alignment for macromolecules. Stretched gels were introduced by Deloche and Samulski [29]. Their alignment strength is solely determined by the amount of mechanical stretching so that arbitrary scaling of RDCs can generally be achieved. Paramagnetic alignment requires a specific ionic binding site for the solute molecule of interest, which either is present naturally or in some cases can be engineered by a paramagnetic tag. The self-aligning paramagnetic ion not only introduces RDCs, but also an effect known as PCS. In the following, some of the renowned alignment media shall be summarized for the most common NMR solvents.

In aqueous solution, most alignment media have been designed for the use of ^{15}N and/or ^{13}C isotope-labeled proteins and nucleic acids, but can also be used for organic molecules at natural abundance. Useful lyotropic mesophases like bicelles, filamentous phage or crystallite suspensions are collected in Ref. [24]. Cross-linked polyacrylamide forms the basis of most stretched aqueous gels used today. Next to the neutral and inert polyacrylamide also copolymers of acrylamide with charged polymeric units have been shown to provide excellent aligning properties (see, e.g. Ref. [30] and references therein). A special gel for partial orientation is represented by gelatin [31], since it is chiral and not covalently cross-linked. Stretching of the gels can be achieved in several ways. Methods range from simple swelling of the dried polymer stick [31] (Fig. 9.6), drying on a glass capillary before reswelling [32], gel compression with a Shigemi plunger [30], to a complex apparatus for squeezing a readily swollen gel into an NMR tube [33]. A promising new approach is the stretching apparatus using a rubber tube to stretch gelatin inside [34]. Paramagnetic tags for proteins (see, e.g. Ref. [35] and references therein) and nucleic acids [36] are readily available. Their use for drugs and drug-like molecules, however, should be strongly limited, since the metal ion often induces conformational changes and paramagnetic relaxation causes line broadening beyond direction.

For aprotic polar organic solvents only few widely applicable alignment media are known. The lyotropic mesophase of poly(γ-benzyl-L-glutamate) (PBLG) with DMF [37] has the disadvantage of a relatively strong minimum alignment. Bicellar

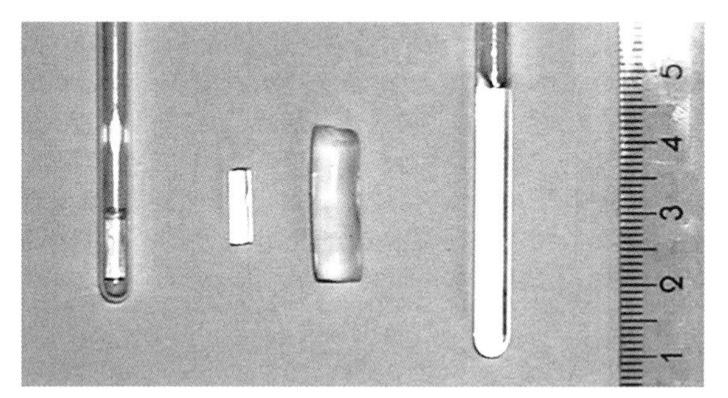

Fig. 9.6 Photograph of cross-linked polymer stick in different states of swelling for the potential use as alignment medium. From left to right: unswollen polymer stick in a standard 5-mm NMR tube, the unswollen polymer stick after polymerization, the polymer stick swollen in chloroform, and the polymer stick swollen inside the NMR tube by adding chloroform to the dry stick. The swollen polymer stick inside the tube is automatically stretched. Molecules diffused into the swollen gel will be partially oriented [42].

lyotropic phases with DMSO as a solvent component [38] provide weak alignment within a narrow but useful temperature range. Most promising are gel based alignment media, like cross-linked poly(acrylonitrile) [147], poly(vinylacetate) [39], or negatively charged PH-polyacrylamide [40], all of which align with DMSO as the solvent. Poly(vinylacetate) was shown to also be compatible with other polar solvents like methanol, acetonitrile, acetone or ethylacetate [39].

Apolar organic molecules can be aligned with a vast number of alignment media. Out of liquid crystalline phases, however, only few systems show sufficiently weak aligning properties for larger organic molecules, e.g. 4-*n*-pentyl-4′-cyanobiphenyl-*d*19 (PCBP) [41] or poly(γ-benzyl-L-glutamate) and poly (γ-ethyl-L-glutamate) (PELG) with the use of variable angle sample spinning. With chloroform, dichloromethane, or tetrahydrofuran as the preferred solvents, cross-linked gels of polystyrene [42, 43], poly(dimethylsiloxane) (PDMS) [44], poly(vinylacetate) [39], poly(methylmethacrylate) (unpublished results), and poly(urethane) (unpublished results) represent arbitrarily scalable alignment media. In this context PDMS cross-linked via accelerated electrons [44] stands out because of the low number of NMR signals which makes it especially useful for solutes at low concentrations.

In all cases, alignment media will interact with the solute. The choice of alignment medium therefore is mainly determined by the solute compatibility with respect to charge, specific hydrophobic interactions or even chemical reactions if, for example, radicals are still present in radically synthesized polymers.

9.2.2.3 **Measurement of RDCs**

If molecules are weakly aligned, RDCs simply contribute to the splitting of scalar couplings. The difference between couplings measured in an isotropic sample and the corresponding partially aligned sample directly results in the RDCs. In principle, most experiments for measuring scalar couplings can be used this way to extract RDCs. In practice, however, the required precision for measuring RDCs is higher than for scalar couplings, where usually deviations on the order of 1 Hz are easily tolerated and the sign information with respect to the coupling is of minor importance. Therefore, existing methods have been strongly revised and extended during the last decade and further techniques can be expected in the future. Since drug-like molecules are usually only available at natural abundance, we focus on this situation and neglect many experiments specifically designed for labeled macromolecules.

The most easily measured RDCs are along one-bond heteronuclear couplings like $^{1}J_{CH}$ or $^{1}J_{NH}$. If the alignment strength is adjusted correctly, RDCs are significantly smaller than the corresponding scalar couplings of known sign, so that a sign-sensitive measurement of RDCs is most easily achieved. Since the distance between directly bound nuclei is usually well-known and fixed, RDCs can also directly be translated into relative angular information. Conventional heteronuclear multiple-quantum correlation (HMQC) or heteronuclear single-quantum correlation (HSQC) experiments can be recorded without heteronuclear decoupling during acquisition and the observed splitting gives the sum of scalar and dipolar couplings (see, e.g. Ref. [42]). More sophisticated methods take into account phase distortions due to mismatched transfer delays, or use spin state selective detection for enhanced resolution [45, 46]. Also quantitative-J [47], J-modulated [48] and J-evolved [49, 50] experiments are approaches with significantly enhanced resolution, although increased measurement times are necessary.

A second class of easily measured RDCs with fixed geometry are along homonuclear $^{2}J_{HH}$ scalar couplings, e.g. in methylene and methyl groups. A number of sophisticated experiments have been developed for the sign-sensitive measurement of these couplings [51, 52], with probably the P.E.HSQC as the most simple experiment suited for the investigation of drug-like molecules [53]. A problem with the extraction of homonuclear two-bond couplings are frequently observed second-order artifacts which often prohibit a simple extraction of coupling constants.

Especially in molecules with few protons available, the measurement of long-range heteronuclear RDCs is of importance. Many important experiments exist, as for example of HMQC/HMBC-type [54, 64] and HSQC-TOCSY-type [55–57] schemes. Only HSQC-TOCSY-type experiments with exclusive correlation spectroscopy (E.COSY) multiplet patterns provide accurate sign-sensitive coupling measurement, but are limited to heteronuclei with a directly attached proton. Additional methods, like HMQC-based experiments, result in accurate determination of long-range couplings also to quaternary carbons, but provide only magnitude information. Therefore the application of the latter method is of limited use when measuring RDCs.

Homonuclear RDCs can be measured using quantitative-J and COSY-type [58, 59] measurement techniques for protons, and INADEQUATE [60] for neighboring carbons at natural abundance. In principle, E.COSY-type experiments provide sign-sensitive information for $^{1}H,^{1}H$ three-spin systems, but the increased linewidths of aligned samples makes the extraction of couplings very difficult. Constant-time-COSY and quantitative-J techniques involve transfer steps to heteronuclei, which makes them insensitive at natural abundance. The insensitivity of INADEQUATE correlating two heteronuclei at natural abundance is well-known, but might partially be overcome by the use of cryogenic probehead technology.

9.2.2.4 Structural Interpretation of RDCs

RDCs for a given interatomic vector depend on its relative orientation with respect to the static magnetic field as described in Section 9.2.2.1. This orientation is averaged over all dynamic processes, including Brownian motion, interactions with the solvent and the alignment medium and conformational changes. If a molecule can be described by a single conformation, the dynamic averaging is identical for all interatomic connectivities and the interpretation of RDCs is straightforward. A very simple example is the distinction of axial versus equatorial protons in six-membered rings. Since all axial $^{1}H—^{13}C$ vectors point into the same direction (or anti-direction), all RDCs being significantly different from these protons directly indicate an equatorial position [61]. More general configurational studies, as for example the identification of diastereotopic protons, can be achieved via the concept of the alignment tensor.

The basic principle of this concept is easily understood: while the magnetic field points along the z-axis in the laboratory frame and the orientation of the molecule changes over time, we can also place ourselves in a frame that is fixed with the orientation of the rigid molecule, where the magnetic field changes its orientation relative to the fixed molecule. The alignment tensor then describes the distribution of the magnetic field in this molecular reference frame (for a readable derivation of the alignment tensor and its matrix representations, see Ref. [62]). From this alignment tensor, all RDCs of a given structural model are easily back-calculated and the comparison with experimental RDCs can verify or falsify the model.

The alignment tensor for conformational and configurational studies is widely used in the field of biomolecular NMR but is also applied to smaller organic compounds. Very nice examples include the assignment of diastereotopic protons in methylene groups [63, 64] or the relative configuration of remote parts of a molecule ([39, 44, 65] and Fig. 9.7), which are usually very difficult to obtain with conventional NMR methods. In these cases, the alignment tensor can be fitted from experimental RDCs from at least five interatomic vectors of different orientation, out of which no more than three vectors lie in a plane. The alignment tensor is then typically derived using singular value decomposition as implemented for example in the program PALES [66].

The full potential of RDCs, however, can be seen by the incorporation of RDC data in structure calculations. Several programs like XPLOR-NIH, DISCOVER or GROMACS allow the incorporation of RDCs as angular or combined angular and distance dependent restraints. Several studies on sugars have been reported (see, e.g. Ref. [43] and references therein) and Fig. 9.8 shows the comparison of three structural models for the backbone of the cyclic undecapeptide cyclosporin A, derived from X-ray crystallography, ROE data in $CDCl_3$ as the solvent, and RDCs and ROEs obtained in a PDMS/$CDCl_3$ stretched gel [22]. Due to the sensitivity to

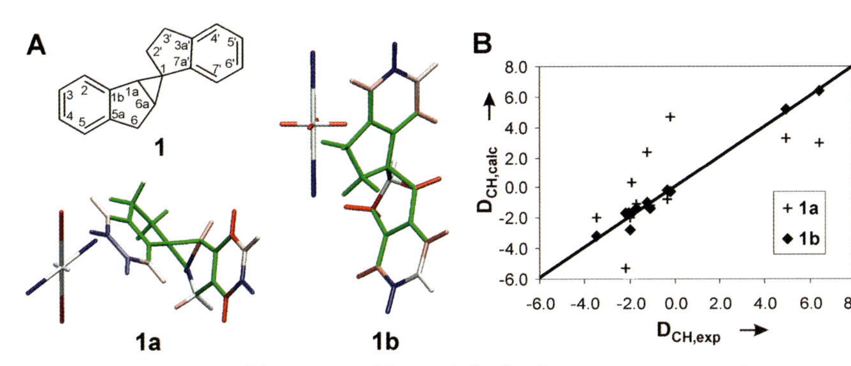

Fig. 9.7 (A) Structures of the two possible diastereomers of the spiroindene **1**. The RDCs determined in a PDMS/$CDCl_3$ gel are color coded to the structures (red, negative RDCs; blue, positive RDCs; green, no RDCs or RDCs not used in analysis). The axis systems to the left of each structure represent the corresponding alignment tensors.
(B) Correspondence of experimental versus back-calculated RDC values for structures **1a** and **1b** [44].

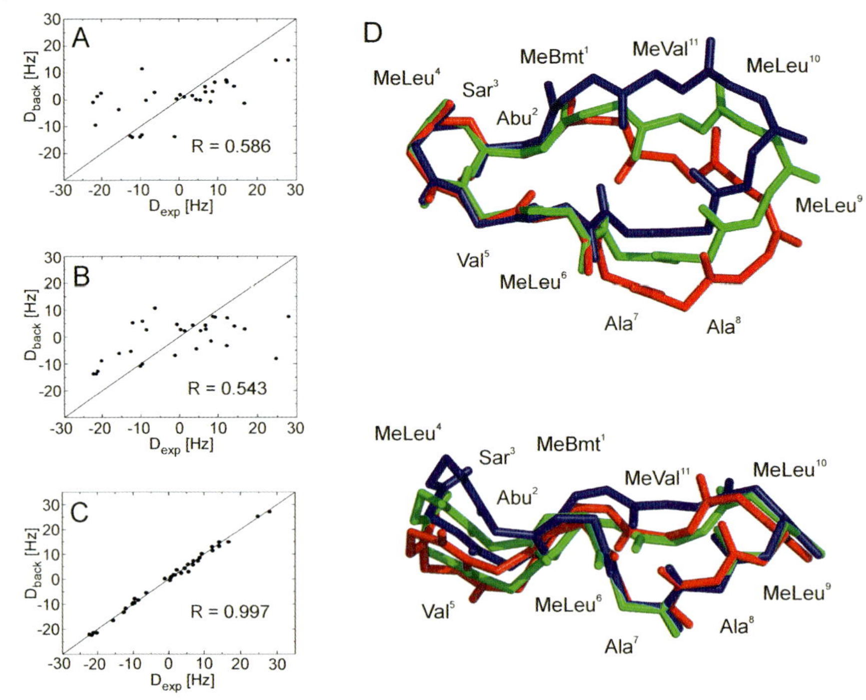

Fig. 9.8 Comparison of experimentally determined and back-calculated one-bond RDCs for the crystal structure (A), the ROE-derived structure (B) and the (RDC+ROE)-refined structure (C) of cyclosporin A. (D) Evaluation of the resulting models for the average structures show a flat geometry in the crystal (red), an already bent backbone for the structure based only on ROE data (green), and a further twist in the RDC-refined backbone (blue), which does not violate any ROE data [22].

angular changes of the backbone relative to the static magnetic field as an external reference, a structural model of the average structure with previously unknown precision can be calculated. If structure calculations are combined with MD simulation, it can be expected that even dynamic models can be obtained with high precision which include conformational changes of the molecule of interest.

A completely different approach for interpreting RDCs is the *in silico* prediction of alignment tensors using various approximations for the interaction between alignment medium and solute. Among these methods, the AP model [67], Chord model [68] and Shape model [69] have been applied to small molecules combined with *ab initio* calculations for dynamical studies. Similar computational method incorporated in PALES have successfully been used to predict the alignment tensors of proteins and nucleic acids, e.g. it was possible to proof a protein dimer with this approach [66]. The TRAMITE approach starts from MD simulations and uses the very fast obtainable moment of inertia tensor as a first order approximation for van der Waals interactions [70]. All methods predict alignment reasonably within their technical limitations, but the high theoretical potential for conformational analysis by computational methods, e.g. the determination of absolute chirality in a chiralorienting medium, are not yet reached.

9.2.3
Other Anisotropic NMR Parameters

RDCs are easy to measure and the interpretation with respect to distance and angular information is readily available. Other anisotropic parameters contain similar information, but are either difficult to obtain experimentally or are more complex in terms of interpretation. In the following, a brief introduction into RQCs, RCSA and PCSs is given.

9.2.3.1 Residual Quadrupolar Coupling (RQCs)

RQCs are only relevant for nuclei with spins of 1 or higher and due to the strong paramagnetic line broadening in other species they are practically limited to deuterium nuclei. RQCs of deuterated solvents are frequently used for the empirically derived determination of alignment strength by measuring a simple ^{2}H-1D experiment. RQCs have the same angular dependence as RDCs. The electric field gradient tensor responsible for the anisotropic interaction with the magnetic field is usually aligned within 1° with the directly attached heteronucleus and therefore also contains the same angular information as one-bond RDCs. However, with 0.015% natural abundance of the deuterium isotope, the measurement is limited to isotopically enriched or highly concentrated samples and no satisfying methods for determining the sign of RQCs are available so far. One application of natural abundance deuterium NMR is the distinction of enantiomers in a chiral alignment medium, which was published by the group around J. Courtieu in a very nice series of papers (see, e.g. Ref. [71] and references cited therein). Quadrupolar couplings as the strongest available anisotropic NMR parameters in this case are well suited to distinguish the slight differences in alignment of an enantiomeric pair.

9.2.3.2 Residual Chemical Shift Anisotropy (RCSA)

In liquid-state NMR spectroscopy only the isotropic component of the chemical shift tensor is measurable. Upon alignment the situation changes and the so-called *zz*-component of the chemical shift tensor includes anisotropic components.

The measurement of chemical shift changes due to partial orientation in an alignment medium is easily done. However, while relative chemical shift positions in a spectrum are well defined, the referencing of chemical shift and the influence of the alignment medium on the chemical shifts have to be taken care of. As nicely summarized in Ref. [26], even molecules with tetrahedral symmetry like tetramethylsilane, tetrachlorocarbon or methane, are partially oriented and show chemical shift changes. Due to the small RCSA observed on these reference molecules, it is usually neglected, but represents a systematic error in RCSA measurements. More restrictive is the influence of the alignment medium as a kind of cosolvent on the chemical shift. Best measurements are therefore obtained using variable angle sample spinning NMR [27, 28] for liquid crystals or variable angle NMR for stretched gels (unpublished results) to scale all anisotropic parameters by the angle of the orientational axis with respect to the static magnetic field. At the magic angle, isotropic chemical shifts are measured and the differences to these shifts represent RCSA values for other angles.

For the interpretation of RCSA values in terms of structural information, good estimates of the chemical shift anisotropy for a given atom have to be available. Although good chemical shift predictions can be achieved today based on empiric and *ab initio* methods, RCSA has only be used in few applications. Examples are the relative orientation of α-helices in transmembrane proteins from so-called PISEMA wheels [72] or the determination of backbone orientation from ^{31}P RCSA data [73]. For drug-like molecules, RCSA seems potentially promising for planar aromatic and olefinic systems with large carbon CSA. Since RDCs for these systems are only available in a single plane, additional information from RCSA might help structure refinement.

9.2.3.3 Pseudo-Contact Shift (PCS)

Paramagnetic ions possess an electron spin and orient themselves in a magnetic field. Therefore molecules containing such ions are automatically partially aligned. In addition to RDCs, RQCs and RCSA, the hyperfine couplings of the nuclei to the electron spin of the paramagnetic center lead to the so-called PCS. The angular dependence and strengths of the PCS is determined by the electron distribution of the paramagnetic center and can generally be used for structure determination. However, few drugs are paramagnetic and applications for this class of molecules are very limited. In biomolecular NMR, instead, specific paramagnetic tags can be engineered into proteins at N- or C-terminal positions or chemically added on single, surface-accessible cysteines [74–77]. As was shown recently, structural information can then be transferred to a bound ligand via transferred PCS as discussed in more detail in Section 9.3.1.3.

9.2.4 Scalar Coupling Constants (J-couplings)

Scalar coupling constants between two nuclei are mediated by electrons via covalent bonds. Depending on the overlap of electron orbitals, the scalar coupling varies in size and sometimes even changes sign. Many relationships have been derived based on empirical observations connecting scalar coupling constants with angular information. Today it is also possible to calculate scalar coupling constants by means of density functional theory to find relationships between couplings and bond angles.

Although easy to measure and generally useful, heteronuclear ^{1}J-coupling constants are rarely applied as angular restraints. One exception are $^{1}J_{C\alpha H\alpha}$-coupling constants in peptides [78]. The vast majority of angular restraints, however, are dihedral angles derived from homonuclear or heteronuclear ^{3}J-couplings. Practically all ^{3}J-couplings obey the empirically derived Karplus relation:

$$^3J = A + B\cos(\phi) + C\cos 2(\phi)$$

with the dihedral angle ϕ between the three bonds constituting the ^3J-coupling and the varying fitting constants *A*, *B* and *C* for a specific structural element. Typical fitting constant values for $^3J_{HNHC}$ can be found in Ref. [79] and for $^3J_{CH}$-couplings in Ref. [80]. Sugars in general can be very well defined by ^3J-coupling constants by the refined Karplus relation which takes into account the influence of neighboring oxygen atoms to the ^3J-coupling [81, 82]. Heteronuclear ^3J-couplings are not used that often although they contain very important configurational and conformational information. Several reviews have been published for different kinds of ^3J-coupling constants [83–85].

For the measurement of long-range homonuclear and heteronuclear coupling constants, a large number of experiments has been developed. The extraction of $^3J_{HH}$-couplings from 1D spectra is routinely used and often just the splitting in the multiplets is indicated as "coupling constants". It should be noted that this is only valid for first order spin systems, e.g. the splittings in the X part of an ABX spin system do not directly yield coupling constants, but the complete spin system including the AB part has to be simulated (see standard NMR textbooks). In case of signal overlap or the measurement of very small coupling constants, 2D methods exhibit distinct advantages. Even if multiplet components are not baseline separated, exact values for couplings can be fitted from COSY spectra calculated in absorption and dispersion mode [86]. E.COSY spectra [59] are used for spin systems involving more than two spins. It is beyond the scope of this article to review those methods and the reader is referred to the literature (e.g. Ref. [85]).

Long-range couplings (over three or more bonds) are only in rare cases directly readable from 1D spectra. If only qualitative values are needed, cross-peak intensities in heteronuclear long-range spectra yield sufficient information. The technology originally developed for carbon detected methods ([87, 88]) is transferred into one of the most often used method for quantitative extraction of heterocoupling via measuring the build-up of cross-peaks depending on coupling evolution times [54]. In systems with more than two spins the heteronuclear analogs of the E.COSY type spectra, e.g. HETLOC [55], yield reliable values of the coupling constants.

Coupling constants are routinely used to determine the side-chain conformation of amino acids in peptides and proteins. Whereas proteins nowadays are almost exclusively studied as ^{13}C- and ^{15}N-labeled isotopomers, peptides usually have these isotopes in natural abundance, i.e. the magnetically active heteronuclei are highly diluted. Most amino acids contain a methylene group at the β-position for which the χ angle is determined by the conformation of the C_α—C_β bond. Two vicinal J_{HH} coupling constants can be measured: H_α to $H_\beta^{pro\text{-}R}$ and H_α to $H_\beta^{pro\text{-}S}$. Usually one can assume that the three staggered conformations are in an energetic minimum state. Two types of vicinal proton coupling constants have to be considered: the antiperiplanar (*trans*) arrangement leads to a large coupling constant of about 12 Hz, whereas the syn-clinal [gauche(+) or gauche(−)] results in ca. 3.5 Hz

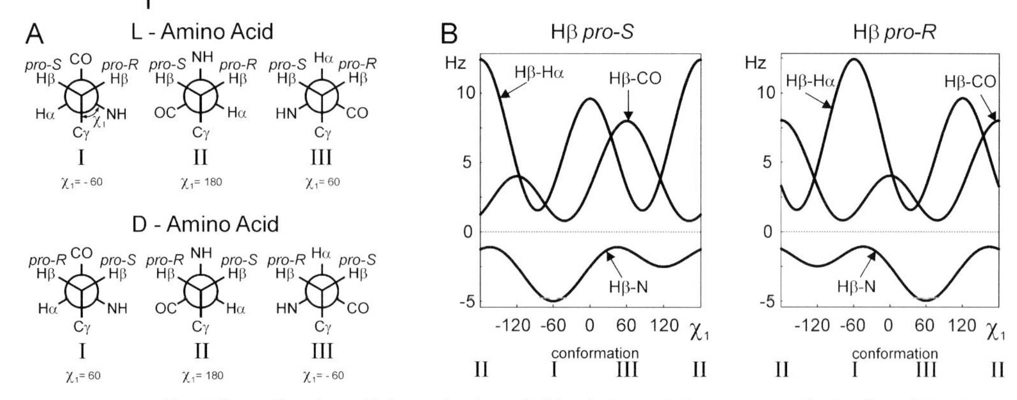

Fig. 9.9 *pro-R* and *pro-S* determination of side-chain methylene protons in the β-position in peptides. (A) The different possible configurations and (B) the Karplus relations for the corresponding $^3J_{H\alpha H\beta}$-, $^3J_{H\beta C'}$- and $^3J_{H\beta N}$-coupling constants [87].

(Fig. 9.9B). Hence, conformation III as shown in Fig. 9.9 A yields two small J-coupling constants and the sum of $^3J_{H\alpha H\beta R}$ and $^3J_{H\alpha H\beta S}$ is about 7 Hz. In conformations I and II of Fig. 9.9(A) the sum is 15.5 Hz. If the three rotamers are in fast exchanging equilibrium, the sum of both vicinal couplings is an indicator of the relative population of III in the equilibrium. In addition, two identical homonuclear couplings indicate that both conformations I and II are equally populated, whereas if they differ (in the most extreme case they are 12 and 3.5 Hz) one of the conformations (I or II) dominates in the equilibrium. Which of the two conformations is the higher populated one can only be decided from the assignment of the diastereotopic β-protons. Sometimes this assignment can be done via NOE/ROE spectra. Since it is strongly recommended not to perform the assignment by NOE parameters which are later on used for the determination of the 3D structure, the measurement of heteronuclear ^3J-coupling should be preferred. If the population of conformation III can be neglected, only the β *pro-R* in L-amino acids is in antiperiplanar arrangement to $^{13}C'$ and exhibits a strong cross-peak in the heteronuclear long-range correlation. In conclusion, the sum of $^3J_{H\alpha H\beta R}$ and $^3J_{H\alpha H\beta S}$ yields the population of III, their difference the preference of I or II and the heteronuclear vicinal coupling decides if I or II is preferred. In ^{15}N-labeled proteins the same procedure can be applied to assign β *pro-R* and β *pro-S* using the large $^{15}NH\beta$ *pro-S* coupling in rotamer I.

Recently it has been found that couplings between ^{15}N and ^{13}C across H-bonds, e.g. in systems containing $^{15}N{-}H\cdots O{=}^{13}C$ units (in proteins), can be directly detected and provide evidence for H-bonds in proteins and nucleic acids [89]. Although this technology is now standard for larger biomolecules it is rarely used for smaller molecules [90].

9.2.5
Cross-Correlated Relaxation (CCR)

A second way of measuring angular information is based on CCR processes. In contrast to empirical relations like the Karplus equation, angular information out of CCR measurements can be directly extracted from fundamental principles.

The determination of dihedral angles via CCR can be explained easily in a simplified form. Considering dipolar relaxation (see also Section 9.2.1), the relaxation of a spin is mediated by the fluctuating magnetic fields caused by an adjacent spin. Usually both spins are considered to fluctuate independently, but CCR indicates that the dipolar coupled spins also might experience fluctuating magnetic fields from another spin pair which influences its relaxation rate. Since both dipolar interactions in a rigid molecule are modulated by the same overall tumbling, they display a very similar time dependence, giving rise to cross-correlation effects. Assuming two internuclear vectors *i,j* and *k,l* as shown in Fig. 9.10, the CCR rate based on the two dipolar relaxing spin pairs is described by the dipole–dipole CCR rate:

$$\sigma^{DD} = \frac{\hbar^2 \mu_0^2 \gamma_i \gamma_j \gamma_k \gamma_l \tau_c}{20\pi^2 r_{ij}^3 r_{kl}^3} \cdot \{3\cos^2\vartheta - 1\}$$

with γ_i, γ_j, γ_k and γ_l as the gyromagnetic ratios, and r_{ij}, r_{kl} the distances between the corresponding nuclei, ϑ being the angle between the two spin pairs as described in Fig. 9.10(A). If we consider fixed distances r_{ij} and r_{kl} and assume that the correlation time of the molecule τ_c is known from other experiments, the only unknown variable in the equation is the dihedral angle ϑ. If we can detect multiple quantum coherences, e.g. double quantum coherences (Fig. 9.10B–D), between the spins *i* and *k*, resulting differences in multiplet component intensities directly translate into ϑ. A similar expression can also be derived for dipole-CSA CCR [91].

The beauty of CCR measurements is that the nuclei *i* and *k* do not need to be neighboring spins, but can be very distant as long as they can be correlated. Based

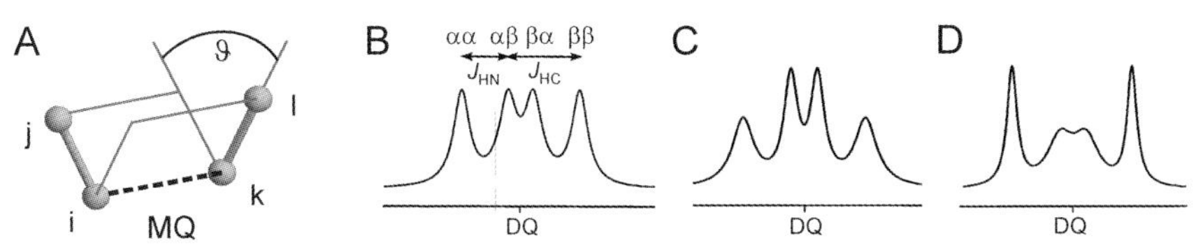

Fig. 9.10 (A) Definition of the dihedral angle ϑ between the spin-pairs *i,j* and *k,l*. (B) Cross-correlated relaxation can be detected by differences in intensities of the multiplet components of double quantum coherences (DQ) between spins *i* and *k*. Multiplets are simulated for H^N,N and Hα,Cα spin pairs in a protein with $\vartheta = 54.7°$ (B), $\vartheta = 0°$ (C) and $\vartheta = 90°$ (D) [145]. MQ, multiple quantum.

on such angular measurements, even a structure based on X-ray crystallography had to be corrected [92]. The usefulness of CCR for structure determination has been demonstrated in many applications (see, e.g. Ref. [93] and references therein), but usually atoms have to be isotopically labeled for effective measurement of multiple quantum multiplet patterns. For this reason it is barely used for the structure determination of drug and drug-like molecules.

9.3
Conformation Bound to the Receptor

Most interesting for pharmaceutical applications, of course, is the conformation of a ligand bound to its receptor. There are several ways of determining the structure of the bound state by NMR. Generally two situations have to be distinguished: tight binding of a ligand to its receptor, and intermediate binding for which the free and bound state are in fast exchange on the NMR timescale. For tight binders the ligand must be assumed to be part of the receptor and the ligand–receptor complex can eventually be solved by biomolecular NMR methods as long as isotope-labeling schemes for the receptor are available and the size of the complex does not exceed limitations with respect to resolution of individual resonances. This situation will not be discussed further here and we refer the interested reader to other comprehensive texts (e.g. Ref. [94]).

Intermediate binders in the fast exchange limit allow the detection of the bound conformation by detecting the free ligand. This can be achieved by using particular NMR interactions that are efficient for the ligand bound to its receptor, but are very inefficient for the free ligand. In isotropic samples the different relaxation properties of a small ligand with a very short correlation time compared to the complex of the ligand bound to a large receptor can be used for detecting exchange-transferred NOE (etNOE) or exchange-transferred CCR (etCCR) (Fig. 9.11A). For samples in which the receptor can be aligned, anisotropic parameters like exchange-

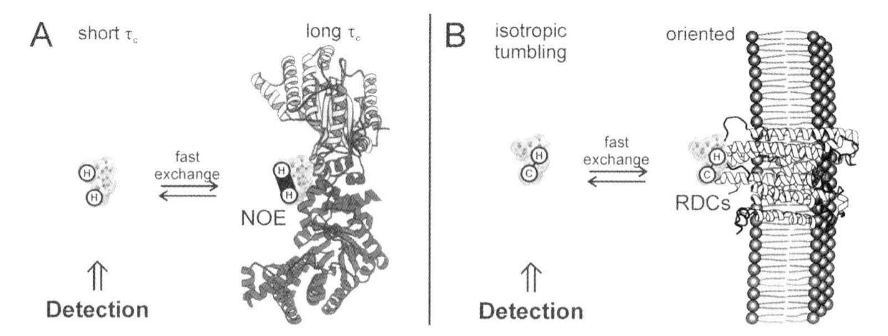

Fig. 9.11 The difference in the strength of relaxation processes (A) or the difference in alignment (B) can be used to detect interactions (e.g. NOE or RDCs, respectively) of the ligand–receptor complex in exchange-transferred experiments.

transferred RDCs (etRDCs) or exchange-transferred PCSs (etPCS) for paramagnetic alignment can be used to define the conformation in the bound state (Fig. 9.11B). In the literature, these exchange-transferred experiments are also referred to as transferred experiments (trNOE, trCCR, trRDCs or trPCS). Finally, saturation transfer difference (STD) and paramagnetic relaxation enhancement (PRE) experiments can be used to qualitatively map the binding surface of the ligand.

All these methods have in common that the receptor is not detected so that no size limit applies. Relaxation-based exchange-transferred experiments actually show best performance for longest correlation times, i.e. very large receptors.

The applicability of exchange-transferred methods is mainly determined by the binding kinetics. It is only possible to determine the conformation of the bound state by measuring the free ligand of the ligand bound to the receptor is released fast enough during the course of an NMR experiment. Therefore the off-rate k_{off} is the main limitation which should generally be larger than approximately $300\,\text{s}^{-1}$ to ensure fast exchange. Assuming diffusion controlled binding with typical on-rates $k_{\text{on}} \approx 10^8$–$10^9\,\text{M}^{-1}\,\text{s}^{-1}$, this implies equilibrium dissociation constants of $K_{\text{D}} \geq 10^{-6}\,\text{M}$. An upper limit for dissociation constants in the millimolar range is mainly given by the solubility of receptor or ligand and the occurrence of nonspecific binding processes which make the determination of a single bound conformation obsolete.

The signal intensity for the bound interaction should be maximized in exchange-transferred experiments. Since usually the receptor has the lowest concentration, it is proportional to the population of the ligand–receptor complex relative to the total concentration of the receptor R, which we call P_{bound}. With the total concentration of the ligand L and the dissociation constant K_{D} assuming a simple ligand–receptor model, it is given by [13]:

$$P_{\text{bound}} = \frac{1}{2}\left\{R + L + K_{\text{D}} - \sqrt{(R + L + K_{\text{D}})^2 - 4RL}\right\}$$

For a receptor concentration of approximately the dissociation constant and a 10-fold excess of ligand $P_{\text{bound}} \approx 1$, i.e. essentially all receptor molecules are complexed with maximum intensity of the desired NMR interactions. Lower receptor concentrations relative to K_{D} can be compensated by a higher ligand excess and might be pushed up to a 1000-fold ligand excess [95], but experiments are easiest applicable with the mentioned 10–20 fold excess.

Since the ligand is in fast exchange with respect to its bound and free state, every molecule experiences an averaging over time where the fraction L

$$\alpha = P_{\text{bound}} \cdot R / L$$

describes the scaling of interactions in the complexed state.

An important aspect of all exchange-transferred experiments is the occurrence of nonspecific binding which contributes equally to the NMR interactions in the bound state and cannot easily be distinguished. Nonspecific interaction is best established by demonstrating competitive binding by a second ligand [96] and must be ruled out in order to reliably being able to deduce a structure of the bound state.

9.3.1
Ligand Conformation

9.3.1.1 Exchange-transferred NOE (etNOE)

Exchange-transferred spectroscopy was introduced with the finding of the etNOE [97] and its theoretical explanation in terms of fast exchange several years later [98] laid the basis for the large variety of applications being present today. The core element of etNOE is the dependence of the cross-relaxation rate σ^{NOE} on the correlation time τ_c. The overall cross-relaxation rate is defined by:

$$\sigma^{NOE}_{overall} = \alpha \cdot \sigma^{NOE}_{bound} + (1-\alpha) \cdot \sigma^{NOE}_{free}$$

with α as defined in the previous section and the NOE cross-relaxation rates for the free and bound state as defined in Section 9.2.1. For a certain mixing time τ_m in an etNOESY experiment, the efficiency of magnetization transfer for small ligands close to the condition $\omega_0\tau_c \approx 1.12$ is very low. For the large ligand–receptor complex, instead, cross-relaxation becomes very efficient (Fig. 9.12 A) and contributions to cross-peaks in etNOESY experiments mainly originate from the bound state despite its relatively low effective mixing time $\alpha\tau_m$. The comparison of a NOESY experiment without and with receptor clearly shows the effect of fast, negative cross-relaxation of the complex (Fig. 9.12B).

Intensive discussions about potential errors of etNOE-derived distances can be found in the literature (see, e.g. Ref. [99] and references therein). Essentially, experimental and systematic errors are identical to conventional NOE-based

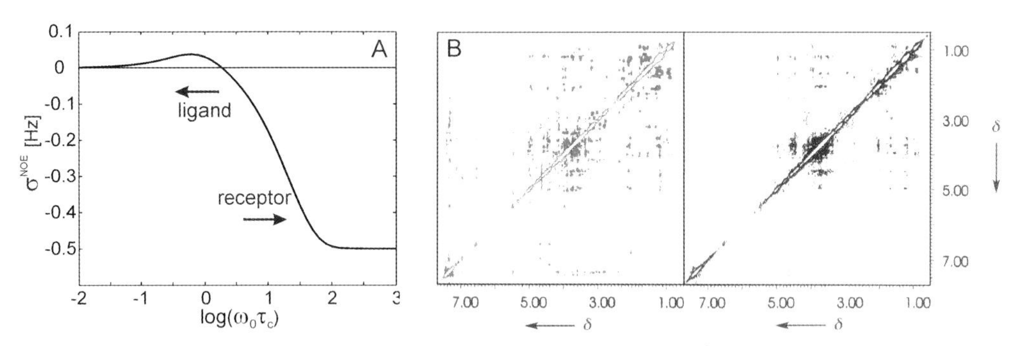

Fig. 9.12 (A) Cross-relaxation rates with respect to the correlation time τ_c and the spectrometer frequency ω_0. (B) While only positive NOE (resulting in cross-peaks with inverted sign compared to the diagonal peaks, left) are present in a mixture of free ligands, the addition of the receptor leads to the occurrence of negative etNOE cross-peaks (with identical sign compared to diagonal peaks, right) which define the conformation in the bound state [146]. In the latter case signals with opposite sign to the diagonal are not shown.

distance determination as discussed in detail in Section 9.2.1. Additional errors might be introduced by residual NOE contributions of the free ligand, nonspecific binding of the ligand and potential intermolecular spin diffusion caused by receptor protons at the binding site. In practice, NOE contributions can easily be identified by the already mentioned NOESY experiment without receptor. Nonspecific binding cannot be distinguished and has to be excluded by competition experiments as mentioned in the previous section. Intermolecular spin diffusion, finally, has been shown in a large number of simulations to be negligible [99, 100].

The basic experimental setup for etNOESY is practically identical to the conventional NOESY experiment shown in Fig. 9.2(B). For suppression of residual receptor signals a relaxation filter can be introduced and the mixing time has to be corrected according to the scaling factor α.

Applications are manifold for etNOE-based structure determination. The majority of etNOE experiments is performed with relatively large, proton-rich and flexible ligands like peptides or sugars, but also other ligands have been used (see Ref. [99] for a list of published applications). On the receptor side almost any class of large soluble proteins has been studied, including antibodies [101] and protein-complexes [102]. The receptor can even be as large as the ribosome [95] or lipid vesicles [103]. For general reviews of the etNOE method, see Refs. [13, 99].

9.3.1.2 **Exchange-transferred RDCs (etRDCs)**

RDCs as anisotropic parameters do not depend on the size of the binding partners, but on the strength of alignment of the participating molecules. To be able to observe etRDCs, the receptor must be significantly aligned while the free ligand behaves basically isotropic. This condition can most easily be fulfilled if the receptor can be considered as being part of the alignment medium. For example, a transmembrane protein can be incorporated into liquid crystalline bicellar phases [104] or a soluble receptor might be covalently attached to a polymer-based alignment medium. With a dissociation constant of $10^{-6}\,\mathrm{M} < K_D < 10^{-3}\,\mathrm{M}$, the interaction strength of the ligand with the receptor as an integral part of the alignment medium can be considered strong compared to the interaction of the ligand with the alignment medium itself.

The measurement of etRDCs is identical to the measurement of conventional RDCs as described in Section 9.2.2. Most easily measured are heteronuclear one-bond RDCs, which are scaled according to:

$$D_{\mathrm{XH}}^{\mathrm{overall}} = \alpha D_{\mathrm{XH}}^{\mathrm{bound}} + (1-\alpha) D_{\mathrm{XH}}^{\mathrm{free}}$$

with the free dipolar coupling negligible small on the order of or less than 1 Hz and the dipolar coupling in the bound state in the kiloHertz range. Since dipolar couplings depend on the angle relative to the static magnetic field, etRDCs not only provide information about the ligand structure in the bound conformation, but also about the relative orientation to the receptor as part of the alignment medium.

The only application known to the authors so far is the transient binding of a selectively labeled transducin undecapeptide to a rhodopsin-containing bicellar alignment medium [104]. This study impressively demonstrates the potential for determining the conformation and orientation of a ligand when bound to an integral membrane receptor. With the biochemical availability of more and more pharmaceutically interesting transmembrane receptors, etRDCs will most certainly become an important technique in rational drug development.

9.3.1.3 Exchange-transferred PCS (etPCS)

As pointed out in Section 9.2.3.3, paramagnetic centers can lead to PCSs due to their spontaneous alignment in magnetic fields. The alignment tensor of a specific paramagnetic ion in most cases is well-known and PCS data can directly be translated into combined angular and distance information. The etPCS requires a receptor that is aligned by a paramagnetic center, while the ligand in its free form is diamagnetic. Similar to etRDCs, the detected free ligand in the end contains the structural information of the complex if the kinetic conditions for exchange-transferred spectroscopy are fulfilled.

The earliest applications of etPCS involve lanthanide shift reagents [105]. Chiral lanthanide shift reagents like $Eu(TFC)_3$ or $Eu(FOD)_3$ [106] are frequently used for the distinction of enantiomers. In such cases the free ligands are the enantiomers and the paramagnetic receptor is the shift reagent. Since the two enantiomers bind either with a different relative orientation to the Eu^{3+} ion or have a different binding affinity to the shift reagent, the resulting etPCS is also different for the enantiomers, leading to separated resonance frequencies.

Biological applications of this phenomenon started only very recently, when binding of thymidine to the receptor ε 186/θ was followed by etPCS [35]. Close to the ligand binding site this specific receptor also has a tight binding site for a paramagnetic lanthanide ion like La^{3+} or Dy^{3+}. With the etPCS obtained, the ligand could be nicely fit into the binding pocket of the already known structure of ε 186/θ.

As mentioned in Section 9.2.3.3, paramagnetic centers can also be engineered into proteins and nucleic acids by chemical or biochemical means. The method therefore is in principle applicable to a wide range of problems.

9.3.1.4 Exchange-transferred CCR (etCCR)

As described in Section 9.2.4.5, CCR can be used to obtain dihedral angular information. Since CCR processes depend linearly on the correlation time, the requirements for exchange-transferred spectroscopy are fulfilled similarly to the etNOE.

The first etCCR application has been reported for a partially ^{13}C—^{15}N-labeled phosphotyrosine peptide derived from interleukin-4 receptor ligated to STAT-6 [107] and subsequent studies involve nucleotide cofactors ligated to human recombinant deoxycytidine kinase [108] and epothilone A bound to tubulin [109]. Since etCCR usually involves isotope-labeling schemes for the ligand, its applicability is limited to specific molecular classes.

9.3.2 Ligand–receptor Binding Surface

9.3.2.1 STD Spectroscopy

The STD experiment is an important method which also relies on the fast exchange of the free and bound state of a ligand [12]. Again, the ligand is detected, but this time the information of interest is caused by saturation transfer of the receptor to the ligand in the bound state. In contrast to the other exchange-transferred experiments, the receptor does not need to have different physical properties compared to the ligand, but is treated differently by the applied pulse sequence. The receptor is selectively saturated by extensive irradiation of CW at a frequency which must not overlap with resonance frequencies of the ligand. Since saturation is distributed via cross-relaxation, also the ligand experiences saturation transfer as soon as it binds to the receptor. By acquiring a spectrum with and without CW irradiation, the difference of the two spectra directly represents the transferred saturation.

Using the heat model introduced in Section 9.2.1, in STD the receptor is selectively heated by CW irradiation. Upon binding, the receptor transfers some of its heat to the ligand and the temperature of the ligand will be hottest at the binding surface. Mapping the heat, i.e. the saturation difference, of the ligand, it is not only possible to identify a binder as such, but also to identify atoms close to the receptor for the potential reconstruction of the binding interface. For medium-sized ligands such as oligosaccharides, it is possible to elucidate the binding site of the ligand because the transfer of saturation is time dependent and most efficient to the part of the molecule which is directly bound to the saturated protein [110].

STD is extensively used as a screening tool for ligands with intermediate binding affinity in high-throughput screening approaches [12]. The identified small ligands are then further used in fragment-based drug design to develop lead structures for drug candidates [111]. Recently, STD has also successfully been applied to membrane-bound receptors [112] and ligand binding to receptors on whole cells [113]. STD is also used in combination with special library designs like the SHAPES approach [114]. Since it is beyond the scope of this article to treat all aspects of STD, the reader is referred to the cited references.

9.3.2.2 Paramagnetic Relaxation Enhancement (PRE)

Dipolar relaxation processes as discussed in Section 9.2.1 are not limited to interactions between two nuclei, but can also occur between a nuclear and an electron spin if a paramagnetic center is close enough in space. Because of the very large gyromagnetic ratio of electrons, relaxation rates are quite strong. As with all dipolar relaxation rates, also relaxation in this case has the typical r^{-6} dependence which, in case of paramagnetic spin labels, leads to significant reductions in signal intensities as far as 20 Å from the paramagnetic center.

A method called SLAPSTIC [115] uses this so-called PRE for detecting ligand binding: With a spin label engineered close to the binding site of a receptor, signal intensities of binding molecules are significantly decreased compared to the free ligand. If the conditions for exchange-transferred spectroscopy are fulfilled, the relative reduction in signal intensities can even be used to extract distances of nuclei to the paramagnetic center. If the position of the spin label is known, distance restrained docking calculations can lead to well-defined structural models of the ligand–receptor complex.

An expansion of the method can be found in second-site SLAPSTIC, where, in favorable cases, a second binding site close to the binding site of interest can be used to infer a spin label via a second ligand [116].

PRE is generally also present in etPCS experiments and the two approaches can be combined to obtain a defined structure of the ligand and its binding surface towards the receptor [35].

9.4 Refinement of Conformations by Computational Methods

The results of NMR measurements have to be converted into a 3D structure. After establishing the constitution by NMR parameters that are transmitted through bond, i.e. J-coupling constants, information about the spatial structure is introduced. Here, mainly distances from NOE build-up rates are used to define the configuration and conformation.

Experimental distances from NOEs/ROEs of small molecules are recommended not to be classified into regions of "small, medium, and large" as it is often done in the structure determination of large molecules. As opposed to macromolecules, the overall correlation time τ_c can be considered constant in small molecules. Thus, it is possible to measure distances in the range between 2 and 5 Å with an accuracy of about 10%. Often distances between protons are almost exclusively used for the structure determination. This leads to the fact that molecules with small numbers of hydrogen atoms are more difficult to determine.

The first step of the structure refinement is the application of distance geometry (DG) calculations which do not use an energy function but only experimentally derived distances and restraints which follow directly from the constitution, the so-called holonomic constraints. Those constraints are, for example, distances between geminal protons, which normally are in the range between 1.7 and 1.8 Å, or the distance between vicinal protons, which can not exceed 3.1 Å when protons are in anti-periplanar orientation.

The second step is the molecular dynamics (MD) calculation that is based on the solution of the Newtonian equations of motion. An arbitrary starting conformation is chosen and the atoms in the molecule can move under the restriction of a certain force field using the thermal energy, distributed via Boltzmann functions to the atoms in the molecule in a stochastic manner. The aim is to find the conformation with minimal energy when the experimental distances and sometimes simultaneously the bond angles as derived from vicinal or direct coupling constants are used as constraints.

Restrained MD (rMD) is followed by the use of MD in explicit solvent, i.e. the conformation as determined above is taken into a box containing many solvent molecules around the molecule. Subsequently, simulated annealing (SA) and energy minimization steps are performed to draw the molecule into the global energy minimum. An MD run (the so-called trajectory) over at least 150 ps to 1 ns is followed and a "mean structure" is calculated from such a trajectory. The conformation must be stable under this condition even when the experimental constraints are removed.

The next step is a careful comparison of the experimental restraints with those from the calculations. Such a check also includes other parameters which are not directly used in the MD calculations, but can be obtained by NMR experiments. When parts of the molecule fulfill the experiments and others do not, it is an indication for flexibility in the part which does not meet the experimental data. To analyze such a behavior, longer MD trajectories or time- or ensemble-averaged distance restraints can be used in the calculations.

In conclusion, the result of a conformational analysis is a structure which is in agreement with the experimental data. There may be other structures which also fulfill the experimental restraints and the "bioactive conformation" may differ in distinct receptors [117]. Thus, the more data are acquired the higher is the probability that the structure is "correct".

In the next sections the different steps of the conformational analysis presented above are described in more detail.

9.4.1
Distance Geometry (DG)

DG was primarily developed as a mathematical tool for obtaining spatial structures when pairwise distance information is given [118]. The DG method does not use any classical force fields. Thus, the conformational energy of a molecule is neglected and all 3D structures which are compatible with the distance restraints are presented. Nowadays, it is often used in the determination of 3D structures of small and medium-sized organic molecules. Compared to force field-based methods, DG is a fast computational technique in order to scan the global conformational space. To get optimized structures, DG mostly has to be followed by various molecular dynamic simulation.

The procedure of DG calculations can be subdivided in three separated steps [119–121]. At first, holonomic matrices (see below for explanation) with pairwise distance upper and lower limits are generated from the topology of the molecule of interest. These limits can be further restrained by NOE-derived distance information which are obtained from NMR experiments. In a second step, random distances within the upper and lower limit are selected and are stored in a metric matrix. This operation is called metrization. Finally, all distances are converted into a complex geometry by mathematical operations. Hereby, the matrix-based distance space is projected into a Cartesian coordinate space (embedding).

9.4.1.1 Distance Matrices

The first step in the DG calculations is the generation of the holonomic distance matrix for all pairwise atom distances of a molecule [121]. Holonomic constraints are expressed in terms of equations which restrict the atom coordinates of a molecule. For example, hydrogen atoms bound to neighboring carbon atoms have a maximum distance of 3.1 Å. As a result, parts of the coordinates become interdependent and the degrees of freedom of the molecular system are confined. The acquisition of these distance restraints is based on the topology of a model structure with an arbitrary, but energetically optimized conformation.

By taking the free rotatability of defined atom bonds into account, the upper and lower limits of the distances can be determined with a tolerance of ±10%. If the lower distance limit is too short, the sum of the van der Waals radii of an atom pair is used for the following calculations. Normally, those limits are very wide. For a further restriction of the distance interval certain geometric computations like the triangle inequality are applied.

At this step, all experimentally derived information like internuclear distances from NOESY/ROESY spectra, can be incorporated into the calculations. Due to experimental errors these restraints should be used with a tolerance of ±10%.

9.4.1.2 Metrization

The structure generating step during a DG run is called metrization. Various methods are known to realize this operation, but in most software tools (e.g. DISGEO [122]) the Havel random metrization is implemented. In this process an atom pair in the holonomic matrix is arbitrarily chosen and a random distance between the upper and lower limit is assigned. By solving the triangle inequality (which states that the sum of the length of two sides of a triangle is always larger than the length of the third side) the holonomic distance limitations of all other atoms inside the molecule are calculated. This procedure is iteratively repeated until all atom distances are allocated. In the end of the computation one obtains a symmetrical matrix of interatomic distances.

Since the starting structure and the initial atom pair was casually selected, distance matrix generation and random metrization should be performed several times in order to get an ensemble of metric matrices.

9.4.1.3 Embedding

The generation of random structures together with the associated metric matrices is followed by the last DG step, the so-called embedding. In this process the structures are projected ("embedded") from the distance space into the Cartesian space. For this, the Eigensystems of the metric matrices are determined. The dimensionality of the Cartesian coordinate space can have any value, but it suggests itself that the transformation should be performed in 3D space [123]. Working with higher dimensioned spaces has the advantage of reduced energy barriers for finding the global energy minimum, e.g. transforming enantiomeric molecules into four dimensions leads to the disappearance of the chiral center [124]. Therefore, no prior knowledge of the stereochemistry of a molecule is needed.

However, the indispensable final reduction of such spaces into the *x*-, *y*- and *z*-dimensions could produce physically meaningless structures.

Due to the fact that the pairwise atom distances inside the upper and lower limits were randomly chosen, the produced embedded structures are often distorted. To overcome this problem further optimization is a must. This can be achieved by the utilization of either distance-dependent or classical force fields.

9.4.2 Molecular Dynamics (MD)

MD [125, 126] has become more and more important in chemistry and biophysics during the last decades. Both enhanced computer performance and improved molecular models enable realistic simulations of the behavior of different kinds of molecules in atomistic detail. In principle, all molecular processes which rely on weak, nonbinding interactions between atoms can be modeled by using computer simulation tools. Since the energies that provoke such processes are in the range of a few tens of kiloJoules per mole the laws of statistical mechanics, i.e. potential energy functions and force fields as part of classical Hamilton operators are applied to describe complete molecular systems. Thus, it is possible to simulate the structure, the dynamics and the motion of small and, if experimental distance restraints are given, of medium-sized and large molecules like peptides and proteins [127].

In this chapter we will focus on the application of MD in structural chemistry. As it was mentioned in Section 9.4.1, 3D structures generated by DG have to be optimized. For this purpose, MD is a well-suited tool. In addition, MD structure calculations can also be performed if no coarse structural model exists. In both cases, pairwise atom distances obtained from NMR measurements are directly used in the MD computations in order to restrain the degrees of motional freedom of defined atoms (rMD; Section 9.4.2.4). To make sure that a calculated molecular conformation is reliable, the time-averaged 3D structure must be stable in a free MD run (fMD; Section 9.4.2.5) where the distance restraints are removed and the molecule is surrounded by explicit solvent which was also used in the NMR measurement. Before both procedures are described in detail the general preparation of an MD run (Section 9.4.2.1), simulations *in vacuo* (Section 9.4.2.2) and the handling of distance restraints in a MD calculation (Section 9.4.2.3) are treated. Finally, a short overview of the SA technique as a special MD method is given in Section 9.4.2.6.

9.4.2.1 Preparation of an MD Simulation

Since several different types of MD software are available, one primarily has to decide which one should be used. While tools like NAMD [128] and AMBER [129] are designed for the simulation of biomolecules (peptides, proteins and nucleic acids), packages like CHARMM [130], GROMOS [131] and GROMACS [132] have a wider range of application. The second step is the selection of an adequate force field. Force fields are empirically developed from known structures or *ab initio* quantum calculations. For example, the OPLS force field [133] was parameterized in particular for amino acids and nucleic acids. Other often used force fields are Amber [134], GROMOS [135], CHARMM [136] (all these force fields are included in the respective software packages), CVFF [137], Tripos [138] and MMFF [139]. Subsequently, a starting conformation of the molecule of interest must be provided. If no coarse model is existent, an arbitrary 3D structure has to be created. This can be done by e.g. commercial software tools like Insight (Accelrys Software) or SYBYL (Tripos).

The following steps contain the build-up of the molecular topology and the choice of the convenient general simulation conditions.

The topology of a molecule includes all information about both the covalent (bond lengths, bond angles, proper and improper dihedral angles) and the non-covalent interactions (Lennard–Jones and Coulomb terms, dipolar reaction field). The binding parameters are not restricted to a single value, but are constrained by a harmonic or trigonometric potential which allows the atoms to move around their equilibrium positions [127]. Care must be taken when the topology is constructed because an inaccurate parameterization could finally result in misleading molecular conformations.

The general setup of an MD run comprehends the system size (dimensions of the simulation box), the temperature and the pressure which affect the molecule(s) and the treatment of the system boundaries. In almost all cases it is recommended to simulate with periodic boundary conditions in order to avoid unwanted surface effects due to the artificial box walls. Periodic boundary conditions are implemented in such a way that the box is completely encircled by identical copies of itself [140, 141].

Finally, the integrating algorithm (e.g. a VERLET leap-frog algorithm [142]) of the Newtonian equations of motion and the duration of the MD run have to be chosen. The integration increments (steps) should not exceed 0.5 fs; otherwise, certain bond vibrations are not covered in the calculations and artificial atom motions are generated. For the computation of the 3D structure of a medium-sized molecule like a peptide a simulation length of about 150 ps to 1 ns is sufficient.

9.4.2.2 MD Simulations *in vacuo*

MD simulations *in vacuo* have the advantage of being much faster than calculations with explicit solvent because fewer interactions have to be computed. However, such trajectories often lead to unreliable or incorrect results. Especially for small molecules which possess a relatively large surface, the solvent may have strong effects on their global conformation. Furthermore, molecules simulated *in vacuo* tend to minimize their overall surface which possibly results in distorted 3D structures. Due to the fact that *in vacuo* no dielectric shielding is applied, intramolecular interactions based on dipoles and charges are overestimated. At last, the formation of intermolecular interactions like hydrogen bonds between the solute and the solvent is completely disregarded. The situation is slightly better when experimental restraints are applied that pull the molecule closer to the real structure. However, to circumvent the above mentioned disadvantages, it is strongly recommended to perform MD simulations of small and medium-sized molecules with explicit solvent even if they require more computational time.

When working with large molecules like proteins – this involves the use of large box sizes and huge amounts of solvent molecules which again causes time-consuming calculations – it is feasible to make use of the "implicit solvent" concept. Instead of using explicit solvent molecules the dielectric constant parameter is adjusted according to the designated solvent. Thus, artificial molecular surface effects as well as unrealistic dipolar and charge-based interactions can be reduced in most instances.

9.4.2.3 Ensemble- and Time-averaged Distance Restraints

As was mentioned before, experimentally (NMR)-derived distance restraints can be applied to MD calculations when a molecule should be enforced to adopt its natural conformation. For this, the force field potentials are extended with energy penalty terms in order to preserve the given distance restrictions. However, such restraints are always averaged over both a tremendous number of molecules and certain time intervals. Since molecules are not static they can change their conformations and not all members of a molecular ensemble exhibit the same structure at a certain point in time. To allow for this aspect, distance information can be implemented as ensemble- or time-averaged restraints [143]. When using ensemble averaging, single representatives are allowed to violate the experimental distance restrictions as long as the whole ensemble is in agreement with the restraints. For this, one has to simulate either a box with several molecules or many identical systems (containing only one molecule) with different starting conditions. On the other hand, the utilization of time-averaged restraints ensures that the distance information is maintained by averaging these restrictions over a certain time period. By all means an average conformation is only meaningful when the recorded trajectory is much longer than the relaxation time of the structure forming process.

9.4.2.4 Restrained MD (rMD)

In the beginning of a rMD the starting structure has to be positioned in the (mostly cubic) simulation box. Afterwards, the cell is filled with the desired type of solvent until the cell has reached the correct density. This can be achieved, for example, by pressure scaling of the box size. Now the experimentally derived distance restraints are included in the topology parameter file. In the following, the solvent molecules are energy minimized while the coordinates of the solute should be kept "frozen" during this process. For this purpose, different energy minimization algorithms have been developed. The steepest descent method should be applied when the starting structure is highly distorted. If the conformation of a molecule is already close to its energy minimum the utilization of the conjugate gradient algorithm is recommended. The first energy minimization step leads to an optimal arrangement of the solvent molecules around the starting structure. Subsequently, the solute is also energy minimized. Now the whole system is slowly heated until the temperature used for the experimental measurements is adopted. According to a Boltzmann distribution, random velocities are assigned to all atoms and the MD run can be started. To ensure that the whole system is well equilibrated the first 50–500 ps of the trajectory (depending on the size of the system) are not comprised in the calculations of the conformation of the solute. During the following 150 ps to 1 ns the coordinates of the molecule of interest are recorded in intervals of about 100 fs. The accumulated 3D structures are now time-averaged whereby one obtains a mean conformation of the solute. Three-dimensional structures achieved by averaging illustrate only a geometrical mean and do not implicitly represent a realistic conformational model. Thus, the averaged structure has to be gently energy minimized in order to eliminate local distortions. The resulting molecular image now displays a 3D structure which should be close to one of the most populated conformations under the experimental conditions used.

9.4.2.5 **Free MD (fMD)**

The last step during the 3D structure generating process includes the simulation of a molecule in a free dynamics run (of course this is only reliable if the calculation is done using explicit solvents) and the comparison of calculated with experimentally derived structural parameters. The procedure explained in Section 9.4.2.4 is exactly repeated, but without distance information effecting the force field. The minimized mean conformation computed in the rMD should serve as the starting structure in the free dynamics run. After a representative conformation was established one has to check if the structure matches the simulated one under restrictive distance conditions. In addition, the pairwise distances within the conformation must be in agreement with the measured NOE/ROE data. If the deviations are too large, both the experimental information and the simulation parameters must be reviewed and, where required, all dynamics calculations have to be repeated. In most cases not all experimentally derived structural parameters are embedded in the rMD. For example, J-coupling constants may directly be used as (soft) constraints in the MD computations, but they can also be utilized to check the reliability of the final results. It is important to note that coupling constants do not have to correlate to the mean structure as described above. Thus, J-couplings have to be taken from the trajectory (in 100-fs to 1-ps intervals) and finally averaged. Hence, the erroneous assumption that the mean structure of an MD represents the reality is avoided. In recent years also orientational restraints (obtained from RDCs and other anisotropic NMR parameters) acquired by NMR experiments in anisotropic environment are utilized for the structure elucidation of a molecule. In our experience it is also recommended to use RDCs to prove the final structure. The consensus of such structure parameters with a simulated conformation provides evidence that the produced structural model is a reliable result of the conformational analysis. If J-couplings and RDCs are directly used as restraints in an MD run they must be soft enough to allow the molecule to pass energy barriers of the restraints during the trajectory.

9.4.2.6 Simulated Annealing (SA)

SA is an often used technique in order to screen the global conformational space of a molecule. The method is applied even when no experimental structure information is available. The preparation of such a simulation is identical with a normal MD run (see Sections 9.4.2.1 and 9.4.2.5). The starting structure can be arbitrarily chosen, i.e. no coarse model conformation needs to be available. By heating up the simulation system to very high temperatures (1000–2000 K), all molecules are allowed to "freely" move on the energy hyper surface. Thus, molecules can partly overcome high energy barriers existent between different conformations. After it spends a certain time (100–500 ps) at high temperature, the whole system is slowly and incrementally cooled down (around 100 K per 200 ps). Each cooling step is followed by a longer period of equilibration (around 200 ps). If the stepwise cooling process is slow enough the molecule is assumed to approach the minimum of the energy hyper surface which represents the thermodynamically most stable conformation. To make the procedure independent of the initial structure the SA process is repeated several times with different starting conformations. Finally, all resulting structures are clustered with respect to the calculated potential energies or, if possible, can be compared with experimental structure information. In addition, it should be noted that SA can be utilized as isolated method or as one preparation step during MD simulations.

9.4.3 Conclusions

As described in Section 9.4, the determination and refinement of molecular conformations comprehends three main methods: DG, MD and SA. Other techniques like Monte Carlo calculations have only a limited applicability in the field of structure elucidation. In principle, it is possible to exclusively make use of DG, MD or SA, but normally it is strongly suggested to combine these methods in order to obtain robust and reliable structural models. Only when the results of different methods match a 3D structure should be presented. There are various ways of combining the described techniques and the procedural methods may differ depending on what kind of molecules are investigated. However, with the flowchart in Fig. 9.13 we give an instruction on how to obtain a reliable structural model.

The starting point of each structure elucidation is the collection of experimental data like NOE/ROE-derived distances, angular restraints from J-couplings or CCRs and RDCs as orientational restraints (1).

If NMR spectra contain a single set of signals, the molecule of interest has either a single low energy conformation or several conformations which are in fast equilibrium (maximal energy barrier height is below 60 kJ M^{-1} [144]). If no other information is available, we start with the assumption of a single preferred structure, which is often incorrectly called the "rigid" conformation. In the next step, experimental information is converted into restraints (2). It is not advisable to add these constraints (J-couplings and RDCs) fully weighted to the MD calculation.

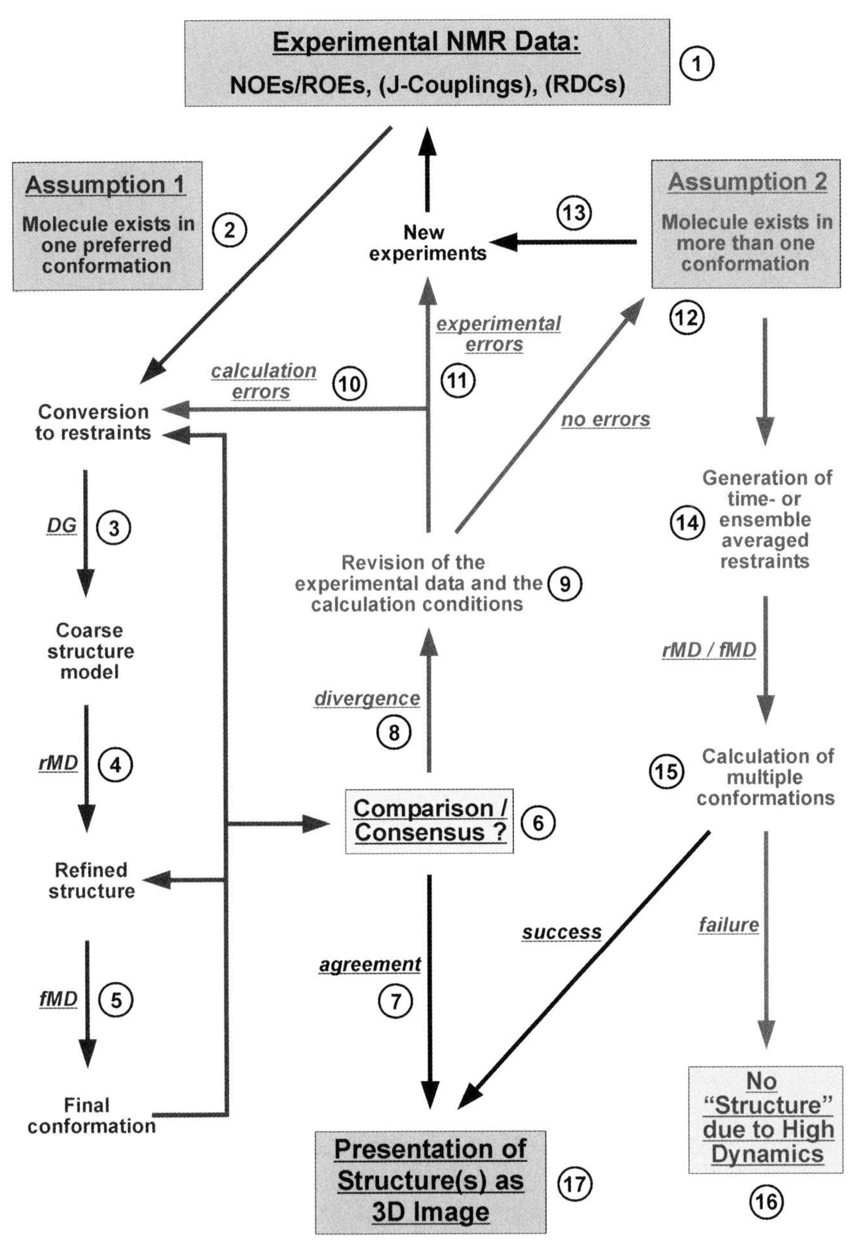

Fig. 9.13 Flowchart presentation of a structure elucidation approach including DG and MD. The steps described in the text above can be followed by the numbers given in this figure.

Instead, one should use them as soft energy terms which allow the molecule to cross energetic barriers during MD (see above). After a set of restraints is generated, the DG computations are performed as described in Section 9.4.1 (3). In order to get an ensemble of different allowed conformations this procedure is repeated several times. The best structural model, i.e. the structure which shows the best agreement with the experimental restraints, serves now as input for a rMD run (4). By averaging all conformations that are present during the trajectory and subsequent energy minimization one gets a refined molecular structure. The rMD run is followed by a free MD simulation by which the final conformational model is produced (5).

At this point it is essential to compare the calculated structure with both the experimental data and the results of the rMD run (6). On the one hand, the interatomic distances of the final model must match the NMR restraints; additionally, the fMD-averaged structure should correspond with the refined conformation obtained by the rMD. Only if the rMD and the fMD simulations result in the same conformational model and no experimental restraints are severely violated the calculated structure can be presented as a 3D image (7).

If the final structure either deviates from the refined model or does not match the NMR restraints (8) one has to revise the experimental data and the parameters used in the DG and MD computations (9). In many cases, mistakes are made when preparing and performing the computational processes (10) or even experimental errors might be present (11). Those errors include a wrong NMR peak assignment, no precise calibration of the NOE/ROE signals, an incorrect conversion of the experimental data to constraints, and a nonfactual parameterization of the rMD and fMD trajectories. In such cases either new calculations or new experiments must be performed.

If the revision of the experiments and the calculation conditions yields neither errors nor inaccuracies one has to change the initial assumption that the molecule of interest has only *one* low-energy structure. Highly flexible molecules like peptides often contain several conformations which are in dynamic exchange (12). If the transition from one conformation to another is slow enough (on the NMR timescale) there are distinguishable sets of signals – one set for each conformation – in the NMR spectra. Thus, the structure elucidation is carried out by taking into account several independent sets of experimental restraints. However, the conformational exchange is often very fast. In this instance, only one averaged NMR signal set can be observed. Since those resonances represent multiple transient molecular conformations it is not possible to assign the NOE/ROE signals to definite, unique interatomic distances [3]. In order to abolish this obstacle one can record new NMR spectra with varied physical conditions (13). The application of different types of solvent, the usage of various pH values and the measurement at lower temperatures may influence both the conformational equilibrium ratio and the exchange kinetics. Thus, in some cases a molecule can be enforced to exhibit one main structure (manipulation of equilibrium) or several clearly distinguishable conformations (slower exchange kinetics) which again results in only one or rather multiple NMR signal sets.

Another sophisticated possibility to handle the problem of conformation-averaged NMR signals is the implementation of time- or ensemble-averaged restraints in MD calculations (14) (see Section 9.4.2.3). The given distance information is either averaged over several molecules or over a certain time interval. Hence, a molecule is not forced to adopt one definite, potentially "meaningless" structure according to the conformation-averaged experimental data. Instead, all conformations which *on average* fulfill the restraints can possibly be observed in a trajectory (15).

If all described procedures fail, no structure determination is possible with the state-of-the-art methods demonstrated in this chapter (16).

In the case of being successful in calculating multiple conformations by using time- or ensemble-averaged MD restraints the solved molecular structures are presented as 3D models and can be deposited in an electronic structure database (17). Finally, it is recommended to provide an accurate explanation of the procedures used for the structure elucidation because the application of different methods (NMR, DG, MD, SA, Monte-Carlo calculations, X-ray crystallography) may result in varying conformational models which do not implicitly display the real state of a molecule. This aspect should be always kept in mind when dealing with structure determination methods.

Acknowledgments

We very much thank Grit Kummerlöwe and Jochen Klages for their immediate support, and Vincent Truffault for helpful suggestions.

References

1 Neuhaus, D., Williamson, M. P. *The Nuclear Overhauser Effect in Structural and Conformational Analysis*, Wiley, New York, **2000**.

2 Kessler, H., Köck, M., Wein, T., Gehrke, M. Reinvestigation of the conformation of cyclosporine A in chloroform. *Helv. Chim. Acta* **1990**, *73*, 1818–1832.

3 Kessler, H., Griesinger, C., Lautz, J., Müller, A., Van Gunsteren, W. F., Berendsen, H. J. C. Conformational dynamics detected by nuclear magnetic resonance NOE values and J coupling constants. *J. Am. Chem. Soc.* **1988**, *110*, 3393–3396.

4 Kessler, H. Peptide conformations. 19. Conformation and biological activity of cyclic peptides. *Angew. Chem. Int. Ed.* **1982**, *21*, 512–523.

5 Loosli, H. R., Kessler, H., Oschkinat, H., Weber, H. P., Petcher, T. J., Widmer, A. Peptide conformations. 31. The conformation of cyclosporin A in the crystal and in solution. *Helv. Chim. Acta* **1985**, *68*, 682–704.

6 Kallen, J., Spitzfaden, C., Zurini, M. G. M., Wider, G., Widmer, H., Wüthrich, K., Walkinshaw, M. D. Structure of human cyclophilin and its binding site for cyclosporin A determined by X-ray crystallography and NMR spectroscopy. *Nature* **1991**, *353*, 276–279.

7 Kessler, H., Gehrke, M., Lautz, J., Köck, M., Seebach, D., Thaler, A. Complexation and medium effects on the conformation of cyclosporin A studied by NMR spectroscopy and molecular dynamics

calculations. *Biochem. Pharmacol.* **1990**, *40*, 169–173.

8 Lautz, J., Kessler, H., Van Gunsteren, W. F., Weber, H. P., Wenger, R. M. On the dependence of molecular conformation on the type of solvent environment: a molecular dynamics study of cyclosporin A. *Biopolymers* **1990**, *29*, 1669–1687.

9 Dechantsreiter, M. A., Planker, E., Mathä, B., Lohof, E., Hölzemann, G., Jonczyk, A., Goodman, S. L., Kessler, H. *N*-methylated cyclic RGD peptides as highly active and selective $\alpha_v\beta_3$ integrin antagonists. *J. Med. Chem.* **1999**, *42*, 3033–3040.

10 Xiong, J. P., Stehle, T., Zhang, R., Joachimiak, A., Frech, M., Goodman, S. L., Arnaout, M. A. Crystal structure of the extracellular segment of integrin $\alpha_v\beta_3$ in complex with an Arg–Gly–Asp ligand. *Science* **2002**, *296*, 151–155.

11 Gottschalk, K. E., Kessler, H. The structures of integrins and integrin–ligand complexes: implications for drug design and signal transduction. *Angew. Chem. Int. Ed.* **2002**, *41*, 3767–3774.

12 Meyer, B., Peters, T. NMR spectroscopy techniques for screening and identifying ligand binding to protein receptors. *Angew. Chem. Int. Ed.* **2003**, *42*, 864–890.

13 Peng, J. W., Moore, J., Abdul-Manan, N. NMR experiments for lead generation in drug discovery. *Prog. Nucl. Magn. Reson. Spectrosc.* **2004**, *44*, 225–256.

14 Kessler, H., Klages, J. Lead search and optimization: NMR in drug discovery. In *Comprehensive Medicinal Chemistry*, Triggle, D. J., Taylor, J. B. (eds.), Elsevier, Oxford, **2006**, Vol. *2*, 901–920.

15 Klages, J., Coles, M., Kessler, H. NMR-based screening: a powerful tool in fragment-based drug discovery. *Mol. Biosyst.* **2006**, *2*, 319–331.

16 Kessler, H., Mronga, S., Gemmecker, G. Multi-dimensional NMR experiments using selective pulses. *Magn. Reson. Chem.* **1991**, *29*, 527–557.

17 Jeener, J., Meier, B. H., Bachmann, P., Ernst, R. R. Investigation of exchange processes by two-dimensional NMR spectroscopy. *J. Chem. Phys.* **1979**, *71*, 4546–4553.

18 Bothner-By, A. A., Stephens, R. L., Lee, J. M., Warren, C. D., Jeanloz, R. W. Structure determination of a tetrasaccharide: transient nuclear Overhauser effects in the rotating frame. *J. Am. Chem. Soc.* **1984**, *106*, 811–813.

19 Thrippleton, M. J., Keeler, J. Elimination of zero-quantum interference in two-dimensional NMR spectra. *Angew. Chem. Int. Ed.* **2003**, *42*, 3938–3941.

20 Bull, T. E. Relaxation in the rotating frame in liquids. *Prog. Nucl. Magn. Reson. Spectrosc.* **1992**, 24.

21 Moseley, H. N. B., Curto, E. V., Krishna, N. R. Complete Relaxation and Conformational Exchange Matrix (CORCEMA) analysis of NOESY spectra of interacting systems; two-dimensional transferred NOESY. *J. Magn. Reson. Ser. B* **1995**, *108*, 243–261.

22 Klages, J., Neubauer, C., Coles, M., Kessler, H., Luy, B. Structure refinement of cyclosporin A in chloroform by using RDCs measured in a stretched PDMS-gel. *ChemBioChem* **2005**, *6*, 1672–1678.

23 Saupe, A., Englert, G. High-resolution nuclear magnetic resonance spectra of oriented molecules. *Phys. Rev. Lett.* **1963**, *11*, 462–464.

24 Bax, A. Weak alignment offers new NMR opportunities to study protein structure and dynamics. *Protein Sci.* **2003**, *12*, 1–16.

25 Simon, B., Sattler, M. *De novo* structure determination from residual dipolar couplings by NMR spectroscopy. *Angew. Chem. Int. Ed.* **2002**, *41*, 437–440.

26 Emsley, J. W., Lindon, J. C. *NMR Spectroscopy Using Liquid Crystal Solvents*, Pergamon Press, Oxford, **1975**.

27 Courtieu, J., Bayle, J. P., Fung, B. M. Variable angle sample spinning NMR in liquid crystals. *Prog. Nucl. Magn. Reson. Spectrosc.* **1994**, *26*, 141–169.

28 Thiele, C. M. Scaling the alignment of small organic molecules in substituted polyglutamates by variable-angle sample spinning. *Angew. Chem. Int. Ed.* **2005**, *44*, 2787–2790.

29 Deloche, B., Samulski, E. T. Short-range nematic-like orientational order in strained elastomers: a deuterium magnetic resonance study. *Macromolecules* **1981**, *14*, 575–581.

30 Cierpicki, T., Bushweller, J. H. Charged gels as orienting media for measurement of residual dipolar couplings in soluble and integral membrane proteins. *J. Am. Chem. Soc.* **2004**, *126*, 16259–16266.

31 Kobzar, K., Kessler, H., Luy, B. Stretched gelatin gels as chiral alignment media for the discrimination of enantiomers by NMR spectroscopy. *Angew. Chem. Int. Ed.* **2005**, *44*, 3145–3147.

32 Sass, H. J., Musco, G., Stahl, S. J., Wingfield, P. T., Grzesiek, S. Solution NMR of proteins within polyacrylamide gels: diffusional properties and residual alignment by mechanical stress or embedding of oriented purple membranes. *J. Biomol. NMR* **2000**, *18*, 303–309.

33 Chou, J. J., Gaemers, S., Howder, B., Louis, J. M., Bax, A. A simple apparatus for generating stretched polyacrylamide gels, yielding uniform alignment of proteins and detergent micelles. *J. Biomol. NMR* **2001**, *21*, 377–382.

34 Kuchel, P. W., Chapman, B. E., Müller, N., Bubb, W. A., Philp, D. J., Torres, A. M. Apparatus for rapid adjustment of the degree of alignment of NMR samples in aqueous media: verification with residual quadrupolar splittings in ^{23}Na and ^{133}Cs spectra. *J. Magn. Reson.* **2006**, *180*, 256–265.

35 John, M., Pintacuda, G., Park, A. Y., Dixon, N. E., Otting, G. Structure determination of protein–ligand complexes by transferred paramagnetic shifts. *J. Am. Chem. Soc.* **2006**, *128*, 12910–12916.

36 Schiemann, O., Piton, N., Mu, Y., Stock, G., Engels, J. W., Prisner, T. F. A PELDOR-based nanometer distance ruler for oligonucleotides. *J. Am. Chem. Soc.* **2004**, *126*, 5722–5729.

37 Panar, M., Phillips, W. D. Magnetic ordering of poly-γ-benzyl-L-glutamate solutions. *J. Am. Chem. Soc.* **1968**, *90*, 3880–3882.

38 Klochkov, V. V., Klochkov, A. V., Thiele, C. M., Berger, S. A novel liquid crystalline system for partial alignment of polar organic molecules. *J. Magn. Reson.* **2006**, *179*, 58–63.

39 Freudenberger, J. C., Knör, S., Kobzar, K., Heckmann, D., Paululat, T., Kessler, H., Luy, B. Stretched poly(vinyl actetate) gels as NMR alignment media for the measurement of residual dipolar couplings in polar organic solvents. *Angew. Chem. Int. Ed.* **2005**, *44*, 423–426.

40 Haberz, P., Farjon, J., Griesinger, C. A DMSO-compatible orienting medium: towards the investigation of the stereochemistry of natural products. *Angew. Chem. Int. Ed.* **2005**, *44*, 427–429.

41 Bendiak, B. Sensitive through-space dipolar correlations between nuclei of small organic molecules by partial alignment in a deuterated liquid solvent. *J. Am. Chem. Soc.* **2002**, *124*, 14862–14863.

42 Luy, B., Kobzar, K., Kessler, H. An easy and scalable method for the partial alignment of organic molecules for measuring residual dipolar couplings. *Angew. Chem. Int. Ed.* **2004**, *43*, 1092–1094.

43 Luy, B., Kobzar, K., Knör, S., Furrer, J., Heckmann, D., Kessler, H. Orientational properties of stretched polystyrene gels in organic solvents and the suppression of their residual ^{1}H NMR signals. *J. Am. Chem. Soc.* **2005**, *127*, 6459–6465.

44 Freudenberger, J. C., Spiteller, P., Bauer, R., Kessler, H., Luy, B. Stretched poly(dimethylsiloxane) gels as NMR alignment media for apolar and weakly polar organic solvents: an ideal tool for measuring RDCs at low molecular concentrations. *J. Am. Chem. Soc.* **2004**, *126*, 14690–14691.

45 Meissner, A., Duus, J. O., Sørensen, O. W. Spin-state-selective excitation. Application for E.COSY-type measurement of J_{HH} coupling constants. *J. Magn. Reson.* **1997**, *128*, 92–97.

46 Ottiger, M., Delaglio, F., Bax, A. Measurement of *J* and dipolar couplings from simplified two-dimensional NMR spectra. *J. Magn. Reson.* **1998**, *131*, 373–378.

47 Vuister, G. W., Bax, A. Quantitative *J* correlation: a new approach for measuring homonuclear three-bond $J(H^NH^\alpha)$ coupling constants in ^{15}N-enriched proteins. *J. Am. Chem. Soc.* **1993**, *115*, 7772–7777.
48 Tjandra, N., Bax, A. Measurement of dipolar contributions to $^1J_{CH}$ splittings from magnetic-field dependence of *J* modulation in two-dimensional NMR spectra. *J. Magn. Reson.* **1997**, *124*, 512–515.
49 Luy, B., Marino, J. P. JE-TROSY: combined *J*- and TROSY-spectroscopy for the measurement of one-bond couplings in macromolecules. *J. Magn. Reson.* **2003**, *163*, 92–98.
50 Furrer, J., John, M., Kessler, H., Luy, B. *J*-Spectroscopy in the presence of residual dipolar couplings: determination of one-bond coupling constants and scalable resolution. *J. Biomol. NMR* **2007**, *37*, 231–243.
51 Carlomagno, T., Peti, W., Griesinger, C. A new method for the simultaneous measurement of magnitude and sign of $^1D_{CH}$ and $^1D_{HH}$ dipolar couplings in methylene groups. *J. Biomol. NMR* **2000**, *17*, 99–109.
52 Nolis, P., Espinosa, J. F., Parella, T. Optimum spin-state selection for all multiplicities in the acquisition dimension of the HSQC experiment. *J. Magn. Reson.* **2006**, *180*, 39–50.
53 Tzvetkova, P., Simova, S., Luy, B. P.E.HSQC: a simple experiment for simultaneous and sign-sensitive measurement of $(^1J_{CH} + D_{CH})$ and $(^2J_{HH} + D_{HH})$ couplings. *J. Magn. Reson.* **2007**, *186*, 193–200.
54 Zhu, G., Renwick, A., Bax, A. Measurement of two- and three-bond ^{1}H-^{13}C *J*-couplings from quantitative heteronuclear *J* correlation for molecules with overlapping ^{1}H resonances, using t_1 noise reduction. *J. Magn. Reson. Ser. A* **1994**, *110*, 257–261.
55 Kurz, M., Schmieder, P., Kessler, H. HETLOC, an efficient method for determining heteronuclear long-range couplings with heteronuclei in natural abundance. *Angew. Chem. Int. Ed.* **1991**, *30*, 1329–1331.
56 Nolis, P., Parella, T. Spin-edited 2D HSQC-TOCSY experiments for the measurement of homonuclear and heteronuclear coupling constants: application to carbohydrates and peptides. *J. Magn. Reson.* **2005**, *176*, 15–26.
57 Kobzar, K., Luy, B. Analyses, extensions and comparison of three experimental schemes for measuring $(^nJ_{CH} + D_{CH})$ couplings at natural abundance. *J. Magn. Reson.* **2007**, *186*, 131–141.
58 Griesinger, C., Sørensen, O. W., Ernst, R. R. Two-dimensional correlation of connected NMR transitions. *J. Am. Chem. Soc.* **1985**, *107*, 6394–6396.
59 Griesinger, C., Sorensen, O. W., Ernst, R. R. Correlation of connected transitions by two-dimensional NMR spectroscopy. *J. Chem. Phys.* **1986**, *85*, 6837–6852.
60 Bax, A., Freeman, R., Frenkiel, T. A., Levitt, M. H. Assignment of carbon-13 NMR spectra via double-quantum coherence. *J. Magn. Reson.* **1981**, *43*, 478–483.
61 Yan, J., Kline, A. D., Mo, H., Shapiro, M. J., Zartler, E. R. A Novel method for the determination of stereochemistry in six-membered chairlike rings using residual dipolar couplings. *J. Org. Chem.* **2003**, *68*, 1786–1795.
62 Kramer, F., Deshmukh, M. V., Kessler, H., Glaser, S. J. Residual dipolar coupling constants: an elementary derivation of key equations. *Conc. Magn. Reson. A* **2004**, *21A*, 10–21.
63 Thiele, C. M., Berger, S. Probing the diastereotopicity of methylene protons in strychnine using residual dipolar couplings. *Org. Lett.* **2003**, *5*, 705–708.
64 Verdier, L., Sakhaii, P., Zweckstetter, M., Griesinger, C. Measurement of long range H,C couplings in natural products in orienting media: a tool for structure elucidation of natural products. *J. Magn. Reson.* **2003**, *163*, 353–359.
65 Aroulanda, C., Boucard, V., Guibe, F., Courtieu, J., Merlet, D. Weakly oriented liquid-crystal NMR solvents as a general tool to determine relative configurations. *Chemistry* **2003**, *9*, 4536–4539.
66 Zweckstetter, M., Bax, A. Prediction of sterically induced alignment in a dilute liquid crystalline phase: aid to protein

structure determination by NMR. *J. Am. Chem. Soc.* **2000**, *122*, 3791–3792.

67 Marcelja, S. Chain ordering in liquid crystals. I. Even–odd effect. *J. Chem. Phys.* **1974**, *60*, 3599–3604.

68 Photinos, D. J., Samulski, E. T., Toriumi, H. Alkyl chains in a nematic field. 1. A treatment of conformer shape. *J. Phys. Chem.* **1990**, *94*, 4688–4694.

69 Patey, G. N., Burnell, E. E., Snijders, J. G., De Lange, C. A. Molecular solutes in nematic liquid crystals: orientational order and electric field gradients. *Chem. Phys. Lett.* **1983**, *99*, 271–274.

70 Azurmendi, H. F., Bush, C. A. Tracking alignment from the moment of inertia tensor (TRAMITE) of biomolecules in neutral dilute liquid crystal solutions. *J. Am. Chem. Soc.* **2002**, *124*, 2426–2427.

71 Canlet, C., Merlet, D., Lesot, P., Meddour, A., Loewenstein, A., Courtieu, J. Deuterium NMR stereochemical analysis of *threo–erythro* isomers bearing remote stereogenic centres in racemic and non-racemic liquid crystalline solvents. *Tetrahedron: Asymmetry* **2000**, *11*, 1911–1918.

72 Lee, S., Mesleh, M. F., Opella, S. J. Structure and dynamics of a membrane protein in micelles from three solution NMR experiments. *J. Biomol. NMR* **2003**, *26*, 327–334.

73 Wu, Z., Tjandra, N., Bax, A. ^{31}P chemical shift anisotropy as an aid in determining nucleic acid structure in liquid crystals. *J. Am. Chem. Soc.* **2001**, *123*, 3617–3618.

74 Wöhnert, J., Franz, K. J., Nitz, M., Imperiali, B., Schwalbe, H. Protein alignment by a coexpressed lanthanide-binding tag for the measurement of residual dipolar couplings. *J. Am. Chem. Soc.* **2003**, *125*, 13338–13339.

75 Pintacuda, G., Moshref, A., Leonchiks, A., Sharipo, A., Otting, G. Site-specific labeling with a metal chelator for protein-structure refinement. *J. Biomol. NMR* **2004**, *29*, 351–361.

76 Haberz, P., Rodriguez-Castaneda, F., Junker, J., Becker, S., Leonov, A., Griesinger, C. Two new chiral EDTA-based metal chelates for weak alignment of proteins in solution. *Org. Lett.* **2006**, *8*, 1275–1278.

77 Su, X. C., Huber, T., Dixon, N. E., Otting, G. Site-specific labeling of proteins with a rigid lanthanide-binding tag. *ChemBioChem* **2006**, *7*, 1599–1604.

78 Mierke, D. F., Grdadolnik, S. G., Kessler, H. Use of one-bond C^{α}–H^{α} coupling-constants as restraints in MD simulations. *J. Am. Chem. Soc.* **1992**, *114*, 8283–8284.

79 Aydin, R., Günther, H. ^{13}C, ^{1}H spin–spin coupling. 9. Norbornane: a reinvestigation of the Karplus curve for $^{3}J(^{13}C, ^{1}H)$. *Magn. Reson. Chem.* **1990**, *28*, 448–457.

80 Bystrov, V. F. Spin–spin coupling and the conformational states of peptide systems. *Prog. Nucl. Magn. Reson. Spectrosc.* **1976**, *10*, 41–81.

81 Haasnoot, C. A. G., De Leeuw, F. A. A. M., De Leeuw, H. P. M., Altona, C. Relationship between proton–proton NMR coupling constants and substituent electronegativities. III. Conformational analysis of proline rings in solution using a generalized Karplus equation. *Biopolymers* **1981**, *20*, 1211–1245.

82 Haasnoot, C. A. G., De Leeuw, F. A. A. M., Altona, C. The relationship between proton–proton NMR coupling constants and substituent electronegativities. I. An empirical generalization of the Karplus equation. *Tetrahedron* **1980**, *36*, 2783–2792.

83 Thomas, W. A. Unravelling molecular structure and conformation – the modern role of coupling constants. *Prog. Nucl. Magn. Reson. Spectrosc.* **1997**, *30*, 183–207.

84 Contreras, R. H., Peralta, J. E. Angular dependence of spin–spin coupling constants. *Prog. Nucl. Magn. Reson. Spectrosc.* **2000**, *37*, 321–425.

85 Eberstadt, M., Gemmecker, G., Mierke, D. F., Kessler, H. Scalar coupling constants – their analysis and their application for the elucidation of structures. *Angew. Chem. Int. Ed.* **1995**, *34*, 1671–1695.

86 Titman, J. J., Keeler, J. Measurement of homonuclear coupling constants from NMR correlation spectra. *J. Magn. Reson.* **1990**, *89*, 640–646.

87 Kessler, H., Griesinger, C., Wagner, K. Peptide conformations. 42. Conformation of side chains in peptides using heteronuclear coupling constants obtained by two-dimensional NMR spectroscopy. *J. Am. Chem. Soc.* **1987**, *109*, 6927–6933.

88 Kessler, H., Griesinger, C., Zarbock, J., Loosli, H. R. Assignment of carbonyl carbons and sequence analysis in peptides by heteronuclear shift correlation via small coupling constants with broad-band decoupling in t_1 (COLOC). *J. Magn. Reson.* **1984**, *57*, 331–336.

89 Grzesiek, S., Cordier, F., Jaravine, V., Barfield, M. Insights into biomolecular hydrogen bonds from hydrogen scalar couplings. *Prog. Nucl. Magn. Reson. Spectrosc.* **2004**, *45*, 275–300.

90 Heller, M., Sukopp, M., Tsomaia, N., John, M., Mierke, D. F., Reif, B., Kessler, H. The conformation of cyclo(D-Pro Ala4): a model for DL4 cyclic pentapeptides. *J. Am. Chem. Soc.* **2006**, *128*, 13806–13814.

91 Tessari, M., Vis, H., Boelens, R., Kaptein, R., Vuister, G. W. Quantitative measurement of relaxation interference effects between ^1H_N CSA and ^1H-^{15}N dipolar interaction: correlation with secondary structure. *J. Am. Chem. Soc.* **1997**, *119*, 8985–8990.

92 Reif, B., Steinhagen, H., Junker, B., Reggelin, M., Griesinger, C. Determination of the orientation of a distant bond vector in a molecular reference frame by cross-correlated relaxation of nuclear spins. *Angew. Chem. Int. Ed.* **1998**, *37*, 1903–1906.

93 Brutscher, B. Principles and applications of cross-correlated relaxation in biomolecules. *Concepts. Magn. Reson.* **2000**, *12*, 207–229.

94 Gemmecker, G. Isotope filter and editing techniques. In *BioNMR in Drug Research (Methods and Principles in Medicinal Chemistry)*, Zerbe, O. (eds.), Wiley-VCH, Weinheim, **2003**, Vol. *16*, 373–390.

95 Verdier, L., Gharbi-Benarous, J., Bertho, G., Mauvais, P., Girault, J. P. Antibiotic resistance peptides: interaction of peptides conferring macrolide and ketolide resistance with *Staphylococcus aureus* ribosomes. Conformation of bound peptides as determined by transferred NOE experiments. *Biochemistry* **2002**, *41*, 4218–4229.

96 Murali, N., Jarori, G. K., Landy, S. B., Rao, B. D. N. Two-dimensional transferred nuclear Overhauser effect spectroscopy (TRNOESY) studies of nucleotide conformations in creatine kinase complexes: effects due to weak nonspecific binding. *Biochemistry* **1993**, *32*, 12941–12948.

97 Balaram, P., Bothner-By, A. A., Breslow, E. Nuclear magnetic resonance studies of interaction of peptides and hormones with bovine neurophysin. *Biochemistry* **1973**, *12*, 4695–4704.

98 Clore, G. M., Gronenborn, A. M. Theory of the time dependent transferred nuclear Overhauser effect: applications to structural analysis of ligand–protein complexes in solution. *J. Magn. Reson.* **1983**, *53*, 423–442.

99 Post, C. B. Exchange-transferred NOE spectroscopy and bound ligand structure determination. *Curr. Opin. Struct. Biol.* **2003**, *13*, 581–588.

100 Zabell, A. P. R., Post, C. B. Intermolecular relaxation has little effect on intra-peptide exchange-transferred NOE intensities. *J. Biomol. NMR* **2002**, *22*, 303–315.

101 Phan-Chan-Du, A., Petit, M. C., Guichard, G., Briand, J. P., Muller, S., Cung, M. T. Structure of antibody-bound peptides and retro-inverso analogues. A transferred nuclear Overhauser effect spectroscopy and molecular dynamics approach. *Biochemistry* **2001**, *40*, 5720–5727.

102 Phan-Chan-Du, A., Hemmerlin, C., Krikorian, D., Sakarellos-Daitsiotis, M., Tsikaris, V., Sakarellos, C., Marinou, M., Thureau, A., Cung, M. T., Tzartos, S. J. Solution conformation of the antibody-bound tyrosine phosphorylation site of the nicotinic acetylcholine receptor β-subunit in its phosphorylated and nonphosphorylated states. *Biochemistry* **2003**, *42*, 7371–7380.

103 Wakamatsu, K., Takeda, A., Tachi, T., Matsuzaki, K. Dimer structure of magainin 2 bound to phospholipid vesicles. *Biopolymers* **2002**, *64*, 314–327.

104 Koenig, B. W., Mitchell, D. C., König, S., Grzesiek, S., Litman, B. J., Bax, A. Measurement of dipolar couplings in a transducin peptide fragment weakly bound to oriented photo-activated rhodopsin. *J. Biomol. NMR* **2000**, *16*, 121–125.

105 Hinckley, C. C. Paramagnetic shifts in solutions of cholesterol and the dipyridine adduct of trisdipivalomethan atoeuropium(III). A shift reagent. *J. Am. Chem. Soc.* **1969**, *91*, 5160–5162.

106 Friebolin, H. *Basic One- and Two-Dimensional NMR Spectroscopy*, Wiley-VCH, Weinheim, **2004**.

107 Blommers, M. J. J., Stark, W., Jones, C. E., Head, D., Owen, C. E., Jahnke, W. Transferred cross-correlated relaxation complements transferred NOE: structure of an IL-4R-derived peptide bound to STAT-6. *J. Am. Chem. Soc.* **1999**, *121*, 1949–1953.

108 Maltseva, T., Usova, E., Eriksson, S., Milecki, J., Földesi, A., Chattopadhayaya, J. An NMR conformational study of the complexes of $^{13}C/^{2}H$ double-labeled 2′-deoxynucleosides and deoxycytidine kinase (dCK). *J. Chem. Soc. Perkin. Trans. 2* **2000**, *2*, 2199–2207.

109 Carlomagno, T., Sanchez, V. M., Blommers, M. J. J., Griesinger, C. Derivation of dihedral angles from CH–CH dipolar–dipolar cross-correlated relaxation rates: a C–C torsion involving a quaternary carbon atom in epothilone A bound to tubulin. *Angew. Chem. Int. Ed.* **2003**, *42*, 2515–2517.

110 Mayer, M., Meyer, B. Characterization of ligand binding by saturation transfer difference NMR spectroscopy. *Angew. Chem. Int. Ed.* **1999**, *38*, 1784–1788.

111 Rees, D. C., Congreve, M., Murray, C. W., Carr, R. Fragment-based lead discovery. *Nat. Rev. Drug Discov.* **2004**, *3*, 660–672.

112 Meinecke, R., Meyer, B. Determination of the binding specificity of an integral membrane protein by saturation transfer difference NMR: RGD peptide ligands binding to integrin $alpha_{IIb}beta_3$. *J. Med. Chem.* **2001**, *44*, 3059–3065.

113 Claasen, B., Axmann, M., Meinecke, R., Meyer, B. Direct observation of ligand binding to membrane proteins in living cells by a saturation transfer double difference (STDD) NMR spectroscopy method shows a significantly higher affinity of integrin $\alpha_{IIb}\beta_3$ in native platelets than in liposomes. *J. Am. Chem. Soc.* **2005**, *127*, 916–919.

114 Fejzo, J., Lepre, C. A., Peng, J. W., Bemis, G. W., Ajay, Murcko, M. A., Moore, J. M. The SHAPES strategy: an NMR-based approach for lead generation in drug discovery. *Chem. Biol.* **1999**, *6*, 755–769.

115 Jahnke, W., Rüdisser, S., Zurini, M. Spin label enhanced NMR screening. *J. Am. Chem. Soc.* **2001**, *123*, 3149–3150.

116 Jahnke, W., Perez, L. B., Paris, C. G., Strauss, A., Fendrich, G., Nalin, C. M. Second-site NMR screening with a spin-labeled first ligand. *J. Am. Chem. Soc.* **2000**, *122*, 7394–7395.

117 Teague, S. J. Implications of protein flexibility for drug discovery. *Nat. Rev. Drug Discov.* **2003**, *2*, 527–541.

118 Blumenthal, L. M. *Theory and Application of Distance Geometry*, Chelsea, New York, **1970**.

119 Braun, W., Go, N. Calculation of protein conformations by proton–proton distance constraints. A new efficient algorithm. *J. Mol. Biol.* **1985**, *186*, 611–626.

120 Kuszewski, J., Nilges, M., Brünger, A. T. Sampling and efficiency of metric matrix distance geometry: a novel partial metrization algorithm. *J. Biomol. NMR* **1992**, *2*, 33–56.

121 Havel, T. F. The sampling properties of some distance geometry algorithms applied to unconstrained polypeptide-chains: a study of 1830 independently computed conformations. *Biopolymers* **1990**, *29*, 1565–1585.

122 Crippen, G. M., Havel, T. F. *Distance Geometry and Molecular Conformation*, Wiley, New York, **1988**.

123 Crippen, G. M. Conformational analysis by energy embedding. *J. Comput. Chem.* **1982**, *3*, 471–476.

124 Beutler, T. C., Van Gunsteren, W. F. Molecular dynamics free energy calculation in four dimensions. *J. Chem. Phys.* **1994**, *101*, 1417–1422.

125 Haile, J. M. *Molecular Dynamics Simulation: Elementary Methods*, Wiley, New York, **1992**.

126 van Gunsteren, W. F., Berendsen, H. J. C. Computer-simulation of molecular-dynamics – methodology, applications, and perspectives in chemistry. *Angew. Chem. Int. Ed.* **1990**, *29*, 992–1023.

127 van Gunsteren, W. F., Bakowies, D., Baron, R., Chandrasekhar, I., Christen, M., Daura, X., Gee, P., Geerke, D. P., Glättli, A., Hünenberger, P. H., Kastenholz, M. A., Oostenbrink, C., Schenk, M., Trzesniak, D., Van der Vegt, N. F. A., Yu, H. B. Biomolecular modeling: goals, problems, perspectives. *Angew. Chem. Int. Ed.* **2006**, *45*, 4064–4092.

128 Phillips, J. C., Braun, R., Wang, W., Gumbart, J., Tajkhorshid, E., Villa, E., Chipot, C., Skeel, R. D., Kale, L., Schulten, K. Scalable molecular dynamics with NAMD. *J. Comput. Chem.* **2005**, *26*, 1781–1802.

129 Case, D. A., Cheatham III, T. E., Darden, T., Gohlke, H., Luo, R., Merz, K. M., Jr., Onufriev, A., Simmerling, C., Wang, B., Woods, R. J. The Amber biomolecular simulation programs. *J. Comput. Chem.* **2005**, *26*, 1668–1688.

130 Brooks, B. R., Bruccoleri, R. E., Olafson, B. D., States, D. J., Swaminathan, S., Karplus, M. CHARMM: a program for macromolecular energy, minimization, and dynamics calculations. *J. Comput. Chem.* **1983**, *4*, 187–217.

131 van Gunsteren, W. F., Billeter, S. R., Eising, A. A., Hunenberger, P. H., Kruger, P., Mark, A. E., Scott, W. R. P., Tironi, I. G., *Biomolecular Simulation: The GROMOS96 Manual and User Guide*, Vdf Hochschulverlag, Zurich, **1996**.

132 van der Spoel, D., Lindahl, E., Hess, B., Groenhof, G., Mark, A. E., Berendsen, H. J. C. GROMACS: fast, flexible, and free. *J. Comput. Chem.* **2005**, *26*, 1701–1718.

133 Jorgensen, W. L., Maxwell, D. S., Tirado-Rives, J. Development and testing of the OPLS all-atom force field on conformational energetics and properties of organic liquids. *J. Am. Chem. Soc.* **1996**, *118*, 11225–11236.

134 Ponder, J. W., Case, D. A. Force fields for protein simulations. *Adv. Protein Chem.* **2003**, *66*, 27–85.

135 Oostenbrink, C., Villa, A., Mark, A. E., Van Gunsteren, W. F. A biomolecular force field based on the free enthalpy of hydration and solvation: the GROMOS force-field parameter sets 53A5 and 53A6. *J. Comput. Chem.* **2004**, *25*, 1656–1676.

136 MacKerell, A. D., Jr., Bashford, D., Bellott, M., Dunbrack, R. L., Evanseck, J. D., Field, M. J., Fischer, S., Gao, J., Guo, H., Ha, S., Joseph-McCarthy, D., Kuchnir, L., Kuczera, K., Lau, F. T. K., Mattos, C., Michnick, S., Ngo, T., Nguyen, D. T., Prodhom, B., Reiher, W. E., Roux, B., Schlenkrich, M., Smith, J. C., Stote, R., Straub, J., Watanabe, M., Wiorkiewicz-Kuczera, J., Yin, D., Karplus, M. All-atom empirical potential for molecular modeling and dynamics studies of proteins. *J. Phys. Chem. B* **1998**, *102*, 3586–3616.

137 Hagler, A. T., Ewig, C. S. On the use of quantum energy surfaces in the derivation of molecular force fields. *Comput. Phys. Commun.* **1994**, *84*, 131–155.

138 Clark, M., Cramer III, R. D., Van Opdenbosch, N. Validation of the general purpose Tripos 5.2 force field. *J. Comput. Chem.* **1989**, *10*, 982–1012.

139 Allinger, N. L., Chen, K., Lii, J. H. An improved force field (MM4) for saturated hydrocarbons. *J. Comput. Chem.* **1996**, *17*, 642–668.

140 Luty, B. A., Van Gunsteren, W. F. Calculating electrostatic interactions using the particle-particle particle-mesh method with nonperiodic long-range interactions. *J. Phys. Chem.* **1996**, *100*, 2581–2587.

141 Hünenberger, P. H., McCammon, J. A. Effect of artificial periodicity in simulations of biomolecules under Ewald boundary conditions: a continuum

electrostatics study. *Biophys. Chem.* **1999**, *78*, 69–88.

142 Verlet, L. Computer "experiments" on classical fluids. I. Thermodynamical properties of Lennard–Jones molecules. *Phys. Rev.* **1967**, *159*, 98–103.

143 Torda, A. E., Brunne, R. M., Huber, T., Kessler, H., Van Gunsteren, W. F. Structure refinement using time-averaged *J*-coupling constant restraints. *J. Biomol. NMR* **1993**, *3*, 55–66.

144 Kessler, H. Detection of intramolecular mobility by NMR spectroscopy. 13. Detection of hindered rotation and inversion by NMR spectroscopy. *Angew. Chem. Int. Ed.* **1970**, *9*, 219–235.

145 Reif, B., Hennig, M., Griesinger, C. Direct measurement of angles between bond vectors in high-resolution NMR. *Science* **1997**, *276*, 1230–1233.

146 Henrichsen, D., Ernst, B., Magnani, J. L., Wang, W. T., Meyer, B., Peters, T. Bioaffinity NMR spectroscopy: identification of an E-selectin antagonist in a substance mixture by transfer NOE. *Angew. Chem. Int. Ed.* **1999**, *38*, 98–102.

147 Kummerlöwe, G., Auernheimer, J., Lendlein, A., Luy, B. Streched poly (acrylonitrile) as a scalable alignment medium for DMSO. *J. Am. Chem. Soc.* **2007**, *129*, 6080–6081.

Part IV
Solubility

10
Drug Solubility in Water and Dimethylsulfoxide

Christopher Lipinski

Abbreviations

DMSO	dimethylsulfoxide
HPLC	high-performance liquid chromatography
HTS	high-throughput screening
MW	molecular weight
NTU	nephelometric turbidity unit
SAR	structure–activity relationship
TFA	trifluoroacetic acid
UV	ultraviolet

10.1
Introduction

This chapter is written for a medicinal chemistry audience. The focus is on the importance of drug solubility in water and in dimethylsulfoxide (DMSO) in the types of activities likely to be of interest to medicinal chemists. The emphasis is on the discovery stage as opposed to the development stage. The reader will find numerous generalizations and rules of thumb relating to solubility in a drug discovery setting.

The solubility of drugs in water is important for oral drug absorption. To simplify, a drug must be soluble in the aqueous contents of the gastrointestinal lumen to be orally absorbed. Drug solubility in DMSO is important in the biology testing of a compound formatted as a DMSO stock solution. Historically the testing has been in high-throughput screening (HTS), although more recently compounds dissolved in DMSO have begun to replace the testing of compounds present in powder form in traditional low-throughput biological assays. Solubility in aqueous media and in DMSO is discussed in the context of similarities, but also important differences.

Molecular Drug Properties. Measurement and Prediction. R. Mannhold (Ed.)

ISBN: 978-3-527-31755-4

10.2
Water Solubility

10.2.1
Where does Drug Poor Water Solubility Come From?

Poor water solubility is an unfortunate fact currently faced by virtually every organization involved in chemical biology/chemical genetics and in drug discovery [1, 2]. Solubility is discussed using two types of concentration units. Micrograms per milliliter (or grams per liter) is the preferred unit used in the pharmaceutical sciences arena. Micromolar or millimolar is the preferred concentration unit used in the biological sciences. In the current era, as a rough guide, one can bin aqueous solubility at room temperature in buffer close to pH 7 into three ranges. Solubility of less than 5 $\mu g\,mL^{-1}$ [10 μM for molecular weight (MW) = 500] is getting into the undesirable low-solubility range. Solubility greater than about 100 $\mu g\,mL^{-1}$ (200 μM for MW = 500) is in a solubility range where few if any solubility related formulation issues may be expected. The range 5–100 $\mu g\,mL^{-1}$ is an intermediate range. These ranges are generalizations and are modified by considerations of drug dose and intestinal permeability.

The following principles define these aqueous solubility generalizations. First, about 80% of the drugs entering clinical study currently are approximately 1 $mg\,kg^{-1}$ in projected human potency [3]. This means roughly a total dose of 100–200 mg is very common. Marketing pressures drive innovator pharmaceutical companies to once a day dosing meaning that the 100–200 mg dose will be given most commonly as a single tablet or capsule. The accepted (US Food and Drug Administration) volume of water swallowed along with the tablet or capsule is 250 mL. Hence the concentration in the gastric lumen that must be achieved if all of the drug is to be dissolved is approximately 100 μM.

To put aqueous solubility into context with regard to DMSO solubility the general range of DMSO stock concentrations used in compound screening repositories is 2–20 mM. These DMSO stock solutions are typically diluted 1000 : 1 with assay buffer to give nominal assay screening concentrations of 2 to 20 μM. Thus it will be seen that the typical requirement for aqueous solubility for oral drug absorption of 100 μM is substantially higher than the concentration range of 2 to 20 μM required for a compound to be soluble in an *in vitro* biology assay. Many compounds are soluble enough to exhibit *in vitro* biological activity, but are not soluble enough to be orally absorbed. The general solubility ranges described above are based on current pharmaceutical industry practices. The histogram of solubility ranges has shifted by 2–3 log molar solubility units towards poorer aqueous solubility over the last decade or more. As a rough generalization, 30–40% of compounds made in a competent organization's drug discovery hit-to-lead and lead optimization chemistry have poor solubility of less than 10 μM.

The generalization as to binned solubility ranges is based on experimental aqueous solubility assays that are primarily intended for early discovery use. Most often these assays employ a drug in DMSO stock solution rather than a powder as the source of the drug and often are termed as "kinetic" solubility assays. The common characteristic of these discovery assays is that to a variable degree they overestimate the drug aqueous solubility compared to the traditional literature thermodynamic aqueous solubility assay.

Using a discovery kinetic solubility assay aqueous solubilities were determined in pH 7 phosphate buffer at Pfizer (Groton, CT, USA) for a large number of compounds separated according to compound source. Compounds achieving phase II status encompassed only about 14% of poorly soluble compounds. Compounds purchased entirely from academic sources exhibited about 30% poor solubility. Compounds prepared as part of a drug discovery effort exhibited 40% poor solubility where the starting point for the medicinal chemistry was an academic compound with poor solubility in the 30% range. Optimization of *in vitro* activity in medicinal chemistry almost invariably worsens aqueous solubility. The most reliable method to improve *in vitro* activity is for a medicinal chemist to increase compound lipophilicity and increased compound lipophilicity almost always decreases aqueous solubility.

10.2.2
Water Solubility is Multifactorial

As a rule-of-thumb experimental solubilities of the same compound from multiple sources under allegedly comparable conditions show a standard deviation when expressed in log molar units of ±0.5 log molar solubility units. Aqueous solubility is only partially determined by compound structure. Summarizing briefly, the single most important factor that is noncompound related that influences aqueous solubility is whether the compound is amorphous or crystalline. Universally, the higher energy amorphous compound form is the more aqueous soluble by typically a factor of 100. This generalization also holds for solubility in DMSO. The amorphous compound can easily be 100 times more soluble in either water or DMSO than any of the crystalline forms. Typically aqueous solubility of different crystalline forms of the same compound varies over a range of 5 or so. It is very likely that this range of crystalline form also holds for solubility in DMSO. However there is no published data on this point.

10.2.3
Water Solubility and Oral Absorption

As a generalization, to be orally well absorbed a compound must be soluble in the contents of the gastrointestinal lumen [4]. Solubility in aqueous buffer is commonly used as a simplifying surrogate for intestinal content solubility. There are rare exceptions to the principle that to be absorbed a compound must be soluble. Solid particles, e.g. starch, can be absorbed. Absorption of very small quantities of even biologically very large compounds can occur via lymphoid tissue, e.g. orally active vaccines. Very lipophilic basic compounds, e.g. certain antimalarials, can be absorbed via the intestinal lymphatics and delivered directly to the heart.

10.2.4
Importance and Guidelines

Figure 10.1 presents the minimum acceptable aqueous thermodynamic solubility for three scenarios. The three scenarios represent different human oral potencies, i.e. the very potent $0.1\,mg\,kg^{-1}$ drugs (left-most three bars), the $1.0\,mg\,kg^{-1}$ drugs (middle three bars) and the $10.0\,mg\,kg^{-1}$ drugs (right-most three bars). The middle triplet of $1.0\,mg\,kg^{-1}$ drugs represents the most common scenario comprising about 80% of cases in current drug discovery. The left-most triplet of very potent $0.1\,mg\,kg^{-1}$ drugs represents 10% or fewer cases in current drug discovery and is not increasing. The right-most triplet of $10.0\,mg\,kg^{-1}$ represents less than 10% of current cases and has been steadily declining over the last decade. Currently, mostly only drugs in the antibiotic and antiviral arena have this low level of oral potency. Each member of a triplet depicts the minimum acceptable aqueous solubility depending on the drug intestinal permeability. The left-most bar of a triplet depicts the solubility required when the permeability is at the lower 10th percentile. With respect to the middle triplet the left-most bar at the permeability 10th percentile might be the case for a drug with peptidomimetic-like structure. For such a compound the aqueous solubility requirement is a little over $200\,\mu g\,mL^{-1}$. The middle bar of the center triplet with solubility of about $50\,\mu g\,mL^{-1}$ represents the solubility required for an average permeability compound at the most common $1.0\,mg\,kg^{-1}$ oral potency range.

It is likely that medicinal chemists might focus on the right-most bar in the left-most triplet depicting a $1.0\,\mu g\,mL^{-1}$ aqueous solubility. This is a frequently encountered rather poor level of aqueous solubility in current drug discovery and the medicinal chemist might argue that indeed this level of solubility might be acceptable. It is true that a solubility of as little as $1.0\,\mu g\,mL^{-1}$ is acceptable for very

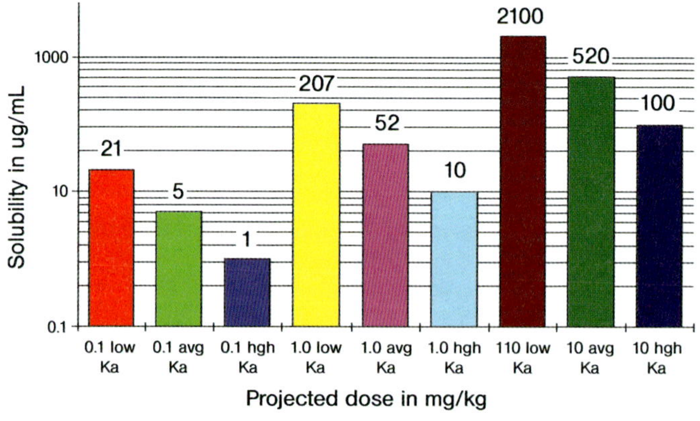

Fig. 10.1 Minimum acceptable solubility (in $\mu g\,mL^{-1}$). Bars show the minimum solubility for low, medium and high permeability (K_a) at a clinical dose. (From Ref. [17].)

potent orally active compounds with the very best permeability characteristics. However, how common is this scenario? The answer is that this is a very uncommon situation. Very potent orally active compounds are discovered by luck, are relatively rare and only a minority of these will have the best permeability properties.

10.2.5
Intestinal Fluid Solubility

Aqueous solubility in buffer is used as a surrogate for the more complex fluid contents of the gastrointestinal lumen [5]. A major factor improving solubility and absorption for some compounds is the presence of surfactant bile acid lecithin-like components secreted from the bile duct. In humans and dogs, bile flow occurs as a bolus following a meal. If the solubility of a compound is improved after a meal due to the presence of bile contents the oral absorption can be improved. This is the "food effect" – oral absorption is better after a meal than on an empty stomach. Formulas for purely synthetic surrogates for fasting and fed gastrointestinal fluid exist, and solubility can be measured in these fluids rather than in aqueous buffer. The food effect may be operative if the primary cause of poor aqueous solubility is because the compound is large and lipophilic. However, the magnitude of the solubility increase (and hence absorption increase) due to bile contents is variable and hard to predict in advance of experiment. A food effect and hence oral absorption increase is unlikely if the primary cause of poor aqueous solubility is strong intermolecular cross-unit cell interactions in the crystal state of the drug. Strong intermolecular cross-unit cell interactions in the crystal state of the drug are typically associated with a high melting point.

It should be noted that it is not particularly desirable to rely on a food effect for improving oral absorption since it places limitations on when a patient can take a drug and increases the variability in drug oral absorption. The reader should also be aware that rodents do not show a "food effect". Rats and mice lack a gall bladder, continually secrete low levels of bile while they nibble on their food throughout the night, and thus do not dump a large bolus of bile stored in the gall bladder at meal time. Hence, a drug subject to the food effect might be expected to show poorer oral absorption in rodents than after a meal in dogs, monkeys and humans.

10.3
Early Discovery Water Solubility and Biological Testing

10.3.1
HTS Application

In general, a compound must be soluble at the assay screening dose to yield reproducible and believable biology testing results. The exception occurs in *in vitro* assays if assay components “solubilize” the compound, e.g. in the presence of serum protein or cell components. Hence it is extremely common, especially in HTS, to prescreen compound collections computationally for aqueous solubility and to include some type of experimental aqueous solubility assay on some or all of the compounds entering an HTS with the goal of identifying or eliminating from assay those compounds with aqueous solubility problems [6]. A related problem relates to quantifying assay compound concentration. In the vast majority of HTS assays the concentration in assay buffer is assumed to be the nominal concentration. For example, diluting a 10 mM compound in DMSO stock concentration by a factor of 1000 into assay buffer is assumed to generate a 10 µM assay buffer concentration. We now know that this assumption is seriously flawed in about 15% of cases [7]. Actual compound concentration is an order of magnitude lower or even very much lower (at the lower limit of quantitation) than the nominal concentration. In the 85% majority of cases the actual concentration is distributed about the nominal concentration, but is irrelevant to the HTS screening result because the HTS screen cannot discern the effect of a concentration factor of 2–3 on assay readout. Generalizing, compounds with poor solubility due to large size and lipophilicity are most successfully predicted computationally. Experience at the Pfizer Groton lab over tens of thousands of compounds tested in a discovery kinetic solubility assay showed that a log $P>5$ was a remarkably effective simple single predictor of poor solubility. Fully 75% of compounds with log $P>5$ had rather poor solubility in pH 7 phosphate buffer of less than 5 $\mu g\,mL^{-1}$. Interestingly this very simple rule held up even when the ionization state of the compound was ignored. The explanation for this phenomenon is given in more detail in the subsequent discussion on ionization.

Compounds with poor aqueous solubility due to strong crystal packing are far more difficult to predict. For these types of compounds predictions are much improved if compound melting points are available. A standard rule-of-thumb is that a 100 °C increase in melting point translates to a 10 times lower aqueous solubility [8]. Unfortunately, in the current era, melting point data has almost completely disappeared from most drug discovery chemistry groups. It is very common for entire discovery medicinal chemistry groups to purify compounds by reverse-phase high-performance liquid chromatography (HPLC) giving predominately amorphous materials with virtually no compounds crystallized in early to mid discovery stages. Predicting compounds insoluble because of strong crystal packing is difficult because prediction of crystal packing strength is itself very difficult.

10.3.2
Improving HTS Assay Quality

Reproducibility in a primary HTS even with a good z' (quality metric) > 0.5 is an issue [9]. The concordance in a triplicate HTS with identical target, identical screening format and identical compounds seldom exceeds 65%. Problems persist post primary HTS. IC_{50} rankings and selectivity panels can be seriously flawed if nominal rather than actual concentration is used. All competent drug discovery organizations resynthesize and retest "actives" before making decisions committing significant resources. Analytical structure confirmation is part of this process. A rule-of-thumb is that 10–50% of reproducible hits from an HTS will not hold up when the compound is resynthesized and carefully retested. The causes of this distressing observation are multifactorial, but the magnitude of the problem reinforces the need for quantitating concentration in the more careful retest stage. In particular, large differences in activity are possible between activity observed using a compound introduced as a DMSO stock concentration and activity observed starting with compound as a crystalline powder. The usual pattern is that activity is lower (sometimes very much lower) using the crystalline powder. Poorer aqueous solubility of the crystalline powder in assay buffer compared to the compound introduced as a DMSO stock solution into the assay buffer is the most common explanation.

10.4 Water Solubility Measurement Technology

10.4.1 Discovery-stage Water Solubility Advantages

Experimental technology to measure aqueous solubility falls into two categories. In one category there are the early discovery aqueous solubility assays (often called kinetic solubility) that usually start with a compound in a DMSO stock solution. The most appropriate use of this type of assay is in early discovery especially when the medicinal chemist has little or no other structure–activity relationship (SAR) input towards changing chemical structure to attain some level of oral absorption. In a second category are so-called thermodynamic aqueous solubility assays which in most respects follow the precepts described for a "proper" solubility in a pharmaceutical sciences textbook. This type of assay may or not be ported to a robot, but regardless of the automation essentially uses a traditional protocol. Discovery kinetic solubility assays are primarily used as SAR inputs to medicinal chemists in early discovery. They tend to be more qualitative than quantitative. So even though results may be reported in a numerical fashion operationally the numerical results are used in binned format, e.g. poor, intermediate, good. Fast assay turnaround time is essential to the value of a discovery solubility assay. Most of the solubility assay value is lost if the assay results come in too slowly to influence the choice of the next compound to make. Experience teaches that chemists do not stop a synthetic sequence because of a solubility result received after synthesis starts. Discovery kinetic solubility assays are very good at categorizing solubility across chemical series, especially if the poor solubility arises from a crystal packing issue. Solubility has the character of a blunt SAR. The best way to avoid a problem is to pick the series for exploration which has the best solubility for the first few compounds. Hence, solubility input is essential in hit to lead chemistry when choices are made as to which series to work on.

10.4.2
Discovery-stage Water Solubility Limitations

Discovery kinetic solubility assays are not a replacement for the thermodynamic solubility assay. These types of assays lose their value in the later stages of discovery. They never should be used to categorize the solubility of a crystalline material and especially should not be used as input to oral absorption estimations in a late discovery setting. Common sense should be used in evaluating a discovery kinetic solubility assay result. *In vivo* experimental data is always more relevant. Measurement on systemic drug levels following *in vivo* oral dosing is always the more relevant (but also more resource intensive). Poor or nonexistent systemic levels in the face of good solubility help in the diagnosis of the problem, which might be due to poor intestinal permeability or a rapid metabolism problem.

10.4.3
In Vivo Dosing Application

Medicinal chemists should be aware that inexperienced biologists can erroneously conclude that poorly aqueous soluble compounds are orally absorbed when compounds are dosed in pharmaceutically unacceptable solvents. Always ask: what is the dosing vehicle? Heroic combinations of DMSO, Cremophor, poly(ethylene glycol), Tween-80 and ethanol are unacceptable and misleading. In case of doubt consult with a pharmaceutical scientist colleague. The reliable standards are an aqueous solution (with perhaps a trace of DMSO) or a suspension (perhaps stabilized with an acceptable quantity of adjuvant, e.g. Tween-80).

10.4.4
In Vivo SAR to Guide Chemistry

Discovery solubility assays provide solubility SAR that is more relevant than thermodynamic solubility data to how biologists actually dose compounds in an early discovery setting. In early discovery biologists do not dose compounds orally in any manner even approximating the conditions of a thermodynamic solubility experiment. If a compound cannot be put into aqueous solution by acceptable pH adjustment, it is put into DMSO as a concentrated stock solution. The gavage vehicle is stirred and the DMSO stock is rapidly added. The DMSO stock plus gavage vehicle is then rapidly dosed to the rodent before the compound precipitates. Typically discovery aqueous solubility assays mimic some or all aspects of this procedure. Contrast this with the conditions of a thermodynamic solubility assay. An excess of well-characterized (from a solid-state perspective) crystalline compound is equilibrated with aqueous buffer with enough equilibration time (often 24–48 h) for equilibrium to occur. The solid phase is separated from the liquid and compound dissolved in the aqueous phase is quantitated. No discovery biologist doses orally with a protocol remotely similar to this.

10.4.5 Discovery Solubility Assay Endpoint Detection

Discovery solubility assays can be subdivided depending on how the solubility point is determined. One type of assay measures compound coming out of solution. The other type of assay measures compound in solution. Both methods have advantages and limitations.

10.4.6 Advantages of Out-of-solution Detection

Measuring compound coming out of solution usually relies on light scattering. Instrumentation is readily available over a wide range of costs. Simple light scattering detectors used by water quality engineers to detect turbidity in drinking water cost less than $1000. These are low capacity, a single test tube at a time. A ultraviolet (UV) or diode array UV machine can be used to detect the absorption due to light scattering either by measuring absorbance at long wavelength (e.g. 600 nm) outside of the usual UV chromophore absorbance range or by capturing the light scattering inverse absorption power curve which generates huge absorbance at low wavelength (e.g. 210 nm). Chemometric techniques (e.g. Fourier transform methods) may be necessary to separate out UV absorbance due to chromophore from UV absorbance due to light scattering. Light scattering can be measured directly using a dedicated plate-based nephelometer. The advantage is that chromophoric absorption or colored material does not interfere with the light scattering signal. Multi-angle light scattering detection as in the commercial solubility scanner originally derived from a fluorescent activated cell sorter provides greater sensitivity and the ability to detect particles in the sub-100 nm size range. By contrast, lower-grade single 90° off-angle light scattering detectors have greatest sensitivity in the size range 500–700 nm.

Measuring compound remaining in solution has the advantage that the endpoint is quantitative and similar to that in a thermodynamic solubility assay. Within reasonable experimental parameters so that Beer's law is followed a UV absorbance is linearly related to concentration in solution. Measuring compound concentration in solution is very well established technology with a wealth of available instrumentation.

10.4.7 Limitations of Out-of-solution Detection

A major disadvantage of using light scattering is that *a priori* one does not know the source of the precipitant. Is it the compound of interest or is it a minor very insoluble impurity? Measuring compound coming out of solution using light scattering has the disadvantage that nephelometry, the measurement of a light scattering signal, is not a quantitative technique. The units measured in a nephelometer are nephelometric turbidity units (NTUs). The magnitude of NTUs are not linearly related to quantity of compound coming out of solution. The NTU signal may decrease with time as particles become larger and agglomerate. The NTU signal strongly depends on particle size. For low-end light scattering detectors the maximum sensitivity occurs when the particle size matches the wavelength of the scattered light. For visible light this translates to a maximum sensitivity for particles in the size range 500–700 nm. There is a general rule-of-thumb relating the cause of poor aqueous solubility to the size of the particle and, hence, the NTU signal. Poorly soluble large lipophilic compounds tend to give larger particles than 1 μm (1000 nm). The sensitivity to detecting precipitation is lower and kinetics seem to be more important in the timing of precipitation. Often the NTU signal rises gradually as the solubility limit is exceeded. Hence, the standard errors in detecting insolubility of large lipophilic compounds can be quite large. By contrast, compounds poorly soluble because of crystal packing tend to precipitate as smaller particles closer to the optimum size for detection. The NTU signal is much more robust and tends to rise very rapidly as the solubility is exceeded. The reproducibility is good in part because the timing of precipitation is much less variable. Nephelometers need to be calibrated using scattering standards that have a limited shelf life.

Measuring compound staying in solution using UV has surmountable limitations not usually encountered in typical analytical assays. If concentration is determined from absorbance at fixed wavelengths there is the danger of a misleading signal arising from a more soluble UV-absorbing impurity. The solution is to match the aqueous UV absorbance pattern with the UV absorbance pattern of a known quantity of a positive standard dissolved in an aqueous organic solution. The assay result is null if the patterns do not match. The UV absorption curve may be pH dependent if the compound contains an ionizable center conjugated to or in close proximity to a UV chromophore. The UV absorption curve patterns between aqueous buffer and calibration standard may not match if the effective pH of aqueous buffer and aqueous organic standard are different. This problem is surmountable, but only with sophisticated software. Commercially available software and hardware that solves both these potential is available for measuring compounds staying in solution by UV.

10.5 Compound Ionization Properties

The simplistic general rule-of-thumb is that only neutral compounds are absorbed via passive processes. This rule is termed the "pH-partition hypothesis". Exceptions are those charged compounds whose crossing of the gastrointestinal wall is facilitated by absorptive biological transporters. An example would be the facilitated absorption of anionic angiotensin-converting enzyme inhibitors and anionic cephalosporins by the nutrient PepT1 transporter. The "pH-partition hypothesis" does not differentiate between anions and cations. Experimentally charged bases (cations) are absorbed better than charged acids (anions). Absorption does occur even for acids existing overwhelmingly as anions at the neutral pH of the upper gastrointestinal tract. A rough rule-of-thumb is that the intestinal permeability will be 3–4 orders of magnitude worse for an anion compared to the free acid. So for acidic drugs with pK_a in the 3–4 range the absorption of acid and anion in the upper jejunum may be fairly similar. The degree of charge localization especially in acids is important in permeability and hence oral absorption. Highly charge delocalized compounds are more permeable and hence more orally available. Extensive literature is available on regional pH gradients in the gastrointestinal tract, and computational programs exist to predict absorption of compounds as a function of compound pK_a, compound aqueous solubility and absorbance site location in the gastrointestinal tract. Generally, simple zwitterionic compounds have their solubility (and absorption) minimum at the isoelectric point midway between the acid and base pK_a. However, structural characteristics can modify the general rule. In particular, charge proximate zwitterions are much better absorbed than charge-separated zwitterions. The truly neutral form of a zwitterionic compound may exist at neutral pH if the zwitterionic acid and base macro pK_as are numerically close (3–4 units apart) and if these bracket neutral pH. The significance is that oral absorption occurs through the truly neutral form and oral absorption is much better than might have been anticipated. An example is the oral absorption of the quinolone class of antibacterials.

10.5.1 Acids

Acids comprise a minority of drugs, 10–12% depending on whether one includes antibacterials. The aqueous solubility of stronger acids (like carboxylic acids) is as one might expect very pH dependent with much higher solubility at pH 7 than at acidic pH. The absorption of very strong acids (like sulfonic acids) is terrible unless the strong acid is very charge delocalized (as in the oxindole nonsteroidal antiinflammatory Tenidap). Any medicinal chemist trying to optimize oral activity in an HTS active sulfonic acid dye is wasting their effort.

Acids consistently display low solubility in DMSO. In our Pfizer studies we found acids to be 4 times more likely to have poor DMSO solubility of less than 5 $\mu g\,mL^{-1}$ than any other functionality. The reason is that DMSO with its very electronegative oxygens is a poor solvator for any compound with partial negative or full anionic character.

10.5.2 Importance and Measurement

Aqueous solubility is not usually considered a priori as a problem in the drug discovery of acidic compounds. More important issues are: (i) the high serum albumin binding of stronger acids, (ii) the very low or nonexistent central nervous system penetration of stronger acids, (iii) the low volumes of distribution of acids limiting these mostly to plasma compartment targets, (iv) the possibility of formation of toxic acylglucuronides and (v) the possibility that smaller acids may be directly cleared through the kidney which effectively eliminates any efficient *in vitro* surrogate for predicting systemic blood levels. Based on this laundry list of potential problems the reader would be correct in guessing that many organizations are wary of working on carboxylic acids. All these potential problems decrease as the acid pK_a weakens so that at weak acid pK_a of 6 and higher there is no relative disadvantage relative to other functionality. At very high concentrations carboxylic acids tend to self-associate as head-to-head dimers, which decreases aqueous solubility and increases apparent lipophilicity of the dimer.

10.5.3 **Bases**

Bases comprise the vast majority of ionizable drugs. The problematical issues found with acids do not exist for bases. Very strongly basic compounds, even if very aqueous soluble, should be avoided because of poor oral absorption. Quaternary compounds are either not absorbed or have low erratic absorption. Basic guanidines tend to exhibit a wide variety of activity leading to selectivity issues. However, the basic pK_a of guanidines is extremely sensitive to the electron-withdrawing character of a substituent. By modifying a nitrogen substituent, pK_a can be varied between 13.5 (very basic) to 0 (totally neutral). As previously discussed, the pH-partition hypothesis is blunted for bases, which is an advantage for oral absorption. At comparable lipophilicity the binding of bases to serum albumin is much less than for acids – an advantage. The volume of distribution of bases tends to be high compared to acids – an advantage in terms of target opportunity. The range of acceptable acidic counterions for bases is large – an advantage in terms of formulation of salts.

10.5.4 **Importance and Measurement**

Solubility is an important consideration for the moderate to weak base or for the base whose free form has very low aqueous solubility (the so called intrinsic solubility). At any pH there is an equilibrium between protonated and free base. If the aqueous solubility of the free base is very low making a salt may not have much value in improving aqueous solubility. The free base in equilibrium with cation precipitates. If the experiment is in buffer the precipitation of free base continues until the free base concentration drops below the solubility limit at the buffer pH. If the salt of a weak base is added to unbuffered water the free base precipitates and the liberated acid counterion acidifies the solution. Free base precipitation continues until a new equilibrium is reached. The message to medicinal chemists is this – never test the solubility of a base in unbuffered water. You may observe what looks like good solubility but the aqueous pH may end up being very, very acidic. Always check solubility in a buffer with enough buffer capacity to be sure the pH does not change.

Bases have the property that it may be important to check the solubility both at neutral pH and at acid pH. The latter measurement is particularly important for a moderately basic compound that is also a low-dose compound with good intestinal permeability. The idea is that a moderate base may be very soluble at low pH, but poorly soluble at neutral pH. If the dose is low the compound dissolves in the acidic gastric lumen and if the permeability is good enough the compound may stay in solution long enough for decent absorption at the neutral pH of the upper gastrointestinal tract before precipitation occurs. For a low-dose 0.1 mg kg^{-1} compound in the upper 10th percentile in intestinal permeability as little as 1 h in the jejunum may be sufficient for acceptable oral absorption. An example of a compound with this type of profile is the moderately basic antipsychotic Ziprasidone.

In discovery the most common base salt is the HCl salt. It is common in biology to use buffers containing chloride and the chloride content of the gastric contents is about 0.15 M. It often happens that the common ion effect of chloride suppresses the solubility of an HCl salt. This is a 100% solvable problem in pharmaceutical sciences.

10.5.5 Neutral Compounds

The prevalence of neutral compounds in drug discovery screening libraries has been growing at an alarming rate. Automated chemistry synthesis takes much of the blame. Adding amine protecting groups to an automated synthesis increases synthetic complexity. Creating diversity by amidation reaction of appendages on a core scaffold gives a neutral linker. Cleanup procedures add to the problem. Reverse-phase HPLC purifications work best on lipophilic neutral compounds. Basic compounds are more difficult to separate because of peak broadening. The peak broadening is often solved by adding trifluoroacetic acid (TFA) to the mobile phase which itself is a very bad approach given the known problem of TFA contamination in enhancing chemistry hydrolytic instability. In addition in long-term cell 24-h culture as little as 10 nM TFA is cytotoxic.

10.5.6 Importance and Measurement

In the current era neutral compounds are important simply because there are so many of them especially in screening collections made in automated chemistry for HTS. Other things being equal, a compound with an ionizable moiety is preferred to a neutral compound. By definition a neutral compound will not give an acid or basic pK_a value. Some essentially neutral compounds can form aqueous unstable salts. This most commonly occurs when an extremely poorly basic compound is dissolved in organic solvent and the salt is precipitatated, e.g. by bubbling in HCl gas.

10.5.7
Zwitterions

True zwitterionic compounds are rare among drugs. The oral absorption of truly zwitterionic compounds is poor unless the compound is a substrate for an absorptive biological transporter as in an α-amino acid which is a substrate for the PepT1 nutrient transporter. The aqueous solubility of a true zwitterionic compound will be at a minimum at the isoelectric point which unfortunately for many compounds happens close to the neutral pH at which oral absorption occurs. Species extrapolation predicting oral absorption and pk/pD from preclinical animal tests to man are difficult for zwitterions.

When the acid and basic pK_as of a zwitterion differ by only 3 or 4 units the compound may exist at isoelectric pH as an equilibrium between the true zwitterion and the formally neutral species. Oral absorption is much improved if this type of equilibrium exits. Charge proximate zwitterions are better absorbed than charge separated zwitterions. When the positive and negative charge centers are close in space there is an overlap in the polarized aqueous salvation shells so that the compound is less polarized than if the charges were far apart.

10.5.8
Importance and Measurement

Zwitterionic character is notable in several therapeutic area series, e.g. in angiotensin-converting enzyme inhibitors, quinolone antibacterials and thrombin inhibitors. The aqueous solubility measurement of zwitterions is very pH dependent as might be expected. The relationship of aqueous solubility to ionization state is extraordinarily complex if the zwitterion is of the type capable of an equilibrium between true zwitterion and formally neutral forms (e.g. as in a quinolone antibacterial). For these types of complex equilibria, salt effects on solubility may be unexpectedly large, e.g. solubility unexpectedly may track with the chaotropic character of the salt.

10.6
Compound Solid-state Properties

10.6.1
Solid-state Properties and Water Solubility

In the current era many medicinal chemists are unaware of the very important role of compound solid state properties on aqueous solubility and therefore to oral absorption. In many organizations compound purification by crystallization has disappeared being replaced by automated reverse-phase HPLC purification. If medicinal chemists isolate a compound as a white powder from evaporation of an HPLC mobile phase they often do not know whether the compound is amorphous or crystalline and why this is important to both aqueous and DMSO solubility. Melting point measurement equipment likely is missing from most medicinal chemistry laboratories. Thus, the metric of a sharp melting point as an indicator of crystallinity is gone. Solid-state infrared spectra have disappeared as discovery tools in reaction monitoring and compound identification being replaced by nuclear magnetic resonance and mass spectroscopy. As a result chemists have lost an easy tool discriminating between amorphous and crystalline materials. The rule-of-thumb is that the higher the energy state of a solid form, the greater is the solubility in any solvent, e.g. water or DMSO. The lower the energy state (the more stable), the lower is the solubility. Compound melting point tracks with energy state. The high energy amorphous and most soluble form typically has a low and very broad melting point range. As the compound is crystallized the melting point rises and sharpens. The compound may crystallize in multiple crystalline forms called polymorphs. Polymorphs with the lowest energies and hence most stable will have the highest melting points and lowest solubilities. The rule-of-thumb relating higher polymorph melting point to lower solubility in drug discovery practice has almost no exceptions. Across all compound types examples do exist where this rule-of-thumb is violated however the exceptions occur when the polymorph melting points are close to the temperature at which the solubilities are measured. Relevant temperatures for solubility in drug discovery are in the range room temperature to 37 °C. Most drug organizations would not develop a compound with melting point of less than 100–110 °C because of concerns about long-term stability. Hence, in drug discovery there is virtually never a drug with melting point close to the temperature of a solubility measurement.

10.6.2 Amorphous

Visual inspection frequently cannot differentiate between an amorphous or crystalline material, e.g. at Pfizer medicinal chemists were required to submit only crystalline and not amorphous compounds to an automated thermodynamic solubility assay. In practice half the white powders that they produced for the assay and that they thought were crystalline were actually amorphous. Prior to 2000 the vast majority of these medicinal chemistry labs had no melting point equipment and it was only in 2000 that the pharmaceutical sciences department started a workshop to teach medicinal chemists the importance of solid state properties, how to crystallize compounds and the importance of salt forms.

Commonly an amorphous solid may have solubility in both aqueous and DMSO that is over 100 times higher than any of the crystalline forms. An amorphous solid can be inadvertently created from a crystalline material, e.g. by heating a crystalline hydrate or solvate for a short time period below its melting point in a drying pistol.

10.6.3 Crystalline

Techniques for differentiating between amorphous and crystalline are: (i) sharp melting point, (ii) sharp peaks in the solid state infrared fingerprint region, (iii) optical birefringence observed when solid is viewed in a phase contrast microscope and (iv) sharp peaks in the powder X-ray diffraction pattern.

Oral absorption from dosing a powder can only be reliably determined if crystalline material is used. Drawing conclusions on oral absorption using amorphous solid is very, very risky. Oral absorption may appear good for amorphous material because the aqueous solubility is high. However, when the compound is crystallized the solubility plummets and so does the oral absorption. The choice of the best orally absorbed compound should always come from a comparison of crystalline compounds. There is nothing wrong with using amorphous compounds throughout most of the discovery process. However, at the late discovery stage crystalline rather than amorphous compounds must be compared when choosing the best orally active compound.

10.6.4 Salt Forms

Different salt forms are prepared to optimize properties such as dissolution rate, solubility, solid-state stability or improve any tendency to hygroscopicity. Most commonly salt screening is performed in late discovery in a pharmaceutical sciences setting. Salt screening has become automated at some well resourced organizations; however the compound sparing efficiency lags far behind a biology HTS screen. So to enable salt screening medicinal chemists will need to prepare compound in the low 100s of milligrams to low gram range.

10.6.5 Ostwald's Rules

For almost all compounds there is a predictability in the sequence of compound isolation [10]. Amorphous material is most likely to be isolated when there is no knowledge of compound isolation conditions. Kinetics predominate in the isolation as in obtaining a solid by vacuum concentration of an HPLC mobile phase. Here the melting point is nonexistent or broad and the solubility is the best. As the chemist learns about isolation conditions, e.g. solvent, temperature and time, the compound is isolated in crystalline form, the isolation is thermodynamically rather than kinetically driven, and the melting point sharpens and rises. As the chemist optimizes isolation conditions a crystalline form with the highest melting point and sharpest melting point is isolated. This crystalline material is usually the purest and unfortunately is the least soluble in aqueous media or DMSO. The sequence described is so predictable and universal that it has been given the name "Ostwald's Rules" after the German chemist describing the phenomenon in the 1880s.

10.6.6 Isolation Procedure Changes

There is a steady decline over 30 years of compounds having melting point data, reaching a nadir in the year 2000, where virtually no compounds were registered with melting points at the Pfizer Groton labs.

10.6.7 Greaseballs

Large and lipophilic poorly water-soluble compounds are sometimes referred to a greaseballs. As previously discussed, the solubility limit for these compounds is more difficult to detect using light scattering equipment than for compounds insoluble because of strong crystal packing forces. At the Pfizer, 45% of water-insoluble compounds were greaseballs as determined over a 6-year time period with over 70 000 compounds tested in a kinetic solubility assay. As previously discussed a calculated log $P > 5$ was predictive of poor aqueous solubility of less than $5\,\mu g\,mL^{-1}$ in pH 7 phosphate buffer at room temperature without regard to compound ionization state.

10.6.8 Properties

Greaseballs will be preferentially selected for in an HTS. The reason is the strong association of better *in vitro* activity with higher lipophilicity. The so-called hydrophobic trap is the endpoint resulting from improving *in vitro* activity by simply increasing lipophilicity. The trap is the end product – a very *in vitro* active compound that is so lipophilic, large and insoluble that aqueous solubility is terrible and oral absorption is nonexistent. The solution to the hydrophobic trap is to never simply choose the most potent series for chemistry pursuit. Rather, a metric-like ligand efficiency should be used that scales *in vitro* activity by compound size. Simplistically, if two compounds have the same IC_{50}, the smaller compound is the better starting point for chemistry SAR.

10.6.9 Measuring and Fixing Solubility

Greaseball solubility is more difficult to quantitate by light scattering. By contrast, greaseball and brickdust poor solubility is comparable in ease of detection when measuring compound in solution as for example by UV. At the Pfizer Groton labs the success in fixing the poor aqueous solubility of a chemical series was about 50%. Those series not fixed in chemistry resulted in clinical candidates requiring a pharmaceutical sciences formulation fix. It should be emphasized that fixing a poor solubility problem in chemistry is by far and away the preferred option. If a poor solubility problem is mostly due to large size and high lipophilicity one cannot expect marked solubility improvement due to chemistry changes unless the compound is significantly downsized. There are more options in formulation sciences for fixing a poorly aqueous soluble greaseball [11] than there are for the poorly soluble compound due to crystal packing – the brickdust compound. Greaseballs are well suited to lipid vehicle formulation approaches and to formulation approaches delivering the compound in amorphous (and more soluble form) in some type of pharmaceutically acceptable matrix. Greaseballs have the advantage that in normal gastrointestinal physiology mechanisms exist for the absorption of lipids.

10.6.10 Brickdust

The poorly soluble compound due to strong inter-molecular interactions in the crystalline state is called a "brickdust" compound. Crystallization of compounds occurs to produce a minimum energy solid. Two mutually antagonistic processes compete to give the low energy solid. The crystalline solid maximizes intermolecular H-bonding and at the same time maximizes density and these processes are mutually opposed. In general all H-bond donors and all or most of the acceptors will be satisfied in the crystalline form. If the H-bonding cannot be satisfied within the compound a hydrate or solvated crystal will form.

10.6.11 Properties

Poorly soluble brickdust compounds are not preferentially selected for by HTS as is the case with greaseballs. However, brickdust compounds are extremely common in certain popular current era targets, e.g. the ATP competitive hinge region binding site in protein kinase inhibitors.

10.6.12 Measuring and Fixing Solubility

Light scattering methods shine for quantitation of brickdust compounds precipitating from solution. As previously discussed brickdust compounds generate very robust and very reproducible light scattering signals. Fixing solubility in brickdust chemistry series within the chemistry lab is particularly important because the solubility formulation fix options are narrower for the brickdust compounds as compared to the greaseball compounds. Obtaining a single crystal X-ray packing diagram for a crystal of a brickdust compound should be the medicinal chemist's first choice in facilitating a chemistry solubility fix. Getting an X-ray structure on a small molecule is an unobvious choice. Asking for the packing diagram is also not customary. The idea is to look at the three-dimensional packing diagram with stereo glasses to see how the compound packs with itself. Where are the H-bonds? Is there some way a strategically placed methyl could disrupt a H-bond and thus improve aqueous solubility?

Brickdust compound poor solubility is harder to fix in formulation than for the greaseball. Lipid formulation fixes are unlikely to work. A poorly aqueous soluble brickdust compound is also likely to be insoluble in a lipid formulation so the lipid formulation fix will not work. Formulation fixes using amorphous delivery technology are also problematic. Many of these methods depend on attaining a high concentration of the drug in a volatile organic solvent. The poorly aqueous soluble brickdust compound has a higher probability of insufficient solubility in the volatile organic solvent. Reduction of particle size to the low nanometer particle size range is a common choice to improve the aqueous solubility and thus to rescue brickdust compounds.

10.6.13 Preformulation Technology in Early Discovery

Should formulation technology be available in early discovery so as to allow biological testing of very poorly water soluble compounds [12]? This is a controversial topic on which reasonable people can disagree. This author's viewpoint is that it is very dangerous to introduce technology into early discovery if it in any way hinders chemistry efforts to fix a solubility problem. The benefit of early discovery formulation, often called preformulation technology, has to be weighed against the adverse effects on chemistry efforts. One should be wary of pharmaceutical science formulation advocates spreading the word to discovery medicinal chemists that formulation technology can fix a solubility problem. The primary responsibility and most cost effective method to solving an aqueous solubility problem is to fix the problem by changing chemistry structure. Formulation technology comes into play only when medicinal chemistry has failed.

10.6.14 Discovery Development Interface Water Solubility

The discovery development interface is important with respect to drug aqueous solubility. In earlier discovery stages kinetic solubility assays are perfectly acceptable even more desirable than the thermodynamic equilibrium solubility assay. As projects approach decision points regarding oral absorption solubility assays must change towards the traditional thermodynamic equilibrium solubility assay. Stated again for emphasis – the early discovery kinetic solubility assay is not a replacement for a traditional thermodynamic equilibrium solubility assay. Accurate oral absorption requires testing of crystalline materials and corresponding accurate solubility testing of crystalline materials.

10.6.15 Thermodynamic Equilibrium Measurements

A thermodynamic equilibrium solubility assay does not mean a traditional low-throughput assay unchanged in design from decades ago. An accurate thermodynamic solubility assay can be ported to a robot to improve efficiency without sacrificing accuracy and scientific rigor. Compound supply and physical form are major challenges facing the porting of thermodynamic equilibrium solubility assays into the late discovery setting. As a project enters the late-stage discovery process there is an aura of excitement. Perhaps the first amorphous compounds are meeting the project product portfolio. Project personnel know that perhaps in a few months a compound will be considered for clinical study. At this time there is a natural tendency for chemists to speed up their chemistry. Yet at this time point the chemists must be convinced to change their work practices. Larger quantities of compound need to be made and these larger quantities must be crystallized. This slows down chemistry output. Yet this change is absolutely necessary to allow for the comparison of the aqueous solubility and oral absorption of crystalline materials.

10.7 DMSO Solubility

Organizations have used compound in DMSO stock solutions in HTS for 15 years. The driver is efficiency. Liquids can be efficiently measured and handled by robotics. With few exceptions this is not the case with solids. Despite a variety of concerns no organization employs a liquid diluent alternative to DMSO, although some organizations introduce small quantities of additives to improve compound solubility. Concerns about compound "integrity" in DMSO have a long history. At earlier time points the concern mostly focused on chemical degradation, i.e. the chemical transformation of the compound in DMSO. Concerns about chemical degradation predominate in older presentations and discussions. We now know that this historical concern with chemical degradation was to a considerable extent incorrect. Currently there is a consensus that although chemical degradation is indeed an issue the predominant problem facing compound storage in DMSO is precipitation [13]. Compound precipitation occurs from the DMSO stock itself especially if it becomes wet and especially if the sample undergoes freeze–thaw cycles. Compound precipitation also occurs as the DMSO stock is diluted into the biological assay buffer.

Why was chemical degradation overemphasized in the earlier years of compound storage in DMSO stock solutions? The cause was an unconscious bias present in analytical chemistry. With respect to compound integrity there are three measurement endpoints. (i) Compound identity focuses on whether the compound in the DMSO stock is the structure shown in the registration system. (ii) Compound purity focuses on whether the material in the DMSO stock is a single component or whether there are multiple impurity/degradation components. (iii) Compound concentration focuses on whether the compound concentration in DMSO solution is experimentally close to the nominal concentration and whether the compound concentration on dilution into assay buffer is experimentally close to the nominal concentration. In high throughput mode analytical chemists routinely focused on measurements (i) and (ii) and their data interpretation was biased by their history with (i) and (ii). However, measurement (iii) is quite difficult to perform in high-throughput mode because it entails quantitation of concentration without using positive standards. As a result many past analytical studies on compound integrity failed to include measurement (iii) – the concentration quantitation. Concentration is needed to interpret an HPLC peak pattern. A compound started storage as a single HPLC peak. After some storage time multiple HPLC peaks appear. There are two possible explanations. The first explanation is that the compound degraded (the chemical structure changed to multiple impurities) and this was the first choice of the analytical chemists because of their past history. The second explanation is that the major component precipitated leaving an HPLC pattern corresponding to a much reduced concentration of the major component plus peaks associated with the minor impurities which were always present in the sample.

The compound in DMSO storage community knew that failed "compound integrity" was associated with multiple freeze–thaw cycles and the widely adopted solution was the replacement of HTS plate-based storage with single use storage tubes. This change minimized the exposure of any one sample to multiple freeze–thaw cycles. This change was successful and "compound integrity" improved. However, the correct change was made for the wrong reason. Single-use storage systems by limiting freeze–thaw cycles primarily improve "compound integrity" by minimizing compound precipitation from the DMSO stock and not by limiting chemical degradation.

10.7.1 Where Does Poor DMSO Solubility Come From?

Precipitation of compound from a DMSO stock is an issue of compound solubility in DMSO [14]. However, operationally, most organizations dissolving compounds in DMSO do not encounter problems in the solubilization process. This apparent paradox is explained by the issue of amorphous compounds. The vast majority of compounds solubilized into DMSO stocks are amorphous and inherently more soluble in DMSO (as previously discussed). The vast majority of the same compounds precipitating from DMSO are crystalline and much less soluble in DMSO (as previously discussed). The problem of compounds precipitating from DMSO occurs because so many compounds entering into screening libraries are amorphous. The precipitation of compounds from DMSO would be far less of a problem if screening compounds were entirely crystalline as they were 20 years ago. Quite simply DMSO insoluble crystalline compounds would never have been successfully dissolved in DMSO in the first place.

10.7.2 DMSO Solubility is Multifactorial

Poorly water-soluble compounds were previously discussed as to the cause of poor water solubility – the large lipophilic greaseballs and the brickdust compounds [15] which are poorly water soluble because of strong intermolecular crystal packing interactions. The ability of DMSO to solubilize a water-insoluble compound depends markedly on whether the water-insoluble compound lies in the greaseball or the brickdust camp.

DMSO is a wonderful solvent for the greaseballs providing they are more than simple hydrocarbons. Simple hydrocarbons (e.g. hexane) have negligible solubility in DMSO. Provided there is some polarization in a greaseball DMSO is an excellent solvent. Dry DMSO has no H-bond donor group so there are no solvent H-bonding networks to disrupt. To dissolve a compound a cavity must be formed in the solvent and cavity formation is more difficult if the solvent exhibits inter-molecular H-bonding. Thus, increased solvent H-bonding decreases the solubilization potential of the solvent (more on this later). The high dielectric constant of DMSO results in favorable interactions with dipoles present in a compound and almost all drugs have one or multiple dipole moments. As previously discussed, DMSO is not such a good solvent for strong acids. However, as previously discussed, acids are only about 10% of drugs. DMSO is a pretty poor solvent for zwitterions. However, as also previously discussed true, zwitterions are rare among drugs. All in all, DMSO is a good solvent for a wide range of organic compound drugs.

DMSO is much less of a good solvent for the poorly water-soluble brickdust compound because the breaking of a strong inter-molecular crystal lattice is pretty well unchanged whether the solvent is water or DMSO. The breaking of the strong crystal lattice is the primary cause of poor water solubility for the pure brickdust compound. The further a poorly water-soluble compound is into the brickdust camp, the less effective is DMSO as a solvent.

10.7.3 DMSO Compared to Water Solubility

Repeating for emphasis – not all poorly water-soluble compounds are equally solubilized by DMSO. The greaseball compounds will be much better solubilized than the brickdust compounds. The most common range of compound in DMSO stock concentrations in compound storage repositories is 2–20 mM. At conferences, organizations employing 2 mM compound in DMSO stock solutions report fewer solubility problems. The downside is that the lower stock concentrations are more limited in utility in some absorption, distribution, metabolism, excretion and toxicity assays that require higher compound concentration.

10.7.4 DMSO Compound Storage Stocks and Compound Integrity

Mention has previously been made of the traditional analytical expertise in quantitating chemical degradation. What is not commonly known is that there are two business reasons for studying chemical degradation attendant to compound storage. In the first business reason the compound store is considered as inventory, and corporate policies and procedures relevant to inventory apply, e.g. inventory shrinkage is a topic of concern. In this scenario standard corporate procedures apply. The issue is addressed by random sampling to determine the magnitude of any potential problem. Thus, for example, inventory shrinkage due to chemical degradation might adversely affect capital budget. Random sampling is a good and appropriate method to address the concern from a business perspective.

Random sampling is not a good method to address the scientific issue. What kinds of chemical functionality are most likely to chemically degrade under certain specified storage conditions. To answer this type of question prioritization by chemistry first principles would make sense. For example, one might want to give priority to weakly or moderately aromatic ring systems put together by a dehydration since such systems might at least in theory be capable of coming apart by hydrolysis. Diluting the theoretically more unstable compounds with lots of perfectly stable phenyl and naphthalenes would not make scientific sense.

The reader should be aware that analytical chemistry studies in the compound in the DMSO storage arena almost exclusively focus on the business inventory issue as opposed to the scientific issue of what chemistry functionality is most susceptible to chemical degradation on storage in DMSO. How likely is it that chemical degradation might occur upon 3 months storage of a compound in DMSO at ambient temperature? Across chemical structures using random sampling the consensus is that chemical degradation is 5% or less at 3 months of room temperature storage provided that samples are not contaminated with TFA.

10.7.5 DMSO Solubility and Precipitation

This section focuses on compound precipitation from compound in DMSO stock solutions. This is an issue of compound solubility in DMSO as well as wet DMSO. This author has not heard any reports at compound management conferences of significant problems initially dissolving compounds in dry DMSO. In the author's experience initial solubilization of compounds in DMSO is seldom a problem. At the author's Pfizer lab over 40 000 compounds were dissolved in DMSO at a 60 mM concentration. Problems dissolving a wide range of structures were so few that they were not worth tracking. However, virtually none of these 40 000 compounds dissolved were crystalline.

10.7.6 DMSO Water Content

The solubility of organic compounds in DMSO is very dependent on the water content of the DMSO. Even small amounts of water markedly depress compound solubility in DMSO. Wet DMSO forms a very structured H-bonding network that is most structured at a water:DMSO mole ratio of 2 : 1 [16]. This mole ratio corresponds to a water content of 33% by weight. The oxygen in DMSO is very electronegative and is a very good H-bond acceptor. The hydrogens in water are excellent H-bond donors. The unfortunate combination of just the wrong bond angles in water and DMSO and the strong H-bond donor and acceptor strengths leads to a highly structured H-bonding network that has been described as ice-1-like in structure. Breaking this H-bond network to create a cavity is energetically unfavorable. A cavity must be formed in a solvent for a compound to dissolve and so even a little water in DMSO greatly depresses compound solubility.

DMSO and water form a solution with nonideal behavior, meaning that the properties of the solution are not predicted from the properties of the individual components adjusted for the molar ratios of the components. The strong H-bonding interaction between water and DMSO is nonideal and is the primary driver for the very hygroscopic behavior of DMSO. Even short exposure of DMSO to humid air results in significant water uptake. Water and DMSO nonideal behavior results in an increase in viscosity on mixing due to the extensive H-bond network.

10.7.7 Freeze–Thaw Cycles

The non ideal behavior of water and DMSO extends to the freezing point of mixtures. Completely dry DMSO freezes at 19 °C. However, 9% by weight water in DMSO depresses the freezing point of the wet DMSO to 4 °C. This is an important temperature because it is the lower temperature limit of the temperature of the nonfreezer part of a lab refrigerator. Typical lab refrigerators as well as home refrigerators have the nonfreezer compartment set at 4–5 °C. The upshot is that if a compound dissolved in wet DMSO is stored in the nonfreezer part of a lab refrigerator it will not be frozen. This is a very bad situation with respect to the compound coming out of solution.

The role of freeze–thaw cycles in harming "compound integrity" is now well understood. Integrity is harmed by compound coming out of solution rather than by something in the freeze–thaw process causing chemical degradation (which chemically does not make much sense). A quartet of factors gives the worst precipitation problem. (i) The problem often starts with a thermodynamically supersaturated solution arising from initial dissolution in DMSO of an amorphous compound. (ii) The problem intensifies as the DMSO takes up water, thereby lowering the compound solubility. (iii) The problem is exacerbated by storing compound in cold wet liquid DMSO, e.g. by storage in the nonfreezer part of a lab refrigerator. (iv) Multiple freeze–thaw cycles compound the problem likely because each thaw cycle at the melting point results in a highly concentrated cold liquid solution of the compound in equilibrium with frozen out solid DMSO. Analytical studies that do not incorporate all these features will underestimate the precipitation problem.

10.7.8 Fixing Precipitation

Understanding what actually happens in freeze–thaw cycles is the clue to fixing the problem. (i) Amorphous DMSO solubility is higher than when the compound is crystalline. Not much can be done about the initial amorphous character of compounds dissolved in DMSO. However, once compounds precipitate they can be redissolved by either contact or noncontact sonication. The highly localized energy input of the exploding micro bubbles formed in sonication-induced cavitation is quite effective in redissolving precipitated crystalline material. This can be done without any thermal adverse effect on compound structural integrity. (ii) A lot can be done to keep the DMSO as dry as possible. (iii) For long-term storage of a compound it is more important that the DMSO matrix is solid (frozen) as opposed to just being colder. At –20 °C DMSO containing 19% by weight water freezes. If DMSO contains 33% by weight water the freezing point is –73 °C (as can be seen from the DMSO–water phase diagram). (iv) Adoption of single use storage tubes reduces the number of freeze–thaw cycles to the minimum of one.

10.7.9 Short-term End-user Storage of DMSO Stocks

How should the end-user of compound in DMSO stocks treat compound DMSO stock solutions? For short-term storage of 3 months or less never put the compound DMSO stock in the nonfreezer part of a lab refrigerator. Store the DMSO stock at ambient temperature away from light and protected from oxygen.

10.8 Conclusions

In the current era, poor compound aqueous solubility is the single most widespread problem hindering compound oral absorption. Poor aqueous solubility of 30–40% of compounds is common in a competent drug discovery organization. Solubility problems relating to compound in DMSO stock solutions are universal across the drug discovery industry. More “compound integrity” problems encountered in compound storage are due to solubility problems than are due to chemical instability problems. Serious problems of precipitation can be expected for about 15% of compound in DMSO stocks diluted into assay buffer.

The bad news is that these problems exist. The good news is that the causes of the problems are mostly known and strategies for mitigating these problems exist.

References

1 Lipinski, C. Drug-like properties and the causes of poor solubility and poor permeability. *J. Pharmacol. Toxicol. Methods* **2000**, *44*, 235–249.

2 Lipinski, C., Lombardo, F., Dominy, B., Feeney, P. Experimental and computational approaches to estimate solubility and permeability in drug discovery and development settings. *Adv. Drug Deliv. Rev.* **1997**, *23*, 3–25.

3 Horspool, K., Lipinski, C. Advancing new drug delivery concepts to gain the lead. *Drug Del. Technol.* **2003**, *3*, 34–44.

4 Lennernäs, H., Abrahamsson, B. The use of biopharmaceutic classification of drugs in drug discovery and development: current status and future extension. *J. Pharm. Pharmacol.* **2005**, *57*, 273–285.
5 Horter, D., Dressman, J. B. Influence of physicochemical properties on dissolution of drugs in the gastrointestinal tract. *Adv. Drug Deliv. Rev.* **2001**, *46*, 75–87.
6 Di, L., Kerns, E. Biological assay challenges from compounds solubility: strategies for bioassay optimization. *Drug Discov. Today* **2006**, *11*, 446–451.
7 Popa-Burke, I., Issakova, O., Arroway, J., Bernasconi, P., Chen, M., Coudurier, L., Galasinski, S., Jadhav, A., Janzen, W., Lagasca, D., Liu, D., Lewis, R., Mohney, R., Sepetov, N., Sparkman, D., Hodge, C. Streamlined system for purifying and quantifying a diverse library of compounds and the effect of compound concentration measurements on the accurate interpretation of biological assay results. *Anal. Chem.* **2004**, *76*, 7278–7287.
8 Sanghvi, T., Jain, N., Yang, G., Yalkowsky, S. Estimation of aqueous solubility by the general solubility equation (GSE) the easy way. *QSAR Comb. Sci.* **2003**, *22*, 258–262.
9 Zhang, J.-H., Chung, T., Oldenburg, K. Confirmation of primary active substances from high throughput screening of chemical and biological populations: a statistical approach and practical considerations. *J. Comb. Chem.* **2000**, *2*, 258–265.
10 Schmalzried, H. On the equilibration of solid phases. Some thoughts on Ostwalds contributions. *Z. Physik. Chem.* **2003**, *217*, 1281–1302.
11 Pouton, C. Formulation of poorly water-soluble drugs for oral administration: Physicochemical and physiological issues and the lipid formulation classification system. *Eur. J. Pharm. Sci.* **2006**, *29*, 278–287.
12 Chaubal, M. Application of drug delivery technologies in lead candidate selection and optimization. *Drug Discov. Today* **2004**, *9*, 603–609.
13 Balakin, K., Savchuk, N., Tetko, I. *In silico* approaches to prediction of aqueous and DMSO solubility of drug-like compounds: trends, problems and solutions. *Curr. Med. Chem.* **2006**, *13*, 223–241.
14 Oldenburg, K., Pooler, D., Scudder, K., Lipinski, C. K. High throughput sonication: Evaluation for compound solubilization. *Comb. Chem. High-Throughput Screen.* **2005**, *8*, 499–512.
15 Rabinow, B. Nanosuspensions in drug delivery. *Nat. Rev. Drug Discov.* **2004**, *3*, 785–796.
16 Catalan, J., Diaz, C., Garcia, F. Characterization of binary solvent mixtures of DMSO with water and other cosolvents. *J. Org. Chem.* **2001**, *66*, 5846–5852.
17 Curatolo, W. Physical chemical properties of oral drug candidates in the discovery and exploratory development settings. *Pharm. Sci. Technol. Today* **1998**, *1*, 387–393.

11
Challenge of Drug Solubility Prediction

Andreas Klamt and Brian J Smith

Abbreviations

AP	ratio of aromatic to total number of heavy atoms
COSMO-RS	conductor-like screening model for realistic solvation
DFT	density functional theory
GCM	group contribution methods
GSE	general solubility equation
MC	Monte Carlo
MD	molecular dynamics
MLR	multiple linear regression
MP	melting point
NN	neural network
PLS	partial least squares
QSPR	quantitative structure–property relationship
RB	number of rotatable bonds
RMSE	root mean square error

Symbols

$\log S_w$	logarithm of aqueous solubility
$\log P_{ow}$	octanol–water partition coefficient

11.1
Importance of Aqueous Drug Solubility

The solubility of drugs, agrochemicals or pesticides is of predominant importance for the physiological availability and hence for the activity of such agents. Since medicinal drugs are the most important and most often considered class of physiological agents, we will use the expression "drug" for all such agents throughout this chapter. Indeed, the expression "solubility" itself is not a well-defined quantity, unless used in context with a solvent medium. Due to the overwhelming importance of the solvent water in all environmental and life processes, the aqueous

Molecular Drug Properties. Measurement and Prediction. R. Mannhold (Ed.)

ISBN: 978-3-527-31755-4

solubility has become a kind of standard reference for solubility and therefore often the specification "aqueous" is omitted when researchers in the pharmaceutical industry discuss aspects of the solubility of drugs. Most drugs are applied orally; hence, they first need to be dissolved in the intestinal fluid before they can pass the intestinal membranes by diffusive or active transport and then be distributed in the body through the mainly aqueous blood serum in order to reach their final destination. Even if not applied orally, the drug first needs to be dissolved in some liquid or gel phase in order to be applied either by injection or by penetration through the skin and then it has to be transported to its final destination. A compound that is highly active and that binds specifically will not become a successful drug without suitable solubility in water for it to be transported to its final destination in the body. Therefore, dissolution, especially aqueous dissolution, is a key primary step for any drug application and hence the measurement as well as the computational prediction of drug solubility is of highest importance for the proper design and development of potent drugs.

Successful drugs with solubilities lower than 10^{-5} or 10^{-6} mol L^{-1}, or in logarithmic units log S_w below –5 or –6, are rarely found, and so this may be viewed as a lower bound for promising drug candidates. On the other hand, almost all drugs have solubilities lower than 0.1 mol L^{-1}, i.e. $\log S_w < -1$. Thus, the range of $-6 < \log S_w < -1$ may be considered as the drug solubility window.

Obviously, the solubility of drugs can and must be measured experimentally, despite the various difficulties and obstacles that have been discussed in detail in Chapter 10 by Lipinski, and which will be further considered from a thermodynamic viewpoint in the next section of this chapter. No regulatory authority would give clearance for a drug without carefully measured and reported solubility data. No authority would accept just calculated or predicted solubility data. So the question is, why there is any need for computational solubility prediction methods? The answer is quite obvious: reliable drug solubility prediction methods are essential during the process of drug design and development, because in the early drug design phase often only virtual compounds are considered by computational drug design techniques or very small amounts of the new drug candidates are synthesized by combinatorial chemistry methods. In both cases there is either no or insufficient substance available for experimental good-quality solubility measurements and hence the only tools for the selection of promising drug candidates with suitable solubility are computational solubility prediction methods, which predict log S_w with reasonable accuracy based only on the chemical structure of the drug candidate. Such purely computational methods often are called *in silico* prediction methods. Any methods making use of other experimental data for the prediction of log S_w are inappropriate in this context, since such data will not be available in the early drug design phase.

The importance of good solubility estimates during the drug design and development process is the driving force for the continuous active development and improvement of computational drug solubility prediction methods by academic and industrial researchers. According to Balakin et al. [1], the number of new methodological publications on drug solubility prediction has been increasing linearly since 1995, with about 24 papers in 2005 and a total number of papers in the order of 300. Apparently, the available methods have not yet reached a degree of maturity which satisfies the needs of the medicinal chemists. A considerable number of reviews on *in silico* drug solubility prediction methods [1–6] appeared in the past 5 years. Apart from some special aspects, all of them consider the different classes of *in silico* solubility prediction methods, and address the most important common features and differences. In this situation it is a difficult task to write another review on drug solubility prediction without just repeating what others have recently written. On the other hand, this volume on drug property measurement and prediction would be incomplete without a chapter on solubility prediction. We therefore decided to focus less on the completeness of the review of the methods and have therefore avoided the more historical aspects. Instead, we use a slightly more fundamental approach and try to shed some new light on the demanding challenge of drug solubility prediction. Nevertheless, it will be unavoidable to give another methodological survey and comparison, and to repeat several findings of previous reviews.

11.2 Thermodynamic States Relevant for Drug Solubility

As a prerequisite for all further considerations it is useful to introduce first a consistent notation and a thermodynamic framework for the different aspects of relevance for aqueous solubility of drugs. First of all we adopt the widely used units convention of moles per liter ($mol\,L^{-1}$) and the abbreviation log S_w for the logarithm of S in these units throughout this article. Nevertheless, one must be aware that many other unit systems are used for drug solubility, such as moles per gram ($mol\,g^{-1}$), milligrams per milliliter ($mg\,mL^{-1}$), moles per moles ($mol\,mol^{-1}$), etc., and one must be very careful in this regard when collecting experimental or calculated solubility values from different sources. Indeed, many errors from unit conversion have made their way into the literature.

Solubility is a thermodynamic property, and as such it depends on the temperature and on the system pressure. Fortunately the pressure dependence of the solubility of liquid and solid compounds is generally very weak, and the pressure range of relevance of drug application is quite limited. Hence a pressure of 1 atm can be generally assumed in drug solubility considerations and the pressure dependence can be ignored otherwise. Temperature is a more critical parameter. Obviously, the temperature of operation of drugs in the human body will be in the narrow range of 35–40 °C, mostly 37 °C. Hence, this should be the reference temperature for drug solubility considerations. Nevertheless, drug solubility is usually measured or calculated at ambient temperature, somewhere in the range between 20 and 25 °C. Most likely this is due to the better feasibility of solubility measurements in a laboratory without the need for heating facilities and thermostats. Fortunately, even in the worst-case scenario that the entire solubility of a sparingly soluble drug with log $S_w = -6$ would be of enthalpic nature, the solubility would only increase by about 0.2 log S_w units with a change in temperature from 25 to 37 °C. In reality, the most significant contributions to the solubility are of an entropic nature, further decreasing the temperature dependence and which can lead to a solubility decrease with raising temperature in some compounds. Since we will see that differences of 0.2 log units are far within the experimental and prediction error, the effect of temperature on drug solubility can be neglected in the typical temperature range of interest, i.e. between 20 and 40 °C. Hence, room temperature of around 25 °C is generally used as the reference temperature for drug solubility and this temperature is assumed unless noted otherwise.

The intrinsic aqueous solubility S_w of a drug X is related to the Gibbs free energy ΔG^X_{sol} of the transfer of the drug X from its lowest free energy crystalline form to a saturated solution of the neutral form of X in water by the simple equation:

$$\log S = \frac{-\Delta G^X_{sol}}{RT \ln 10} = \frac{\mu^X_{cryst} - \mu^X_{aq,sat}}{RT \ln 10} \tag{1}$$

where μ denotes the pseudo-chemical potential according to the definition by Ben-Naim [7]. This transfer is schematically illustrated by the bold arrow in Fig. 11.1. One big problem of drug solubility measurement and prediction arises from the fact that for typical drug compounds there are often several other states of the drug in both the solid and aqueous phases competing with the two states that define the intrinsic solubility. In the solid state there may be a number of different polymorphs, i.e. crystal forms. There is no guarantee that the drug crystallizes into its thermodynamically most stable form, i.e. into the form with the lowest chemical potential. Often the formation of the different crystal forms is controlled by dynamic aspects and less stable polymorphs can be formed during a solubility measurement. The equilibration between different polymorphs is often an extremely slow process that may never occur by itself. Hence, the experimentally measured solubility of such polymorphs will be higher than the true solubility. Typically, the spontaneously forming polymorphs of drugs are less than 4 kJ mol^{-1} higher in free energy than the most stable one, resulting in log S_w differences of up to 0.7 log units. Since a demanding polymorph analysis is rarely done in drug solubility measurements, the noise in experimental log S_w data arising from the polymorphic ambiguity must be assumed to be in the range of around 0.3–0.5.

There are other solid states which sometimes confuse the measurement and definition of solubility. The drug may crystallize as a hydrate, i.e. under inclusion of water molecules. If the hydrate form is more stable than the pure form it may be difficult to measure the intrinsic solubility of the drug at all. Often drugs tend to precipitate in an amorphous form, often under the inclusion of impurities. As with metastable polymorphs, such amorphous precipitates may lead to erroneously high solubility measurements. Commercially, drugs are often crystallized in salt form, e.g. as the hydrochloride salt, a cation with a chloride anion. In these co-crystallized salts, a much lower solubility than the intrinsic solubility will typi-

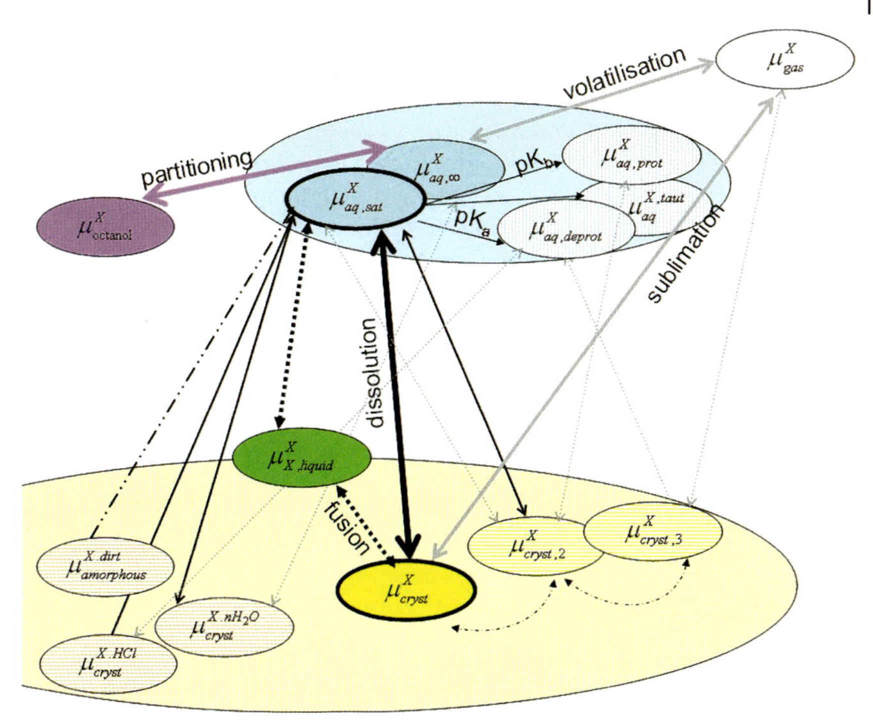

Fig. 11.1 States and equilibria relevant for solubility prediction.

cally be measured. There are many drugs that can only be crystallized in an ionized form as salts. For these the definition of an intrinsic, neutral solubility is not possible at all. Although these classes are very important, they are usually disregarded in drug solubility prediction methods due to the extreme additional difficulties arising from the computational treatment of the ions. We will follow this convention in this article.

It is not only the solid state of a drug that suffers from ambiguities, but also the aqueous state. The state relevant for the intrinsic solubility is the state of the saturated solution of the neutral species. Since most aqueous drug solubilities are small, direct interactions of the drug molecules are usually rare. Hence, this state is usually very similar to the state of the drug at infinite dilution in water. Most computational methods disregard saturation effects. Usually this is a good approximation, but one should keep in mind that this approximation may result in some moderate, but systematic errors at the upper end of the solubility scale.

More important are the ambiguities arising from acid–base equilibria, since many drugs contain several ionizable groups. At neutral pH, an acidic or basic drug will exist in water almost exclusively in its deprotonated state until a concentration of 10^{-pKa} or 10^{-pKb} mol L^{-1} is reached, respectively. Typical pK_a and pK_b values are in the range 4–5 for carboxylic acid groups or 9–11 for amine groups, which requires experimental solubility values to be carefully corrected to the effective concentration of the neutral species. This either requires knowledge of the pK_a or the experiment must be performed in a buffered aqueous solution that suppresses the ionization of the drug. Unfortunately, the presence of the buffer may change slightly the solubility, and the measurement is no longer for the aqueous solubility. Nevertheless, if the buffer is used moderately, the error introduced by the presence of the buffer should be small. Often solubilities are reported without ionization corrections, which may lead to large differences of reported solubility values in the literature. The situation becomes more complicated for compounds forming tautomers or zwitterions, e.g. amino acids. If the zwitterionic state is the most stable one, solubility can only be measured with respect to this state, unless the small concentration of the unionized form is measured, e.g. by spectroscopic means. Computational methods will usually consider only the neutral form and hence will not be able to treat the solubility of zwitterionic drugs.

Given the large number of different solid and aqueous states a drug may occupy, there is an even larger number of possible thermodynamic pseudo-equilibria between these states which may be erroneously measured or interpreted as the solubility of a drug. Experimental problems arising from difficulties in the detection of the often very small amounts of drug dissolved in water, together with the sometimes very slow solvation kinetics, are the major reasons for the large scatter in reported experimental drug solubilities. Various analyses on large collections of compounds have led to the identical finding that the experimental root mean square error (RMSE) [8] of log S_w due all of these inconsistencies is about 0.6. For the remainder of the chapter we will ignore all alternative aqueous and crystalline states, and focus on the still demanding problem of the transfer from a well-defined crystalline state to a well-defined dilute aqueous state of the neutral drug.

The process of dissolving a compound in water sometimes is conceptualized as a process of sublimation followed by hydration from the gas phase, i.e. reverse volatilization. We consider this thermodynamic cycle provides little utility in the context of drugs. Due to the very low vapor pressures of drug-like compounds the gaseous state of these compounds is hardly accessible experimentally, as are the room temperature vapor pressures or sublimation free energies. Equally, the corresponding free energies of hydration of the gaseous drug are unavailable for most drugs. Theoretically and computationally the usage of the gas phase as an intermediate reference point is analogous to going from London to Paris via the South Pole, because the two condensed states of the crystalline drug and the aqueous drug solution have much more in common than either state has with the gas phase. In both of the latter states the drug will experience strong H-bond and Coulomb interactions, as well as dispersive van der Waals interactions. While these interactions will be different but of the same order of magnitude in both states, they are essentially absent in the gas phase. Hence, we can benefit from a considerable amount of error cancellation if we avoid the gaseous state. Therefore, this thermodynamic pathway will not be considered further in this review.

Although not directly related to aqueous solubility, the state of the drug dissolved in another solvent, most notably in 1-octanol, has been used often in the computation of aqueous solubility. Hansch [9] introduced the simple linear relationship of

$$\mathrm{Log}\,S = 0.5 - \log P_{\mathrm{ow}}^{\mathrm{X}} \tag{2}$$

where $P_{\mathrm{ow}}^{\mathrm{X}}$ is the octanol–water partition coefficient. This has been further refined by Yalkowski [10] and published as the "general solubility equation" (GSE) in the form

$$\mathrm{Log}\,S = 0.5 - \log P_{\mathrm{ow}}^{\mathrm{X}} - 0.01(\mathrm{MP} - 25) \tag{3}$$

where MP is the melting point of the compound (°C). The term added by Yalkowski accounts for ΔG_{fus} because according to Walden's rule [11] a larger nonsymmetric compound has roughly an entropy of fusion of 50–60 J mol^{-1} K^{-1}, which translates into a slope of around 0.01 if converted to log units K^{-1}. The GSE, first formulated some 25 years ago, remains of comparable accuracy for drug solubility prediction as many of the modern approaches for drug solubility prediction [12]. While for log $P_{\mathrm{ow}}^{\mathrm{X}}$ a number of good *in silico* prediction methods are available (*vide infra*), the GSE requires the experimental MP as input, because no reliable prediction methods for MPs of drug-like compounds are yet available. This makes it almost useless in drug screening. Nevertheless, the GSE is widely used for compounds with known MP and it is used in combination with a crude MP guess of 125 °C [4] in industry. Despite its name and its apparent empirical success the GSE must clearly be classified as a very empirical equation. From a theoretical point of view it can be justified if the chemical potential of the compound in octanol is similar to the one in the neat liquid state. This may be the case if the polarity of the drug is close to that of octanol, i.e. if $\log P_{\mathrm{ow}}^{\mathrm{X}}$ is close to $\log P_{\mathrm{ow}}^{\mathrm{octanol}}$, which is about 3. Hence, we should at most expect GSE to be applicable for compounds in the range of $2 < \log P_{\mathrm{ow}}^{\mathrm{X}} < 4$, and indeed restrictions like this have been reported. The GSE benefits from the fact that this is a very typical log P_{ow} range for drug compounds.

The last state in Fig. 11.1 that has not yet been discussed is the state of the neat liquid compound X. For liquid compounds this is the relevant initial state for solubility, but almost all drug-like compounds are solid at room temperature. In this case the neat liquid is a virtual state of a supercooled liquid which can hardly be accessed experimentally. However, it is an interesting intermediate state because it allows us to split the calculation of solubility into two separate steps, which are conceptually and for some methods computationally easier to handle than the complete step from the crystalline state of the drug to the liquid state of the drug dissolved in water. In the first step we only have to transfer the compound from its neat crystalline state to its neat liquid state. The free energy of this fusion transfer is usually called ΔG_{fus} (or $\Delta G_{\mathrm{cryst}}$ if considered in the opposite direction). The second step is a transfer between two liquid states, the supercooled liquid drug and the drug in aqueous solution. Thus, in the first step the system itself is kept constant and its phase transition is considered. The second step is a transfer of the compound between two liquid environments of the same state of aggregation. We consider this route as most suited for the understanding and for a rational prediction of drug solubility. Therefore, we will analyze the different methods used for drug solubility prediction under the consideration of their ability to describe these two steps. Since this will be relatively short and general, we will start with a section on the prediction of ΔG_{fus}.

11.3 Prediction of ΔG_{fus}

For *in silico* drug solubility prediction, the rigorous prediction of ΔG_{fus} requires the prediction of the crystal structure of the drug molecule without any experimental information. Crystal structure determination means to predict the space group (of the 230 theoretically possible and about 10 more or less common space groups of organic crystals), the number of molecules in the unit cell, the positions and conformations of the molecules in the unit cell, and finally of the lattice parameters. Despite almost two decades of active research and a number of success stories for the prediction of the crystal structure of smaller and mostly rigid organic compounds, there is little hope that the crystal structure of drug-sized and drug-like molecules will be predictable by any computational method within the next 10 years, and perhaps for even longer. A recent review of this field has been given by Price [13].

Even if we could predict the crystal structure, for the prediction of ΔG_{fus} we would need a method which is able to handle the liquid and the crystalline state of the compound on the same footing, because any change in the method would introduce a large error in the free energy difference of both states. However, the methods for the treatment of the vibrational and motional degrees of freedom are very different for liquid-phase and fluid-phase simulations. Hence, we must conclude that the starting point for rational drug solubility prediction is currently inaccessible and that the situation appears hopeless. There are a few empirical considerations that can help us to overcome this deadlock. On the one hand, according to Walden's rule the entropy of fusion ΔS_{fus} is known to be in the order of 50–60 J mol^{-1} K^{-1} for drug-like molecules. On the other hand, most drug-like compounds have MPs in the order of 75–200 °C, rarely up to 250 °C. Under the assumption of a rather temperature independent ΔH_{fus} and ΔS_{fus}, the equilibrium condition of $\Delta G_{fus} = \Delta H_{fus} - T\Delta S_{fus}$ at the MP leads to a range of $2.5 < \Delta G_{fus} < 12.5$ kJ mol^{-1}. Thus, using a constant estimate of 7.5 kJ mol^{-1} for dG_{fus} should furnish log S_w with a maximum error of roughly 0.9 log units and probably within 0.7 log units RMSE arising from ΔG_{fus} for real drugs. A similar procedure was proposed by Delaney [4, 14] who reported that within Syngenta a MP of 125 °C is usually used in combination with Walden's rule. Unfortunately, in a very recent study on a diverse set of 26 orally administered drugs Wassvik et al. [15] report an average ΔS_{fus} of $83\,J\,mol^{-1}\,K^{-1}$, ranging from 56 to $114\,J\,mol^{-1}\,K^{-1}$, and an average MP of 177 °C. These results suggest a larger inaccuracy from the application of simple estimates based on Walden's rule.

Beyond such mean value estimates one can try to access ΔG_{fus} by quantitative structure–property relationship (QSPR) methods. There are a few rather obvious factors which should influence ΔG_{fus}. Larger molecules should have larger ΔG_{fus} than smaller ones, those with more H-bonding should have larger ΔG_{fus} than less H-bonding ones and also rigidity should give rise to larger ΔG_{fus}. Klamt et al. [16, 17] proposed a QSPR relation for ΔG_{fus} at 25 °C as:

$$\Delta G_{fus}^{X} \cong \mathrm{Max}(0; -12.2V^{X} + 0.76N_{ringatom}^{X} - 0.54 * \mu_{W}^{X}) \quad (4)$$

where V^X is the volume of drug X, $N_{ringatom}$ is the number of atoms involved in rings and μ_w^X is the conductor-like screening model for realistic solvation (COSMO-RS) [18–20] chemical potential of the drug in water, which is used as a combined polarity and H-bond descriptor in this situation. While this QSPR equation catches about 60%–80% of the variance of ΔG_{fus} on drug-like datasets, it possesses no scientifically sound basis for the calculation for ΔG_{fus}. For example, it completely fails for long alkanes, which are solid at room temperature, while Eq. (4) predicts them as liquids.

Delaney also indirectly introduced a ΔG_{fus} expression in his solubility model ESOL [14]:

$$\log S_{w}^{X} \cong 0.066\mathrm{RB} - 0.74\mathrm{AP} - 0.0062\mathrm{MW} - 0.62\log P_{ow} + 0.16 \quad (5)$$

where RB is the number of rotatable bonds and AP is the ratio of aromatic to total number of heavy atoms. Assuming the validity of the GSE equation (GSE1) this would correspond to an expression

$$\Delta G_{fus}^{X} \cong 0.09\mathrm{RB} - \mathrm{AP} - 0.0085\mathrm{MW} + 0.52\log P_{ow} + 0.46 \quad (6)$$

This also involves flexibility and rigidity descriptors (RB and AP), a size descriptor (MW) and a polarity descriptor (log P_{ow}). It appears that these kinds of descriptors are essential for ΔG_{fus}.

Delaney [4, 14] and Klamt [16] argued that for drug-like compound datasets only about 20% of the variance of log S_w arises from ΔG_{fus}. This is further confirmed by the study of Wassvik et al. [15] in which 77% of the variance is due to the solubility of the supercooled liquid. Hence, applying crude estimates by mean values or by QSAR approaches we can reasonably expect that the inaccuracies introduced in drug solubility prediction by our theoretical ignorance of ΔG_{fus} is less than, or at least not much bigger than, the inaccuracies introduced by the estimates of the larger part, i.e. the liquid solubility, and by the experimental difficulties in solubility measurement.

11.4
Prediction of Liquid Solubility with COSMO-RS

COSMO-RS [18–20] will be used here to elucidate the nature of the free energy of liquid solubility, which is the free energy of transfer of a molecule from its liquid or supercooled liquid state to its state in a solvent, i.e. in our case in water. For this model – and as we believe for the general understanding of solvation – a graphical analogy as shown in Fig. 11.2 can be very useful. The starting point of our consideration shall be the noninteracting molecule X in a perfect vacuum. This state is most easily accessible by computational means and therefore considered as the reference state for almost all computational chemistry methods. Nowadays good quality quantum chemical calculations, e.g. density functional theory (DFT) calculations, can be performed even for drug sized molecules, yielding geometries, electron densities and reference energies of good quality. For the following let us consider this clean reference state as the South Pole of the world of physical states of molecules. From this reference point the ideal gas phase can be

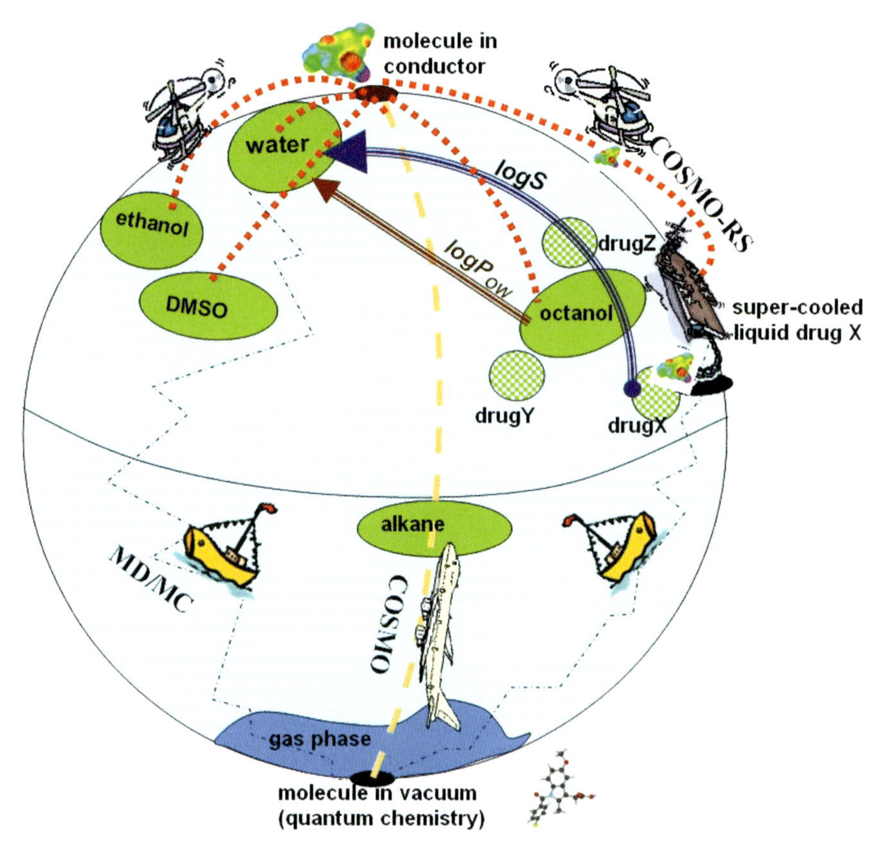

Fig. 11.2 Schematic illustration of the states and modes of transfer relevant for solubility prediction.

easily accessed by addition of translational and rotational free energy contributions, and hence the gas phase is a kind of Antarctic of our globe. Nevertheless, if we want to go from the isolated molecule in vacuum to a condensed state, we have to bring a large number of molecules from infinity to nearest neighbor distance. Along this path we have to integrate the long-ranged Coulomb interactions and the polarizability responses to these interactions. This is a major burden for all computational methods.

In 1995, one of the authors (A.K.) introduced the state of a molecule embedded in a perfect conductor as an alternative reference state, which is almost as clean and simple as the vacuum state. In this state the conductor screens all long-range Coulomb interactions by polarization charges on the molecular interaction surface. Thus, we have a different reference state of noninteracting molecules. This state may be considered as the North Pole of our globe. Due to its computational accessibility by quantum chemical calculations combined with the conductor-like screening model (COSMO) [21] we will denote this as the COSMO state.

Liquid solvents in this picture may be considered as islands in the northern hemisphere, because the state of molecules in a liquid solvent is quite definitely closer to the COSMO state than to the vacuum. The only exception may be alkane solvents, which are located somewhere close to the equator due to their fully nonpolar character. Solids may be considered as sunken islands and their depths below sea level may be considered as ΔG_{fus}. As discussed before, the methods to explore this depth are rather limited, but we can be quite sure that in general the depth below sea level will be much smaller than the distance of the islands from the North Pole or from each other. We now explore the methods to go from the sea level position of any island to the North Pole or *vice versa*. Given such a method we will be able to transfer a compound from any liquid or supercooled liquid state to any other such state.

The COSMO-RS method starts from quantum chemical calculations in the gas phase, i.e. at the South Pole, which are carried to the state of a molecule in a conductor, i.e. to the North Pole, by DFT calculations in combination with COSMO, which is an efficient variant of the old idea of continuum solvation models. Given the noninteracting molecules in a conductor, COSMO-RS introduces molecular interactions in the condensed phase as local contact interactions of surface segments, which result from the removal of the conductor between two closely contacting molecules. COSMO-RS quantifies the interactions based on the local polarization charge densities σ and σ' of the interacting segments, which are known for each molecular surface segment from the prior DFT/COSMO calculation. While van der Waals interactions can be well described by simple element-specific surface tensions $\tau(e)$, which are considered as part of the energy of the compound in the conductor, the deviations of the interactions of a molecule between the conductor and real solvent molecules are thus described as electrostatic misfit interactions of the polarization charge densities:

$$E_{misit}(\sigma, \sigma') = a_{eff} \frac{\alpha'}{2}(\sigma + \sigma')^2 \tag{7}$$

and as H-bond interactions:

$$E_{\mathrm{HB}} = (\sigma, \sigma') = a_{\mathrm{eff}} c_{\mathrm{HB}}(T) \min(0, \sigma\sigma' + \sigma_{\mathrm{HB}}^2) \tag{8}$$

The H-bond threshold σ_{HB} ensures that only segment pairs with strong and opposite polarity make H-bonds, and the temperature-dependent coefficient expresses the fact that the formation of H-bonds releases a huge amount of enthalpy, but also has a considerable penalty due to the enthalpy arising from the strong distance and orientation demands of H-bonds. For the following statistical thermodynamics step we need to represent each solute and solvent molecule by its surface histogram $p^{X}(\sigma)$ with respect to the polarization charge density σ, which is called the σ-profile further on. The generation of the σ-profile from the polarization charges on the COSMO surface is shown schematically in Fig. 11.3.

Next, the statistical thermodynamics of the pairwise surface segment interactions can be performed exactly, i.e. without any additional approximations beyond the assumption of surface pair formation, using:

$$\mu_S(\sigma) = -RT \ln\left\{\int p_S(\sigma, \sigma') \exp\left(\frac{\mu_S(\sigma') - E_{\mathrm{misfit}}(\sigma, \sigma') - E_{\mathrm{HB}}(\sigma, \sigma')}{RT}\right) d\sigma'\right\} \tag{9}$$

where $\mu_S(\sigma)$ is the chemical potential of a surface segment of polarity σ in solvent S. $p_S(\sigma)$ is the normalized σ-profile of the liquid system, i.e. of the solvent, which in the case of a pure solvent is just the σ-profile of the of the solvent divided by

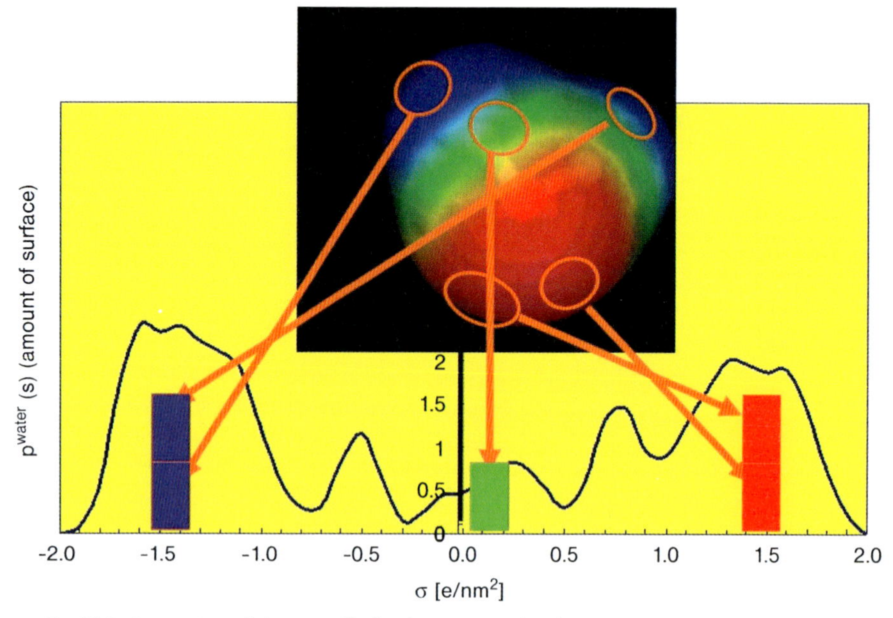

Fig. 11.3 Generation of the σ-profile for the water molecule.

surface area A^S and in the general case of a mixture, is trivially given by the normalized mole-fraction-weighted σ-profile:

$$p_S(\sigma) = \sum_{i=1}^{n} x_i p^i(\sigma) \Big/ \sum_{i=1}^{n} x_i A^i \tag{10}$$

The segment chemical potential $\mu_S(\sigma)$is also called the σ-potential of a solvent. It is a specific function expressing the affinity of a solvent S for solute surface of polarity σ. Typical σ-profiles and σ-potentials are shown in Fig. 11.4. From the σ-potentials it can clearly be seen that hexane likes nonpolar surfaces and increasingly dislikes polar surfaces, that water does not like nonpolar surfaces (hydrophobic effect), but that it likes both H-bond donor and acceptor surfaces, that methanol likes donor surfaces more than does water, but acceptors less, and many other features.

Given the σ-potential of a solvent S, the chemical potential of a molecule X in S can be essentially expressed as a surface integral of this σ-potential $\mu_S(\sigma)$ over the surface of the solute:

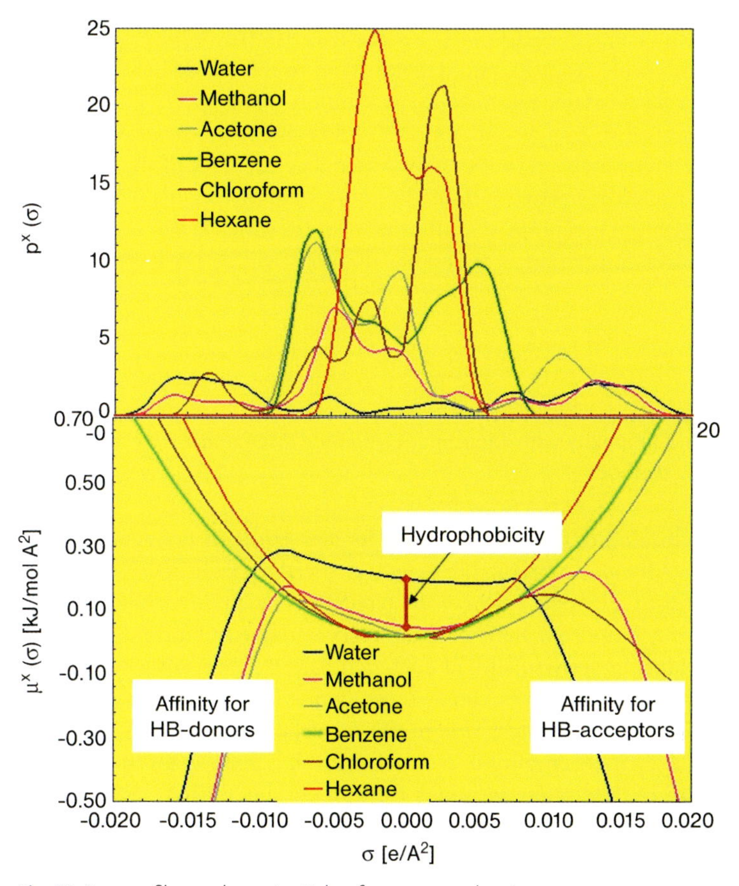

Fig. 11.4 σ-profiles and σ-potentials of common solvents.

$$\mu_S^X = \oint\!\!\oint_X \mu_S(\sigma(\underline{r}))d^2\underline{r} + \mu_{comb}(X,S) = \int \mu_S(\sigma)p^X(\sigma)d\sigma + \mu_{comb}(X,S) \tag{11}$$

With the help of the σ-profile the surface integral can be elegantly transformed into a σ-integral (right side in Eq. 11), but we should keep in mind that the chemical potential of a solute in a solvent is essentially a surface integral of a solvent specific function over the surface of the solute. This fact is important for the analysis of the problem of solubility prediction.

The "combinatorial term" $\mu_{comb}(X,S)$ in Eq. (11) denotes a usually small correction for size and somewhat also for shape differences of the solutes and solvents, which is reasonably well understood from chemical engineering models. It is an empirical function, depending only on the surface areas and volumes of solute and solvent molecules, which are reasonably known from the COSMO cavities. The error arising from the approximations made in the combinatorial term can be expected to be much less than 1 kJ mol^{-1}, or 0.17 log units, and hence can be safely neglected in the context of drug solubility estimation.

As a result of Eq. (11) we are able to calculate the chemical potential of any molecule X in any liquid system S, relative to the chemical potential in a conductor, i.e. at the North Pole. Hence, COSMO-RS provides us with a vehicle that allows us to bring any molecule from its liquid state island to the North Pole and from there to any other liquid state, e.g. to aqueous solution. Thus, given a liquid, or a reasonable estimate of ΔG_{fus} of a solid, COSMO-RS is able to predict the solubility of the compound in any solvent, not only in water. The accuracy of the predicted ΔG of transfer of molecules between different liquid states is roughly 0.3 log units (RMSE) [19, 22] with the exception of amine systems, for which larger errors occur [16, 19]. Quantitative comparisons with other methods will be presented later in this article.

11.5
Prediction of Liquid Solubility with Molecular Dynamics (MD) and Monte Carlo (MC) Methods

The fluid-phase simulation approach with the longest tradition is the simulation of large numbers of the molecules in boxes with artificial periodic boundary conditions. Since quantum chemical calculations typically are unable to treat systems of the required size, the interactions of the molecules have to be represented by classical force fields as a prerequisite for such simulations. Such force fields have analytical expressions for all forces and energies, which depend on the distances, partial charges and types of atoms. Due to the overwhelming importance of the solvent water, an enormous amount of research effort has been spent in the development of good force field representations for water. Many of these water representations have additional interaction sites on the bonds, because the representation by atom-centered charges turned out to be insufficient. Unfortunately it is impossible to spend comparable parameterization work for every other solvent and solute. Therefore, one needs to be aware that the quality of the force field representation of other molecules is much less accurate, perhaps with the exception of alkanes and other nonpolar compounds. Special care must be taken for drug-like molecules with all their rather demanding functional groups and heterocycles. Given a reasonable force field for all involved compounds and their interactions, an initial configuration of the ensemble is generated and its energy is evaluated. Starting from this, new configurations are iteratively generated and evaluated, either by following the classical Newton's laws of motion, or by supervised random choices. The first class of methods is summarized under the name MD and the others are called MC simulations. Finally, after achievement of thermodynamic convergence, the free energy and the enthalpy of the system is evaluated by appropriate averages over a large number of configurations. If, as in the case of solubility calculation, only energetic aspects and no dynamic aspects are of interest, MC simulations are considerably more efficient, but they still require at least several central processing unit hours of computation for a reasonably converged calculation of the free energy or chemical potential of a drug sized molecule in water. Anderson and Siepmann [23] recently presented a review on the usage of MC methods for the evaluation of "aqueous solubility" with respect to the gas phase. Almost all work done so far is for atoms or small molecules of size up to tetrahydrofurane or acetonitrile that are considered as large molecules in this context. Only one study is mentioned that deals with log P_{ow} calculation for two larger and truly drug-like molecules, and which included the evaluation of the chemical potential in water [24]. The CHARMM force field [25] was used based on DFT charges. Such prior quantum chemical calculations are standard for MC simulations of electronically more demanding molecules, e.g. of drugs, because simpler, more empirical atomic partial charge assignments usually fail to reproduce the electrostatics with sufficient accuracy. In the picture of the globe in Fig. 11.2 this means that the quantum chemical information about the molecule in vacuum, i.e. at the South Pole, is shipped by classical boats towards the solvated states, including the aqueous state. This is still a rather time-consuming journey which has not yet proven to be reliable for drug-sized molecules.

Although not really evaluating the chemical potential of drugs in water, we should mention here the BOSS and QikProp methods by Jorgensen et al. [26–28] based on MC simulation of drugs in water using the OPLS force field [29]. BOSS samples the different contributions to the interaction energy as well as other quantities of the drug in water during a MC simulation and subsequently derives 17 descriptors for the drug interactions in solvents. These are used to generate QSAR expressions for all kinds of solvation-related properties, one of them being log S_w. It should be noted that BOSS never simulates the solute in the neat liquid phase nor has any specific terms for ΔG_{fus}. Since the BOSS simulations takes about 2 h per compound, Jorgensen et al. developed the QikProp method, which replaces the rigorous MC sampling by an empirical surface analysis of the solute based on a similar molecular representation as it was used in the MC simulations. Thus, QikProp no longer requires a real molecular simulation and as a result it takes about a second to compute the properties of any one compound.

11.6
Group–Group Interaction Methods

After this side track let us return to the methods of thermodynamic evaluation of the chemical potential of solutes in water and in their neat liquid. Models of pairwise group interactions, mostly called group contribution methods (GCMs), such as UNIFAC or ASOG [30–32], have been developed in the chemical engineering arena for over 30 years to infer the activity and hence the chemical potential of solutes in solvents and solvent mixtures. It is worth noting that these chemical engineering GCMs treat solute and solvent by group decomposition, while GCMs typically used in drug design only consider a decomposition of the solute, while the solvent system is kept fixed; the CLOGP [33, 34] method for calculating the octanol–water partition coefficient is one such example. Using a large quadratic matrix of group–group interaction parameters with several thousand entries adjusted to experimental phase equilibrium data, the pairwise group interaction methods perform a fast empirical evaluation of the statistical thermodynamics of the interacting groups and end up with the activity coefficient for each molecule in pure and mixed solvents at variable temperature. It should be noted that for low solubilities the inverse activity coefficient is just the solubility. Hence, these methods are basically suited to calculate solubility within a proven fluid phase thermodynamics framework. They have been widely used in chemical engineering thermodynamics, but they have problems when it comes to molecules with several functional groups and intramolecular interactions of such groups, e.g. by electronic push–pull effects or by H-bonding. As an example, the dramatic variation of the polarity of the aromatic nitro group in the series *p*-dinitrobenzene, nitrobenzene and *p*-nitroaniline cannot be reflected by these GCMs, because the interaction parameters of an aromatic nitro group are considered as independent of the chemical environment. The introduction of more or larger groups in order to take such effects into account is prohibitive, because the number of adjustable parameters increases as the square of the number of groups. Another reason for the failure of these GCMs is the usage of the approximate statistical thermodynamics instead of an exact quasichemical treatment. This approximation does not correctly reflect the strong increase of chemical potential towards infinite dilution [35] and hence it has problems to correctly describe small solubility values as are typical for drugs in water. While being computationally very fast, these two shortcomings clearly disqualify chemical engineering GCMs for the evaluation of the chemical potential of drugs in water and in their supercooled liquid state.

11.7
Nonlinear Character of Log S_w

Almost all widely used, reliable prediction models for logarithmic partition coefficients, and especially for the octanol–water partition coefficient log P_{ow}, are linear regression models with respect to fragment counts, atom types, bond types or other unimolecular descriptors of the compound X. This can be well understood, since according to Eq. (11) the logarithmic partition coefficient of a compound X between two phases S and S′ is given as a surface integral:

$$\log P^{X}_{S,S'} = \frac{\mu^{X}_{S'} - \mu^{X}_{S}}{RT\ln 10} = \frac{1}{RT\ln 10}\Big(\oiint_{X}(\mu_{S'}(\sigma(\underline{r})) - \mu_{S}(\sigma(\underline{r})))\mathrm{d}^2\underline{r} + \mu_{\mathrm{comb}}(X,S') - \mu_{\mathrm{comb}}(X,S)\Big)$$
$$= \frac{1}{RT\ln 10}\left(\sum_{\alpha\in X}\oiint_{\alpha}(\mu_{S'}(\sigma(\underline{r})) - \mu_{S}\sigma(\underline{r})))\mathrm{d}^2\underline{r} + \Delta\mu_{\mathrm{comb}}(X,S',S)\right) \tag{12}$$

where α denotes the atoms of compound X. Ignoring the small and probably rather constant contribution from the combinatorial term, we realize that log $P_{SS'}$ can be written as a sum of atom contributions. Hence, to first order it is plausible to assume that the individual atom contributions are characteristic for each atom type. Therefore, log $P_{SS'}$ is well accessible by linear regression with respect to atom type counts, fragment or group counts and with respect to other descriptors, which themselves can be roughly considered as surface integrals or as linear combinations of atomic contributions.

The situation is very different for the solubility of a compound X, e.g. a supercooled liquid drug, in any solvent S, which can be written as:

$$\log S^{X}_{S} = \frac{\mu^{X}_{X} - \mu^{X}_{S}}{RT\ln 10} = \frac{1}{RT\ln 10}\Big(\oiint_{X}(\mu_{X}(\sigma(\underline{r})) - \mu_{S}(\sigma(\underline{r}))\mathrm{d}^2\underline{r} + \mu_{\mathrm{comb}}(X,X) - \mu_{\mathrm{comb}}(X,S)\Big) \tag{13}$$

No decomposition into atom contribution is possible, because X not only occurs as the solute, but also as the solvent in μ^{X}_{X}. In other words, since the drug is in a dual role of solute and solvent, it is not possible to dissect the contribution of individual atoms or groups of atoms in the drug from their role in the solvent state from that in the solute. Reiterating this important point, log S_w cannot be assumed to be accessible by linear regression with respect to composition descriptors, as is the case for log P_{ow}. Moreover, it can hardly be – as often assumed – a linear function of log P_{ow}, because the latter is linear with respect to composition. Thus, any account of log S_w requires the nonlinear character to be taken into account.

As an example of this nonlinear character we may consider two pairs of compounds, naphthalene versus quinoline and indole versus benzimidazole (Fig. 11.5). In both pairs of compounds the second differs from the first by a mutation of an aromatic -CH group to an aromatic nitrogen, which introduces a strong H-bond acceptor into the molecule. In quinoline, which has no H-bond donor, the acceptor has no favorable interaction partner in the supercooled liquid or crystalline state, while it can make strong H-bonds with the solvent in water. Therefore, log S_w of quinoline is about 2 log units higher [35, 36] than that of naphthalene, i.e. the introduction of the H-bond acceptor strongly increases solubility in this

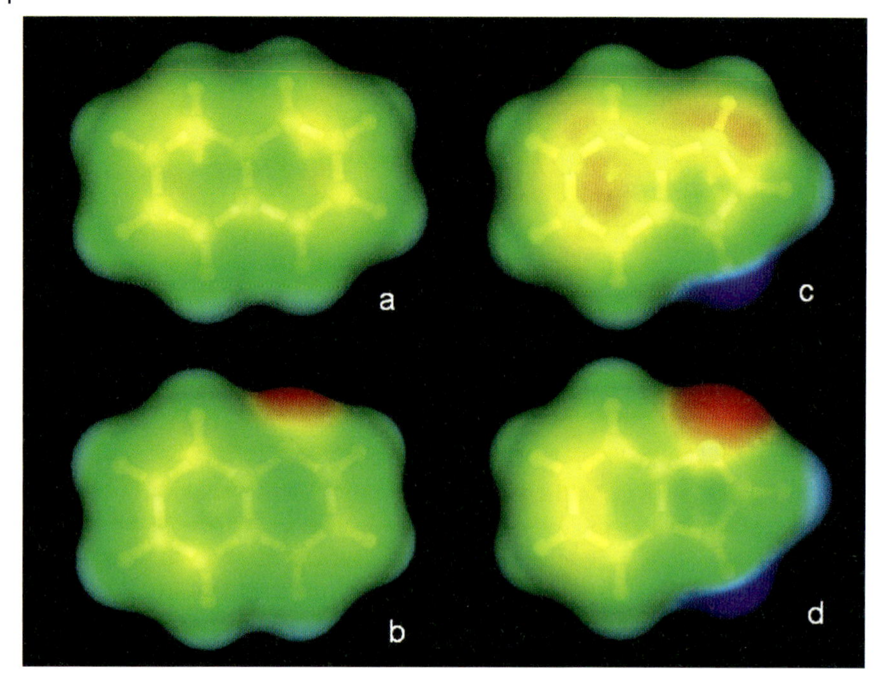

Fig. 11.5 COSMO surfaces for two pairs of compounds differing by the mutation of an aromatic CH- group to an aromatic nitrogen: (a) naphthalene, (b) quinoline, (c) indole and (d) benzimidazole. The influence of the aromatic nitrogen on the aqueous solubility is very dissimilar in both cases.

case. This corresponds to the usually observed trend, since in most drugs we have more H-bond acceptors than donors and, hence, the introduction of additional acceptors usually leads to an increase in solubility since the additional H-bond acceptor cannot find good interactions in the pure state. However, if – by the mutation to benzimidazole – we introduce the additional H-bond acceptor in indole, which already has a strong H-bond donor, a small solubility decrease of around –0.25 is observed [36, 37], because the additional H-bond acceptor stabilizes the pure compound state through interactions with the H-bond donor more strongly than the aqueous solution phase. Hence, the analogous chemical replacement in one case leads to a strong increase in solubility, while in the other case to a slight decrease; clearly, the direction of the change depends on the overall composition of the compound. The difference between the two changes, i.e. the $\Delta\Delta\log S_w$, is around 2.25 log units. Such nonlinear behavior is very hard to model unless one calculates the sophisticated thermodynamic equilibria between all the groups in the liquid and aqueous states of a drug. Therefore, most of the simpler methods fail to reproduce such trends. Almost all of the methods tested by us correctly predict the increase by about 2 log units for the naphthalene to quinoline mutation, but predict a similar increase for the indole to benzimidazole mutation. Only COSMO-RS – due to its inherent thermodynamics – is able to treat this nonlinear behavior correctly ($\Delta\Delta\log S_w = 2.33$ [38]). As expected, log P_{ow} shows a linear behavior, i.e. the effect of the introduction of the additional H-bond acceptor is a decrease of log P_{ow} by 0.8 and 1.2 log units, respectively, for the two mutations. This trend is well reflected by standard group additivity methods.

The nonlinear character of log S_w has not often been discussed previously. Nevertheless, Jorgensen and Duffy [26] argued the need for a nonlinear contribution to their log S_w regression, which is a product of H-bond donor capacity and the square root of H-bond acceptor capacity divided by the surface area. Indeed, for the example above their QikProp method partially reflects for this nonlinearity by predicting a much smaller solubility increase for the indole to benzimidazole mutation (0.45 versus 1.82 [39, 40]). Abraham and Le [41] introduced a similar nonlinearity in the form of a product of H-bond donor and H-bond acceptor capacity while all logarithmic partition coefficients are linear regressions with respect to their solvation parameters. Nevertheless, Abraham's model fails to reflect the test case described above. It yields changes of 1.8(1.5) and 1.7(1.7) [42] for the mutations described above.

11.8 QSPRs

The concept of functional groups in chemistry is pivotal to our understanding of physicochemical behavior. It is not surprising, then, that we might hope that various properties of molecules might be able to be described based on this concept. From this simple premise the world of QSPRs emerges. This field of study cannot be adequately reviewed here and, further, Dearden [3] has recently reviewed the important features of deriving a QSPR for the prediction of aqueous solubility. The most critical requirement of any QSPR is that it be predictive; it is not sufficient that QSPR be able to reproduce the training data. To this issue, it is very important that the QSPR not be over-fit. Like many physicochemical properties, solubility tends to be an information-deficient property. Thus, the number of compounds whose solubility is accurately determined experimentally and published is limited to at most a few thousand, and question marks have to be put to several of these data due to the reasons mentioned above, i.e. due to unreported protonation and tautomeric states, polymorphism, etc. Often the number of compounds used to derive a QSPR is limited to a few hundred. It is, therefore, imperative that the number of descriptors used in the QSPR be restricted to a very small number to ensure the predictive nature of the model be retained, even at the cost of a more precise model. Although there is no formal ratio of observations to descriptors needed to ensure a predictive model, it is not unusual to expect ratios of at least 10 to 20.

The process of defining any QSPR model involves three fundamental components: (i) a set of descriptors, (ii) a method to select the most appropriate descriptors, and (iii) the experimental data to train and test the model. It is important to note here that none of these components are unique and many models can be derived to describe the same property. Thus, there exist in the literature many different models that can be applied to predict aqueous solubility. Equally important to note is the lack of any requirement for the descriptors to provide any intuition into the solubility process. Indeed, many descriptors currently in use have no physical interpretation, most notably the BCUT descriptors, which are just eigenvalues of an abstract adjacency matrix [43].

There are many different methods for selecting those descriptors of a molecule that capture the information that somehow encodes the compounds solubility. Currently, the most often used are multiple linear regression (MLR), partial least squares (PLS) or neural networks (NN). The former two methods provide a simple linear relationship between several independent descriptors and the solubility, as given in Eq. (14). This equation yields the independent contribution, b_i, of each descriptor, D_i, to the solubility:

$$\log S_w = a + \Sigma b_i D_i \tag{14}$$

It is usual to have the coefficient of determination, r^2, and the standard deviation or RMSE, reported for such QSPR models, where the latter two are essentially identical. The r^2 value indicates how well the model fits the data. Given an r^2 value close to 1, most of the variation in the original data is accounted for. However, even an r^2 of 1 provides no indication of the predictive properties of the model. Therefore, "leave-one-out" tests of the predictivity are often reported with a QSAR, where sequentially all but one descriptor are used to generate a model and the remaining one is predicted. The analogous statistical measures resulting from such leave-one-out cross-validation often are denoted as q^2 and s_{PRESS}. Nevertheless, care must be taken even with respect to such predictivity measures, because they can be considerably misleading if clusters of similar compounds are in the dataset.

A problem of all such linear QSPR models is the fact that, by definition, they cannot account for the nonlinear behavior of a property. Therefore, they are much less successful for log S_w as they are for all kinds of logarithmic partition coefficients.

NN can be used to select descriptors and to produce a QSPR model. Since NN models can take into account nonlinearity, these models tend to perform better for log S_w prediction than those refined using MLR and PLS. However, to train nonlinear behavior requires significantly more training data that to train linear behavior. Another disadvantage is their black-box character, i.e. that they provide no insight into how each descriptor contributes to the solubility.

11.9 Experimental Solubility Datasets

Refinement of a QSPR model requires experimental solubilities to train the model. Several models have used the dataset of Huuskonen [44] who sourced experimental data from the AQUASOL [45] and PHYSPROP [46] databases. The original set had a small number of duplicates, which have been removed in most subsequent studies using this dataset, leaving 1290 compounds. When combined, the log S_w

of the compounds in this dataset ranges from −11.62 to +1.58. The properties of drugs (or even drug-like molecules) tend to differ somewhat from those of the broader class of organic compounds. This difference stems from the requirement for these compounds to have favorable pharmacokinetic properties. Huuskonen has also compiled a database of log S_w of 211 drugs, covering the range −5.60 to +0.55; this dataset contained 85 compounds from the original compilation above [44]. Again, it should be kept in mind that aqueous solubility can be measured using a variety of different experimental techniques, sometimes resulting in a large variation in the reported solubilities for any one compound [47]. Thus, any single measurement of solubility may be accompanied by large errors of sometimes 1 or even 2 log units. As an example, the Huuskonen dataset includes the compound natamycin (CAS-RN: 7681-93-8) and reports a log S_w of −3.1. A quick internet search yields values ranging from −2.1 to −4.1. More importantly, Suloff [48] reports it to be a zwitterion in water, which is quite plausible. Hence it is almost impossible to predict such high solubility from the neutral structure given in the Huuskonen dataset. Many pharmaceutical companies maintain solubility data on their own compounds. In some instances researchers from these laboratories use this data to construct QSPR models. The content of these datasets, including reliability, are usually not disclosed. It is widely assumed that the larger the number of compounds used to train the QSPR the more reliable the model should become; Hansen et al. [49] extracted 4548 compounds with drug-like properties from the PHYSPROP database in their recent study. However, we must keep in mind that for big datasets the number of questionable solubility values and of wrongly reported structures usually increases, and this may ameliorate any advantage of the bigger datasets.

It is usual in developing a QSPR to split the database into two. One part is used for training the model, while the other part is used to validate the model. This goes to the predictability of the model; the model is assumed to be predictive if it can predict the solubility of the validation set. Since the validation set is used intimately with the training set to refine the model, it is questionable if this partitioning is warranted [50]. This partitioning is particularly questionable where the available experimental data is sparse.

With all QSPR studies it is not possible to separate the influence of the data used to train the model and the computational approach used to derive the model from the final model. Ideally, the QSPR should be sufficiently general to be applied to any compound that is reasonably represented by the data used to derive the model.

To examine these effects we can look to the recent work by McElroy and Jurs [51], who derived several QSPR models using different databases, using both MLR and computational NN models to determine the optimal model in each situation. The compounds constituting the first dataset all contained at least one nitrogen atom. The compounds in the second dataset contained at least one oxygen, but no nitrogen atoms. Datasets one and two contained 176 and 223 compounds, respectively. The third dataset contained all compounds in datasets one and two. The best performing relationship from the MLR analysis for the three datasets contained 7, 11 and 11 descriptors, respectively. The three models shared only one common descriptor, a topological descriptor describing the distance between 1°

and 4° carbon atoms; 83 of the compounds in dataset 3 contained both 1° and 4° carbon atoms. All three models contained a descriptor that relied on the H-bonding characteristics of the compounds. The number of descriptors required to adequately represent the data also differed in each model. Each model had comparable performance, with r^2 values of 0.75, 0.92 and 0.86 for datasets 1, 2 and 3, respectively. Critically in this regard, the model derived from dataset 3 did not contain descriptors formed by the simple combination of descriptors found for datasets 1 and 2.

NN models for the three datasets contained the same number of descriptors as the MLR models, yet no more than two descriptors in each model were the same in both NN and MLR models. No descriptor was found in common with all models, although, each model contained a descriptor that relied on H-bonding in some manner. Nonlinear modeling from the NN approach gave better representation of the data than the linear models from MLR; the r^2 value for the three datasets was 0.88, 0.98 and 0.90, respectively.

From the results described above it is clear that a different QSPR model can be obtained depending on what data is used to train the model and on the method used to derive the model. This state of affairs is not so much a problem if, when using the model to predict the solubility of a compound, it is clear which model is appropriate to use. The large disparity between models also highlights the difficulty in extrapolating any physical significance from the models. Common to all models described above is the influence of H-bonding, a feature that does at least have a physical interpretation in the process of aqueous solvation.

The question of selecting the most appropriate method for any one compound has been addressed recently by Kühne et al. [52]. Initially several different methods are used to predict the solubility of a reference library of compounds. A subset of compounds from this reference library that are most similar to the compound of interest is identified and the method with the smallest sum of errors in the predicted solubility for this subset is chosen to predict the solubility. Dearden [3] considered whether a consensus approach could improve prediction over any one method. While the predictions from certain pairs of methods could be combined with improved results, some combinations led to poorer performance than either method alone. Chen et al. [53] were able to achieve improved correlation with their QSPR model using different QSPRs for different classes of compounds. Thus, while each QSPR used the same set of eight descriptors, the contribution of each descriptor changed according to the compound type. Each group had 82–101 compounds and achieved an r^2 of 0.86–0.92.

11.10 Atom Contribution Methods, Electrotopological State (E-state) Indices and GCMs

Many researchers have developed models under the assumption that in some way, although not always linearly, each atom in the molecule contributes to the overall solubility. E-state indices [54] that take into account the immediate environment of the atoms in a molecule, are often used to encode this atom-specific influence on aqueous solubility. The features often used to distinguish atoms are the connectivity to other atoms, connectivity to hydrogen atoms and valency. Atoms with no E-state index defined are assumed to contribute nothing to the solubility.

Hou et al. [50] created a QSPR using 79 of these atom descriptors, but needed to augment these with an additional descriptor to account for errors associated with hydrophobic carbon atoms, and a correction factor equal to the square of the molecular weight. Using the Huuksonen 1290 compound dataset for training, the QSPR had an r^2 of 0.91. Tetko et al. [55] used 30 E-state indices and the molecular weight, trained against a subset of the Huuskonen dataset, retaining 412 for validation purposes, and achieved an r^2 of 0.86 using MLR. Using a NN the r^2 improved to 0.95.

While E-state indices capture a description of the immediate environment of an atom in a molecule, often it is a group of connected atoms that impart a certain property. For example, a carbonyl group attached to a nitrogen atom in an amide is far less reactive than a carbonyl attached to an oxygen atom in an ester; the E-state index of the carbon of the carbonyl is the same, yet a group description might be quite different in the amide or ester. The contribution of each group identified in the molecule is combined to yield a predicted solubility. Unlike E-state indices, any one atom can participate in several group contributions. For large molecules the number of groups present can be large, and the resulting predicted solubility can be distorted toward very large (positive or negative) values. To compensate for this failing, Klopman and Zhu [56] proposed a method to scale predictions that fall toward these extreme values using a heuristic process of stereographic projection, which reduces the predictions to the range of log S_w typically observed for drug-like compounds. In this model 192 group-contribution descriptors were required to adequately describe the solubility of 1168 compounds. Notably, the ratio of observations to variables is very small (6.1) in this example and the question of over-fitting is worth bearing in mind.

Others who have applied the GCM are Wakita et al. [57], who were more economical in the choice of 40 descriptors to fit the solubilities of 307 organic liquids, with a resulting r^2 of 0.96. The group of Yalkowsky have developed their AQUAFAC [47] approach and used 44 descriptors trained against 970 compounds, a ratio of 22. The r^2 for this model was 0.98, with an RMSE of 0.43. For solids, the MP is also required, highlighting the connection with Yalkowsky's earlier GSE. While the types of fragments used in the QSPR are similar in these two studies, the AQUAFAC approach utilizes the connectivity of these fragments as an additional level of specificity. Thus, a methyl group attached to a sp^3 or sp^2 neighbor will contribute differently to the solubility.

11.11
Three-dimensional Geometry-based Models

Most of the models and descriptors discussed so far are based on the two-dimensional representation of the compounds, i.e. on their structural formula.

The advantage of such two-dimensional QSPR approaches is their simplicity in representing the compounds of interest and the uniqueness of the descriptors, provided we exclude tautomerism. Some solubility models and descriptors require the three-dimensional structure to determine geometrical or electronic features, such as molecular volume, or molecular moments, and other parameters relevant for intermolecular interactions. Within these models are the above-mentioned BOSS and QikProp models, which obtain the descriptors from more or less rigorous force field modeling of the solute–water interactions. Often more demanding quantum chemical calculations for the individual compounds are performed in order to get information about the charge distribution and the electrostatic properties of the entire compound or of the individual atoms in the compound. Such information is later used to build a solubility QSPR model, mostly by linear regression. The advantage of such three-dimensional descriptors is that they can often better account for specific differences resulting from intramolecular interactions, such as H-bonding or electronic push–pull interactions, e.g. those displayed in Fig. 11.3. Disadvantages of such approaches are the higher computational demand and, especially, the conformational ambiguity. The choice of different geometrical conformations leads to different results, but it is not simple to give rules how to generate a unique conformation for such methods. Some models and descriptors are based on atomic partial charges. The problem of calculating atomic charges is particularly insidious, since there is no unique way of determining these charges, even if one resorts to a quantum mechanical description. Despite the ambiguity, many different charge models exist, including empirical methods, such as Gasteiger [58], and quantum mechanical methods, such the semiempirical AM1 and PM3 methods. Thus, for example, McElroy and Jurs [51] used charges calculated using the AM1 Hamiltonian based on geometries minimized using the PM3 Hamiltonian.

11.12 Conclusions and Outlook

A very large number of empirical relationships for predicting solubility have appeared over the past few decades. The advent of sophisticated modeling tools has seen a large increase in the number and type of approaches used. The variety of methods used to perform regression analysis, the plethora of potential descriptors and the emergence of larger collections of experimental data for training has produced an explosion of new models. Most of them are computationally very efficient. These models are routinely able to predict log S_w with a standard deviation of 0.6–1.0 log units of the experimental values upon which they are trained, which is not too far from the intrinsic experimental error of the presently available broader solubility datasets, which is generally assumed to be no smaller than 0.5–0.7 log units.

Nevertheless, while for log P_{ow} a few widely accepted standard models such as CLOGP or LOGKOW [34, 59] have emerged, which are all linear models based on atom or fragment contributions with some additional correction terms, unfortunately no such convergence to any accepted set of descriptors or any mathematical model can be observed for log S_w. Also, almost none of the methods undergo regular improvements. It appears that computational chemists can continue this approach of developing new solubility models from scratch for decades without achieving convergence. We are convinced that this is due to the inherent nonlinear character of log S_w compared to the linear character of log P_{ow}. For this reason linear modeling of log S_w will always yield very different and biased models depending on the dataset used for the training. Nonlinear models are definitely required, but it is questionable whether NN techniques will ever be able to yield reliable and structurally extrapolative models, because the rules of nonlinearity are very hard to learn just from a dataset without a physicochemical model. Considering the fact that around 20 000 quite good experimental data are required to yield widely usable and accurate linear log P_{ow} models, we can be quite sure that the number of good quality data required for training unsupervised nonlinear log S_w models will be in the order of several hundred thousand. Given the enormous costs of highly accurate solubility measurements, this must be considered unrealistic at present.

In this situation the only rational way to proceed is to make use of the fundamental laws of physical chemistry and thermodynamics as much as possible. The pathway using the supercooled state of the drug as intermediate state and splitting log S_w into two contributions, i.e. one smaller contribution arising from the free energy of fusion, and a large contribution from the solubility of the supercooled drug, appear to be the only sensible way for a reasonably rational approach. For the fusion part no theoretically fundamental approach is to be seen and more work will be required to improve the empirical models. Fortunately, ΔG_{fus} data can be obtained independently of solubility measurements from MPs and melting enthalpies. For the solubility of the supercooled liquid drug a consistent thermodynamic modeling of all the interactions in the aqueous and neat liquid systems is required. Since force field-based simulation approaches cannot currently handle such liquid drug systems with the required accuracy and within acceptable timings, and since the chemical engineering group–group interaction models fail with respect to the description of the extremely broad variety of functional groups and intramolecular interactions in drug systems, the COSMO-RS approach appears to be the most promising method in this regard. Although, admittedly, in its present state it cannot be proven to be more accurate than most empirical methods, it does have the potential to be rationally improved. Potential improvements to the presently available COSMO-RS solubility prediction model include a more accurate fusion term, improvements to the COSMO-RS interactions themselves, especially those of amines, and more reliable experimental data. A dataset of a few hundred drug-like compounds with reliable, highly accurate intrinsic solubilities, characterized polymorphic forms, and measured MPs and melting enthalpies will be required for the development of more accurate COSMO-RS-based models, as they were published recently for 26 drugs [15]. A broader study of this kind is being performed presently at the Unilever Centre for Molecular Science Informatics in

Cambridge in collaboration with Pfizer which should be published in the near future [60]. Efficiency concerns may arise with respect to the time-consuming quantum chemical basis of COSMO-RS, but a highly efficient short-cut of COSMO-RS has already been developed in form of the COSMOfrag approach [61].

Finally, it should be noted that COSMO-RS might also open a path for the prediction of the solubility of drug salts, which are not properly addressed by any current model, despite the fact that a large number of commercial drugs are administered as salts. Since the supercooled state of a drug salt is an ionic liquid and since COSMO-RS has been widely proven to be able to treat ionic liquid thermodynamics [62–64], the missing smaller part is again the free energy of fusion. Work will have to be done to obtain reasonable approximations for this component in drug salts.

References

1 Balakin, K. V., Savchuk, N. P., Tetko, I. V. *In silico* approaches to prediction of aqueous and DMSO solubility of drug-like compounds: trends, problems and solutions. *Curr. Med. Chem.* **2006**, *13*, 223–241.

2 Jorgensen, W. L., Duffy, E. M. Prediction of drug solubility from structure. *Adv. Drug Deliv. Rev.* **2002**, *54*, 355–366.

3 Dearden, J. C *In silico* prediction of aqueous solubility. *Expert Opin. Drug Discov.* **2006**, *1*, 31–52.

4 Delaney, J. S. Predicting aqueous solubility from structure. *Drug Discov. Today* **2005**, *10*, 289–295.

5 Bondesson, L. Microscopic views of drug solubility. *Thesis*, Universitetsservice, Stockholm, **2006**.

6 Johnson, S. R., Zheng, W. Recent progress in the computational prediction of aqueous solubility and absorption. *AAPS J.* **2006**, *8*, 27 – 40.

7 Ben Naim, A. (ed.), *Solvation Thermodynamics*, Plenum, New York, **1987**.

8 For a normal (Gaussian) error distribution, the RMSE is by a factor of $\sqrt{2}$ larger than the mean absolute error, also denoted as mean unsigned error. The error distribution of log Sw prediction methods appears to be somewhat less inhomogeneous than a Gaussian distribution and typically leads to a ratio of RMSE/mean absolute error of about 1.25 [1]. This should be kept in mind when reading papers on drug solubility prediction and comparing error data.

9 Hansch, C., Quinlan, J. E., Lawrence, G. L. The linear free-energy relationship between partition coefficients and the aqueous solubility of organic liquids. *J. Org. Chem.* **1968**, *33*, 347–350.

10 Yalkowsky, S. H., Valvani, S. C. Solubility and partitioning I: solubility of nonelectrolytes in water. *J. Pharm. Sci.* **1980**, *69*, 912–922.

11 Walden, P. Z. Über die Schmeltzwärme, spezifische Kohäsion und Molekulargröße bei der Schmeltztemperatur. *Z. Elektrochem.* **1908**, *14*, 713–728.

12 Sanghi, T., Jain, N., Yang, G., Yalkowsky, S. H. Estimation of aqueous solubility by the general solubility equation (GSE) the easy way. *QSAR Comb. Sci.* **2002**, *22*, 258–262.

13 Price, S. L. The computational prediction of pharmaceutical crystal structures and polymorphism. *Adv. Drug Deliv. Rev.* **2004**, *56*, 301–319.

14 Delaney, J. S. ESOL: estimating aqueous solubility directly from molecular structure. *J. Chem. Inf. Comput. Sci.* **2004**, *44*, 1000–1005.

15 Wassvik, C. M., Holmen, A. G., Bergström, C. A. S., Zamora, I., Artursson, P. Contribution of solid-state properties to the aqueous solubility of drugs. *Eur. J. Pharm. Sci.* **2006**, *29*, 294–305.

16 Klamt, A., Eckert, F., Hornig, M., Beck, M., Bürger, T. Prediction of aqueous solubility of drugs and pesticides with COSMO-RS. *J. Comp. Chem.* **2002**, *23*, 275–281.

17 In Ref. [16] the terms "fusion" and "crystallization" are mixed up. The equation given here is corrected for this mistake.

18 Klamt, A. Conductor-like screening model for real solvents: a new approach to the quantitative calculation of solvation phenomena. *J. Phys. Chem.* **1995**, *99*, 2224–2235.

19 Klamt, A., Jonas, V., Bürger, T., Lohrenz, C. Refinement and parameterization of COSMO-RS. *J. Phys. Chem.* **1998**, *102*, 5074–5085.

20 Klamt, A. (ed.), *COSMO-RS: From Quantum Chemistry to Fluid Phase Thermodynamics and Drug Design*, Elsevier, Amsterdam, **2005**.

21 Klamt, A., Schüürmann, G. COSMO: a new approach to dielectric screening in solvents with explicit expressions for the screening energy and its gradient. *J. Chem. Soc. Perkin Trans.* **1993**, *2*, 799–805.

22 Putnam, R., Taylor, R., Klamt, A., Eckert, F., Schiller, M. Prediction of infinite dilution activity coefficients using COSMO-RS. *Ind. Eng. Chem. Res.* **2003**, *42*, 3635–3641.

23 Anderson, K. E., Siepmann, J. I. Molecular simulation approaches to solubility. In *Developments and Applications in Solubility*, Letcher, T. (ed.), IUPAC, Oxford, **2007**, pp. 171–184.

24 Lyubartsev, A. P., Jacobsson, S. P., Sundholm, G., Laaksonen, A. Water/octanol systems. A expanded ensemble molecular dynamics simulation study of log *P* parameters. *J. Phys. Chem. B* **2001**, *105*, 7775–7782.

25 Mackerell, A. D., Bashford, D., Bellott, M., Dunbrack, R.L., Jr., Evanseck, J., Field, M.J., Fischer, S., Gao, J., Guo, H., Ha, S., Joseph-McCarthy, D., Kuchnir, L., Kuczera, K., Lau, F.T.K., Mattos, C., Michnick, S., Ngo, T., Nguyen, D.T., Prodhom, B., Reiher, W.E., III, Roux, B., Schlenkrich, M., Smith, J., Stote, R., Straub, J., Watanabe, M., Wiorkiewicz-Kuczera, J., Yin, D., Karplus, M. All-atom empirical potential for molecular modeling and dynamics studies of proteins. *J. Phys. Chem. B* **1998**, *102*, 3586–3616.

26 Jorgensen, W. L., Duffy, E. M. Prediction of drug solubility from Monte Carlo simulations. *Bioorg. Med. Chem. Lett.* **2000**, *10*, 1155–1158.

27 Duffy, E. M., Jorgensen, W. L. Prediction of properties from simulations: free energies of solvation in hexadecane, octanol, and water. *J. Am. Chem. Soc.* **2000**, *122*, 2878–2888.

28 QikProp, version 2.2, Schrödinger, LLC, New York, NY, **2005**.

29 Jorgensen, W. L, Tirado-Rives, J. The OPLS force field for proteins. Energy minimizations for crystals of cyclic peptides and crambin. *J. Am. Chem. Soc.* **1988**, *110*, 1657–1666.

30 Fredenslund, A., Gmehling, J., Rasmussen, P. (eds.), *Vapor Liquid Equilibria Using UNIFAC*, Elsevier, Amsterdam, **1977**.

31 Gmehling, J., Lohmann, J., Jakob, A., Li, J., Joh, R. A modified UNIFAC (Dortmund) model. 3. Revision and extension. *Ind. Eng. Chem. Res.* **1998**, *37*, 4876–4882.

32 Tochigi, K., Minami, S., Kojima, K. Prediction of vapour–liquid equilibria with chemical reaction by analytical solutions of groups. *J. Chem. Eng. Jpn.* **1977**, *10*, 349–354.

33 Hansch, C., Leo, A. J. (eds.) *Substituent Parameters Correlation Analysis in Chemistry and Biology*, Wiley, New York, **1979**.

34 CLOGP program, BioByte Corp., Claremont, CA, USA.

35 Klamt, A., Krooshof, G. J. P., Taylor, R. The surface pair activity coefficient equation: alternative to conventional activity coefficient models. *AICHE J.* **2002**, *48*, 2332–2349.

36 US Environmental Protection Agency, http://www.epa.gov/ttn/atw/hlthef/quinolin.html.

37 http://us.chemcas.com/specification_d/chemicals/supplier/cas/Benzimidazole.asp.

38 COSMOtherm, release C21_01_06, BP-TZVP_01_06 parameterization, COSMOlogic GmbH & Co. KG.

39 QikProp, version 2.5, Schrödinger LLC, New York, NY, USA.

40 We thank Matthew Repasky from Schrödinger Inc. for providing us with the QikProp results.

41 Abraham, M. H., Le, J. The correlation and prediction of the solubility of compounds in water using an amended solvation energy relationship. *J. Pharm. Sci.* **1999**, *88*, 868–880.

42 Values calculated based on original and interpolated Abraham descriptors. Values in brackets are calculated based on Abraham parameters regressed from COSMO σ-moments [65].

43 Burden, F. R. Molecular identification number for substructure searches. *J. Chem. Inf. Comput. Sci.* **1989**, *29*, 225–227.

44 Huuskonen, J. Estimation of aqueous solubility for a diverse set of organic compounds based on molecular topology. *J. Chem. Inf. Comput. Sci.* **2000**, *40*, 773–777.

45 Yalkowsky, S. H., Dannefelser, R. M. The Arizona Database of Aqueous Solubility, College of Pharmacy, University of Arizona, Tuscon, AZ, **1990**.

46 Syracuse Research Corporation. Physical/Chemical Property Database (PHYSPROP), SRC Environmental Center, Syracuse, NY, **1994**.

47 Myrdal, P. B., Manka, A. M., Yalkowsky, S. H. AQUAFAC 3: aqueous functional group activity coefficients; application to the estimation of aqueous solubility. *Chemosphere* **1995**, *30*, 1619–1637.

48 Suloff, E. C Comparative study of semisynthetic derivative of natamycin and the parent antibiotic on the spoilage of shredded cheddar cheese. *Thesis*, Virginia Polytechnic Institute and State University, Blacksburg, VA, **1999**, http://scholar.lib.vt.edu/theses/available/etd-120399-111506/unrestricted/thesisetd.pdf.

49 Hansen, N. T., Kouskoumvekaki, I., Jørgensen, F. S., Brunak, S., Jónsdóttir, S. O. Prediction of pH-dependent aqueous solubility of druglike molecules. *J. Chem. Inf. Model.* **2006**, *46*, 2601–2609.

50 Hou, T. J., Xia, K., Zhang, W., Xu, X. J. ADME evaluation in drug discovery. 4. Prediction of aqueous solubility based on atom contribution approach. *J. Chem. Inf. Comput. Sci.* **2004**, *44*, 266–275.

51 McElroy, N. R., Jurs, P. C. Prediction of aqueous solubility of heteroatom-containing organic compounds from molecular structure. *J. Chem. Inf. Comput. Sci.* **2001**, *41*, 1237–1247.

52 Kühne, R., Ebert, R-U., Schüürmann, G. Model selection based on structural similarity – method description and application to water solubility prediction. *J. Chem. Inf. Model.* **2006**, *46*, 636–641.

53 Chen, X-Q., Cho, S. J., Li, Y., Venkatesh, S. Prediction of aqueous solubility of organic compounds using a quantitative structure–property relationship. *J. Pharm. Sci.* **2002**, *91*, 1838–1852.

54 Hall, L. H., Kier, L. B. Electrotopological state indices for atom types: a novel combination of electronic, topological and valence state information. *J. Chem. Inf. Comput. Sci.* **1995**, *35*, 1039–1045.

55 Tetko, I. V., Tanchuk, V. Y., Kasheva, T. N., Villa, A. E. P. Estimation of aqueous solubility of chemical compounds using E-state indices. *J. Chem. Inf. Comput. Sci.* **2001**, *41*, 1488–1493.

56 Klopman, G., Zhu, H. Estimation of the aqueous solubility of organic molecules by the group contribution approach. *J. Chem. Inf. Comput. Sci.* **2001**, *41*, 439–445, 1096–1097.

57 Wakita, K., Yoshimota, M., Miyamoto, S., Watanabe, H. A method for calculation of the aqueous solubility of organic compounds by using new fragment solubility constants. *Chem. Pharm. Bull.* **1986**, *34*, 4663–4681.

58 Gasteiger, J., Marsili, M. Iterative partial equalization of orbital electronegativity – a rapid access to atomic charges. *Tetrahedron* **1980**, *36*, 3219–3228.

59 Meylan, W. M., Howard, P. H. Atom/fragment contribution method for estimating octanol–water partition coefficients. *J. Pharm. Sci.* **1995**, *84*, 83–92.

60 Private communication, Glen, R. C. Unilever Centre for Molecular Science Informatics, **2006**.

61 Hornig, M., Klamt, A. COSMOfrag: a novel tool for high-throughput ADME property prediction and similarity screening based on quantum chemistry. *J. Chem. Inf. Model.* **2005**, *45*, 1169–1177.

62 Diedenhofen, M., Eckert, F., Klamt, A. Prediction of infinite dilution activity coefficients of organic compounds in ionic liquids using COSMO-RS. *J. Chem. Eng. Data* **2003**, *48*, 475–479.

63 York, C., Kristen, C., Pieraccini, D., Stark, A., Chiappe, C., Beste, Y. A., Arlt, W. Tailor-made ionic liquids. *J. Chem. Thermodynam.* **2005**, *37*, 537–558.

64 Banerjee, T., Singh, M. K., Khanna, A. Prediction of binary VLE for imidazolium based ionic liquid systems using COSMO-RS. *Ind. Eng. Chem. Res.* **2006**, *45*, 3207–3219.

65 Zissimos, A. M., Abraham, M. H., Klamt, A., Eckert, F., Wood J. A comparison between the two general sets of linear free energy descriptors of Abraham and Klamt. *J. Chem. Inf. Comput. Sci.* **2002**, *42*, 1320–1331.

Part V
Lipophilicity

12
Lipophilicity: Chemical Nature and Biological Relevance

Giulia Caron and Giuseppe Ermondi

Abbreviations

2D	two-dimensional
ADMET	absorption, distribution, metabolism, excretion and toxicity
BLW-ED	block-localized wave function energy decomposition
hERG	human ether-a-go-go-related gene
QSAR	quantitative structure–activity relationship

Symbols

pK_a	ionization constant
$\log P$	partition coefficient
$\log D$	distribution coefficient

12.1
Chemical Nature of Lipophilicity

Lipophilicity is a molecular property widely exploited in the pharmaceutical and many other industries. Since its popular definition (a chemical is lipophilic if it dissolves much more easily in lipids than in water, whereas is hydrophilic if the reverse is true, Fig. 12.1) is intuitively perceived, one could argue that lipophilicity is a very simple concept and that no chemistry is required for its understanding. This is entirely false and some essential chemical concepts must be clearly understood before attempting a dive into the abyss of lipophilicity.

12.1.1
Chemical Concepts Required to Understand the Significance of Lipophilicity

12.1.1.1 Molecular Charges and Dipoles

In a molecule, the presence of charges is the result of the formation of bonds that cause an electron flow from the original atoms to the new bonded atoms, and thus

Molecular Drug Properties. Measurement and Prediction. R. Mannhold (Ed.)

ISBN: 978-3-527-31755-4

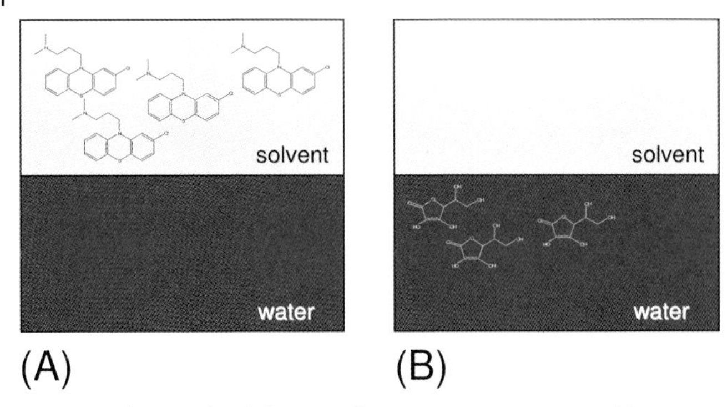

Fig. 12.1 The popular definition of lipophilicity. (A) A chemical is lipophilic if it dissolves much more easily in lipids than in water (e.g. chlorpromazine). (B) A substance is hydrophilic if it dissolves much more easily in water than in lipids (e.g. ascorbic acid).

alters the original electronic distribution. The relationship between electron distribution and molecular charges is well known in computational chemistry and its full analysis is beyond the scope of this chapter. However, some essential notions must be fixed before discussing lipophilicity factorization (Sections 12.1.3 and 12.1.4).

The electronic properties of a molecule are defined through the electron distribution around positively charged nuclei. Detailed information about the electron distribution can either be obtained via experimental results, e.g. X-ray diffraction studies, or through calculations using quantum mechanical methods [1]. Computational procedures provide only the electron density distribution throughout three-dimensional space. For the purpose of obtaining an operative definition of atomic charge, point charges located at the centre of the atom positions are mostly needed. To achieve the transformation, electron density must be converted into the so-called partial or point charges [1]. Numerous methods, both quantum chemical and empirical, have been proposed for partitioning the electron density distribution among the atoms of a molecule [2].

An example of quantum mechanical schemes is the oldest and most widely used Mulliken population analysis [1], which simply divides the part of the electron density localized between two atoms, the overlap population that identifies a bond, equally between the two atoms of a bond. Alternatively, empirical methods to allocate atomic charges to directly bonded atoms in a reasonable way use appropriate rules which combine the atomic electronegativities with experimental structural information on the bonds linking the atoms of interest. A widely used approach included in many programs is the Gasteiger–Hückel scheme [1].

Two additional concepts of relevance and related to charges are the dipole moment and the induced dipole.

The permanent dipole moment $\mathbf{p}$ (Eq. 1, vector entities are in bold type) between two equal but opposite charges, separated by a distance r, is defined as the product of the charge q and the distance $\mathbf{r}$:

$$\mathbf{p} = q\mathbf{r} \tag{1}$$

The direction of the dipole moment is from negative to positive sign.

The induced dipole, $\mathbf{p}_{\text{ind}}$, is a dipole produced by an external electric field $\mathbf{E}'$:

$$\mathbf{E}' = \alpha \mathbf{p}_{\text{ind}} \tag{2}$$

the polarizability α is the ease of distortion of the electron cloud of a molecular entity by an electrical field. In ordinary usage, the term refers to the 'mean polarizability', i.e. the average over three rectilinear axes of the molecule [3].

The link between lipophilicity and point charges is given by intermolecular electrostatic interactions (Sections 12.1.1.2, 12.1.3 and 12.1.4 address this topic) and ionization constants. The mathematical relationships between lipophilicity descriptors and pK_as are discussed in detail in Chapter 3 by Alex Avdeef. Here, we recall how pK_a values are related to the molecular electron flow by taking the difference between the pK_a of aromatic and aliphatic amines as an example. The pK_a of a basic compound depends on the equilibrium shown in Fig. 12.2(A). A chemical effect produces the stabilization or destabilization of one of the two forms, the free energy difference (ΔG) decreases or increases and, consequently,

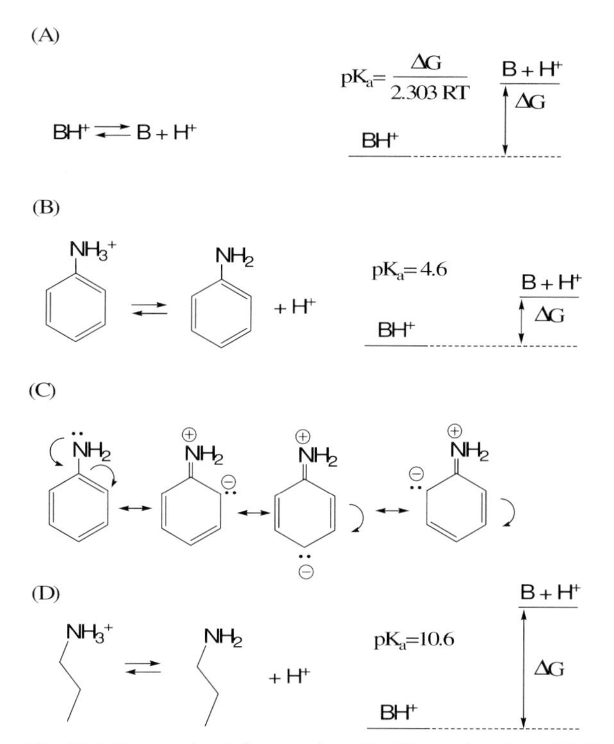

Fig. 12.2 Factors that influence the pK_a. (A) The acid–basic equilibrium governed by the K_a. (B) The acid–base equilibrium in the case of aniline. (C) The structure of aniline and its resonance canonical forms. (D) The acid–base equilibrium in the case of *n*-propylamine.

the pK_a changes in the same direction. For aniline (Fig. 12.2B) the neutral form is stabilized by the resonance effect, which is due to the electron density of the unshared pair that is spread over the ring (Fig. 12.2C). Conversely, protonation does not allow resonance to occur, since the electron pair is involved in the formation of the bond with the acidic hydrogen atom. For the aliphatic amine, no resonance effect takes place either on the neutral or on the ionic form (Fig. 12.2D) and thus, at least to a first approximation, an aliphatic amine is more basic than an aromatic one.

12.1.1.2 **Intermolecular Forces**

In the absence of solvent molecules, the intermolecular forces governing the molecular interactions are essentially of an electrostatic nature and depend on the presence of electrical charges and dipoles in the molecules [3, 4].

The main intermolecular interactions governing chemical processes have been discussed elsewhere [3] and are listed in Table 12.1. The interaction between these charged entities can be treated in term of the electrostatic interactions on the basis of the Coulomb law (T1 in Table 12.1) to obtain the equations of the corresponding potentials T1–T4. T1 can be attractive or repulsive depending on the sign of the two charges involved in the interaction, whereas T2–T4 are always attractive interactions. In general, T1–T4 are the products of three factors: one depends on particle properties (e.g. the charge), the second depends on some constant (e.g. ε_0, k, in certain cases the temperature T) and the third depends on a certain power of the inverse of the distance. There is a fifth attractive contribution present in all molecules due to the presence of an instantaneous electric dipole originated by the continuous motion of the electrons. The interaction between these instantaneous dipoles (called dispersion energy) is attractive and inversely proportional to the sixth power of the intermolecular separation (Table 12.1). The dispersion energy cannot be analyzed by classical mechanics since its origin is purely quantum mechanical. Finally, as the atoms get too close, at some point there is a strong repulsion from overlapping electron clouds and Pauli's exclusion principle whereby filled electron shells of an atom cannot accommodate any more electrons. The repulsive interactions between electron densities are named short-range interactions and simply define the molecular volume.

The nomenclature and equations determining these interactions are often derived by simple systems, not always realistic, as point charges and thus they can be considered as simple interactions. In "real" molecules these interactions are present, but other interactions emerge from the growing complexity of the system. These emerging interactions can be viewed as a combination of the simple forces (combined in Table 12.1). H-bonding and interactions involving π systems represent two important cases (Table 12.1), which are treated in more detail elsewhere [3].

12.1.1.3 **Solvation and Hydrophobic Effect**

The intermolecular interactions listed in Table 12.1 are influenced by the solvent. It is well known that in a medium, Coulomb interactions are reduced

Tab. 12.1 Main intermolecular forces in chemistry.

Simple forces	
Category	***Potential[1] and approximate energy of interaction***
Ion – ion	(T1) $\frac{1}{4\pi\varepsilon_0} \cdot \frac{q_1 q_2}{r}$
Ion – permanent dipole	(T2) $-\frac{1}{(4\pi\varepsilon_0)kT} \cdot \frac{q^2 p^2}{r^4}$
Van der Waals forces	
Permanent dipole – permanent dipole (Keesom forces)	(T3) $-\frac{2}{3}\frac{1}{(4\pi\varepsilon_0)^2 kT} \cdot \frac{p_1^2 p_2^2}{r^6}$
Permanent dipole – induced dipole	(T4) $-\frac{1}{(4\pi\varepsilon_0)^2} \cdot \frac{p^2 \alpha_p}{r^6}$
Induced dipole – induced dipole (Dispersion forces)	Always present and attractive. Depends on $-1/r^6$. Quantum mechanical origin
Repulsive forces	Always present. Quantum mechanical origin. Repulsion derived from overlapping electron clouds.
Combined intermolecular forces	
Category	***Description of the simple forces involved***
Cation-π and π–π	Attractive interactions involving π systems. The interaction energy depends on both the nature of the π system and the nature of the cation. When the ligand is a metal cation, electrostatic forces dominate the interaction. When the ligand is a non-polar molecule (hydrocarbons, etc.) the dispersive interactions dominate. A combination of electrostatic and dispersive forces governs the interaction when the ligand is polar.
Normal and reinforced H-bond	The H-bond is an intermediate range intermolecular interaction between an electron-deficient hydrogen and a region of high electron density. H-bonds result from an electrostatic attraction between a hydrogen atom bound to an electronegative atom X (usually N or O) and an additional electronegative atom Y or a π-electron system.

1 q_i is the charge of the ion i, p_i is the dipole moment of the dipole i, α_p is the polarizability, r is the distance between objects, k is the Boltzmann constant, ε_0 is the permittivity of the free space, T is the temperature; the subscript i is omitted where unnecessary.

because of the presence of solvent molecules. The effect of the medium on Coulomb interactions can be considered by substituting the permittivity of free space in Eqs. T1–T4 (Table 12.1) with the dielectric constant of the medium. The dielectric constant of water is about 80, whereas it is 1 in vacuum. Intermediate dielectric constants of 1–20 are assumed for other organic solvents (octanol = 10, cyclohexane = 2). All electrostatic interactions are affected by changes in the dielectric constant, but the effect is particular evident in stronger interactions, i.e. ion–ion. Since H-bonds, to a first approximation, comprise an important ion–ion contribution, they are also particularly sensitive to this effect.

In a recent paper, Mo and Gao [5] used a sophisticated computational method [block-localized wave function energy decomposition (BLW-ED)] to decompose the total interaction energy between two prototypical ionic systems, acetate and methylammonium ions, and water into permanent electrostatic (including Pauli exclusion), electronic polarization and charge-transfer contributions. Furthermore, the use of quantum mechanics also enabled them to account for the charge flow between the species involved in the interaction. Their calculations (Table 12.2) demonstrated that the permanent electrostatic interaction energy dominates solute–solvent interactions, as expected in the presence of ion species (76.1 and 84.6% for acetate and methylammonium ions, respectively) and showed the active involvement of solvent molecules in the interaction, even with a small but evident flow of electrons (Fig. 12.3). Evidently, by changing the solvent, different results could be obtained.

Apart from acting on the Coulomb interactions, the second effect of the solvent on intermolecular interactions is related to the hydrophobic effect. Briefly, because hydrocarbon molecules (but more generally an apolar moiety of a chemical) are not solvated in water owing to their inability to form H-bonds with water molecules, the latter become more ordered around the hydrocarbons. The resulting increase in solvent structure leads to a higher degree of order in the system than in bulk water and thus a loss of entropy. When hydrocarbon molecules come together they squeeze out the ordered water molecules that lie between them. Since the displaced water is no longer a boundary domain, it reverts to a less-ordered structure, which results in an entropy gain [4].

12.1.2
Lipophilicity Systems

Lipophilicity represents the affinity of a molecule or a moiety for a lipophilic (= fat-loving) environment and is commonly measured by the partition coefficient, P_{aaa} (where aaa represent a generic biphasic system, e.g. oct indicates the standard octanol–water). P is valid for a single electrical species, to be specified (P^{N} for neutral forms and P^{I} for ionized species). The distribution coefficient, expressed as D^{pH}, is a pH-dependent descriptor (Eq. 3) for ionizable solutes and results from the weighted contributions of all electrical forms present at this pH:

Tab. 12.2 Combined *ab initio* quantum and molecular mechanical simulations of solvation of acetate and methylammonium ions in aqueous solution.

Force[1]	*Methylammonium*	*Acetate*
H-bonds[2]	3.5	5.6
Permanent electrostatic[3]	84.6%	76.1%
Polarization[3]	13.0%	20.0%
solvent[4]	86.0%	71.0%
solute[4]	14.0%	29.0%
Charge transfer[3]	2.4%	3.9%
Electron flow[5]	0.025*e* (gained)	0.022*e* (lost)

1 Energy decomposition based on BLW-ED theory [5].
2 Average number of molecules able to form H-bonds with the ion considered.
3 Percentage contribution to total energy.
4 Contribution of solvent and solute polarization to overall polarization effects showing that polarization of the solvent is the main effect but the solute itself is polarized by the aqueous environment.
5 Electron flow across the solute–solvent surface.

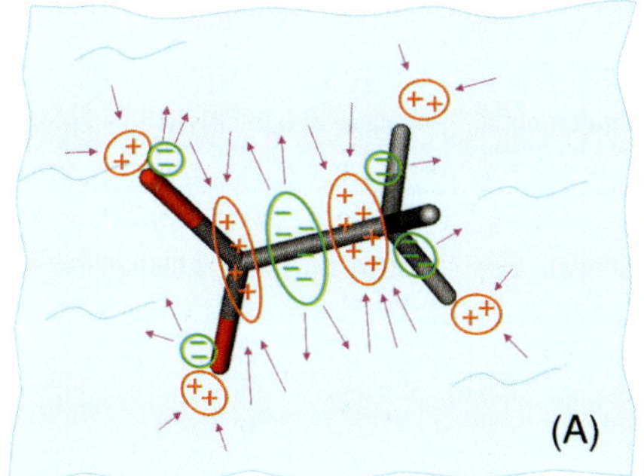

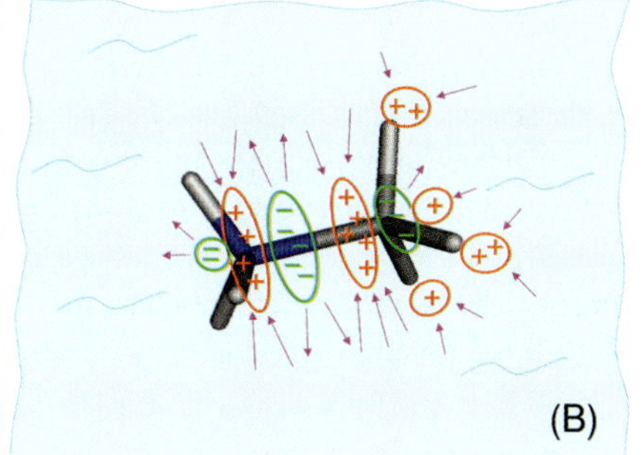

Fig. 12.3 Schematic representation of the charge flow due to polarization effects in the case of (A) acetate (ion loses about 0.022*e*) and (B) methylammonium (ion gains about 0.025*e*) ions in aqueous solution. Purple arrows represent the negative charge flow between ions and water (cyan background). Green regions denote electron loss (minus), whereas bright orange regions denote electron gain (plus).

$$D^{\mathrm{pH}} = f^{\mathrm{N}}P^{\mathrm{N}} + \sum(f^{\mathrm{I}}P^{\mathrm{I}}) \tag{3}$$

where f^{N} and f^{I} are the respective molar fractions of the neutral and ionized forms. Major insights about the practical use of these two descriptors are given in Chapter 16 by Lombardo et al.

Today, lipophilicity can be determined in many systems that are classified by the characteristics of the nonaqueous phase. When the second phase is an organic solvent (e.g. *n*-octanol), the system is isotropic, when the second phase is a suspension (e.g. liposomes), it is anisotropic, and when the second phase is a stationary phase in liquid chromatography, it is an anisotropic chromatographic system [6]. Here, we discuss the main aspects of isotropic and anisotropic lipophilicity and their biological relevance; the chromatographic approaches are investigated in the following chapter by Martel et al.

In very simple terms, an isotropic system shows a more evident separation of the phases than an anisotropic one, but the spatial distribution of ionic charges in the nonaqueous phase is the true factor discriminating isotropic from anisotropic systems (the aqueous phase is always the same, independent of the system). In the former, partial or point charges are carried by salt buffers and thus they have no defined location (see Section 12.1.1.3). In anisotropic media such as liposomes, ionic charges have a fixed location on the polar head of phospholipids. Since each anisotropic system has its own phospholipids features (they may be organized in liposomes, free, linked to high-performance liquid chromatography columns or to other supports [7]), each system shows its own charge features. This is one reason why no standard anisotropic system has yet been defined.

12.1.3
Determination of Log *P* and Log *D*

Log P and log D can be experimentally measured and computationally calculated. Both measurements and calculations can be made by a variety of methods, most of which are quite simple to perform (see following chapters). Our experience recommends, if possible, the use of both procedures. In fact the combination of theory (i.e. how things should be) with practice (i.e. how things are) enables both a better set-up of experiments and the identification of the best predictive method to be used for the chosen dataset of compounds.

12.1.4
Traditional Factorization of Lipophilicity (Only Valid for Neutral Species)

Log P factorization (the decomposition of an object into a product of other objects, or factors, which when multiplied together give the original) demonstrates that lipophilicity is governed by the laws of chemistry (see Section 12.1.1) and thus provides a qualitative chemical insight into partitioning data [8].

Abraham's equation (Eq. 4) is the most common tool to factorize log P^{N} (called simply log P in Eq. 4):

$$\log P = c + eE + sS + aA + bB + vV \tag{4}$$

where E is the excess molar refraction, S is the dipolarity/polarizability, A is the H-bond acidity, B is the H-bond basicity, V is the molar volume of the solute, c is a constant, and e, s, a, b and v are the regression coefficients of the equation and thus vary with the system investigated [9].

As the solute descriptors (E, S, A, B and V) represent the solute influence on various solute–solvent phase interactions, the regression coefficients e, s, a, b and v correspond to the complementary effect of the solvent phases on these interactions. As an example, consider the product aA in Eq. (4). Since A is the H-bond acidity of the solute, a is the H-bond basicity of the system. In other words, the intermolecular forces discussed in Sections 12.1.1.2 and 12.1.1.3 are present in all Abraham's log P factorization equations, with the exception of those interactions involving ions. This is the reason why Abraham's equations are valid for neutral species only.

As previously mentioned the regression coefficients c, s, a, b and v vary with the partitioning system investigated (e.g. a is 0.034 for octanol–water and –3.45 for dodecane–water as reported by ADME Boxes software version 3.5.), whereas E, S, A, B and V are constant for a given solute (Fig. 12.4 shows the value of A, 0.78, for atenolol). This means that the balance between intermolecular forces varies with the system investigated as would be expected from a careful reading of Section 12.1.1.3. This can also be demonstrated by using a completely different approach to factorize log P, i.e. a computational method based on molecular interaction fields [10]. Volsurf descriptors [11] have been used to calculate log P of neutral species both in n-octanol–water and in alkane–water [10].

A comparative analysis of coefficients and descriptors clarifies the relationship between lipophilicity and hydrophobicity (V in Eq. 4 is the molar volume which assesses the solute's capacity to elicit nonpolar interactions (i.e. hydrophobic forces) which, as also clearly stated in the International Union of Pure and Applied Chemistry definitions [3] are not synonyms but, when only neutral species are concerned, may be considered as interchangeable. In the majority of partitioning systems, the lipophilicity is chiefly due to the hydrophobicity, as is clearly indicated by the finding that the product of numerical values of the descriptors V and of the coefficient v is larger in absolute value than the corresponding product of other couples of descriptors/coefficients [9]. This explains the very common linear rela-

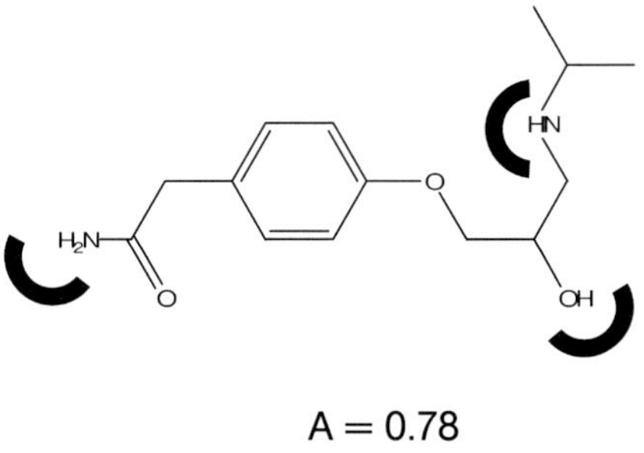

HN
O
H_2N
O
OH

A = 0.78

Fig. 12.4 Atenolol: the molecular moieties that determine the value of A as calculated (0.78) by ADME Boxes software version 3.5 are evidenced in blue.

tionships found among log P^N values determined in different partitioning systems. Moreover, the dominant effect of hydrophobicity outlines the relevance of Δ descriptors (e.g. $\Delta \log P_{aaa-bbb}$ is the difference between the log P values of the same compound in the same electrical state, measured in the systems aaa–water and bbb–water) which cancel the contributions of some parameters (i.e. hydrophobicity) and give relevance to e.g. H-bonding properties; see also Chapter 16.

12.1.5
General Factorization of Lipophilicity (Valid For All Species)

We have not yet introduced the influence of the presence of point charges on the lipophilicity of a chemical. Nevertheless, Sections 12.1.1.2 and 12.1.1.3 do warn that the lipophilic behavior of an ionized molecule might be very different from that of its parent neutral compound. Indeed, in order to investigate the balance of forces governing the lipophilicity of ionized species we must do without Abraham's equations, since they do not exist when ions are considered. Recently, Abraham et al. also demonstrated what had long been perceived intuitively – descriptors for ions are not the same as those for nonelectrolytes [12].

To shed light on this topic, we propose a general equation (Eq. 5) that can be applied to all electrical species and all partitioning systems [3]:

$$\log P = \nu V - \Lambda + I + IE \tag{5}$$

where ν is a constant, V is the molar volume related to the hydrophobic effect as in Eq. (4), Λ accounts for the polar interactions between solute and system, I accounts for ionic interactions between solute and system, and IE represents the intramolecular effects of the solutes which is mainly due to conformational flexibility.

The I term is of particular relevance since, in anisotropic media such as liposomes and artificial membranes in chromatographic processes, ionic charges are located on the polar head of phospholipids (see Section 12.1.2) and thus able to form ionic bonds with ionized solutes, which are therefore forced to remain in the nonaqueous phase in certain preferred orientations. Conversely, in isotropic systems, the charges fluctuate in the organic phase and, in general, there are no preferred orientations for the solute. Given this difference in the I term (but also the variation in polar contributions, less evident but nevertheless present), it becomes clear that log P^I in anisotropic systems could be very different from the value obtained in isotropic systems.

Equation (5) is of more academic than practical use, but demonstrates the critical role that the introduction of a molecular charge has on the lipophilicity of drugs and drug candidates, which often bear an ionization center and may thus be partially or completely ionized at physiological pH.

Some may argue that, in practical terms, we do not often have to do with log P^I, but rather with log D^{pH}. This is true, but for the use of log D^{pH} in mathematical relations with biological data (see Section 12.2) we need to know which species is responsible for the numerical value of the descriptor. In particular for compounds that are more than 90% ionized [6], but also for less ionized ones, we must be aware that D^{pH} is mainly the expression of P^{I} (Eq. 3), thus the ionized species is responsible for the relationship with biological data and not the neutral species.

12.2 Biological Relevance of Lipophilicity

The attraction of lipophilicity in medicinal chemistry is mainly due to Corwin Hansch's work and thus it is traditionally related to pharmacodynamic processes. However, following the evolution of the drug discovery process, lipophilicity is today one of the most relevant properties also in absorption, distribution, metabolism, excretion and toxicity (ADMET) prediction, and thus in drug profiling (details are given in Chapter 2).

12.2.1 Lipophilicity and Membrane Permeation

Permeability is a kinetic property expressed by the permeability coefficient (centimeters per second), a number indicating the rate at which molecules pass from aqueous solution across a membrane to another solution on the other side. Permeability is a molecular property used to screen for more complex absorption processes (i.e. *in vitro* permeability is measured to estimate *in vivo* absorption).

Lipophilicity is intuitively felt to be a key parameter in predicting and interpreting permeability and thus the number of types of lipophilicity systems under study has grown enormously over the years to increase the chances of finding good mimics of biomembrane models. However, the relationship between lipophilicity descriptors and the membrane permeation process is not clear. Membrane permeation is due to two main components: the partition rate constant between the lipid leaflet and the aqueous environment and the flip-flop rate constant between the two lipid leaflets in the bilayer [13]. Since the flip-flop is supposed to be rate limiting in the permeation process, permeation is determined by the partition coefficient between the lipid and the aqueous phase (which can easily be determined by log D) and the flip-flop rate constant, which may or may not depend on lipophilicity; and if it does so depend, on which lipophilicity scale should it be based?

It is our opinion that, among isotropic systems, alongside the standard octanol–water, the alkane–water system (partitioning between water and different alkanes is relatively independent of the alkane used [14]) is the only system that can be successfully used in ADMET prediction, because of its completely different nature from octanol–water. The situation is much more confused for anisotropic systems (see Ref. [7] for a brief review) since no standard system has been defined to date.

However, it is clear that anisotropic systems become relevant in the presence of ionic species (see Section 12.1.4).

To sum up, lipophilicity is only one component of permeability, and thus any relationships found between passive permeation and $\log D_{\text{aaa}}^{\text{pH}}$ are reliable for the investigated series of compounds, but cannot be used to make general predictions.

12.2.2 Lipophilicity and Receptor Affinity

The relevance of lipophilicity in pharmacodynamics is due to the fact that inter- and intramolecular interactions governing lipophilicity (Sections 12.1.1.2 and 12.1.1.3) are of the same nature as those that govern drug recognition and binding to biological sites of action [3, 4, 15].

Two-dimensional (2D) quantitative structure–activity relationship (QSAR) studies have in recent years shown the great importance of hydrophobic effects (see Section 12.1.3 for the relation between lipophilicity and hydrophobicity) expressed as CLOGP (the calculated log *P* by the Hansch and Leo approach) in chemical–biological interactions [16]. In particular, considering data for purified enzymes and receptors with varying degrees of purity, C-QSAR (the database containing 2D QSAR models collected from a careful check of the literature) now contains 2129 examples of which 1164 (55%) lack hydrophobic terms in the form of log *P* or π (the calculated octanol–water partition coefficient of the substituent). For the more complex systems, from organelles to whole organisms, C-QSAR has 3677 examples, of which 2937 (80%) contain hydrophobic terms. Of 709 examples of receptors, only 300 (42%) contain a log *P* or π term [16].

Some considerations may be made from these data. If activity depends on membrane passage, it might generally be hypothesized that the lipophilic character of chemicals could help them to cross cell membranes (see Section 12.2.1) and thus the presence of the Clog *P* term in a QSAR equation might be related to pharmacokinetic aspects. If activity does not depend on membrane passage, the presence of the Clog *P* term would suggest that hydrophobic interactions between ligand and receptors are expected to occur. Conversely, the lack of such a term would suggest that the crucial reaction could occur on the cell surface or that active transport could be involved [16], or else the interaction with a receptor could mainly be of a polar/ionic nature. Finally, the absence of the Clog *P* term in a QSAR model could also be ascribed to the limited variation of the hydrophobic contribution all along the investigated series of compounds.

However, Clog *P* and, more generally, lipophilicity descriptors referring to octanol–water are not the only lipophilicity parameters to be taken into account. As mentioned above, isotropic and anisotropic lipophilicity values gave rise to two different lipophilicity scales for ionized compounds and thus it is recommended to test both of them (after checking the absence of any colinearity) when looking for a QSAR model involving ions.

12.2.3
Lipophilicity and the Control of Undesired Human Ether-a-go-go-related Gene (hERG) Activity

The pharmaceutical industry has recently begun to consider an addition to ADMET profiling, based on the ability of a compound to bind and inhibit the hERG potassium channel [17]. This interest arises from the observation that at least five blockbuster drugs have recently been withdrawn from the market due to reports of sudden cardiac death [17]. A recent paper by Jamieson et al. critically analyses the state of the art of hERG optimization [18] and lists the most common strategies adopted by various research groups to limit hERG activity. Since QSAR data often show series-dependent correlation between potency of hERG block and measures of lipophilicity, the control of log *P* is one of the most widely used strategies in hERG optimization. This is supported by recently published hERG homology models and mutagenesis data suggesting that a lipophilic ligand-binding site exists that is accessed from the intracellular domain. When the polarity of a drug molecule is increased, interaction with the lipophilic cavity is destabilized (Fig. 12.5) [18].

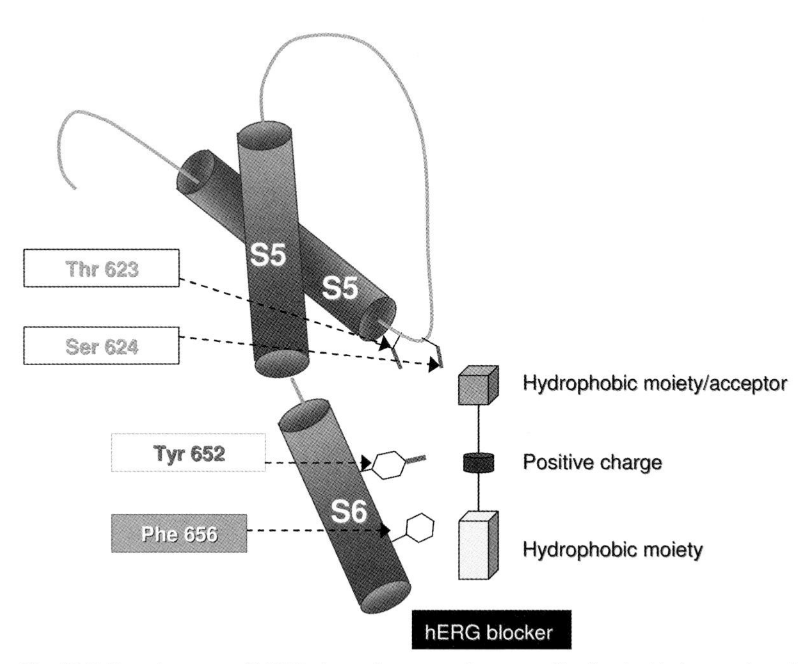

Fig. 12.5 Key elements of hERG channel topology are schematically illustrated using one subunit of the X-ray structure of bacterial KvAP [17]. A hERG blocker is represented according to published evidence [17]. The relevance of hydrophobic interactions between the channel and the blocker justifies the use of log *P* control as one of the most widely used strategies in hERG optimization.

The Jamieson paper reports the results of a number of studies, some successful, others not. Failures can be ascribed to the difficulties encountered in log *P* control. The first evident trouble concerns the choice of the lipophilicity descriptor; many prefer log *P*, but this choice is questionable as has been outlined by Lombardo (see Chapter 16). Secondly, variations in lipophilicity profile influence not only hERG activity, but also target selectivity and also ADMET properties. Lipophilicity is a bulk property and its modification can involve different moieties of the molecules. Once the chemical modulation has been designed, but before moving to the bench, the research group should predict the consequences of this change on each step of the drug's action, but unfortunately this is not always done.

12.3 Conclusions

In recent years we have assisted in the revolution of medicinal chemistry mainly due to the elucidation of the human genome, along with new automated experimental technologies and progress in computer sciences. This revolution demonstrated that a successful drug candidate requires not only good pharmacodynamic properties, but also a suitable pharmacokinetic profile. Curiously, log *P* (and log *D*), in addition to its original application to generate pharmacodynamic 2D QSAR models, is now a key parameter in characterizing ADMET profiles of drugs and drug candidates as well. This finding confirms once again that lipophilicity is a modern medicinal chemistry property since its chemical nature enables us to rationalize a number of biological events, and that lipophilicity is not a very simple concept and a lot of chemistry is required for its understanding.

Acknowledgments

G.C. and G.E. are indebted to the University of Turin for financial support. We also thank Pharma Algorithm (http://www.ap-algorithms.com) for the complimentary copy of the ADME Boxes software (version 3.5, release date November 2006).

References

1 Höltje, H.-D., Sippl, W., Rognan, D., Folkers, G. *Molecular Modeling*, 2nd edn., Wiley-VCH, Weinheim, **2003**.

2 Gross, K. C., Seybold, P. G., Hadad, C. M. Comparison of different atomic charge schemes for predicting pK_a variations in substituted anilines and phenols. *Int. J. Quant. Chem.* **2002**, *90*, 445–458.

3 Caron, G., Scherrer, R. A., Ermondi, G. Lipophilicity, polarity and hydrophobicity. In *ADME/Tox Approaches*, Van de Waterbeemd, H., Testa, B. (vol. eds.), Vol. 5 in *Comprehensive Medicinal Chemistry*,

2nd edn., Taylor, J. B., Triggle, D. J. (eds.), Elsevier, Oxford, **2006**, pp. 425–452.

4 Ermondi, G., Caron, G. Recognition forces in ligand–protein complexes. *Biochem. Pharmacol.* **2006**, *72*, 1633–1645.

5 Mo, Y., Gao, J. Polarization and charge-transfer effects in aqueous solution via *ab initio* QM/MM simulations. *J. Phys. Chem. B Lett.* **2006**, *110*, 2976–2980.

6 Caron, G., Ermondi, G. New insights into the lipophilicity of ionized species. In *Pharmacokinetic Profiling in Drug Research: Biological, Physicochemical and Computational Strategies*, Testa, B., Kraemer, S., Wunderli-Allenspach, H., Folkers, G. (eds.), VHCA, Zurich, **2006**, pp. 165–185.

7 Hartmann, T., Schmitt, J. Lipophilicity – beyond octanol/water: a short comparison of modern technologies. *Drug Discov. Today Technol.* **2004**, *1*, 431–439.

8 Bodor, N., Buchwald, P. Molecular size based approach to estimate partition properties for organic solutes. *J. Phys. Chem. B* **1997**, *101*, 3404–3412.

9 Abraham, M. H., Ibrahim, A., Zissimos, A. M., Zhao, Y. H., Comer, J. E., Reynolds, D. Application of hydrogen bonding calculations in property based drug design. *Drug Discov. Today* **2002**, *7*, 1056–1063.

10 Caron, G., Ermondi, G. Calculating virtual log *P* in the alkane/water system ($\log P_{\mathrm{alk}}^{\mathrm{N}}$ and its derived parameters $\Delta\log P_{\mathrm{oct-alk}}^{\mathrm{N}}$ and $\log D_{\mathrm{alk}}^{\mathrm{pH}}$. *J. Med. Chem.* **2005**, *48*, 3269–3279.

11 Cruciani, G., Crivori, P., Carrupt, P. A., Testa, B. Molecular fields in quantitative structure–permeation relationships: the VolSurf approach. *J. Mol. Struct. Theochem.* **2000**, *503*, 17–30.

12 Abraham, M. H., Zhao, Y. H. Determination of solvation descriptors for ionic species: hydrogen bond acidity and basicity. *J. Org. Chem.* **2004**, *69*, 4677–4685.

13 Thomae, A. V., Wunderli-Allenspach, H., Krämer, S. D. Permeation of aromatic carboxylic acids across lipid bilayers: the pH-partition hypothesis revisited. *Biophys. J.* **2005**, *89*, 1802–1811.

14 Abraham, M. H., Chadha, H. S., Whiting, G. S., Mitchell, R. C. Hydrogen bonding. 32. An analysis of water–octanol and water–alkane partitioning and the Δlog *P* parameter of Seiler. *J. Pharm. Sci.* **1994**, *83*, 1085–1100.

15 Taillardat-Bertschinger, A., Carrupt, P. A., Barbato, F., Testa, B. Immobilised artificial membrane HPLC in drug research. *J. Med. Chem.* **2003**, *46*, 655–665.

16 Hansch, C., Kurup, A., Garg, R., Gao, H. Chem-bioinformatics and QSAR: a review of QSAR lacking positive hydrophobic terms. *Chem. Rev.* **2001**, *101*, 619–672.

17 Aronov, A. M. Predictive *in silico* modeling for hERG channel blockers. *Drug Discov. Today* **2005**, *10*, 149–155.

18 Jamieson, C., Moir, E. M., Rankovic, Z., Wishart, G. Medicinal chemistry of hERG optimizations: highlights and hang-ups. *J. Med. Chem.* **2006**, *49*, 5029–5046.

13
Chromatographic Approaches for Measuring Log *P*

Sophie Martel, Davy Guillarme, Yveline Henchoz, Alexandra Galland, Jean-Luc Veuthey, Serge Rudaz, and Pierre-Alain Carrupt

Abbreviations

ADMET	absorption, distribution, metabolism, elimination and toxicity
CE	capillary electrophoresis
CHI	chromatographic hydrophobicity index
IAM	immobilized artificial membrane
LC	liguid chromatography
LEKC/VEKC	liposome/vesicular electrokinetic chromatography
LSER	linear solvation energy relationship
LSS	linear solvent strength
MEEKC	microemulsion electrokinetic chromatography
MEKC	micellar electrokinetic chromatography
MS	mass spectrometry
ODP	octadecyl polyvinyl
ODS	octadecylsilane
OS	octylsilane
PC	phosphatidylcholine
QSAR	quantitative structure–activity relationship
RPLC	reversed-phase liquid chromatography
SDS	sodium dodecyl sulfate
UV	ultraviolet

Symbols

S	slope
V_D	system dwell volume
$\log P_{oct}$	octanol–water partition coefficient

Molecular Drug Properties. Measurement and Prediction. R. Mannhold (Ed.)

ISBN: 978-3-527-31755-4

13.1
Introduction

Many modern drug discovery methods were developed in the last 15 years to obtain drug candidates with better pharmacokinetic profile in particular in terms of absorption, distribution, metabolism, elimination and toxicity (ADMET) properties [1–3]. Among them, chromatographic methods represent a good alternative to the traditional shake-flask method (for reviews, see, e.g. Refs. [4, 5]). This chapter presents a brief overview of the main features of reversed-phase liquid chromatography (RPLC) (isocratic condition and gradient elution) and capillary electrophoresis (micellar electrokinetic chromatography (MEKC), microemulsion electrokinetic chromatography (MEEKC) and liposome/vesicular electrokinetic chromatography (LEKC/VEKC)) methods used for lipophilicity determination of neutral or neutral form of ionizable compounds ($\log P_{oct}$); estimation of distribution coefficients of ionizable compounds ($\log D_{oct}$) is reviewed in Chapter 16. The determination of lipophilicity using thin-layer chromatography will not be covered here considering the recent excellent review of Gocan et al. [4].

Relationships between lipophilicity and retention parameters obtained by RPLC methods using isocratic or gradient condition are reviewed. Advantages and limitations of the two approaches are also pointed out, and general guidelines to determine partition coefficients in 1-octanol–water are proposed. Finally, more recent literature data on lipophilicity determination by capillary electrophoresis of neutral compounds and neutral forms of ionizable compounds are compiled. Quotation is restricted to key references for every method presented – an exhaustive listing is only given for the last few years.

13.2
Lipophilicity Measurements by RPLC: Isocratic Conditions

RPLC methods were largely used for the determination of $\log P_{oct}$, as illustrated by many well-documented reviews [4–9], due to their well-known advantages even if criticized as not truly replacing shake-flask values [10]. These methods are characterized by their good throughput, the small amounts of compounds required and their general insensitivity to impurities or degradation products which might affect bulk partitioning or analysis. Furthermore, a broader range of $\log P$ can be estimated by RPLC than by the shake-flask method. Considering the large amount of literature on RPLC determination of lipophilicity indices, we restrict this part of the chapter to a brief overview of frequently used technical and experimental conditions, and extend it only in two directions, i.e. the ability of the retention on large number of stationary phases, including immobilized artificial membranes (IAM), to predict $\log P_{oct}$ and some guidelines to select the most useful experimental conditions.

13.2.1
Main Features of RPLC Approaches

13.2.1.1 Principles of Lipophilicity Determination

Based on the partitioning of the solute between a polar mobile and an apolar stationary phase (RPLC), the chromatographic data were expressed as retention factors ($\log k$) given by:

$$\log k = \log[(t_r - t_0)/t_0] \tag{1}$$

where t_r and t_0 are the retention time of the solute and a nonretained compound, respectively.

Retention parameters can be correlated to $\log P_{oct}$ values by:

$$\log P_{oct} = a \log k + b \tag{2}$$

However, some authors also used additional terms (correction term or additional variable) and therefore, the general correlation is given by:

$$\log P_{oct} = a \log k + bX + c \tag{3}$$

where $\log k$ is the retention factor of the solute, $\log P_{oct}$ is the partition coefficient of the solute, X is an additional term, and a, b and c are the linear regression constants.

In practice, coefficients a and b (and c) must be determined for every couple of a stationary phase and a mobile phase (and sometime for a given class of compounds) using retention factors of reference compounds of known $\log P_{oct}$. Then this calibration equation can be used to determine $\log P_{oct}$ of novel compounds.

13.2.1.2 Retention Factors Used as RPLC Lipophilicity Indices

With most reversed phases, pure aqueous solutions as mobile phases cannot be used due to the collapse of the hydrophobic side-chains and/or to the very long elution times. Therefore, an organic modifier has to be added to avoid this phenomenon. However, the use of an organic modifier (typically methanol or acetonitrile) leads to a retention factor measured in a mixture of water and organic solvent. Although the retention factor obtained at a single mobile phase composition, i.e. isocratic $\log k$, can be used as lipophilicity parameter, its value depends on the percentage of organic modifier in the mobile phase. It has been pointed out that at certain modifier concentrations, a solute can be more retained than another one and the situation can be reversed at a different concentration [11].

Thus, as a first approximation, it has been proposed to use $\log k_w$ obtained by extrapolation of a series of isocratic $\log k$ values against the percentage of the organic modifier used in the mobile phase. The relationship between isocratic $\log k$ and organic modifier concentration depends on the organic modifier used (methanol or acetonitrile). In fact, using methanol, $\log k$ values are linearly correlated to the organic modifier percentage (ϕ) in the mobile phase as described by:

$$\log k = \log k_w - S\phi \qquad (4)$$

where $\log k_w$ is the retention factor extrapolated to 100% water as mobile phase ($\phi = 0$) and S (the slope) is a constant for a given solute and a given LC system.

Using acetonitrile, the correlation between isocratic $\log k$ and organic modifier percentages in the mobile phase (ϕ) can be described by the general quadratic equation:

$$\log k = \log k_w + B\phi + A\phi^2 \qquad (5)$$

where $\log k_w$ is the retention factor extrapolated to 100% water as mobile phase ($\phi = 0$), and A and B are constants for a given solute and LC system.

13.2.2 Relation Between Log k_w and Log P_{oct} Using Different Conventional Stationary Phases

Considering recent reviews [4–9] and Chapter 16 devoted to the estimation of log D, this section will focus only on the relationships between lipophilicity (log P_{oct}) and retention factors ($\log k_w$) obtained on a number of stationary phases recently explored for neutral compounds and neutral form of ionizable compounds using the general equations (Eqs. 2 and 3) describing relationships between lipophilicity (log P_{oct}) and retention factor ($\log k_w$). A detailed presentation of stationary phases explored and the associated literature is given in the supplementary material (see Section 13.5).

13.2.2.1 Conventional Apolar Stationary Phases

A summary of the linear relationships existing between log P_{oct} and $\log k_w$ can be found in the supplementary material (see Section 13.5). A part of these relationships cannot be used for the prediction of log P_{oct} values due the low range of lipophilicity covered by the solutes or the weak statistics resulting from the low number of solutes measured and their limited chemical diversity.

Even if it is evident that the retention mechanism cannot exactly model the partitioning process in shake-flask experiments, a number of acceptable correlations between $\log k_w$ and log P_{oct} have been established. Stationary phases for which the slope of the linear relationships is close to unity are promising phases for the estimation of log P_{oct} for new drugs, as the slope is assumed to reflect the degree of similarity between the chromatographic partition system and the 1-octanol–water system [12]. A slope close to 1 indicates that the two processes (partitioning and retention) are homoenergetic, i.e. the Gibbs energy changes are the same [13, 14].

It has been largely demonstrated that the extrapolated logarithm of the retention factors for a pure aqueous mobile phase gives a better correlation with lipophilicity than isocratic retention factors [6, 15], particularly for noncongeneric compounds probably because $\log k_w$ could decrease compounds discrimination due to the H-bonding effect except for very strong H-bond acceptors [9, 16, 17]. However, even for series of congeneric compounds, differences in H-bond capacity could influence the correlations according to the stationary phase employed [6].

At the beginning of the 1980s, several types of stationary phases were investigated, in particular silica gel-bonded phases such as octadecylsilane (ODS). Such stationary phases were largely used, in spite of the poor batch-to-batch reproducibility and the large number of free silanol groups present. The high level of free silanol can lead to specific interactions such as ionic interactions with the solutes especially when they contain a basic nitrogen [18–20]. Therefore ODS stationary phases discriminate compounds according to three classes, i.e. H-bond acceptors, H-bond donors and non-H-bonders. Therefore, great progress has been realized on silica-based stationary phases using different approaches (such as end-capping, encapsulation etc.) as reviewed by Vervoort et al. [21] and Stella et al. [20].

Studies of octylsilane (OS) phases, deactivated by end-capping, have shown that such stationary phases lead to a discrimination between compounds according to their H-bond donor capacity, as the stationary phase presents strong accessible H-bond acceptor groups (-Si-O-Si-) [22, 23]. For OS phases with a uniform matrix of cross-linked polysiloxane alkyl groups, relatively low correlations between $\log k_w$ and $\log P_{oct}$ were found.

Polymer-based stationary phases such as octadecyl polyvinyl-alcohol copolymer (ODP) offer good alternatives to ODS to assess lipophilicity, particularly of basic compounds due to their stability on a large pH range (1–12) and the residual silanol groups which are effectively hindered. Bidentate silane phases such as Zorbax™ Extend-C_{18} demonstrated to accurately estimate $\log P_{oct}$ especially for basic compounds when using high pH [24]. On the other hand, a study conducted with a high-density bonding stationary phase [20] such as luna™ C_{18} shows that this stationary phase discriminates compounds according to their H-bonding ability [17].

Promising stationary phases such as LC-ABZ$^+$Plus or Discovery™ RP-Amide-C_{16} were developed where silanophilic interactions are absent due to electrostatic coating [24]. These polar-embedded phases have been considered better than end-capped ones for lipophilicity estimation. First, the deviation of isocratic $\log k$ observed when the percentage of organic modifier decreases is less pronounced, which results in a decreased extrapolation error [25], and thus an increased correspondence between extrapolated and nonextrapolated $\log k_w$ values (as discussed above). Second, they can be used with pure aqueous mobile phases since the polar groups reduce the hydrophobic collapse generally observed on traditional columns. The above-mentioned phases exhibit good correlation between $\log k_w$ and $\log P_{oct}$ for series of simple compounds which cover a broad property space. Furthermore it has been shown by linear solvation energy relationship (LSER) analysis that the retention mechanism which occurs on such stationary phases is closer to the partition mechanism in traditional shake-flask method than other stationary phases [24].

Chemically bonded stationary phases, e.g. alkylamide silica reversed phases, were also developed. Despite a generally good stability and good quality of resolution and less interactions with free silanol groups, correlations between $\log P_{oct}$ and $\log k_w$ are relatively poor compared to a number of other stationary phases [26]. Finally, monolithic silica stationary phases have also been applied for lipophilicity determination of a series of β-blockers [27].

A solution explored to definitely eliminate the problem of interactions with free silanol groups was the use of non-silica based stationary phases. Many studies recently reviewed [9] were conducted on this class of stationary phases to test their usefulness for lipophilicity prediction. A few of these phases including polybutadiene-coated alumina [4] and octadecyl-bonded alumina show interesting results. Conversely, several of them show poor correlations or correlations not better than those obtained on ODS stationary phases between the retention parameter and $\log P_{oct}$.

13.2.2.2 IAMs

Recently, new types of stationary phases based on IAMs were developed as putatively better mimics of biomembranes [19]. IAM-phases are monolayers of phospholipids covalently bound to silica particles [28]. IAM columns are based on phosphatidylcholine (PC) linked to a silica propylamine surface. As, after coupling of the PC molecules on the surface, residual free propylamino groups still remained, the capacity to predict lipophilicity in 1-octanol–water system was evaluated using IAM columns with end-capped free amino groups (mainly IAM.PC.DD, IAM.PC.DD2, IAM.PC.MG). The relationships between lipophilicity parameters determined on IAM columns ($\log k_{IAMw}$ or $\log k_{IAM}$ depending on the compound's elution time) and $\log P_{oct}$ strongly varied with the class of compounds studied [19]. These studies revealed that partitioning in 1-octanol–water and IAM retention are governed by a different balance of intermolecular interactions [29, 30], especially when ionizable compounds are included. The special behavior of basic or acidic compounds mainly related to different electrostatic interactions [30] are beyond the scope of this chapter.

For neutral compounds, the plots of $\log P_{oct}$ against $\log k_{IAM}$ presented slopes higher than unity for the three columns (see the supplementary material in Section 13.5) suggesting also real, but smaller, differences in the balance of intermolecular interactions between the two systems. A detailed analysis identified polar interactions (H-bond and electrostatic interactions) between phospholipids and small polar neutral compounds ($\log P_{oct} < 2$) as being responsible for these deviations and established that for hydrophobic compounds ($\log P_{oct} > 2$) the relationships between $\log P_{oct}$ and $\log k_{IAM}$ increased significantly when hydrophobic interactions are the dominant intermolecular interactions [19].

Recently, other IAM stationary phases based on different types of phospholipids like phosphatidylserine (IAM.PS), phosphatidylethanolamine (IAM.PE) or sphingomyelin (IAM.SM) were also studied for their ability to estimate lipophilicity. For the same set of neutral compounds it has been demonstrated that the retention mechanism was not influenced by the type of phospholipids used to form the IAM stationary phase (see also the supplementary material in Section 13.5): LSER analyses showed that the solute's size and H-bond acceptor basicity are the two predominant factors in this mechanism, confirming the dual retention mechanisms of polar and nonpolar neutral compounds [31].

13.2.3 Some Guidelines for the Selection of Adequate Experimental Conditions

13.2.3.1 Organic Modifiers

The different organic modifiers used to derive the most suitable mobile phases lead to different parameters namely isocratic $\log k$ and extrapolated $\log k_w$. The extrapolation method has no reality in terms of chromatographic behavior of solutes. However, mainly by correlation with $\log P_{oct}$ (Eqs. 2 and 3) several studies have demonstrated the interest of these extrapolated retention factors as predictors of the lipophilicity of solutes.

By definition, extrapolated $\log k_w$ should be independent from the organic modifier used. However, several studies on different stationary phases showed statistical differences for this parameter when using either methanol or acetonitrile [7, 32]. This phenomenon is probably due to the stationary phase solvation which differs with the organic modifier. However, recent studies on Discovery™ RP-amide-C_{16} established a good correlation between nonextrapolated $\log k_w^{ne}$, retention factors directly measured in 100% aqueous mobile phase and extrapolated $\log k_w^{e}$, retention factors obtained by linear extrapolation of isocratic $\log k$ measured with different methanol–aqueous mobile phases [33]. Moreover quadratic extrapolated $\log k_w$ values obtained from $\log k$ measured for the same compounds with acetonitrile–aqueous mobile phases were also well-correlated with non extrapolated $\log k_w^{ne}$, suggesting that extrapolated $\log k_w$ using different organic modifiers could be used with some specific stationary phases [34] in quantitative structure–property relationship analyses.

It has been demonstrated that the linear (using methanol) or quadratic (using acetonitrile) relationship was limited to a restricted range of solvent percentage. In particular, when using methanol, a curvature is observed for very low volume fractions of organic modifier. Therefore, the range 30–70% is preferred [11].

Some advice can be formulated for the choice of organic modifier. (i) Acetonitrile as an aprotic solvent cannot interact with residual silanols, whereas the protic methanol can. Thus, when measuring retention factors, methanol is the cosolvent of choice, as it reduces the secondary interactions between the solutes and the free silanol groups. (ii) For the study of the performance of new stationary phases one should use acetonitrile, as the effects of free silanol groups are fully expressed [35]. (iii) Acetonitrile with its better elution capacity can be considered as the best organic modifier for lipophilicity measurements of highly lipophilic compounds with adequate stationary phases [36].

13.2.3.2 Addition of 1-Octanol in the Mobile Phase

It has been shown for many RPLC methods that correlations between log P_{oct} and retention parameters were improved by separating compounds in two classes, i.e. H-bond acceptor and donor compounds. Minick et al. [23] propose to add 0.25% (v/v) of 1-octanol in the organic portion of the mobile phase (methanol was preferred in this study) and to prepare the aqueous portion with 1-octanol-saturated water to minimize this discrimination regarding H-bond properties. For a set of heterogenous neutral compounds. the addition of 0.25% (v/v) of 1-octanol in methanol and the use of water-saturated 1-octanol to prepare mobile phase improve the correlation between log k_w obtained on the LC-ABZ column and log P_{oct} [13].

The same conclusions emerged using a Discovery™ RP-amide-C_{16} as stationary phase. The use of 1-octanol increased the correlation between extrapolated log k_w and log P_{oct} for a set of diverse compounds. Furthermore, comparing the two equations obtained with [33] and without [24] 1-octanol showed that slopes of the linear correlations were slightly closer to unity when using 1-octanol. However, preliminary studies evidenced that the presence of 1-octanol in the mobile phase requires careful experimental designs, especially with longer conditioning times at least for Discovery™ RP-amide-C_{16} as stationary phase because the retention factors seem to decrease steadily before reaching a plateau [37].

13.2.3.3 Column Length

In recent years, the need to screen a large number of compounds for high-throughput screening has initiated the use of shorter columns [13, 38]. Their use presents three main advantages: (i) a significant reduction of the analysis time, (ii) a diminution of organic modifier concentration even for lipophilic compounds and (iii) a broadening of log P_{oct} range determination.

For example, with a LC-ABZ 50×4.6 mm stationary phase, Lombardo et al. [13] used 60–70% MeOH to predict log P_{oct} between 3 and 5, whereas using a LC-ABZ 150×4.6 mm stationary phase, the determination of log P_{oct} of less lipophilic compounds (log $P_{oct} < 4$) required 50–80% MeOH. Moreover, according to stationary phases, retention factors can be obtained directly with an aqueous mobile phase with an acceptable analysis time for quite lipophilic compounds (log P_{oct} up to 4) as recently shown with a Discovery™ RP-amide-C_{16} 20×4 mm column [33]. Conversely, nonextrapolated values can be obtained only for a limited range (log P_{oct} less than 1.5) when using Discovery™ RP-amide-C_{16} 150×4 mm columns. Thus, in this case, the analysis time of 26 compounds including conditioning time, solvent change and data processing was reduced from 24 h using a 150×4.6 mm column to 14 h using a 20×4.6 mm column with an increase in quality of results (no extrapolation required). It was also verified on this latter stationary phase that retention factors did not significantly differ between the 150×4.6 and the 20×4 mm column [33].

Actually, the decrease in column length allows to reach log P_{oct} values up to 7.8 using for example an ODP-50 cartridge with a reasonable analysis time [38]. Moreover the use of acetonitrile as organic modifier and a Discovery™ RP-amide-C_{16}

20×4 mm column has recently shown promising results for the determination of $\log P_{oct}$ of very lipophilic compounds [36].

13.2.4 Limitations of the Isocratic Approach for log *P* Estimation

In summary, the estimation of $\log P$ data from isocratic RPLC with extrapolation to pure water elution ($\log k_w$) is a well-accepted method for the determination of lipophilicity. However, isocratic procedures remain tedious and time consuming because numerous chromatographic runs are needed (i.e. between six and eight) within a wide range of isocratic mobile phase conditions (i.e. generally from 30 to 70% of organic modifiers). In addition, some compounds are either insufficiently retained in isocratic conditions and the predictive models are characterized by a too small window of $\log P_{oct}$ values. Hence, the isocratic approach is only recommended for $\log P_{oct}$ between 0 and 6. Alternatively, the use of the organic modifier percentage required to achieve an equal distribution of a compound between the mobile and stationary phase, i.e. ϕ_0, was introduced to replace $\log k_w$. With ϕ_0, a better correlation with $\log P_{oct}$ was observed using a Lichrosorb RP-18 for structurally diverse sets of compounds [39].

13.3 Lipophilicity Measurements by RPLC: Gradient Approaches

Gradient elution procedures have been proposed to substantially speed up the procedures of lipophilicity characterization and to extend the range of measurable $\log P_{oct}$. Several authors assessed comparisons between isocratic and gradient modes [40–43]. Gradient retention times were in good agreement with ϕ_0 values obtained in isocratic mode using 76 diverse drugs [40]. Therefore, gradient elution represents an attractive alternative over isocratic RPLC methods to increase the throughput and extend the investigated $\log P_{oct}$ range. This part will review the potential benefits and limitations of gradient elution to determine chromatographic parameters and related indices that have to be used for $\log P_{oct}$ determination in RPLC.

13.3.1 Gradient Elution in RPLC

Gradient elution in RPLC is performed by a gradual change of the mobile phase composition during the chromatographic run. The gradient elution in RPLC is generally obtained by adding an organic solvent (solvent B, e.g. methanol or acetonitrile) to water or buffer solution (solvent A). As a result, the apparent analyte's retention decreases during the analysis and polar as well as nonpolar compounds can be retained without excessive elution time. Therefore, when samples present a broad polarity range and require different chromatographic conditions, gradient elution is the recommended solution to solve the "general elution problem". Gradient elution in RPLC is compatible with the analysis of very complex samples and can be performed with one generic method to elute variety of compounds [44–46].

Since the first report in 1979 [47] of gradient RPLC use for log *P* determination about 40–50 fundamental and applied articles have been published and recently reviewed [5].

13.3.2 Significance of High-performance Liquid Chromatography (HPLC) Lipophilicity Indices

13.3.2.1 General Equations of Gradient Elution in HPLC

When a limited range of organic modifier proportions within the mobile phase is considered, the retention changes in RPLC with mobile phase composition are generally adequately described by the linear Soczewinski–Snyder model (Eq. 4) adapted to both isocratic and gradient mode [5].

In 1996, Snyder and Dolan elaborated the linear solvent strength (LSS) approach allowing the evaluation of $\log k_w$ from a single gradient run and its precise determination from two gradient runs [48]. From a practical point of view, LSS is the most useful approach to describe theoretical aspects and LSS gradients are convenient for optimization studies. Several commercially available optimization software packages which are able to predict resolution or retention in RPLC are currently based on this approach.

Retention time in a linear-gradient separation (t_r) can be expressed as follow:

$$t_r = (t_0/b)\cdot\log(2.3\cdot k_0\cdot \mathrm{b}+1)+t_0+t_D \tag{6}$$

where k_0 is the k value at the beginning of the gradient (for $\phi=\phi_0$), t_0 is the column dead time (min) which depends on the column dead volume and mobile phase flow rate ($t_0=V_0/D$), t_D is the system dwell time for gradient elution (min). This fundamental parameter to obtain accurate results could be experimentally determined using a 10-min gradient of 0.1% acetone in methanol, according to a procedure described by Dolan [49].

The gradient steepness parameter, b, is obtained by:

$$b = (t_0\cdot\Delta\phi\cdot S)/t_G \tag{7}$$

where t_G is the gradient time from the beginning to the end of the gradient (minutes) and $\Delta\phi$ is the change in ϕ during the gradient [equal to $(\%B_{\mathrm{final}}-\%B_{\mathrm{initial}}]/100)$ ranging from 0 to 1. The parameter b should ideally remain constant throughout the gradient run and has the same value for all the compounds eluted during the chromatographic process.

13.3.3 Determination of $\log k_w$ from Gradient Experiments

Due to instrumental or laboratory variability, the information given by the experimental gradient retention times (t_r) is not sufficient to obtain accurate and repeatable $\log P$ values [38]. As described by Eqs. (4), (6) and (7) the lipophilicity measurement in gradient mode is therefore based on the determination of $\log k_0$ and S because both terms are necessary to obtain $\log k_w$.

13.3.3.1 From a Single Gradient Run

Assuming that S is roughly constant for structurally similar compounds (typical value of $S=4$ for samples with molecular weight less than 1000) [48], Eq. (6) allows us to estimate $\log k_w$ from a single gradient run, taking into account both the parameters of experimental conditions (t_G, V_o, F, V_D, ϕ, ϕ_0) and the experimentally observed retention time of the compound in gradient mode (t_r). This rapid determination method was principally described and used by Valko et al. [11, 40]. A fast gradient (total gradient time below 15 min) was applied in order to limit the influence of the S parameter on compounds retention time. Estimate on the basis of a single initial gradient run is however limited by the assumption of a constant S and of errors not larger than the S variability (±0.8 units) according to Ref. [48].

13.3.3.2 From Two Gradient Runs

More accurate values of S and $\log k_w$ could be obtained from two gradient runs. For this purpose, two gradients with a different t_G are necessary (20 and 60 min are acceptable for a conventional LC column geometry) [9]. This methodology was recently reviewed by Poole et al. [7].

13.3.3.3 With Optimization Software and Two Gradient Runs

As previously described, Eq. 6 contains two constants characteristic of the system and the sample, k_0 and S, which can be determined by two chromatographic runs differing only in t_G. These two values allow to calculate $\log k_w$ using Eq. 4. However, because there is no empirical solution, values of $\log k_w$ and S have to be computed by iteration. Such procedures are included in several commercially available LC software packages, such as Drylab (Rheodyne, CA, USA), Chromsword (Merck, Darmstadt, Germany), ACD/LC simulator (Advanced Chemical Development, Toronto, Canada) or Osiris (Datalys, Grenoble, France). This approach was comprehensively described and successfully applied for accurate $\log P$ determination of several solutes with diverse chemical structures [9, 12, 43, 50].

13.3.4 Chromatographic Hydrophobicity Index (CHI) as a Measure of Hydrophobicity

The ϕ_0 scale approach introduced in 1993 by Valko and et al. [39] for the isocratic determination of hydrophobicity was extended to the CHI in gradient mode [40].

The CHI parameter approximates the percentage of organic modifier in the mobile phase for eluting the compounds and can be used for high-throughput determination of physicochemical properties (50–100 compounds per day). CHI is a system property index, and depends on the nature of the stationary phase and the organic modifier as well as the pH of the mobile phase for ionizable compounds.

13.3.4.1 Experimental Determination of CHI

The CHI can be obtained without preliminary method development directly from a single fast-gradient run with a cycle time less than 15 min with a 150-mm column [40] or 5 min with 50-mm column [42]. In this case, the obtained retention time, t_r, is expressed within an organic phase concentration (ϕ_0) scale using a calibration set of compounds. CHI value can be obtained from:

$$\mathrm{CHI} = At_r + B \tag{8}$$

where A and B are derived from the calibration set of analytes (i.e. around 10 compounds).

The values ϕ_0 for the calibration standards are estimated by plotting $\log k$ values as a function of organic modifier concentrations assessed from at least three isocratic experiments and obtained with [39]:

$$\phi_0 = -\log k_w / S \tag{9}$$

In case of fast gradient (below 15 min), S could be considered constant for all the investigated molecules and will only have a small influence on the retention time of the compounds. Thus, the gradient retention times, t_r, of a calibration set of compounds are linearly related to the ϕ_0 values [39]. Moreover, Valko et al. also demonstrated that the faster the gradient was, the better the correlation between t_r and ϕ_0 [40]. Once the regression model was established for the calibration standards, Eq. 8 allowed the conversion of gradient retention times to CHI values for any compound in the same gradient system. Results are then suitable for interlaboratory comparison and database construction. The CHI scale (between 0 and 100) can be used as an independent measure of lipophilicity or also easily converted to a $\log P$ scale.

13.3.4.2 Advantages/Limitations of CHI

The CHI index is reportedly a relevant parameter in quantitative structure–activity relationship (QSAR) studies [41]. With this approach, $\log P$ could be determined in the range $-0.45 < \log P < 7.3$ [40]. CHI was determined for more than 25 000 compounds with excellent reproducibility (within ±2 index units) and reported in a GlaxoSmithKline database [11]. Two main drawbacks were identified using this approach: (i) the assumptions used in Ref. [7], i.e. that S is constant for all compounds and that the system dwell volume V_D is excluded in calculations, yield some discrepancies in the resulting $\log P$, and (ii) the set of gradient calibration standards is not absolute and needs to be carefully selected to perfectly align ϕ_0 values with CHI scales [41].

13.3.5
Experimental Conditions and Analysis of Results

13.3.5.1 Prediction of log *P* and Comparison of Lipophilicity Indices

As in isocratic mode, the estimate of log *P* is indirect and based on the construction of a linear retention model between a retention property characteristic of the solute ($\log k_w$) and a training set with known $\log P_{oct}$ values. To assess the most performing procedures, the three hydrophobicity indexes (ϕ_0, CHI and $\log k_w$) were compared on the basis of the solvation equation [41]. These parameters were significantly inter-related with each other, but not identical. Each parameter was related to log *P* with r^2 values between 0.76 and 0.88 for the 55 tested compounds; fitting quality associated with the compound nature.

The results obtained in gradient mode by Kaliszan et al. for $\log k_w$ determination are based on a relatively complex mathematical treatment, using RPLC optimization software. As described, log *P* values were determined in the range $-2 < \log P_{oct} < 7$. The correlation coefficient between $\log k_w$ and log *P* was satisfactory, and included in the range $0.94 < r^2 < 0.98$.

Concerning the CHI approach developed by Valko et al., a much larger set of compounds was investigated (about 25 000). The CHI measurement takes only 5–15 min and covers a range of 7–8 log *P* units according to the nature of the stationary and mobile phases. Simple data processing can be used to convert the gradient retention time into CHI. Correlation coefficients between CHI and log *P* are lower ($0.81 < r^2 < 0.88$) than for $\log k_w$, but CHI can also be used successfully as an independent measurement of lipophilicity or improved using solvatochromic parameters [42]. In several studies, Valko revealed that the major difference between log *P* and CHI lipophilicity scales was their sensitivity towards the H-bond acidity of the compounds [7, 11]. To tackle this problem, CHI values of uncharged compounds were measured and the H-bond acidity term computed from *in silico* methods (e.g. ABSOLV; Sirius Analytical Instruments, East Sussex, UK). As expected, correlating log *P* with CHI and $\Sigma\alpha 2H$ improved the prediction of log *P* ($r^2 = 0.94$) for the 86 tested compounds. The main limitation of this approach is associated with the quality of H-bond molecular descriptors [51, 52].

In another study, Valko et al. presented correlations between various retention data based on 62 compounds [41]. log *P* and $\log k_w$ were highly correlated ($r^2 = 0.88$) while correlations between CHI and log *P* ($r^2 = 0.85$) and between ϕ_0 and log *P* ($r^2 = 0.76$) were less significant. On the opposite, Poole et al. concluded that isocratic separations provided the most accurate results for log *P* values in the range −0.1 to 4 while gradient elution was more suitable for fast separations of compounds with large log *P* values [43]. The standard error measured for a set of 30 log *P* values was 0.135 log units in the isocratic mode ($r^2 = 0.99$) and 0.288 in the gradient mode ($r^2 = 0.94$). It has to be noted that the gradient procedure was applied for lipophilicity determination of very hydrophobic compounds such as tetrachlorobenzyltoluenes [53], with an average log *P* higher than 7 or 55 aurones and flavonoides with log *P* values between 3 and 8 [54]. Lipophilicity of peptides with log *P* values as low as −10 was also investigated with a gradient approach [55]. Even if a controversy remains for selecting isocratic or gradient procedures, several authors chose the gradient approach for QSAR studies [53–57].

13.3.6
Approaches to Improve Throughput

A major disadvantage of gradient elution in terms of fast analysis remains the time to adequately equilibrate the chromatographic column between two experiments. However, Carr et al. recently demonstrated an excellent repeatability (±0.002 min in retention time) obtained with two column volumes of re-equilibration instead of the usual 10 column volumes when a small amount of ancillary solvent (1–3% of 1-butanol or 1-propanol) is added to the mobile phase [45, 58].

In order to speed up the log *P* estimate by gradient RPLC, several solutions were investigated, i.e. fast gradient elution obtained with the help of monolithic supports or short columns and the use of mass spectrometry (MS).

13.3.6.1 Fast Gradient Elution in RPLC

The initial configuration proposed by Valko et al. for log *P* gradient determination was based on a gradient cycle time of about 15 min with a 150-mm column [40]. This procedure was modified by Mutton, who stated that resolution could be maintained when the gradient time and/or column length were reduced or the flow rate increased [59, 60].

In 2001, Valko et al. reduced the column length to only 50 mm and increased the flow rate to 2 mL min^{-1} [42]. The gradient time was diminished to 2.5 min with a gradient cycle time of 5 min. Measurement of CHI and evaluation of log *P* were excellent with a 3-fold improved productivity. In these conditions, the system dwell volume (V_D) becomes essential and only dedicated chromatographic devices with V_D lower than 0.8 mL can be used [42]. Special attention should be paid to the injected volume, which must remain lower than 3 µL to avoid any overloading or extra-column volume contributions.

Donovan and Pescatore described another fast-gradient approach with very short columns (20 × 4.6 mm internal diameter) packed with a porous polymer (known as ODP columns) [38]. This chromatographic support presents a high chemical stability and can be used at pH 2, 10 or 13 to analyze neutral analytes. This procedure allowed a relatively high flow rate (2 mL min^{-1}) and a gradient from 10 to 100% methanol in only 7 min. The mathematical treatment was simplified and based on the direct transformation of retention time to log *P*. For this purpose, two standards (toluene and triphenylene) were used to minimize retention time variations from run-to-run and instrument-to-instrument, and to facilitate the scaling of gradient retention time. This method was described to provide fair accuracy and good precision (equal or better than ±0.01 log *P* units) for compounds with log *P* values between 2 and 6. However, several drawbacks were identified: (i) it is difficult to separate the two standards from the compounds of interest with the relatively low efficiency value afforded by the short column length (N_{max} = 1500 theoretical plates), (ii) the limited column resistance leads to retention changes and variability in log *P* determination, and (iii) due to the column capacity, a sample overloading could induce inaccurate retention times.

Nevertheless, this method was successfully applied by Gulyaeva et al. for the log *P* and log *D* determination of 15 β-sympatholytic drugs [56]. Another study by Welerowicz and Buszewski compared the lipophilicity values of β-blockers obtained with a column made of a monolithic-silica C_{18} with a conventional porous silica particles C_{18} as reference material [27]. A modified method was used for evaluating log *P* with two main differences: (i) log k_g was considered rather than retention times, and (ii) benzene and butyl-benzene were used as calibration compounds. In this study, chromatographic experiments were 10 times faster with the monolithic column and results were equivalent to those obtained with the silica-based columns. This approach could be further optimized with faster gradient since flow rate should be increased by a factor 3 or 7 compared to conventional C_{18} supports [61, 62] and gradient time reduced by the same factor [63] to fully exploit the potential of monolithic supports.

13.3.6.2 **Use of MS Detection**

Camurri and Zaramella adapted the methodology of Valko et al. [40] with MS detection [64]. In the experimental set-up the ultraviolet (UV) detector was substituted by a single-quadrupole MS instrument. The MS detection allows a reduction of the analysis time because CHI can be simultaneously determined for a mixture of molecules. Thirty-two drugs, as the maximum number of channels available on the MS unit, were analyzed and the throughput increased by the same factor. CHI values determined by both LC-UV and LC-MS differed by only 0–3 CHI units (CHI scale between 0 and 100). Another advantage using LC-MS methodology concerned compounds with poor UV absorption. In 2004, Valko reported that Comgenex (Budapest, Hungary) adopted this method for the rapid determination of lipophilicity [11].

Some studies reported the possible sample lipophilicity, integrity and purity determination with gradient techniques and MS detection for several structurally diverse drugs [65, 66]. The clear advantages of the MS detection method are limited by the fact that each analyte should possess a distinct *m*/*z* value. Therefore, isobaric compounds cannot be distinguished unless they are chromatographically separated or high-resolution MS is used since it is able to assess highly accurate *m*/*z* measurements. Another limitation concerns compound polarity because the most widely used ionization source (electrospray ionization) is not adapted for nonpolar compounds. Therefore, dual ionization sources (electrospray ionization and atmospheric pressure chemical ionization) could be a solution to overcome this issue.

13.3.7
Some Guidelines for a Typical Application of Gradient RPLC in Physicochemical Profiling

In summary, gradient RPLC is a fast and convenient method for the estimation of lipophilicity. An additional advantage of the gradient system over isocratic determination is the extended range of measurable log *P*; in isocratic mode, determination could be assessed in the log *P* range of 0–6, while in gradient mode, the range could be extended from 2 to 8. Two approaches to determine log *P* in gradient mode are mainly used, i.e. the determination of $\log k_w$ or CHI. The latter is based on a single fast-gradient experiment and represents the fastest approach. The method using $\log k_w$ is more accurate, but requires two gradient runs and a relatively complex mathematical treatment, generally performed with LC optimization software. To fully benefit from this methodology some conditions have to be fulfilled.

13.3.7.1 A Careful Selection of Experimental Conditions

- *Stationary phase.* Supelcosil C_{18} ABZ (Supelco Scientific, Bellefonte, PA, USA) was the most often employed support and gave the best correlations. This stationary phase should be selected in a first instance with a geometry adapted to the application: for conventional gradient experiments, supports of 150 × 4.6 mm, 5 µm represent a good choice while a shorter column (i.e. 50 mm or lower) with smaller particle size (i.e. 3–3.5 µm) must be preferentially selected for fast gradient analysis.
- *Flow rate.* Its value must be selected according to the column geometry and back-pressure limitations of the system. For 4.6 mm internal diameter columns (5 µm), 1 mL min^{-1} is adapted for a conventional length, while for short columns packed with smaller particles the flow rate can be increased.
- *Injected quantity.* When column length or internal diameter is reduced, the injected volume should be decreased proportionally to avoid any overloading of the column and variability of retention times.
- *Gradient profile.* The gradient profile must be as wide as possible since structure and retention of studied compounds are generally unknown. Silica-based materials are, however, not adapted to work with pure water or pure organic modifier. For long-term stability of the support, a 5–95% organic modifier gradient range should be preferred. The gradient time must be set according to the flow rate and column geometry. For this purpose, several rules and equations were proposed by Dolan and Snyder [48, 63]. For a good repeatability of the gradient without adaptation of mobile composition, gradient re-equilibration time must be equal at least to 10 column volumes.
- *pH.* pH is one of the most critical parameters for a successful determination of the lipophilicity of the neutral form of ionizable compounds. For this reason, pH must be higher than $pK_a + 2$ for bases and lower than $pK_a - 2$ for acidic compounds. Some problem of column

instability could be observed at extreme pH conditions with basic compounds and new chromatographic supports compatible with high pH applications could be chosen.

- *Chromatographic system*. Generally, a conventional HPLC instrument is sufficient. However, it is important to accurately estimate the column dead volume V_0 and the system dwell volume V_D. V_0 depends on the column geometry and could be estimated mathematically with the following equation ($V_0 = \pi d_c^2 / 4L\varepsilon$) or approximated by the injection of an unretained compound (i.e. uracil, $NaNO_3$ or $LiNO_3$). V_D is specific of the used chromatographic system and should be experimentally obtained with the method described by Dolan [49]. For fast-gradient experiments, high detector acquisition rates and a low system dwell volume are mandatory.
- *Detection*: UV detection represents the less expensive and the most widespread approach but is limited to analytes possessing chromophoric moieties. When the analytical cost is not an issue, MS detection should be used preferentially because of its quasi-universality and selectivity.

13.3.7.2 General Procedure for $\log k_w$ Determination

Two gradients are necessary for this approach. A first gradient should have a slope between 1 and 4% min^{-1} and the second gradient should possess exactly the same boundaries (initial and final compositions), but with a slope 3 times higher. The selection of the operating conditions should be made with an optimization software. Retention times of the analyte must be monitored for both gradients to find the corresponding model for each compound, according to Eq. 4. The lipophilicity of unknown compounds can thus be determined from the plot of $\log k_w$ versus $\log P$.

13.3.7.3 General Procedure for CHI Determination

In this approach, the mathematical treatment is simplified. It is recommended to perform a fast gradient with short columns and high flow rate. Only a single gradient is necessary to determine CHI. Retention time of the compounds is obtained and the mathematical treatment is achieved as described above with the help of several standards.

13.4 Lipophilicity Measurements by Capillary Electrophoresis (CE)

CE was early identified as a powerful experimental technique to monitor the partitioning of neutral solutes and the neutral form of ionizable compounds [67]. In the search of partitioning parameters closer than $\log P_{oct}$ to the pharmacokinetic behavior of drug compounds [68] several different CE experimental conditions were explored and recently reviewed in detail [5, 69].

Lipophilicity determination by CE exhibits the same advantages as all chromatographic methods, i.e. insensitivity to impurities, fast analysis time, low cost and potential for automation. In addition, for high-throughput screening, the CE methods appear very attractive because of the much lower sample consumption than traditional RPLC approaches. However, CE methods request special treatments to handle ionizable compounds [69] since the migration of all electrical forms of an ionizable compound has to be taken into account to derive pertinent physicochemical profiles (pK_a and $\log P$) for this large class of solutes. Detailed discussion of ionizable compounds is out of the scope of this chapter. Here, only the key features and the most interesting recent results applied to neutral compounds will be reported for three experimental modes, i.e. MEKC, MEEKC and VEKC/LEKC.

13.4.1 MEKC

MEKC is a CE mode based on the partitioning of compounds between an aqueous and a micellar phase. This analytical technique combines CE as well as LC features and enables the separation of neutral compounds. The buffer solution consists of an aqueous solution containing micelles as a pseudo-stationary phase. The composition and nature of the pseudo-stationary phase can be adjusted but sodium dodecyl sulfate (SDS) remains the most widely used surfactant.

Similarly to RPLC, there is a strong linear correlation between the logarithm of the retention factor measured by MEKC ($\log k$) for neutral solutes or the neutral form of ionizable compounds and $\log P_{oct}$. As shown in the supplementary material (see Section 13.5), numerous papers reported good correlations between $\log P_{oct}$ and $\log k$.

Practically, the retention factor k of a neutral solute can be calculated from MEKC measurements:

$$k = \frac{t_r - t_{EOF}}{\left(1 - \frac{t_r}{t_{MC}}\right) \cdot t_{EOF}} \tag{10}$$

where t_r, t_{EOF} and t_{MC} are the migration time of the solute, the electroosmotic flow marker and the micelle marker, respectively. Highly water-soluble neutral compounds such as acetone or dimethylsulfoxide are used as electroosmotic flow markers, whereas highly water-insoluble compounds such as dodecaphenone are used as micelle markers.

QSAR studies showed high correlations between bioactivity and $\log k$ in MEKC using bile salt surfactants and mixed bile salt systems [70, 71], and micellar pseudo-stationary phases are considered to mimic better biological membranes than 1-octanol or RPLC stationary phases [69]. However, the success of MEKC to estimake $\log P_{oct}$ remains limited. Indeed, congeneric behavior was observed for several groups of compounds mainly with SDS micelles but also with other types of surfactants [70, 72]. This congeneric behavior was confirmed by LSER results to be due to the differences in H-bond donor capacity and dipolarity/polarizability of the various surfactants systems used compared with 1-octanol–water system [7, 73].

13.4.2 MEEKC

MEEKC is a CE mode similar to MEKC, based on the partitioning of compounds between an aqueous and a microemulsion phase. The buffer solution consists of an aqueous solution containing nanometer-sized oil droplets as a pseudo-stationary phase. The most widely used microemulsion is made up of heptane as a water-immiscible solvent, SDS as a surfactant and 1-butanol as a cosurfactant. Surfactants and cosurfactants act as stabilizers at the surface of the droplet.

Similarly to MEKC, strong linear correlations between the $\log k$ measured by MEEKC and the $\log P_{oct}$ were reported by several authors as shown in the supplementary material (see Section 13.5).

Practically, the retention factor k can be calculated from MEEKC measurements in the same way as from MEKC measurements (Eq. 10). Calibration curves of $\log P_{oct}$ against $\log k$ can be constructed with a universal set of standard compounds and then successfully employed for the $\log P_{oct}$ estimation [74–80]. Microemulsion systems of low and high pH are used for the $\log P_{oct}$ determination of the neutral form of acids and bases, respectively. Separations at low pH require either negative polarity [81], or negatively coated capillaries [82–85] to overcome the low electro-osmotic flow. As for rapid approaches, pressure-assisted MEEKC can improve the throughput to two samples per hour using a single capillary system [75] and 96-capillary multiplexed instruments can improve throughput to 46 samples per hour [78, 79]. A miniaturization attempt by microchip MEEKC with indirect fluorimetric detection also provided an accurate determination of $\log P_{oct}$ for several analytes [80].

The SDS/1-butanol/heptane microemulsion system provided a better estimation of $\log P_{oct}$ than the SDS micellar system. LSER analyses confirmed this microemulsion system to be more similar to phospholipids vesicles than SDS micellar systems [86]. Moreover, QSAR studies using electrophoretic migration data achieved by MEEKC provided better correlations than those using conventional $\log P_{oct}$ [86–88]. The MEEKC method is becoming a more and more important method to estimake $\log P_{oct}$ due to higher throughput, lower sample consumption and higher accuracy than LC [89].

13.4.3 LEKC/VEKC

Other electrokinetic chromatography techniques using liposomes (LEKC) and vesicles (VEKC) have also been applied for $\log P_{oct}$ estimation. Liposomes and vesicles are organized structures containing continuous bilayers of monomers enclosing an aqueous core region.

Linear correlations between the log k measured by LEKC/VEKC and the log P_{oct} were reported by few authors [77, 90, 91].

The log k obtained by MEKC, MEEKC and LEKC were also compared with membrane permeability reference data by Örnskov et al. [92]. An improved correlation was obtained in the order: MEKC > MEEKC > LEKC. Thus, LEKC appears to provide experimental conditions that mimic more closely physiological membranes [93]. However, liposomes and vesicles remain unstable and difficult to prepare reproducibly. Their use is then devoted to some particular applications.

13.5 Supplementary Material

In order to offer a permanently updated information, we have created a website devoted to lipophilicity measurements via chromatographic methods. Literature and main experimental data associated with this chapter are accessible at http://www.unige.ch/pharmacochimie.

References

1 Testa, B., Krämer, S. D., Wunderli-Allenspach, H., Folkers, G. (eds.). *Pharmacokinetic Profiling in Drug Research. Biological, Physicochemical, and Computational Strategies*, Wiley-VCH, Weinheim and VHCA, Zurich, **2006**.

2 Martel, S., Castella, M. E., Bajot, F., Ottaviani, G., Bard, B., Henchoz, Y., Gross Valloton, B., Reist, M., Carrupt, P. A. Experimental and virtual physicochemical and pharmacokinetic profiling of new chemical entities. *Chimia* **2005**, *59*, 308–314.

3 Lombardo, F., Gifford, E., Shalaeva, M. Y. *In silico* ADME prediction: data, models, facts and myths. *Mini-Rev. Med. Chem.* **2003**, *3*, 861–875.

4 Gocan, S., Cimpan, G., Comer, J. Lipophilicity measurements by liquid chromatography. *Adv. Chromatogr.* **2006**, *44*, 79–176.

5 Nasal, A., Kaliszan, R. Progess in the use of HPLC for evaluation of lipophilicity. *Curr. Comput-Aided Drug Des.* **2006**, *2*, 327–340.

6 Van de Waterbeemd, H., Kansy, M., Wagner, B., Fischer, H. Lipophilicity measurement by reversed-phase high performance liquid chromatography. In *Lipophilicity in Drug Action and Toxicology*, Pliska, V., Testa, B., Van de Waterbeemd, H. (eds.), VCH, Weinheim, **1996**, pp. 73–87.

7 Poole, S. K., Poole, C. F. Separation methods for estimating octanol–water coefficients. *J. Chromatogr. B* **2003**, *797*, 3–19.

8 Kaliszan, R., Haber, P., Baczek, T., Siluk, D., Valko, K. Lipophilicity and pK_a estimates from gradient high-performance liquid chromatography. *J. Chromatogr. A* **2002**, *965*, 117–227.

9 Nasal, A., Siluk, D., Kaliszan, A. Chromatographic retention parameters in medicinal chemistry and molecular pharmacology. *Curr. Med. Chem.* **2003**, *10*, 381–426.

10 Sangster, J. *Octanol–Water Partition Coefficients: Fundamentals and Physical Chemistry*, Wiley, Chichester, **1997**.

11 Valko, K. Application of high-performance liquid chromatography based measurements of lipophilicity to model biological distribution. *J. Chromatogr. A* **2004**, *1037*, 299–310.

12 Kaliszan, R., Nasal, A., Markuszewski, M. J. New approaches to chromatographic determination of lipophilicity of xenobiotics. *Anal. Bioanal. Chem.* **2003**, *377*, 803–811.

13 Lombardo, F., Shalaeva, M. Y., Tupper, K. A., Gao, F., Abraham, M. H. ElogPoct: a tool for lipophilicity determination in drug discovery. *J. Med. Chem.* **2000**, *43*, 2922–2928.

14 Lombardo, F., Shalaeva, M. Y., Tupper, K. A., Gao, F. ElogDoct: a tool for lipophilicity determination in drug discovery. 2. Basic and neutral compounds. *J. Med. Chem.* **2001**, *44*, 2490–2497.

15 Kaliszan, R. Quantitative-structure–retention relationships applied to reversed-phase high-performance liquid chromatography. *J. Chromatogr.* **1993**, *656*, 417–436.

16 Yamagami, C. Recent advances in reversed-phase-HPLC techniques to determine lipophilicity. In *Pharmacokinetic Optimization in Drug Research. Biological, Physicochemical, and Computational Strategies*, Testa, B., Van de Waterbeemd, H., Folkers, G., Guy, R. H. (eds.), VHCA, Zurich, **2001**, pp. 383–400.

17 Du, C. M., Valko, K., Bevan, C., Reynolds, D., Abraham, M. H. Rapid method for estimating octanol–water partition coefficient ($\log P_{OCT}$) from isocratic RP-HPLC and a hydrogen bond acidity term (A). *J. Liquid Chromatogr. Rel. Technol.* **2001**, *24*, 635–649.

18 Stella, C., Seuret, P., Rudaz, S., Carrupt, P. A., Gauvrit, J. Y., Lanteri, P., Veuthey, J. L. Characterization of chromatographic supports for the analysis of basic compounds. *J. Sep. Sci.* **2002**, *25*, 1351–1363.

19 Taillardat-Bertschinger, A., Carrupt, P. A., Barbato, F., Testa, B. Immobilized artificial membrane (IAM)-HPLC in drug research. *J. Med. Chem.* **2003**, *46*, 655–665.

20 Stella, C., Rudaz, S., Veuthey, J. L., Tchapla, A. Silica and other materials as supports in liquid chromatography. Chromatographic tests and their importance for evaluating these supports. Part I. *Chromatographia* **2001**, *53*, S-113–S-131.

21 Vervoort, R. J. M., debets, A. J. J., Claessens, H. A., Cramers, C. A., De Jong, G. J. Optimization and characterization of silica-based reversed-phase liquid chromatographic systems for the analysis of basic pharmaceuticals. *J. Chromatogr. A* **2000**, *897*, 1–22.

22 Vallat, P., Fan, W., El Tayar, N., Carrupt, P. A., Testa, B. Solvatochromic analysis of the retention mechanism of two novel stationary phases used for measuring lipophilicity by RP-HPLC. *J. Liquid Chromatogr.* **1992**, *15*, 2133–2151.

23 Minick, D. J., Frenz, J. H., Patrick, M. A., Brent, D. A. A comprehensive method for determining hydrophobicity constants by reversed-phase high-performance liquid chromatography. *J. Med. Chem.* **1988**, *31*, 1923–1933.

24 Stella, C., Galland, A., Liu, X., Testa, B., Rudaz, S., Veuthey, J. L., Carrupt, P. A. Novel RPLC stationary phases for lipophilicity measurement: solvatochromic analysis of retention mechanisms for neutral and basic compounds. *J. Sep. Sci.* **2005**, *28*, 2350–2362.

25 Tate, P. A., Dorsey, J. G. Column selection for liquid chromatographic estimation of the k'_w hydrophobicity parameter. *J. Chromatogr. A* **2004**, *1042*, 37–48.

26 Buszewski, B., Gadza-la-Kopciuch, R. M., Markuszewski, M. I., Kaliszan, R. Chemically bonded silica stationary phases: synthesis, physicochemical characterization, and molecular mechanism of reversed-phase HPLC retention. *Anal. Chem.* **1997**, *69*, 3277–3284.

27 Welerowicz, T., Buszewski, B. The effect of stationary phase on lipophilicity determination of β-blockers using reverse-phase chromatographic systems. *Biomed. Chromatogr.* **2005**, *19*, 725–736.

28 Pidgeon, C., Venkataram, U. V. Immobilized artificial membrane chromatography: supports composed of membrane lipids. *Anal. Biochem.* **1989**, *176*, 36–47.

29 Vrakas, D., Giaginis, C., Tsantili-Kakoulidou, A. Different retention behavior of structurally diverse basic and neutral drugs in immobilized artificial membrane and reversed-phase high performance liquid chromatography:

comparison with octanol–water partitioning. *J. Chromatogr. A* **2006**, *1116*, 158–164.

30 Taillardat-Bertschinger, A., Marca-Martinet, C. A., Carrupt, P. A., Reist, M., Caron, G., Fruttero, R., Testa, B. Molecular factors influencing retention on immobilized artificial membranes (IAM) compared to partitioning in liposomes and *n*-octanol. *Pharm. Res.* **2002**, *19*, 729–737.

31 Galland, A. Towards the validation of *in silico* models and physicochemical filters to identify and characterize new chemical entities, *PhD Thesis*, University of Lausanne, **2004**.

32 Baczek, T., Markuszewski, M., Kaliszan, R. Linear and quadratic relationships between retention and organic modifier content in eluent in reversed phase high-performance liquid chromatography: a systematic comparative statistical study. *J. High Resolut. Chromatogr.* **2000**, *23*, 667–676.

33 Ayouni, L., Cazorla, G., Chaillou, D., Herbreteau, B., Rudaz, S., Lanteri, P., Carrupt, P. A. Fast determination of lipophilicity by HPLC. *Chromatographia* **2005**, *62*, 251–255.

34 Martel, S., Carrupt, P. A. Retention factors in RP-HPLC as lipophilicity indices: comparison between extrapolated and non-extrapolated $\log k_w$. Unpublished results.

35 Stella, C., Seuret, P., Rudaz, S., Tchapla, A., Gauvrit, J. Y., Lanteri, P., Veuthey, J. L. Simplification of a chromatographic test methodology for evaluation of base deactivated supports. *Chromatographia* **2002**, *56*, 665–671.

36 Martel, S., Moccand, C., Gross Valloton, B., Carrupt, P. A. The lipophilicity of cyclosporine A revisited using a RP-HPLC approach suitable to handle highly hydrophobic compounds. Unpublished results.

37 Girod, L., Martel, S., Carrupt, P. A. The lipophilicity of zwitterionic compounds by RP-HPLC: the marked influence of 1-octanol in mobile phase. Unpublished results.

38 Donovan, S. F., Pescatore, M. C. Method for measuring the logarithm of the octanol–water partition coefficient by using short octadecyl-poly(vinyl alcohol) high-performance liquid chromatography columns. *J. Chromatogr. A* **2002**, *952*, 47–61.

39 Valko, K., Slegel, P. New chromatographic hydrophobicity index (ψ_0) based on the slope and the intercept of the $\log k'$ versus organic phase concentration plot. *J. Chromatogr.* **1993**, *631*, 49–61.

40 Valko, K., Bevan, C., Reynolds, D. Chromatographic hydrophobicity index by fast-gradient RP-HPLC: a high-throughput alternative to $\log P/\log D$. *Anal. Chem.* **1997**, *69*, 2022–2029.

41 Du, C. M., Valko, K., Bevan, C., Reynolds, D., Abraham, M. H. Rapid gradient RP-HPLC method for lipophilicity determination: a solvation equation based comparison with isocratic methods. *Anal. Chem.* **1998**, *70*, 4228–4234.

42 Valko, K., Du, C. M., Bevan, C., Reynolds, D. P., Abraham, M. H. Rapid method for the estimation of octanol/water partition coefficient ($\log P_{oct}$) from gradient RP-HPLC retention and a hydrogen bond acidity term ($\Sigma\alpha_2^H$). *Curr. Med. Chem.* **2001**, *8*, 1137–1146.

43 Dias, N. C., Nawas, M. I., Poole, C. F. Evaluation of a reversed-phase column (Supelcosil LC-ABZ) under isocratic and gradient elution conditions for estimating octanol–water partition coefficients. *Analyst* **2003**, *128*, 427–433.

44 Gagliardi, L., De Orsi, D., Manna, L., Tonelli, D. Simultaneous determination of antioxidants and preservatives in cosmetics and pharmaceutical preparations by reversed phase HPLC. *J. Liquid Chromatogr. Rel. Technol.* **1997**, *20*, 1979–1808.

45 Schellinger, A. P., Carr, P. W. Isocratic and gradient elution chromatography: a comparison in terms of speed, retention reproducibility and quantitation. *J. Chromatogr. A* **2006**, *1109*, 253–266.

46 Dolan, J. W. Starting out right. Part VI. The scouting gradient alternative. *LC. GC Eur.* **2000**, June, 388–394.

47 Veith, G. D., Austin, N. M., Morris, R. T. A rapid method for estimating $\log P$ for organic chemicals. *Water Research*, **1979**, pp. 43–47.

48 Snyder, L. R., Dolan, J. W. Initial experiments in high-performance liquid chromatographic method development. I. Use of a starting gradient run. *J. Chromatogr. A* **1996**, *721*, 3–14.

49 Dolan, J. W. Gradient transfer does not have to be problematic. *LC GC North Am* **2006**, *24*, 458–466.

50 Kaliszan, R., Haber, P., Baczek, T., Siluk, D. Gradient HPLC in the determination of drug lipophilicity and acidity. *Pure Appl. Chem.* **2001**, *73*, 1465–1475.

51 Zissimos, A. M., Abraham, M. H., Barker, M. C., Box, K. J., Tam, K. Y. Calculation of Abraham descriptors from solvent–water partition coefficients in four different systems; evaluation of different methods of calculation. *J. Chem. Soc.* **2002**, *3*, 470–477.

52 Caron, G., Rey, S., Ermondi, G., Crivori, P., Gaillard, P., Carrupt, P. A., Testa, B. Molecular hydrogen-bonding potentials (MHBPs) in structure–permeation relations. In *Pharmacokinetic Optimization in Drug Research: Biological, Physicochemical and Computational Strategies*, Testa, B., Van de Waterbeemd, H., Folkers, G., Guy, R. H. (eds.), Wiley-VCH, Weinheim and VHCA, Zurich, **2001**, pp. 525–550.

53 Paschke, A., Manz, M., Schüürmann, G. Application of different RP-HPLC methods for the determination of the octanol/water partition coefficient of selected tetrachlorobenzyltoluenes. *Chemosphere* **2001**, *45*, 721–728.

54 Hallgas, B., Patonay, T., Kiss-Szikszai, A., Dobos, Z., Hollosy, F., Eros, D., Orfi, L., Keri, G., Idei, M. Comparison of measured and calculated lipophilicity of substituted aurones and related compounds. *J. Chromatogr. B* **2004**, *801*, 229–235.

55 Kaliszan, R., Baczek, T., Cimochowska, A., Juszczyk, P., Wisniewska, K., Grzonka, Z. Prediction of high-performance liquid chromatography retention of peptides with the use of quantitative structure–retention relationships. *Proteomics* **2005**, *5*, 409–415.

56 Gulyaeva, N., Zaslavsky, A., Lechner, P., Chlenov, M., Chait, A., Zaslavsky, B. Relative hydrophobicity and lipophilicity of β-blockers and related compounds as measured by aqueous two-phase partitioning, octanol–buffer partitioning, and HPLC. *Eur. J. Pharm. Sci.* **2002**, *17*, 81–93.

57 Gulyaeva, N., Zaslavsky, A., Lechner, P., Chlenov, M., McConnell, O., Chait, A., Kipnis, V., Zaslavsky, B. Relative hydrophobicity and lipophilicity of drugs measured by aqueous two-phase partitioning, octanol–buffer partitioning and HPLC. A simple model for predicting blood–brain distribution. *Eur. J. Med. Chem.* **2003**, *38*, 391–396.

58 Schellinger, A. P., Stoll, D. R., Carr, P. W. High speed gradient elution reversed-phase liquid chromatography. *J. Chromatogr. A* **2005**, *1064*, 143–156.

59 Mutton, I. M. Use of short columns and high flow rates for rapid gradient reversed-phase chromatography. *Chromatographia* **1998**, *47*, 291–298.

60 Mutton, I. M. *Handbook of Analytical Separations*, Valko, K. (ed.), Elsevier, Amsterdam, **2000**, pp. 20–52.

61 Cabrera, K. Applications of silica-based monolithic HPLC columns. *J. Sep. Sci.* **2004**, *27*, 843–852.

62 Van Nederkassel, A. M., Aerts, A., Dierick, A., Massart, D. L., Vander Heyden, Y. Fast separations on monolithic silica columns: method transfer, robustness and column ageing for some case studies. *J. Pharm. Biomed. Anal.* **2003**, *32*, 233–249.

63 Dolan, J. W., Snyder, L. R. Maintaining fixed band spacing when changing column dimensions in gradient elution. *J. Chromatogr. A* **1998**, *799*, 21–34.

64 Camurri, G., Zaramella, A. High-throughput liquid chromatography/mass spectrometry method for the determination of the chromatographic hydrophobicity index. *Anal. Chem.* **2001**, *73*, 3716–3722.

65 Kerns, E. H., Di, L., Petusky, S., Kleintop, T., Huryn, D., McConnell, O., Carter, G. Pharmaceutical profiling method for lipophilicity and integrity using liquid chromatography–mass spectrometry. *J. Chromatogr. B* **2003**, *791*, 381–388.

66 Wilson, D. M., Wang, X., Walsh, E., Rourick, R. A. High throughput log *D* determination using liquid chromatography-mass spectrometry. *Comb. Chem. High-Throughput Screen.* **2001**, *4*, 511–519.

67 Taillardat-Bertschinger, A., Carrupt, P. A., Testa, B. The relative partitioning of neutral and ionised compounds in sodium dodecyl sulfate micelles measured by micellar electrokinetic capillary chromatography. *Eur. J. Pharm. Sci.* **2002**, *15*, 225–234.

68 Testa, B., Crivori, P., Reist, M., Carrupt, P. A. The influence of lipophilicity on the pharmacokinetic behavior of drugs: concepts and examples. *Perspect. Drug Discov. Des.* **2000**, *19*, 179–211.

69 Jia, Z. Physicochemical profiling by capillary electrophoresis. *Curr. Pharm. Anal.* **2005**, *1*, 41–56.

70 Yang, S., Bumgarner, J. G., Kruk, L. F. R., Khaledi, M. G. Quantitative structure–activity relationships studies with micellar electrokinetic chromatography. Influence of surfactant type and mixed micelles on estimation of hydrophobicity and bioavailability. *J. Chromatogr. A* **1996**, *721*, 323–335.

71 Detroyer, A., Vander, H. Y., Cambre, I., Massart, D. L. Chemometric comparison of recent chromatographic and electrophoretic methods in a quantitative structure–retention and retention–activity relationship context. *J. Chromatogr. A* **2003**, *986*, 227–238.

72 Trone, M. D., Leonard, M. S., Khaledi, M. G. Congeneric behavior in estimations of octanol–water partition coefficients by micellar electrokinetic chromatography. *Anal. Chem.* **2000**, *72*, 1228–1235.

73 Trone, M. D., Khaledi, M. G. Statistical evaluation of linear solvation energy relationship models used to characterize chemical selectivity in micellar electrokinetic chromatography. *J. Chromatogr. A* **2000**, *886*, 245–257.

74 Kibbey, C. E., Poole, S. K., Robinson, B., Jackson, J. D., Durham, D. An integrated process for measuring the physicochemical properties of drug candidates in a preclinical discovery environment. *J. Pharm. Sci.* **2001**, *90*, 1164–1175.

75 Jia, Z., Mei, L., Lin, F., Huang, S., Killion, R. B. Screening of octanol–water partition coefficients for pharmaceuticals by pressure-assisted microemulsion electrokinetic chromatography. *J. Chromatogr. A* **2003**, *1007*, 203–208.

76 Lucangioli, S. E., Kenndler, E., Carlucci, A., Tripodi, V. P., Scioscia, S. L., Carducci, C. N. Relation between retention factors of immunosuppressive drugs in microemulsion electrokinetic chromatography with biosurfactants and octanol–water partition coefficients. *J. Pharm. Biomed. Anal.* **2003**, *33*, 871–878.

77 Lucangioli, S. E., Carducci, C. N., Scioscia, S. L., Carlucci, A., Bregni, C., Kenndler, E. Comparison of the retention characteristics of different pseudostationary phases for microemulsion and micellar electrokinetic chromatography of betamethasone and derivatives. *Electrophoresis* **2003**, *24*, 984–991.

78 Wehmeyer, K. R., Tu, J., Jin, Y., King, S., Stella, M., Stanton, D. T., Kenseth, J., Wong, K. S. The application of multiplexed microemulsion electrokinetic chromatography for the rapid determination of log P_{ow} values for neutral and basic compounds. *LC GC North Am* **2003**, *21*, 1078–1088.

79 Wong, K. S., Kenseth, J., Strasburg, R. Validation and long-term assessment of an approach for the high throughput determination of lipophilicity (log P_{OW}) values using multiplexed, absorbance-based capillary electrophoresis. *J. Pharm. Sci.* **2004**, *93*, 916–931.

80 Tu, J., Halsall, H. B., Seliskar, C. J., Limbach, P. A., Arias, F., Wehmeyer, K. R., Heineman, W. R. Estimation of log P_{ow} values for neutral and basic compounds by microchip microemulsion electrokinetic chromatography with indirect fluorimetric detection (muMEEKC-IFD). *J. Pharm. Biomed. Anal.* **2005**, *38*, 1–7.

81 Gluck, S. J., Benkö, M. H., Hallberg, R. K., Steele, K. P. Indirect determination of octanol–water partition coefficients by microemulsion electrokinetic chromatography. *J. Chromatogr. A* **1996**, *744*, 141–146.

82 Poole, S. K., Durham, D., Kibbey, C. Rapid method for estimating the octanol–water partition coefficient ($\log P_{ow}$) by microemulsion electrokinetic chromatography. *J. Chromatogr. B* **2000**, *745*, 117–126.

83 Caliaro, G. A., Herbots, C. A. Determination of pK_a values of basic new drug substances by CE. *J. Pharm. Biomed. Anal.* **2001**, *26*, 427–434.

84 Ostergaard, J., Hansen, S. H., Larsen, C., Schou, C., Heegaard, N. H. Determination of octanol–water partition coefficients for carbonate esters and other small organic molecules by microemulsion electrokinetic chromatography. *Electrophoresis* **2003**, *24*, 1038–1046.

85 Poole, S. K., Patel, S., Dehring, K., Workman, H., Dong, J. Estimation of octanol–water partition coefficients for neutral and weakly acidic compounds by microemulsion electrokinetic chromatography using dynamically coated capillary columns. *J. Chromatogr. B Anal. Technol. Biomed. Life Sci.* **2003**, *793*, 265–274.

86 Ishihama, Y., Oda, Y., Uchikawa, K., Asakawa, N. Evaluation of solute hydrophobicity by microemulsion electrokinetic chromatography. *Anal. Chem.* **1995**, *67*, 1588–1595.

87 Ishihama, Y., Oda, Y., Asakawa, N. Hydrophobicity of cationic solutes measured by electrokinetic chromatography with cationic emulsions. *Anal. Chem.* **1996**, *68*, 4281–4284.

88 Ishihama, Y., Oda, Y., Asakawa, N. A hydrophobicity scale based on the migration index from emulsion electrokinetic chromatography of anionic solutes. *Anal. Chem.* **1996**, *68*, 1028–1032.

89 Berthod, A., Carda-Broch, S. Determination of liquid–liquid partition coefficients by separation methods. *J. Chromatogr. A* **2004**, *1037*, 3–14.

90 Razak, J. L., Cutak, B. J., Larive, C. K., Lunte, C. E. Correlation of the capacity factor in vesicular electrokinetic chromatography with the octanol:water partition coefficient for charged and neutral analytes. *Pharm. Res.* **2001**, *18*, 104–111.

91 Klotz, W. L., Schure, M. R., Foley, J. P. Rapid estimation of octanol–water partition coefficients using synthesized vesicles in electrokinetic chromatography. *J. Chromatogr. A* **2002**, *962*, 207–219.

92 Örnskov, E., J. Gottfries, M. Erickson, S. Folestad. Experimental modelling of drug membrane permeability by capillary electrophoresis using liposomes, micelles and microemulsions. *J. Pharm. Pharmacol.* **2005**, *57*, 435–442.

93 Huie, C. W. Recent applications of microemulsion electrokinetic chromatography. *Electrophoresis* **2006**, *27*, 60–75.

14
Prediction of Log *P* with Substructure-based Methods

Raimund Mannhold and Claude Ostermann

Abbreviations

ADMET	absorption, distribution, metabolism, excretion and toxicity
AFC	atom/fragment contribution
BBB	blood–brain barrier
CASE	computer automated structure evaluation
EVA	experimental value adjusted
HTS	high-throughput screening
IC	isolating carbon
MAE	mean absolute error
MLP	molecular lipophilicity potential
MLR	multivariate linear regression
QSAR	quantitative structure–activity relationship

Symbols

$\log P$	octanol–water partition coefficient
CM	magic constant

14.1
Introduction

Lipophilicity is the measure of the partitioning of a compound between a lipidic and an aqueous phase [1]. The terms "lipophilicity" and "hydrophobicity" are often used inconsistently in the literature. Lipophilicity encodes most of the intramolecular forces that can take place between a solute and a solvent. Hydrophobicity is a consequence of attractive forces between nonpolar groups and thereby is a component of lipophilicity [2]. Lipophilicity is one of the most informative physicochemical properties in medicinal chemistry and since long successfully used in quantitative structure–activity relationship (QSAR) studies. Its

Molecular Drug Properties. Measurement and Prediction. R. Mannhold (Ed.)

ISBN: 978-3-527-31755-4

important role in governing pharmacokinetic and pharmacodynamic events has been extensively documented [3–16].

Hansch and Leo [13] described the impact of lipophilicity on pharmacodynamic events in detailed chapters on QSAR studies of proteins and enzymes, of antitumor drugs, of central nervous system agents as well as microbial and pesticide QSAR studies. Furthermore, many reviews document the prime importance of log *P* as descriptors of absorption, distribution, metabolism, excretion and toxicity (ADMET) properties [5–18]. Increased lipophilicity was shown to correlate with poorer aqueous solubility, increased plasma protein binding, increased storage in tissues, and more rapid metabolism and elimination. Lipophilicity is also a highly important descriptor of blood–brain barrier (BBB) permeability [19, 20]. Last, but not least, lipophilicity plays a dominant role in toxicity prediction [21].

The quantitative descriptor of lipophilicity, the partition coefficient *P*, is defined as the ratio of the concentrations of a neutral compound in organic and aqueous phases of a two-compartment system under equilibrium conditions. It is commonly used in its logarithmic form, log *P*. Whereas 1-octanol serves as the standard organic phase for experimental determination, other solvents are applied to better mimic special permeation conditions such as the cyclohexane–water system for BBB permeation. Measurement of log *P* is described in Chapters 12 and 13 as well as in Ref. [22].

The extent of existing experimental log *P* data is negligible compared to the huge number of compounds for which such data are needed. Thus, methods deriving log *P* from molecular structure are highly desired. The first published method for calculating log *P* from structure was based on a "substitution" procedure and was developed with substituent π constants for aromatic rings in mind [23]. Of course this method was limited to deriving a new log *P* from a "parent" structure whose log *P* was already known. Rekker and his group [24–27] were the first to publish a new and more general procedure assigning "fragmental constants" to structural moieties.

Methods for calculating log *P* [28–32] are either substructure- or property-based. Substructure-based approaches cut molecules into fragments or down to the single-atom level; summing the substructure contributions gives the final log *P*. Property-based approaches utilize descriptions of the entire molecule including molecular lipophilicity potentials (MLP), topological indices or molecular properties like charge densities, volume and electrostatic potential to quantify log *P*.

In this chapter, we describe substructure-based log *P* calculation approaches comprising fragmental and atom-based methods; property-based approaches are discussed in Chapter 15.

14.2
Fragmental Methods

Fragmental methods (Table 14.1) cut molecules down into fragments and apply correction factors in order to compensate for intramolecular interactions. Fragmental methods work according to the general formula given in Eq. (1). The first term considers the contribution of fragment constants, f_i, and the incidence of this fragment, a_i, in the query structure; the second term considers the contribution of the correction factor, F_j, and its frequency, b_j:

$$\log P = \sum_{i=1}^{n} a_i f_i + \sum_{j=1}^{m} b_j F_j \tag{1}$$

Defining fragments larger than single atoms guarantees, that significant electronic interactions are comprised within one fragment; this is a prime advantage of using fragments. On the other hand, fragmentation can be arbitrary and missing fragments may prevent calculation. These are the main disadvantages.

Among fragmental methods Σf, KLOGP and KOWWIN are *reductionistic* approaches, i.e. fragment and correction factor coefficients are derived by multiple regression of experimental data. Fragment values and interaction factors are identified and evaluated concurrently. CLOGP and ACD/LogP are based on the principle of *constructionism*. The basic fragment values are derived from measured $\log P$ data of simple molecules, then the remaining fragment set is constructed. These methods systematically interpret and generalize all the possible increments. AB/LogP combines the advantages of both *reductionistic* and *constructionistic* approaches by using hierarchical cluster analysis.

14.2.1 Σf System

Fragmental constants were derived from experimental $\log P$ values for about 100 simple organics via Free–Wilson-type regression analyses and fine-tuned by a stepwise enlargement of the dataset. This resulted in a first Σf system [24–27] comprising 126 fragment values. The fragmentation procedure leaves functional groups with direct resonance interaction intact and generates fragments ranging from atoms over substituents to complicated, in particular heterocyclic rings; all fragments are differentiated according to their aliphatic or aromatic attachment.

Tab. 14.1 Fragmental methods.

Program	*Provider*	*Internet access*
KLOGP	MULTICASE	http://www.multicase.com
KOWWIN	Syracuse Research Corporation	http://www.syrres.com/esc/est_soft.htm
	US EPA	http://www.epa.gov/opptintr/exposure/pubs/episuite.htm
CLOGP	Daylight	http://www.daylight.com
	Biobyte	http://www.biobyte.com
ACD/LogP	Advanced Chemistry Development	http://www.acdlabs.com
AB/LogP	Advanced Pharma Algorithms	http://www.ap-algorithms.com

Regression analyses revealed systematic differences between experimental log *P* and log *P* calculations based on the summation of fragment values. These differences could be attributed to chemical characteristics of the molecules, which in turn allowed the definition of correction rules such as chain conjugation, electronegativity facing bulk or the proximity effect, which describes the presence of electronegative centers in a molecule separated by one or two carbons. Correction values needed for log *P* calculation were shown to represent multiples of a constant value of 0.289, which is known as the "magic constant" (CM).

Although the method operated well, a number of intriguing points remained: (i) the bad fit of aliphatic hydrocarbon log *P* with Σf values, (ii) the irregular fit of log *P* for simple halo-alkanes with calculation data, and (iii) the correction factor of −0.46 for structures with electronegativity facing alkyl bulk and the impossibility of connecting this correction with CM.

To treat these problems, the revision of the Σf system [33] gave great care to a correct tracing of CM. A set of 15 structure pairs ($C_6H_5 \pm (CH_2)1$ or $2 \pm X$ versus $C_6H_5 \pm X$), with X representing an electronegative substituent, revealed a fairly constant difference between fX_{ar} and fX_{al} of 0.87 (±0.06), close to 3 times the original CM of 0.289. It became more and more clear, however, that a CM value of 0.87/4 rather than 0.87/3 would be preferable for renewing the *f* system and the value of CM was revised to 0.219.

- *Multihalogenation.* Nevertheless, correction rules of the revised approach raised problems in the calculation of multi-halogenated structures and demanded the following update: a halogenation pattern with two halogens on the same C atom requires one extra CM in the calculation; a halogenation pattern with three geminal halogens demands four extra CM. For per-halogenated compounds there are no satisfactory rules for CM application.
- *Decoupling of resonance.* In aromatic rings, neutral (e.g. alkyl) groups in an *ortho* position to another substituent that can undergo resonance interaction may perform a decoupling of resonance with regard to the aromate. This will convert the lipophilicity contribution of the aromatic substituent to a more aliphatic value. The difference between aliphatic and aromatic fragment values is related to multiples of CM depending on the resonance power of the substituent: $OCH_3 = -5$ CM; COOH or $CONH_2 = -4$ CM; C=O, CONH or NHCO = −3 CM; $NH_2 = -2$ CM.
- *Resonance interaction.* The combination of two groups like nitro or carbonamide on a phenyl ring in *para* or *meta* position triggers a resonance interaction responsible for increased log *P* values (1–3 CM). As subrules could not be developed, an averaged correction of 2 CM was proposed.

The actual Σf system [34] lists 13 correction rules and 169 fragment values, including 14 new heterocyclic fragments as well as doubly and triply halogenated methyls. A typical calculation is depicted in Fig. 14.1 for quinidine. The Σf system is the only fragment method allowing manual log *P* calculation. Computerized

QUINIDINE ($C_{20}H_{22}N_2O_2$)	fragments		
	1 quinolinyl (- 1 H)	+ 1.617	
	1 O (armatic)	- 0.450	
	1 OH (aliphatic)	- 1.448	
	1 N (aliphatic)	- 2.074	
	sum	**- 2.355**	
	CH residual		
	$C_{11}H_{18}$	+ 4.893	
	sum	**+ 2.583**	
	corrections		
	proximity effect	+ 0.438	(+ 2 CM)
	electronegativity facing bulk	- 0.438	(- 2 CM)
	O-C-Ar	+ 0.219	(+ 1 CM)
experimental log P = 3.33	**calculated log P**	**2.76**	

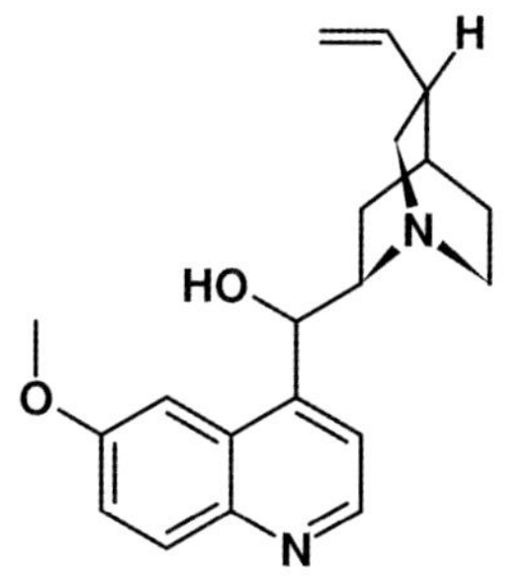

Fig. 14.1 Log *P* calculation for quinidine with the fragmental method according to Rekker and Mannhold. Calculation starts with the gross formula $C_{20}H_{24}N_2O_2$; definition of fragments leaves the CH residual; calculation is finalized by application of correction rules using the so-called magic constants (CM).

versions like SANALOGP_ER, based on an extended, revised Σf system with 302 fragmental constants [35], allow the calculation of larger databases.

14.2.2 KLOGP

The reductionistic KLOGP approach [36] was derived via regression analysis from a database containing 1663 diverse organics, which stem from Refs. [37–41]. A set of basic group parameters was defined and a program was developed to automatically identify the occurrence of each basic group parameter in the compounds. The basic group parameters consist of two types: (i) heavy atoms with both their hybridization and the number of hydrogens attached to them, and (ii) fundamental functional groups. In some cases, the nearest heavy atoms are also specified.

The initial calculation model was established with the basic group contribution parameters without considering interactions between these groups. Sixty-four fundamental group parameters were finally used in regression analysis. Log *P* calculations using only group contributions were satisfactory for simple compounds, whereas complex compounds showed high deviations. This was attributed to unparametrized interactions between substructures. The Computer Automated Structure Evaluation (CASE) methodology [42] was then used to identify the substructures responsible for large calculation errors. This analysis led to the identification of 25 statistically significant fragments. Apart from the CASE correction factors, five additional correction parameters were defined accounting for estimation errors seen in alkanes, unsaturated hydrocarbons and compounds with folding capability. In addition to the basic group parameters, the significant fragments identified by CASE were then used for developing the final log *P* estimation model:

$$\log P = a + \sum b_i B_i + \sum c_j C_j \tag{2}$$

where a, b_i and c_j are regression coefficients, B_i is the number of occurrences of the *i*th basic group, and C_j is the number of occurrences of the *j*th correction factor identified by CASE.

The final log *P* model was obtained by correlating a total of 94 parameters with the log *P* values through MLR analysis. Correction fragments identified by the CASE program can be classified into four categories: (i) tautomerization effects, (ii) dipolar ion effects, (iii) proximity effects and (iv) conjugated multiheteroatomic effects.

The revised version of KLOGP [43] is built on a training set of 8320 compounds and a test set of 1667 compounds. In total, 153 basic parameters were selected based on the diversity of the atoms and their environment inside the molecule. Every non-hydrogen atom and specific functional groups like carbonyl or nitro are treated as a "center" for a basic parameter. The bonds connected to these centers are also described, e.g. cyclic, aliphatic, nitro group binding with aromatic ring. For ring atoms, following specifications account for the environments of center atoms: (i) the types of ring structures are classified as normal (six-membered) or specified as three-membered, four-membered and so on, (ii) atoms included in more than one ring are distinguished from atoms included in one ring, and (iii) if any center atom has a lone pair, it is necessary to account for it whether there is a resonance effect between this atom and its neighbors. In addition, a set of 14 surface area and molecular volume descriptors were used. Via MULTICASE (upgrade of CASE), 41 substructures were identified and used as correction factors in building the final log *P* model:

$$\log P = C_0 + \sum_{i=1}^{n} C_i (1 - H_i) G_i \tag{3}$$

Beyond the parameter sets, described above, a steric index, H_i, is introduced, which represents the steric hindrance of the *i*th atom by other atoms in the molecule. By definition, the *H* value of the *i*th atom ranges from a minimum value of 0 to a maximum value of 1 proportional to its shielding by all other atoms of the molecule. Thus, a functional group next to large substituents will weakly contribute to the estimated log *P*. Including the steric index yields a small, but significant improvement in model accuracy.

In the recently published structural analog approach [44], log *P* of unknown chemicals is calculated from the known experimental log *P* of their closest structural analogs. The contribution of the differing molecular parts is estimated from a compilation of fragment contributions. Such a strategy promises an enhancement of prediction accuracy.

14.2.3 KOWWIN

Meylan and Howard [45–47] developed the atom/fragment contribution (AFC) method, that also employs a *reductionist* approach. About 9500 compounds out of their database ($n = 13\,062$) are common with the CLOGP-Starlist [48]; 2473 mostly simple compounds are used in the training set for deriving fragment coefficients and correction factors. The remaining 10 589 compounds are used in the validation set containing simple, moderate and complex molecules.

In total, 150 atom/fragments are used. In general, each non-hydrogen atom in a structure is core for a fragment; the exact fragment is determined by the type of the atoms connected to the core. Several functional groups, such as carbonyl, nitro, nitrate, and cyano, are treated as cores. Connections to each core are either general or specific; specific connections take precedence over general connections. For example, aromatic carbon, oxygen and sulfur atoms have nothing but "general" connections (the fragment and its value is the same no matter what is connected to the atom). In contrast, there are several aromatic nitrogen fragments.

Then, 250 correction factors, derived from the training set, were added to the AFC method. There are two groups of correction factors: (i) aromatic ring substituent positions and (ii) miscellaneous factors. Collectively, there are correction factors for various steric interactions, H-bondings and effects from polar substructures. Fragment and correction factor coefficient values were derived with two separate MLR analyses. The first regression (1120 compounds) correlated atom/fragment values of compounds that are adequately estimated by fragments alone. The second regression (1231 compounds) correlated correction factors using the difference between experimental log *P* and the log *P* estimated by fragments alone. Regression analyses yield the following general equation for estimating log *P* via the AFC method:

$$\log P = \sum (f_i n_i) + \sum (c_j n_j) + 0.229 \qquad (4)$$

where $\Sigma(f_i n_i)$ is the summation of f_i (the fragment coefficient) times n_i (the number of times the fragment occurs in the structure) and $\Sigma(c_j n_j)$ is the summation of c_j (the correction factor coefficient) times n_j (the number of times the factor occurs in the structure); 0.229 is the constant value generated by the multiple linear regression.

The model built on the current AFC training data set (2473 compounds) has the following statistics: $r^2 = 0.98$; $s = 0.22$; mean absolute error (MAE) = 0.16. The validation set (10 589 compounds) demonstrates the predictive capabilities of the AFC method with compounds it has not seen: $r^2 = 0.943$; $s = 0.473$; MAE = 0.354.

A strength of the AFC method is its ability to generate both "neutral" and "zwitterionic" log *P* estimates for amino acids and other selected zwitterions. The inherent weakness of any estimation method is predicting structure types not included during method training. The AFC method is no exception.

KOWWIN, presented so far, calculates log *P* from scratch as the sum of all fragment and correction values. The experimental value adjusted (EVA) approach is an extension to the scratch method; the estimate begins with the experimental log *P* of a structural analog of the query compound. The analog is then modified by subtracting and adding fragments and factors to build the query compound. The estimate then becomes the sum of the experimental log *P* and the value of the fragment/factor modifications. Estimation from measured values of similar analogs is more accurate than estimating log *P* from scratch.

14.2.4 CLOGP

The CLOGP approach of Corwin Hansch and Albert Leo [48–52] is based on the principles of *constructionism*. The basic fragmental values were derived from measured log *P* data of simple molecules, then the remaining fragment set was constructed. Great emphasis was given to carefully-measured values for molecular hydrogen, methane and ethane, in order to derive fragment constants for carbon and hydrogen that would be free of obscuring interactions. For more complex hydrocarbons, whose measured values were not the sum of fragment values, the differences were defined in terms of universally applicable correction factors. Within CLOGP they are subgrouped into structural factors and interaction factors (Table 14.2). A brief overview of fragmentation rules and correction factors is given below.

Tab. 14.2 Correction factors in CLOGP.

STRUCTURAL FACTORS	*INTERACTION FACTORS*
Bonds	**Aliphatic proximity, measured topologically**
Chain Bonds	Halogen versus Halogen (X vs. X)
Ring Bonds	H-Polar versus H-Polar (Y vs. Y)
Branch Bonds	Halogen versus H-Polar (X vs. Y)
Branching at Isolating Carbons	**Electronic effects through π-bonds**
Chain Branch	Fragment Valence Type
Group Branch	Extension of Aromaticity
	Sigma/Rho Fragment Interaction
	Special ortho effects
	Crowding
	Intramolecular Hydrogen Bonding

14.2.4.1 Fragmentation Rules

The definition of isolating carbon (IC) atom is a main concept within fragmentation. Such a carbon is not doubly or triply bonded to a heteroatom; it may be bonded to a heteroatom by a single or an aromatic bond. Although the hydrophobic value of an IC is constant, several types must still be distinguished. The degree to which they delocalize electrons in polar fragments attached to them has a great influence on log *P*. The presently identified types of ICs are aliphatic, benzyl, vinyl, styryl and aromatic.

A fragment is an atom or group of atoms bounded by ICs and all except hydrogen are considered polar. A fragment may have many internal bonds, but those connecting it to ICs are called valence bonds. Valence bonds are most often single, but can be aromatic. Polar fragments can interact in various ways. To quantitate this interaction, several types of polar fragments need to be defined: (i) X = any halogen and (ii) Y = all non-X fragments; these are further subdivided according to sensitivity to halogen interaction as Y-1, Y-2 and Y-3 containing -OH or not.

14.2.4.2 Structural Factors

Special attention is dedicated to bonds. CLOGP needs to know the number and types of certain bonds in the solute structure. Generally, the effect of all bonds within any fragment is reflected by the fragment value and so it is not necessary to keep track of them. Bonds which need to be identified are: (i) chain bonds, i.e. nonring bonds between ICs plus any valence bonds to fragments, (ii) ring bonds, i.e. nonaromatic ring bonds between ICs plus any valence bonds to fragments, and (iii) branch bonds, i.e. chain bonds emanating from branched fragments and counted to the last IC preceding any polar fragment. A separate count of each of these bond types must be made and a negative correction applied. For chain bonds only, this correction applies to bonds after the first in each chain. For example, there is no net bond correction for ethane but there is one for propane.

14.2.4.3 Interaction Factors: Aliphatic Proximity

- *Halogen versus halogen (X versus X)*. The positive correction for this interaction is attributed to dipole shielding and is limited to halogens on the same (geminal) or adjacent (vicinal) ICs.
- *H-polar versus H-polar (Y versus Y)*. The negative sign on Y-fragments may be due to their structure-breaking and thus their cavity-reducing ability in the water phase. Y-fragments appear to eliminate the cavity requirement for two or more ICs they are attached to. If two Y-fragments are located on the same or adjacent ICs, some of this cavity reduction is counted twice. Thus, a positive correction factor is applied when the topological separation is less than three ICs. CLOGP does not apply the same correction for every Y-C-Y or Y-C-C-Y; correction is proportional to the degree of hydrophilic character involved.
- *Halogen versus H-Polar (X versus Y)*. Consideration of this interaction is limited to that which takes place across single bonds. It is therefore,

probably due to an inductive or field effect. In evaluating X-C-Y correction, all halogens can be treated alike. However, there are several levels of sensitivity shown by Y-type fragments. The most sensitive class, Y-3, is restricted to -SO_2-R. Y-2 consists of —CONH—R, —O—R, —S—R and —NH—R; and Y-1 of all other H-polar fragments. CLOGP makes no separation of Y types to make the X—C—C—Y correction, but needs to distinguish fluorine from the other halogens.

14.2.4.4 Interaction Factors: Electronic Effects through π-Bonds

- *Fragment valence type*. All fragments (X or Y) are assigned the most negative values when bonded to aliphatic ICs – this is the "base" or "intrinsic" level. If the fragment value when attached to a vinyl IC has not been measured it can be estimated as the average of the base and aromatic-bonded values. Likewise the value for the styryl-attached fragment can be estimated as two-thirds the way from the base to the aromatic value. CLOGP only makes these estimations when measured values are not present in the database.
- *Extension of aromaticity*. The extension of the aromatic ring system through fusion (as in naphthalene) or direct substitution (as in biphenyl) appears to increase log *P*, especially if the heteroaromatic atom is next to the juncture. If the ring-joining carbons are attached only to other aromatic carbons, electron delocalization is minimal; the correction is +0.10 for each IC. If the ICs are also attached to a polar (fused-in) fragment, such as in quinoline, the correction is +0.31.
- *The σ / ρ fragment interaction*. When two or more X- and/or Y-type fragments are attached to an aromatic ring system, correction can be calculated by a method very similar to Hammett calculations of the electronic effects in other equilibria, such as acid ionization. This requires the assignment of a measure of electronic strength (σ) and susceptibility (ρ). Regarding electronic effects on partitioning equilibria, a few fragments appear to act bidirectionally, and require both σ and ρ values, although they cannot, of course, act upon themselves.

14.2.4.5 Interaction Factors: Special *Ortho* Effects

Crowding of certain fragment types can effectively lower their aromatic-attached values. This holds for fragments attached to the aromatic ring through a hetero-atom possessing an electron pair, such as -$NHCOCH_3$. A reasonable explanation is that the lone pair can no longer remain in the ring plane, making the fragment attachment aliphatic rather than aromatic. The magnitude of the correction appears to depend on both steric and electronic (field) effects.

H-bonding occurs intramolecularly between *ortho*-substituted donors and acceptors like in *o*-nitrophenol. In terms of H-bonding between aromatic *ortho* substituents, the octanol–water system seems to be sensitive to a very restricted class. The only clear-cut cases result from a carbonyl group directly attached to the ring acting as acceptor and a directly attached -OH or -NH- acting as donor. In all observed cases, the correction is close to +0.63.

CLOGP is the most frequently used log *P* calculation program. Recent versions include the FRAGCALC algorithm [53] for calculating fragment values from scratch. It is based on a test set of 600 dependably measured fragments having only aliphatic or aromatic bonds.

14.2.5
ACD/LogP

The ACD/LogP algorithm [54, 55] is based on log *P* contributions of separate atoms, structural fragments and intramolecular interactions between different fragments. These contributions have been derived from an ACD/Labs internal database of over 18 400 structures for which one or more experimental log *P* values have been published. The log *P* increments are stored in the internal databases of log *P* contributions:

The database of "Fragmental Contributions" contains increments for over 1200 different functional groups. They differ by their chemical structure, attachment to the hydrocarbon skeleton (aliphatic, vinylic or aromatic), cyclization (cyclic or noncyclic), and aromaticity (nonaromatic, aromatic or fused aromatic). Fragmentation rules are based on Hansch-Leo's approach, but differ in some respects from the definition in CLOGP. First, C sp^2 attached to two aromatic heteroatoms or to C sp are not ICs. For example, 2-pyridinyl or acetylenyl carbons are included in larger functional groups in ACD/LogP. This simplifies the analysis of intramolecular interactions. Second, hydrogens are never detached from ICs. This enlarges the list of fragmental increments, but eliminates the need for several structural correction factors.

The database of "Carbon Atom Contributions" contains increments for different types of carbons that are not involved in any functional group. They differ by their state of hybridization (sp, sp^2, sp^3), number of attached hydrogens or branching (primary, secondary, tertiary and quaternary), cyclization (cyclic and noncyclic) and aromaticity (nonaromatic, aromatic and fused aromatic).

The database of "Intramolecular Interaction Contributions" contains increments for over 2400 different types of pairwise group interactions. They differ by the type of the interacting terminal groups and the length and type of the fragmental system in between the interacting groups (aliphatic, aromatic and vinylic). Intramolecular Interaction Contributions comprise the main subgroups of aliphatic and aromatic interactions. Great care has been taken to generalize these interactions and to reflect their bidirectional character, in order to avoid too many specialized interaction rules.

If, during log *P* calculation, fragmental or intramolecular interaction contributions are not found in the internal databases, they are calculated by special "secondary" algorithms. In such cases the calculated log *P* values are provided with larger uncertainty limits (±0.6 or greater).

ACD/LogP calculations involve following steps: (i) structure fragmentation and assignment of f constants (missing fragments are estimated by atomic increments similar to Ghose–Crippen), (ii) assignment of implemented F_{ij} constants (missing interfragmental interactions are calculated by a polylinear expression similar to the Hammett–Palm equation), and iii) summation of the implemented and estimated f and F_{ij} constants. ACD/LogP uses the following equation:

$$\log P = \sum f_i + \left(\sum Q_j\right) + \sum \text{aliph-}F_{ijk} + \sum \text{vinyl-}F_{ijk} + \sum \text{arom-}F_{ijk} \quad (5)$$

where f_i = fragmental increments, Q_j = increments of “superfragments” (very occasional use), F_{ijk} = increments of interactions between any two (ith and jth) groups separated by k aliphatic, vinylic or aromatic atoms.

The careful analysis of fragmental (f_i) and aliphatic interaction (aliph-F_{ijk}) increments [54] is a strong point of ACD/LogP. Fragmental increments are represented as a sum of smaller (atomic and bond) increments and aliphatic interactions as a product of Hammett-type constants ($F_{ij} = \rho_k \tau_i \tau_j$). These “secondary” relations provide (i) the assurance that the original increments are consistent and have no “chance” values, and (ii) the possibility to estimate new increments if they are missing in the original dataset. A weak point is in the lack of similar treatment for vinylic (vinyl-F_{ijk}) and especially aromatic (arom-F_{ijk}) interactions. ACD/LogP uses over 1000 increments of aromatic interactions, many of which stem from single point determinations.

Another powerful feature of ACD/LogP is the ability to train the algorithm to improve log P prediction accuracy. Training is done in two different ways. The first option is to build a User Database containing representative compounds from the chemical classes of interest. Such a database can be used for algorithm training for special chemical classes. Secondly, an “Accuracy Extender” option allows users to define their own new fragments and aromatic or aliphatic interactions, and assign corresponding values to these increments to increase the accuracy of prediction for specific classes.

14.2.6
AB/LogP

In some instances, the general procedure of AB/LogP [56] resembles that of ACD/LogP: (i) ICs are defined, (ii) the remaining parts of the molecule are treated as “fragments” (which in most cases correspond to common functional groups), (iii) interactions between any two functional groups are identified and (iv), AB/LogP allows the user to define “superfragments” when building custom algorithms. Steps of the AB/LogP approach are illustrated in Fig. 14.2.

Step 1. ICs are defined as any carbons containing no double or triple bonds to heteroatoms. ICs isolate functional groups from each other according to the classical Hammett concept, so that the treatment of complex structures is simple.

Step 2. Whatever remains after identifying ICs represents functional groups. The functional groups are differentiated according to branching (hydrogenation),

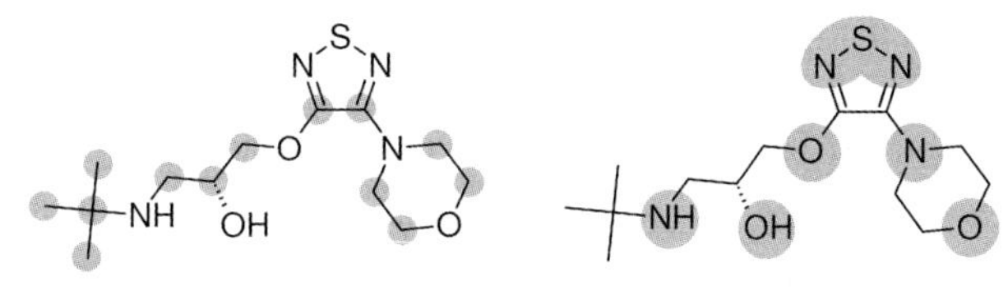

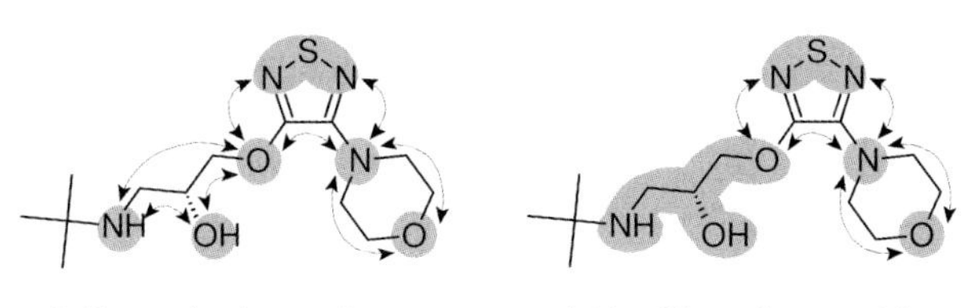

Fig. 14.2 Fundamental steps of the AB/LogP approach are illustrated for the complex chemical structure timolol. This graph is taken from Ref. [56] with kind permission of the copyright owners.

cyclization and aromaticity. In addition, they depend on conjugation with IC atoms (aromatic, vinylic and aliphatic). Increments of the missing functional groups can be estimated via special algorithms.

Step 3. Correction factors are responsible for deviations from simple group additivity. In most cases correction factors reflect internal (electronic, steric and H-bonding) interactions between polar functional groups. Figure 14.2 describes them as two-way arrows between any two functional groups, thereby reflecting the bidirectional nature of interactions (interaction between the ith and jth fragments separated by the kth type of skeleton) as expressed in:

$$Int_{ijk}(X_i - Z_k - Y_j) = \sum(\rho_{xz}\sigma_y + \rho_{yz}\sigma_x + \rho_z\sigma_x\sigma_y) + (H_B + E_S)_{ortho} \quad (6)$$

Here summation is done over all types of interactions, including inductive, resonance and mixed (which is possible due to nonlinear term). The σ constants denote electronic (Hammett-type) constants, whereas E_S and H_B denote steric and H-bonding interaction increments which mostly occur in the *ortho* position.

Step 4. The last step implies the definition of "superfragments" as parts of a molecule that are not subject to the IC-based fragmentation. AB/LogP normally does not use superfragments, but it allows the user to define them when building custom algorithms. They represent a simple way to account for various specific effects, such as ionization, stereochemical and other types of three-dimensional interactions. The program superimposes superfragments on a structure before applying fragmentation rules, so they appear as "exceptions" to the general rules. In AB/LogP superfragments can also be regarded as special cases of Free–Wilson-type fragmentation which is based on the manual definition of certain skeleton(s).

Based on the above discussed steps, AB/LogP uses the following equation:

$$\log P = \left(\sum Q_k\right) + \sum IC_i + \sum F_j + \sum Int_{ijk} \qquad (7)$$

where Q_k = increments of superfragments (only occasional use), IC_i = increments of isolating carbons, F_j = increments of functional groups and Int_{ijk} = increments of interactions.

A major problem of any fragmental method is that it produces a large number of variables which are difficult to optimize. There exist different possible ways for optimizing increments. The constructionist approach implies the stepwise construction of a model by analyzing separate classes of compounds. After each step, the obtained increments are generalized and used as fixed constants in the next step. This approach, as used in CLOGP and ACD/LogP, leads to a good understanding of all structural factors that affect the property being investigated, but requires a very long development time. The reductionist approach implies an "upfront" reduction of parameters, with optimization of the remaining increments in terms of simple MLR. The advantage is that analysis is rapid and requires smaller datasets when compared to the constructionist approach. The disadvantage is that the interpretation of obtained increments cannot be done in terms of the Hammett concept. This diminishes the ability to predict the performance of calculation algorithms when dealing with new compounds.

Similarity clustering implies an automated generalization of increments using a similarity hierarchical clustering procedure, followed by the optimization of the generic increments. This procedure combines the advantages of both the constructionist and reductionist approaches, and is a central method in AB/LogP.

Generalization by hierarchical clustering analysis was based on similarity of H-bonding and electronic interactions. A similarity key was derived from analysis of 10000 Abraham's β (H-accepting) parameters. The latter are known to play a dominant role in determining $\log P$ values. The obtained similarity key is implemented as a standard function in AB/LogP. Generic increments are obtained by MLR of the following equation:

$$f(X) = \sum IC_i + \sum F_j + \sum \{Int_{ijk}\}_{\text{clusters}} \qquad (8)$$

"Clust Int_{ijk}" indicates the number of clusters in Eq. (8). The increments can be automatically converted into calculation algorithms. This procedure dramatically reduces the development time of new algorithms. The procedure described above allowed to reduce the number of increments leading to (i) avoiding many single-point determinations, (ii) increased statistical significance, (iii) increased predicting power (as defined by "leave-one-out" cross-validation) and (iv) avoiding manual errors in generalized increments.

An instructive feature is the graphical display of $\log P$ distribution, as illustrated in Fig. 14.3 for diazepam.

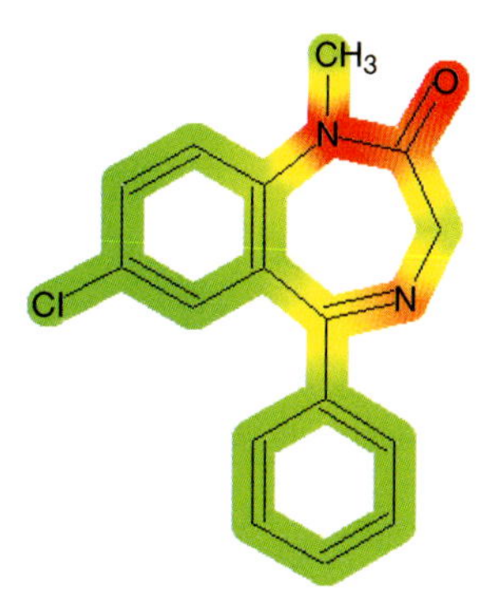

Fig. 14.3 Log *P* distribution on diazepam. Green denotes lipophilic regions and red hydrophilic regions.

14.3 Atom-based Methods

Atom-based methods [40, 57–68] cut molecules down to single atoms and commonly do not apply correction rules. According to Eq. (9) they work by summing the products of the contribution of an atom type *i* times the frequency of its presence in a query molecule:

$$\log P = \sum n_i a_i \tag{9}$$

where n_i = number of atoms of type i and a_i = contribution of an atom of type i.

Since the partition coefficient is not a simple additive property, the constitutive feature is covered by classifying huge numbers of atom types according to structural environment. An advantage of atom-based methods is that ambiguities are avoided; a shortcoming is the failure to deal with long-range interactions, e.g. as found in *p*-nitrophenol.

Several atom-based approaches form the basis of frequently used software packages such as MOLCAD, TSAR, PROLOGP, ALOGP98 and XLOGP (Table 14.3). MOLCAD, TSAR and PROLOGP are based on the original Ghose–Crippen approach [40, 61–63]. ALOGP98 is based on a refined version [66]. XLOGP is the only atom-additive method applying corrections [67, 68].

14.3.1 Ghose–Crippen Approach

The group of Crippen [40, 61–63] has described the development of a purely atom-based procedure, which exclusively applies atomic contributions and avoids correction factors:

Tab. 14.3 Atom-based methods.

Program	***Provider***	***Internet access***
MOLCAD	Tripos	http://www.tripos.com
TSAR	Accelrys	http://www.accelrys.com
PROLOGP	Compudrug	http://www.compudrug.com
ALOGP98 in Cerius^2	Accelrys	http://www.accelrys.com
XLOGP 2.0	Luhua Lai	lai@ipc.pku.edu.cn

$$\log P = \sum a_k N_k \tag{10}$$

where N_k is the occurrence if the kth atom type.

Carbon, hydrogen, oxygen, nitrogen, sulfur and halogens are classified into 110 atom types; after several revisions, the number of atom classifications has increased to 120 obtained from a training set of 893 structures ($r^2 = 0.86$, $s = 0.50$). Hydrogen and halogens are classified by the hybridization and oxidation state of the carbon they are bonded to; carbons are classified by their hybridization state and the chemical nature of their neighboring atoms. The complexity of classification is attested by a total of 44 carbon types alone. A typical calculation example is depicted in Table 14.4 for quinidine. The original Ghose–Crippen approach underlies the MOLCAD and TSAR software. MOLCAD is available from Tripos (http://www.tripos.com) and TSAR from Accelrys (http://www.accelrys.com).

Model building within PROLOGP was based on nearly 13000 experimental log P values. The software includes three calculation algorithms. Two of them are linear models: one is based on the Rekker fragmental method, the other on the Ghose–Crippen atomic method, while the third one is a feedforward neural network model that is able to recognize hidden and nonlinear relationships between chemical structure and log P [65]. The latter method involves forming the next member of a new class of pseudo-linear algorithms, where the precision of the nonlinear approaches is combined with the transparency of earlier linear methods. A further option of PROLOGP combines the optimum results obtained by the different models. PROLOGP is part of the Pallas package and available from Compudrug (http://www.compudrug.com).

ALOGP98 [66] is a refinement of the original Ghose–Crippen approach aimed at considering earlier criticisms, in particular the chemical sense of atomic contributions. The new version comprises 68 atomic definitions obtained via SMARTS from Daylight. The chemical interpretation of the atomic definitions is improved by constraining several carbon atom types to have positive contributions to log P in the fitting process. The training set was expanded to the 9000 structures in the POMONA database: a standard deviation of 0.67 is reported. ALOGP98 is implemented in the software package Cerius^2 and is available from Accelrys (http://www.accelrys.com).

Tab. 14.4 Log *P* calculation for quinidine via the Ghose–Crippen approach.

Description	*Frequency*	*Contribution*
C in CH_2R_2	2	−0.9748
C in CHR_3	2	−0.7266
C in CH_3X	1	−1.0824
C in CH_2RX	2	−1.6740
C in CHR_2X	2	−1.0420
C in $=CH_2$	1	−0.1053
C in =CHR	1	−0.0681
C in R -- CH -- R	4	+0.0272
C in R -- CR -- R	2	+0.3200
C in R -- CX -- R	2	−0.2066
C in R -- CH -- X	1	+0.0598
H attached to C^0_{sp3} with no X to next C	1	+0.4410
H attached to C^1_{sp3} or C^0_{sp2}	16	+5.3488
H attached to C^2_{sp3}, C^1_{sp2} or C^0_{sp}	1	+0.3161
H attached to heteroatom	1	−0.3260
H attached to C^0_{sp3} with one X to next C	5	+1.8475
O in alcohol	1	+0.1402
O in Al–Al, Ar_2O, R •• O •• R or R–O–C=X	1	+0.2712
N in Al_3N	1	+0.3954
N in R -- N -- R or R -- N -- X	1	−0.1106
	log P	**2.852**

Log *P* calculation for quinidine with the atom contribution method according to Ghose and Crippen. *R*: group connected to C; *X*: heteroatom; =: double bond; --: aromatic bond; ••: aromatic single bond (e.g. C=N in pyrrole); subscripts give the hybridization state and superscripts the formal oxidization number. For the quinidine structure see Fig. 14.1.

14.3.2
XLOGP

XLOGP [67, 68] is a further atom-additive method, as expressed by its almost exclusive use of atomic contributions. However, in contrast to pure atom-based methods correction rules are defined, to account for intramolecular interactions, which is typical for fragmental methods.

In the original XLOGP approach [67], atom types for carbon, nitrogen, oxygen, sulfur, phosphorus and halogen atoms are defined by their hybridization states and their neighboring atoms. Seventy-six basic atom types and four pseudoatom types for functional groups (cyano, isothiocyano, nitroso and nitro) gave a total of 80 descriptors in atom classification. The following four correction rules account for intramolecular interactions: (i) the number of “hydrophobic carbons” (sp^3 and sp^2 carbon without any attached heteroatom), (ii) an indicator variable of amino acids, (iii) presence of intramolecular H-bonds and (iv) two corrections for “poly-halogenation”.

The revised version [68] was developed with experimental log P data from Ref. [48] for a training set of 1853 compounds. Ninety atom types are used to classify atoms in neutral organic compounds. The classification scheme differentiates atoms according to: (i) element, (ii) hybridization state, (iii) accessibility to the solvent (represented by the number of attached hydrogens), (iv) nature of the neighboring atoms and (v) adjacency to π systems. Thus, atoms belonging to the same atom type generally have similar solvent accessible surfaces and charge densities. This establishes the rough theoretical support for the assumption that a certain type of atom has a specific contribution to the partition coefficient. An additional atom type for hydrogens is not used because they are included in the atom classification scheme implicitly. In other words, the atom types in this method are "united" atoms that already include the attached hydrogen atoms.

XLOGP uses correction factors to compensate for intramolecular group–group interactions. Since the atom types defined in XLOGP take the neighboring atoms into account, these correction factors aim at 1–3, 1–4 or even further intramolecular interactions. Ten correction factors are derived to correctly handle hydrophobic carbon, internal H-bond, halogen 1–3 pair, aromatic nitrogen 1–4 pair, *ortho* sp^3 oxygen pair, *para* donor pair, sp^2 oxygen 1–5 pair, α-amino acid, salicylic acid and *p*-amino sulfonic acid. The model for log P calculation includes additive (atom types) and constitutive (correction factors) terms:

$$\log P = \sum a_i A_i + \sum b_j B_j \tag{11}$$

where A_i is the occurrence of the ith atom type and B_j is the occurrence of the jth correction factor; a_i is the contribution of the ith atom type and b_j is the contribution of the jth correction factor.

The contributions of each atom type and correction factor are obtained by using Eq. (11) to perform multivariate regression analysis on the training set. The regression analysis yields $n = 1853$, $r = 0.973$, $s = 0.349$, $F = 312.4$. The standard deviation of 0.349 log units is within the experimental error range of roughly 0.4 log units. Internal prediction was performed via leave-one-out cross-validation, which yields a correlation coefficient (r) between the experimental and the predicted log P values of 0.966 and a standard deviation in prediction of 0.373 log units.

XLOGP, version 2.0, is written in C++. The program reads the query compound (represented in SYBYL/MOL2 format), performs atom classification, detects correction factors, and then calculates the log P value. Due to its simple methodology the program is quite fast. It can process about 100 medium-sized compounds per second on an SGI O2/R10000 workstation.

14.4 Predictive Power of Substructure-based Approaches

Log P prediction software has to be validated by comprehensive comparison of calculated versus experimental values. Several such comparisons have appeared

Tab. 14.5 Validity check of substructure-based calculation approaches.

		KOWWIN	CLOGP	ACD/LogP	AB/LogP	ALOGP98	XLOGP
Entire set (284)	MAE	0.55	0.49	0.59	0.53	0.64	0.73
	acceptable	176	189	180	188	154	145
	disputable	62	54	50	58	76	62
	unacceptable	46	41	54	38	54	77
Star (234)	MAE	0.48	0.43	0.49	0.44	0.59	0.67
	acceptable	157	166	164	163	136	128
	disputable	47	45	35	50	60	51
	unacceptable	30	23	35	21	38	55
Non-Star (50)	MAE	0.88	0.78	1.06	0.99	0.89	1.00
	acceptable	19	23	16	25	18	17
	disputable	15	9	15	8	16	11
	unacceptable	16	18	19	17	16	22

in the literature [69–77]. Here we present a validity check of six substructure-based approaches (Table 14.5). Experimental log P values ($n = 284$) stem from a compilation of Alex Avdeef [78] and 234 of these test molecules are also part of the BioByte StarList [79]. StarList values were intensively used to develop many log P prediction methods. Molecules outside the StarList may represent relatively new chemical classes and may be more challenging for prediction methods. Thus, the entire database ($n = 284$), molecules from the StarList ($n = 234$) and the remaining Non-Star set ($n = 50$) were considered separately.

Validity is compared as follows: (i) the MAE for the differences between experiment and calculation is given, and (ii) differences ($\Delta\log P$) between measured and calculated data in the range of 0.00 to ±0.49 are qualified as acceptable, $\Delta\log P$ values of ±0.50 to ±0.99 are viewed as disputable and differences exceeding ±0.99 as unacceptable. For both the entire set and the StarList molecules a superiority of fragment- over atom-based methods is observed; CLOGP and AB/LogP perform best. In the case of the Non-Star set, predictive power is significantly worsened by roughly a factor of two for all six software programs. Fragment- and atom-based methods, in particular ALOGP98, exhibit rather similar performance for the Non-Star set.

Acknowledgments

We want to thank Alex Avdeef for supply with an updated log P database. Thanks are also due to Pranas Japertas (Pharma Algorithms, Canada) for calculating the test set with AB/LogP and to Remy Hoffmann (Accelrys, France) for doing such calculations with ALOGP98.

References

1 Taylor, P. J. Hydrophobic properties of drugs. In *Quantitative Drug Design*, Ramsden, C. A. (ed.), Pergamon Press, Oxford, **1990** Vol. *4*, pp. 241–294.

2 Van de Waterbeemd, H., Testa, B. The parameterization of lipophilicity and other structural properties in drug design. *Adv. Drug Res.* **1987**, *16*, 85–225.

3 Kubinyi, H. Lipophilicity and drug activity. *Prog. Drug Res.* **1979**, *23*, 97–198.

4 Dearden, J. C. Partitioning and lipophilicity in quantitative structure–activity relationships. *Environ. Health Perspect.* **1985**, *61*, 203–228.

5 Smith, D. A., Jones, B. C., Walker, D. K. Design of drugs involving the concepts and theories of drug metabolism and pharmacokinetics. *Med. Res. Rev.* **1996**, *16*, 243–266.

6 Lipinski, C. A., Lombardo, F., Dominy, B. W., Feeney, P. J. Experimental and computational approaches to estimate solubility and permeability in drug discovery and development settings. *Adv. Drug Deliv. Rev.* **1997**, *23*, 3–25.

7 Van de Waterbeemd, H., Smith, D. A., Beaumont, K., Walker, D. K. Property-based design: optimization of drug absorption and pharmacokinetics. *J. Med. Chem.* **2001**, *44*, 1313–1333.

8 Van de Waterbeemd, H., Gifford, E. ADMET *in silico* modelling: towards prediction paradise? *Nat. Drug Discov. Rev.* **2003**, *2*, 192–204.

9 Lombardo, F., Gifford, E., Shalaeva, M. Y. *In silico* ADME prediction: data, models, facts and myths. *Minirev. Med. Chem.* **2003**, *3*, 861–875.

10 Van de Waterbeemd, H., Lennernäs, H., Artursson, P. (vol. eds.). *Drug Bioavailability (Methods and Principles in Medicinal Chemistry)*, Mannhold, R., Kubinyi, H., Folkers, G. (eds.), Wiley-VCH, Weinheim, **2003**.

11 Smith, D. A., Van de Waterbeemd, H., Walker, D. K. (vol. eds.). *Pharmacokinetics and Metabolism in Drug Design (Methods and PrIncíples In Medicinal Chemistry)*, Mannhold, R., Kubinyi, H., Folkers, G. (eds.), Wiley-VCH, Weinheim, **2006**.

12 Van de Waterbeemd, H., Testa, B. (vol. eds.). *ADME/Tox Approaches*, Vol. 5 in *Comprehensive Medicinal Chemistry*, 2nd edn., Taylor, J. B., Triggle, D. J. (eds.), Elsevier, Oxford, **2007**.

13 Hansch, C., Leo, A. J. *Exploring QSAR: Fundamentals and Applications in Chemistry and Biology, Vol. 1*, American Chemical Society, Washington, DC, **1995**.

14 Pliska, V., Testa, B., Van de Waterbeemd, H. (vol. eds.), *Lipophilicity in Drug Action and Toxicology*, Vol. 4 in *Methods and Principles in Medicinal Chemistry*, Mannhold, R., Kubinyi, H., Timmerman, H. (eds.), VCH, Weinheim, **1996**.

15 Testa, B., Van de Waterbeemd, H., Folkers, G., Guy, R. (eds.). *Pharmacokinetic Optimization in Drug Research*, Wiley-VCH, Weinheim and VHCA, Zurich, **2001**.

16 Testa, B., Krämer, S. D., Wunderli-Allenspach, H., Folkers, G. (eds.). *Biological and Physicochemical Profiling in Drug Research*, Wiley-VCH, Weinheim and VHCA, Zurich, **2006**.

17 Testa, B., Crivori, P., Reist, M., Carrupt, P.-A. The influence of lipophilicity on the pharmacokinetic behavior of drugs: concepts and examples. *Perspect. Drug Discov. Des.* **2000**, *19*, 179–211.

18 Van de Waterbeemd, H., Smith, D. A., Jones, B. C. Lipophilicity in PK design: Methyl, Ethyl, futile. *J. Comput.-Aided Mol. Des.* **2001**, *15*, 273–286.

19 Crivori, P., Cruciani, G., Carrupt, P.-A., Testa, B. Predicting blood–brain barrier permeation from three-dimensional molecular structure. *J. Med. Chem.* **2000**, *43*, 2204–2216.

20 Atkinson, F., Cole, S., Green, C., Van de Waterbeemd, H. Lipophilicity and other parameters affecting brain penetration. *Curr. Med. Chem. Central Nervous System Agents* **2002**, *2*, 229–240.

21 Cronin, M. T. D. The role of hydrophobicity in toxicity prediction. *Curr. Comput.-Aided Drug Des.* **2006**, *2*, 405–413.

22 Sangster, J. *Octanol–Water Partition Coefficients: Fundamentals and Physical Chemistry*, Wiley, New York, **1997**.

23 Fujita, T., Iwasa, J., Hansch, C. A new substituent constant, pi, derived from

partition coefficients. *J. Am. Chem. Soc.* **1964**, *86*, 5175–5180.

24 Nys, G. G., Rekker, R. F. Statistical analysis of a series of partition coefficients with special reference to the predictability of folding of drug molecules. Introduction of hydrophobic fragmental constants (*f*-values). *Chim. Therap.* **1973**, *8*, 521–535.

25 Nys, G. G., Rekker, R. F. The concept of hydrophobic fragmental constants (*f*-values). II. Extension of its applicability to the calculation of lipophilicities of aromatic and hetero-aromatic structures. *Chim. Therap.* **1974**, *9*, 361–375.

26 Rekker, R. F. *The Hydrophobic Fragmental Constant. Its Derivation and Application*, Elsevier, Amsterdam, **1977**.

27 Rekker, R. F., De Kort, H. M. The hydrophobic fragmental constant; an extension to a 1000 data point set. *Eur. J. Med. Chem.* **1979**, *14*, 479–488.

28 Carrupt, P.-A., Testa, B., Gaillard, P. Computational approaches to lipophilicity: methods and applications. *Rev. Computat. Chem.* **1997**, *11*, 241–315.

29 Mannhold, R., Dross, K. Calculation procedures for molecular lipophilicity: a comparative study. *Quant. Struct.-Act. Relat.* **1996**, *15*, 403–409.

30 Mannhold, R., Cruciani, G., Dross, K., Rekker, R. F. Multivariate analysis of experimental and calculative descriptors for molecular lipophilicity. *J. Comput.-Aided Mol. Design* **1998**, *12*, 573–581.

31 Buchwald, P., Bodor, N. Octanol–water partition: searching for predictive models. *Curr. Med. Chem.* **1998**, *5*, 353–380.

32 Klopman, G., Zhu, H. Recent methodologies for the estimation of *n*-octanol/water partition coefficients and their use in the prediction of membrane transport properties of drugs. *Minirev. Med. Chem.* **2005**, *5*, 127–133.

33 Rekker, R. F., Mannhold, R. *Calculation of Drug Lipophilicity. The Hydrophobic Fragmental Constant Approach*, VCH, Weinheim, **1992**.

34 Mannhold, R., Rekker, R. F., Dross, K., Bijloo, G., De Vries, G. The lipophilic behaviour of organic compounds: 1. An updating of the hydrophobic fragmental constant approach. *Quant. Struct.-Act. Relat.* **1998**, *17*, 517–536.

35 Petelin, D. E., Arslanov, N. A., Palyulin, V. A., Zefirov, N. S. Extended parameterization of Rekker's *f*-system for drug lipophilicity calculation. In *Proc. 10th Eur. Symp. on Structure–Activity Relationships*, Prous, Barcelona, **1995**, abstr. B263.

36 Klopman, G., Li, J. W., Wang, S., Dimayuga, M. Computer automated log *P* calculations based on an extended group contribution approach. *J. Chem. Inf. Comput. Sci.* **1994**, *34*, 752–781.

37 Hansch, C., Leo, A. J. *Substituent Constants for Correlation Analysis in Chemistry and Biology*, Wiley, New York, **1979**.

38 Leo, A. J., Hansch, C., Elkins, D. Partition coefficients and their uses. *Chem. Rev.* **1971**, *71*, 525–616.

39 Suzuki, T., Kudo, Y. Automated log *P* estimation based on combined additive modeling methods. *J. Comput.-Aided Mol. Des.* **1990**, *4*, 155–198.

40 Viswanadhan, V. N., Ghose, A. K., Revankar, G. R., Robins, R. K. Atomic physicochemical parameters for three dimensional structure directed quantitative structure–activity relationships. 4. Additional parameters for hydrophobic and dispersive interactions and their application for an automated superposition of certain naturally occurring nucleoside antibiotics. *J. Chem. Inf. Comput. Sci.* **1989**, *29*, 163–172.

41 Viswanadhan, V. N., Reddy, M. R., Bacquet, R. J., Erion, M. D. Assessment of methods used for predicting lipophilicity: application to nucleosides and nucleoside bases. *J. Comput. Chem.* **1993**, *14*, 1019–1026.

42 Klopman, G. MULTICASE 1. A hierarchical computer automated structure evaluation program. *Quant Struct-Act Relat.* **1992**, *11*, 176–184.

43 Zhu, H., Sedykh, A. Y., Chakavarti, S. K., Klopman, G. A new group contribution approach to the calculation of log *P*. *Curr. Comput.-Aided Drug Des.* **2005**, *1*, 3–9.

44 Sedykh, A. Y., Klopman, G. A structural analogue approach to the prediction of the octanol–water partition coefficient. *J. Chem. Inf. Model.* **2006**, *46*, 1598–1603.

45 Meylan, W. M., Howard, P. H. Atom/fragment contribution method for estimating octanol–water partition coefficients. *J. Pharm. Sci.* **1995**, *84*, 83–92.

46 Meylan, W. M., Howard, P. H., Boethling, R. S. Improved method for estimating water solubility from octanol/water partition coefficient. *Environ. Toxicol. Chem.* **1996**, *15*, 100–106.

47 Meylan, W. M., Howard, P. H. Estimating log *P* with atom/fragments and water solubility with log *P*. *Perspect. Drug Discov. Des.* **2000**, *19*, 67–84.

48 Hansch, C., Leo, A. J., Hoekman, D. *Exploring QSAR: Hydrophobic, Electronic, and Steric Constants, Vol. 2*, American Chemical Society, Washington, DC, **1995**.

49 Leo, A. J., Jow, P. Y. C., Silipo, C., Hansch, C. Calculation of hydrophobic constant (log *P*) from π- and *f*-constants. *J. Med. Chem.* **1975**, *18*, 865–868.

50 Leo, A. J. Some advantages of calculating octanol–water partition coefficients. *J. Pharm. Sci.* **1987**, *76*, 166–168.

51 Leo, A. J. Hydrophobic parameter: measurement and calculation. *Methods Enzymol.* **1991**, *202*, 544–591.

52 Leo, A. J. Calculating log P_{oct} from structure. *Chem. Rev.* **1993**, *93*, 1281–1306.

53 Leo, A. J., Hoekman, D. Calculating log P_{oct} with no missing fragments; the problem of estimating new interacting parameters. *Perspect. Drug Discov. Des.* **2000**, *18*, 19–38.

54 Petrauskas, A. A., Kolovanov, E. A. ACD/LogP method description. *Perspect. Drug Discov. Des.* **2000**, *19*, 99–116.

55 Petrauskas, A. A., Kolovanov, E. A. ACD approaches for phys-chem data prediction. In 13th Eur. Symp. on Quantitative Structure–Activity Relationships, Düsseldorf, 2000, abstr. book p. 4.

56 Japertas, P., Didziapetris, R., Petrauskas, A. A. Fragmental methods in the design of new compounds. Applications of the advanced algorithm builder. *Quant. Struct.-Act. Relat.* **2002**, *21*, 23–37.

57 Broto, P., Moreau, G., Vandycke, C. Molecular structures, perception, autocorrelation descriptor and SAR studies; system of atomic contributions for the calculation of the octanol–water partition coefficient. *Eur. J. Med. Chem.* **1984**, *19*, 71–78.

58 Convard, T., Dubost, J. P., Le Solleu, H., Kummer, E. SmilogP: a program for a fast evaluation of theoretical log *P* from the Smiles code of a molecule. *Quant. Struct.-Act. Relat.* **1994**, *13*, 34–37.

59 Ghose, A. K., Crippen, G. M. The distance geometry approach to modeling receptor sites In: *Comprehensive Medicinal Chemistry. The Rational Design, Mechanistic Study and Therapeutic Application of Chemical Compounds*, Hansch, C., Sammes, P. G., Taylor, P. J. (eds.), Pergamon Press, Oxford, **1990**, Vol. *4*, pp. 715–733.

60 Viswanadhan, V. N., Erion, M. D., Reddy, M. R. GLOGP: a new algorithm for the estimation of "log P" for organic and biological molecules. Poster at *Gordon Research Conf.*, Department of Chemistry, Gensia Inc., San Diego, CA, **1995**.

61 Ghose, A. K., Crippen, G. M. Atomic physicochemical parameters for three-dimensional structure-directed quantitative structure–activity relationships. I. partition coefficients as a measure of hydrophobicity. *J. Comp. Chem.* **1986**, *7*, 565–577.

62 Ghose, A. K., Crippen, G. M. Atomic physicochemical parameters for three-dimensional structure directed quantitative structure–activity relationships. II. modeling dispersive and hydrophobic interactions. *J. Chem. Inf. Comp. Sci.* **1987**, *27*, 21–35.

63 Ghose, A. K., Pritchett, A., Crippen, G. M. Atomic physicochemical parameters for three-dimensional structure directed quantitative structure–activity relationships III: modeling hydrophobic interactions. *J. Comp. Chem.* **1988**, *9*, 80–90.

64 Ghose, A. K., Viswanadhan, V. N., Wendoloski, J. J. Prediction of hydrophobic (lipophilic) properties of small organic molecules using fragmental methods: an analysis of

ALOGP and CLOGP methods. *J. Phys. Chem. A* **1998**, *102*, 3762–3772.

65 Molnár, L., Keserü, G. M., Papp Á., Gulyás, Z., Darvas, F. A neural network based prediction of octanol–water partition coefficients using atomic5 fragmental descriptors. *Bioorg. Med. Chem. Lett.* **2004**, *14*, 851–853.

66 Wildman, S. A., Crippen, G. M. Prediction of physicochemical parameters by atomic contribution. *J. Chem. Inf. Comput. Sci.* **1999**, *39*, 868–873.

67 Wang, R., Fu, Y., Lai, L. A new atom-additive method for calculating partition coefficients. *J. Chem. Inf. Comp. Sci.* **1997**, *37*, 615–621.

68 Wang, R., Gao, Y., Lai, L. Calculating partition coefficient by atom-additive method. *Perspect. Drug Discov. Des.* **2000**, *19*, 47–66.

69 Mannhold, R., Dross, K., Rekker, R. F. Drug lipophilicity in QSAR practice. I. A comparison of experimental with calculative approaches. *Quant. Struct.-Act. Relat.* **1990**, *9*, 21–28.

70 Rekker, R. F., Mannhold, R., ter Laak, A. M. On the reliability of calculated log *P*-values: Rekker-, Hansch/Leo- and Suzuki-approach. *Quant. Struct.-Act. Relat.* **1993**, *12*, 152–157.

71 Kühne, R., Rothenbacher, C., Herth, P., Schüürmann, G. Group contribution methods for physicochemical properties of compounds. In *Software Development in Chemistry*, Jochum, C. (eds.), GDCh, Frankfurt, **1994**, Vol. 8, pp. 207–224.

72 Schüürmann, G., Kühne, R., Ebert, R.-U., Kleint, F. Multivariate error analysis of increment methods for calculating the octanol/water partition coefficient. *Fresenius Environ. Bull.* **1995**, *4*, 13–18.

73 Mannhold, R., Rekker, R. F., Sonntag, C., Ter Laak, A. M., Dross, K., Polymeropoulos, E. E. Comparative evaluation of the predictive power of calculation procedures for molecular lipophilicity. *J. Pharm. Sci.* **1995**, *84*, 1410–1419.

74 Ghose, A. K., Viswanadhan, V. N., Wendoloski, J. J. Prediction of hydrophobic (lipophilic) properties of small organic molecules using fragmental methods: an analysis of ALOGP and CLOGP Methods. *J. Phys. Chem.* **1998**, *A 102*, 3762–3772.

75 Petrauskas, A. A., Kolovanov, E. A. Neural nets versus fragments: SciLogP ULTRA versus ACD/LogP. In *13th Eur. Symp. on Quantitative Structure–Activity Relationships*, Düsseldorf, 2000, abstr. book p. 84.

76 Erös, D., Kövesdi, I., Orfi, L., Takács-Novák, K., Acsády, G., Kéri1, G. Reliability of log *P* predictions based on calculated molecular descriptors: a critical review. *Curr. Med. Chem.* **2002**, *9*, 1819–1829.

77 Mannhold, R., Petrauskas, A. Substructure versus whole-molecule approaches for calculating log *P*. *QSAR Comb. Sci.* **2003**, *22*, 466–475.

78 Avdeef, A. *Absorption and Drug Development – Solubility, Permeability, and Charge State*, Wiley-Interscience, Hoboken, NJ, **2003**.

79 Leo, A. J. *The MedChem Database*, BioByte Corp. and Pomona College, Daylight Chemical Information Systems, Mission Viejo, CA, **2003**.

15
Prediction of Log *P* with Property-based Methods

Igor V. Tetko and Gennadiy I. Poda

Abbreviations

2D, 3D	two-, three-dimensional
AAM	arithmetic average model
AE	average error
CLIP	calculated lipophilicity potential
DFT	density functional theory
HINT	hydrophobic interactions
LSER	linear solvation energy relationship
MAE	mean absolute error
MD	molecular dynamics
MLP	molecular lipophilicity potential
OPS	optimal prediction space
QSAR	quantitative structure–activity relationships
RMSE	root mean square error

Symbols

log P	partition coefficient
log D	distribution coefficient

15.1
Introduction

The importance of methods to predict log P from chemical structure was described in Chapter 14, which is focused on fragment- and atom-based approaches. In this chapter property-based approaches are reviewed, which comprise two main categories: (i) methods that use three-dimensional (3D) structure representation and (ii) methods that are based on topological descriptors.

Molecular Drug Properties. Measurement and Prediction. R. Mannhold (Ed.)

ISBN: 978-3-527-31755-4

15.2
Methods Based on 3D Structure Representation

For two relatively immiscible solvents log P can be considered [1] proportional to the molar Gibbs free energy of transfer between octanol and water:

$$-2.303RT\log P = \Delta G_{\mathrm{oct}} - \Delta G_{\mathrm{w}} = \Delta G_{\mathrm{ow}} = \Delta G_{\mathrm{ow}}^{\mathrm{el}} + \Delta G_{\mathrm{ow}}^{\mathrm{nel}} \quad (1)$$

where ΔG_{w} and ΔG_{oct} are the solvation free energies of the solute in water and octanol, respectively, and $\Delta G_{\mathrm{ow}}^{\mathrm{el}}$, $\Delta G_{\mathrm{ow}}^{\mathrm{nel}}$ correspond to electrostatic and nonpolar terms (cavity formation and van der Waals dispersion) of solute–solvent interactions.

As theoretical prediction of the Gibbs energy in Eq. (1) is difficult, one can consider physical effects influencing the partition of molecules amid both phases and derive a restricted set of experimental parameters governing this process. Such approaches are exemplified in Section 15.2.1 by linear solvation energy relationship (LSER), SLIPPER and SPARC approaches. The measurements of experimental values cannot be used to derive log P for large datasets. However, one can expect that parameters that capture the physical effects important for the Gibbs energy such as molecular size and atomic charges should be able to provide reliable models for prediction of this property. These assumptions are the main driving forces for a number of studies described in Section 15.2.2. Continuum solvation models simplify the problem of Gibbs energy calculations by considering the solution as continuous medium. These models were originated by Born in 1920 [2] and mainly cover $\Delta G_{\mathrm{ow}}^{\mathrm{el}}$, i.e. the largest component of the solvation free energy. Three continuum solvation models are reviewed in Section 15.2.3. A computationally demanding method to calculate Gibbs free energy was published by Duffy and Jorgensen [3], who applied a molecular dynamics (MD) approach and used explicit water molecules in their model. This model and its simplification, QikProp, are considered in Section 15.2.4. The analysis of 3D approaches is completed in Section 15.2.5 with methods modeling the molecular lipophilicity potential (MLP) surrounding the query molecule.

15.2.1
Empirical Approaches

15.2.1.1 LSER

An empirical solution of Eq. (1) consists of analysis of the solvation process of the target molecule in solute, finding descriptors, which govern each phase and using them to calculate log P. This was done, for example, in the LSER approach which considered that the process of any solvation involves (i) endoergic creation of a cavity in the solvent and (ii) incorporation of the solute in the cavity with consequent setting up of various solute–solvent interactions [4–6]. Each of these steps requires a relevant solute descriptor and five parameters, also known as Abraham's descriptors, were selected [4–6]. A model to predict log P using these descriptors was calculated [6]:

$$\log P = 0.08 + 0.58E - 1.09S + 0.03A - 3.40B + 3.81V \quad (2)$$

$$n = 584, \text{SD} = 0.13$$

In this equation E (R_2) is the excess molar refraction, $S(\pi_2^{\text{H}})$ is the solute dipolarity-polarizability, $A(\Sigma\alpha_2^{\text{H}})$ and $B(\Sigma\beta_2^{\text{H}})$ are the solute H-bond acidity and basicity, respectively, and V is the McGowan characteristic volume (in $\text{cm}^{-3}\,\text{mol}^{-1}/100$). The solute size, V, (molecule favors octanol) together with solute H-bond basicity, B, (favors water) are the dominating parameters of this equation. The use of $B_0(\Sigma\beta_2^{\text{O}})$ resulted in equation

$$\log P = 0.088 + 0.562E - 1.054S + 0.032A - 3.460B_{\text{o}} + 3.814V \quad (3)$$

$$n = 613, \text{SD} = 0.116$$

with higher accuracy of log P prediction [7]. While V and E can be easily calculated from molecular structure, the other parameters, S, A, B and B_{o}, are normally obtained from experiments. This provided serious limitations on the applicability of the method to predict new series of compounds. A fragmental-based approach to calculate these parameters has been developed [8]. The calculated values of Abraham's descriptors were used [9] to develop a new equation

$$\log P = 0.315 + 0.962E - 0.841S + 0.241A - 2.506B_{\text{o}} + 2.647V \quad (4)$$

$$n = 8844, \text{RMSE} = 0.674$$

for prediction of compounds from the BioByte StarList [10]. Despite a change in the regression coefficients, the same terms, solute size and H-bond basicity, were calculated as the most important to predict log P. This method is available within the ABSOLV program [11] which uses either the original Eq. (3) or a new equation developed with Abraham's descriptors and over 15 000 compounds with experimental log P values (P. Japertas, personal communication):

$$\log P = 0.395 + 0.738E - 0.586S - 0.338A - 2.972B + 2.74V \quad (5)$$

Another implementation of this method [8] is available in the ChemProp software developed by the group of Gerrit Schüürman [12].

15.2.1.2 **SLIPPER**

The importance of molecular size and H-bond strength for octanol–water partitioning is used by the SLIPPER model [13, 14], which calculates lipophilicity for

2850 simple compounds using only two terms, the polarizability, α, and the H-bond acceptor strength, ΣC_a:

$$\log P = 0.267\alpha - \sum C_a \qquad (6)$$

$$n = 2850, \mathrm{SD} = 0.23$$

Notice, that the polarizability is highly intercorrelated with other size-related parameters, such as molecular volume, molecular weight, surface area and refractivity. Either of these terms could be used in Eq. (6), but the best results were obtained using polarizability. The second term reflects intermolecular interactions during the solvation process and was calculated with the HYBOT (Hydrogen Bond Thermodynamics) program, that was calibrated via experimentally determined enthalpies and free energies of H-bond formation between 650 donors and 2250 acceptors in different solvents [15]. Thus, the terms in Eq. (6) and the two dominating terms in Eqs. (2)–(5) describe the same physical effects.

The application of Eq. (6) to predict lipophilicity for compounds with several functional groups runs into problems. The difficulties are associated with intramolecular interactions, which could not be addressed by additive schemes as used in the SLIPPER model. Therefore, the authors correct the log *P* prediction of a given molecule according to the lipophilicity values of the nearest neighbors by using cosine similarity measures and molecular fragments [13, 14].

15.2.1.3 **SPARC**

The SPARC (Sparc Performs Automated Reasoning in Chemistry) approach was introduced in the 1990s by Karickhoff, Carreira, Hilal and their colleagues [16–18]. This method uses LSER [19] to estimate perturbed molecular orbitals [20] to describe quantum effects such as charge distribution and delocalization, and polarizability of molecules followed by quantitative structure–activity relationship (QSAR) studies to correlate structure with molecular properties. SPARC describes Gibbs energy of a given process (e.g. solvation in water) as a sum of:

$$\Delta G = \Delta G_{\text{dispersion}} + \Delta G_{\text{induction}} + \Delta G_{\text{dipole-dipole}} + \Delta G_{\text{H-bond}} + \Delta G_{\text{other}} \qquad (7)$$

Depending on the nature of a given molecule, one or several terms will be considered in the calculation. SPARC does not calculate log *P* directly, but uses the calculated activities at infinite dilution of the molecular species in both phases as

$$\mathrm{Log\,P} = \log \gamma_0^{\infty} - \log \gamma_w^{\infty} + \log R_m \qquad (8)$$

where γ^{∞} are the activities at infinite dilution of the compound of interest in each phase and $R_m = -0.82$ is the ratio of the molecularities of the two phases, i.e. coefficients that convert the mole fraction concentration to moles per liter for water and "wetted" octanol. To calculate individual terms SPARC breaks molecules into functional fragments. Calculation of molecular properties is performed using a linear combination of the fragment contributions:

$$\chi_j^0(\text{molecule}) = \sum_i (\chi_j^0 - A_i) \quad (9)$$

where χ_j^0 are the intrinsic fragmental contributions (usually tabulated values) and A_i are adjustments related to steric perturbations of the analyzed molecule caused by continuous structural elements of the molecule. Both χ_j^0 and A_i are empirically trained either on the investigated or related properties. For octanol–water partitioning these properties include activity coefficient, solubility, gas chromatography retention time, Henry's constant and distribution coefficients [21]. SPARC calculates all required molecular descriptors, such as volume, polarizability, dipole and H-bonding parameters, but it can also use the values provided by the user. These same mechanistic models are also used to calculate other physical properties, i.e. vapor pressure, solubility, Henry's constant, etc.

15.2.2 Methods Based on Quantum Chemical Semiempirical Calculations

15.2.2.1 Correlation of Log *P* with Calculated Quantum Chemical Parameters

Semiempirical quantum chemical methods are used for fast calculation and optimization of molecular structures (in vacuum or gas phase) since the second half of the last century. In 1969, Rogers [22] pioneered studies to empirically correlate log *P* with molecular properties contributing to the Gibbs free energy. Quantum chemical calculations (topology-based MINDO/3 and Hückel-type) were applied for log *P* prediction by Klopman [23]. Afterwards, the BLOGP method, involving 18 parameters, was developed using the AM1 methodology [24, 25]. The calculated quantum chemical parameters usually were limited to a sum (absolute or squared) of atomic charges of specific atoms, dipole moments and parameters reflecting molecular shape. Despite a strong educational impact these approaches could not be used for log *P* predictions of drug-like molecules. Indeed, in those studies only a few hundreds of compounds from structurally simple chemical classes were used.

Two models of practical interest using quantum chemical parameters were developed by Clark et al. [26, 27]. Both studies were based on 1085 molecules and 36 descriptors calculated with the AM1 method following structure optimization and electron density calculation. An initial set of descriptors was selected with a multiple linear regression model and further optimized by trial-and-error variation. The second study calculated a standard error of 0.56 for 1085 compounds and it also estimated the reliability of neural network prediction by analysis of the standard deviation error for an ensemble of 11 networks trained on different randomly selected subsets of the initial training set [27].

15.2.2.2 QLOGP: Importance of Molecular Size

The importance of molecular size for log *P* prediction was demonstrated by Buchwald and Bodor [28]. Molecular size determines the $\Delta G_{\text{ow}}^{\text{nel}}$ term in Eq. (1), i.e. the energy required to create a cavity for the solute in the solvent. The calculation of the molecular volume, v, was performed following geometry optimization using molecular mechanics or the semiempirical AM1 method [29]. A preliminary analysis of molecules lacking strong polar H-bonding atoms gave

$$\log P = -0.042(\pm 0.061) + 0.033(\pm 0.0005)v \quad (10)$$

$$n = 142, r = 0.98$$

with an intercept close to zero. The residuals for other molecules were surprisingly equal within different classes of monosubstituted compounds. The final equation was

$$\log P = -0.032(\pm 0.0002)v - 0.723(\pm 0.007)n + 0.010(\pm 0.0007)vI \quad (11)$$

$$n = 320, r = 0.989, s = 0.214$$

where n is a positive integer, which increases in an additive manner by each functional molecular moiety (see below); I is an indicator variable, which equals 1 for saturated unsubstituted hydrocarbons and 0 for others. QLOGP, developed using Eq. (11), recognizes 40 functional groups able to form H-bonds, such as —OH, —O—, —CN, —CON< and others. The groups contribute values from 1 to 4, depending on their type and attachment. For example, the —OH group contributes 1 or 2 depending on aromatic ring or alkyl chain attachment, respectively. The authors pointed out that a change in n by one unit is related to an average free energy change of around 4.2 kJ mol^{-1} for an H-bond in water, thus suggesting a relation to changes occurring in H-bonding during the octanol–water transport. Indeed, although n is only an integer, the authors found a very good correlation of this descriptor with the H-bond acceptor basicity $\Sigma\beta_2^H$ used in LSER Eq. (2)

$$\sum \beta_2^H = 0.114 + 0.118n \quad (12)$$

$$n = 257, r = 0.939, s = 0.07$$

for compounds from Kamlet's original article [4]. QLOGP was successfully applied to predict the log P of proteins [30].

15.2.3 Approaches Based on Continuum Solvation Models

15.2.3.1 GBLOGP

Totrov [31] developed a model to estimate electrostatic solvation transfer energy ΔG_{el}^{ow} in Eq. (1) based on the Generalized Born approximation, which considers the electrostatic contribution to the free energy of solvation as:

$$\Delta G_{ow}^{el} = -0.5(1/\varepsilon_w - 1/\varepsilon_{oct})\sum q_i q_j / f_{GB} \quad (13)$$

where ε_w and ε_{oct} are dielectric constants, q_i and q_j are partial charges, and f_{GB} is a function that interpolates between the so-called "effective Born radius", when the distance between atoms is small and r_{ij} is large. Thus, the Generalized Born approximation reduces the complex multicenter nature of solvation electrostatics to the pairwise interactions of atoms and self-contribution terms but requires the calculation of Born radii, which themselves depend on the overall shape of the molecules.

The author assumed that the Born radii of atoms can be estimated from the solvent exposure factors for sampling spheres around the atoms. Two spheres were used in a five-parameter equation to calculate the Born radii. The parameters of the equation were estimated using numerical calculations from X-ray protein structures for dihydrofolate reductase. In addition to ΔG^{el}_{ow} the author also considered the ΔG^{nel}_{ow} term accounting for cavity formation and dispersion of the solute–solvent interactions as:

$$\Delta G^{nel}_{ow} = b^{ow} + \gamma^{ow} ASA + \sum \delta^{ow}_i \tag{14}$$

where δ^{ow}_i are six atomic correction factors, ASA is water accessible surface area, and b^{ow} and γ^{ow} are adjustable parameters. The author also noticed that the microscopic dielectric constant of octanol in Eq. (13) can be different from its macroscopic values and thus its value should be adjusted. All nine parameters of Eq. (14) were estimated using dataset of 81 molecules and calculated an RMSE of 0.23. The prediction power of the method was evaluated for 19 drugs (RMSE=0.96, n=19) originally used by Moriguchi [32].

15.2.3.2 COSMO-RS (Full) Approach

The COSMO-RS (Continuum Solvation Model for Real Solvents) approach of Klamt considers interactions in a liquid system as contact interactions of the molecular surfaces [33–35]. The analysis starts with a density functional theory (DFT)/COSMO (Conductor-like Screening Model) [33] calculation to get the total energy E_{COSMO} and the polarization (or screening) charge density σ on the surface of a molecule. Once calculated, COSMO files for each molecule are stored for future use. The interaction of molecules is modeled as an ensemble of pairwise interacting molecular surfaces. The interaction energies, such as H-bonding and electrostatic terms, are written as pairwise interactions of the respective polarization charge densities:

$$E_{es}(\sigma,\sigma') = \alpha/2(\sigma+\sigma')^2 \tag{15}$$

$$E_{hb}(\sigma,\sigma') = c_{hb} \min\{0, \sigma\sigma' + \sigma^2_{hb}\} \tag{16}$$

where constants α, c_{hb} and σ_{hb} were adjusted to fit a large amount of thermodynamic data, and σ and σ' refer to solute and solvent, respectively. The ensemble of interacting molecules is replaced by the corresponding system of surface segments (both for solute and solvent) and the system is solved under the condition that the surfaces of both solute and solution interact with each other. The resulting function $\mu_s(\sigma)$ describes the solvent behavior regarding electrostatics, H-bond affinity and hydrophobicity, and covers both enthalpy and entropy of the solvation. The integration of the $\mu_s(\sigma)$ over the surface of a compound X gives a chemical potential of X in solute S. This allows us to derive models for different solvent–water partition systems, including octanol, benzene, hexane, etc., or their mixtures. Calculation of such properties using the COSMOtherm program requires only fractions of a second.

15.2.3.3 **COSMOfrag (Fragment-based) Approach**

The DFT/COSMO calculations are the rate-limiting part of the method and can easily take a few hours for molecules with up to 40 heavy atoms on a 3-GHz computer [36]. To overcome speed limitations, the authors developed the COSMOfrag method. The basic idea of this method is to skip the resource-demanding quantum chemical calculations and to compose σ profiles of a new molecule from stored σ profiles of precalculated molecules within a database of more than 40 000 compounds. A comparison of the full and fragment-based versions for log P prediction was performed using 2570 molecules from the PHYSPROP [37]. RMSE values of 0.62 and 0.59 were calculated for the full COSMO and COSMOfrag methods, respectively [36].

15.2.3.4 ***Ab Initio* Methods**

Increasing computer power enabled the use of extremely resource-demanding approaches, such as *ab initio* methods for log P prediction. Geometry optimization and calculation of the electrostatic potential of 74 molecules at the Hartree–Fock/6-31G* level using GAUSSIAN 94 were done already 10 years ago by Haeberlein and Brinck [38]. They calculated about 100 theoretical descriptors reflecting bulk, cavity, dipolarity, polarizability and H-bonding potential of molecules, and derived a three-parameter equation involving surface area, polarity and H-bonding terms with $r=0.979$, $s=0.32$, $n=74$.

Recently, Chuman et al. [39] directly estimated electrostatic solvation energy using *ab initio* MO-self-consistent reaction field calculations within the COSMO model [33]. Molecular geometry of compounds was optimized in each solvent phase using the Hartree–Fock calculation at the 3-21G* level. For molecules with multiple local energy minima the conformational analysis was performed in each phase and the resulting conformation was used to evaluate the total electrostatic energy E_{sol} in each of the solvent phases using the COSMO model and then ΔE_{ow} between n-octanol and water phases was calculated. A two-descriptor model:

$$\log P = a\Delta E_{ow} + bASA + c \tag{17}$$

was considered. The second term, the water accessible surface area, ASA, covered the effect of the solute size and thus the entropic effect. The analysis was performed for a set of 155 small molecules (less than 30 atoms) subdivided on groups of nonhydrogen bonders, hydrogen acceptors, amphitropics and (dia)azine molecules. The use of Eq. (17) provided reasonably good correlation for each subset with the best results for nonhydrogen bonders (n=69, s=0.16) and the lowest accuracy for amphitropic molecules (n=50, s=0.50). The final equation:

$$\log P_{\text{sol/w}} = -0.776\Delta E_{\text{oct}} + 0.0266ASA - 0.760I_{\text{HAc}} + 0.564I_{\text{sol}} - 0.421 \quad (18)$$

$$n = 208, r = 0.972, s = 0.28$$

included two indicator variables: I_{HAc} was used to differentiate non H-bonders from the rest of the molecules and I_{sol} was used for predicting octanol or chloroform–water systems. A similar accuracy of prediction s=0.31 was demonstrated for 51 compounds not included in the model development. The discussion of possible reasons why the present procedure failed to elucidate log P values of a heterogeneous set of compounds amid others reasons included limitations of the COSMO theory (particularly for donor molecules) and insufficient modeling of the entropy of partitioning. The author also concluded that the molecular orbital-related procedures published to date have been too optimistic in prediction of log P values and require further elaboration.

15.2.3.5 **QuantlogP**

QuantlogP, developed by Quantum Pharmaceuticals, uses another quantum-chemical model to calculate the solvation energy. As in COSMO-RS, the authors do not explicitly consider water molecules but use a continuum solvation model. However, while the COSMO-RS model simplifies solvation to interaction of molecular surfaces, the new vector-field model of polar liquids accounts for short-range (H-bond formation) and long-range dipole–dipole interactions of target and solute molecules [40]. The application of QuantlogP to calculate log P for over 900 molecules resulted in an RMSE of 0.7 and a correlation coefficient r^2 of 0.94 [41].

15.2.4 Models Based on MD Calculations

MD simulations in explicit solvents are still beyond the scope of the current computational power for screening of a large number of molecules. However, mining powerful quantum chemical parameters to predict log P via this approach remains a challenging task. QikProp [42] is based on a study [3] which used Monte Carlo simulations to calculate 11 parameters, including solute–solvent energies, solute dipole moment, number of solute–solvent interactions at different cutoff values, number of H-bond donors and acceptors ($HBDN$ and $HBAC$) and some of their variations. These parameters made it possible to estimate a number of free energies of solvation of chemicals in hexadecane, octanol, water as well as octanol–water distribution coefficients. The equation calculated for the octanol–water coefficient is:

$$\log P = 0.0145ASA - 0.731HBAC - 1.064NA + 1.172NN - 1.772 \quad (19)$$

$$n = 200, \text{RMSE} = 0.55, r^2 = 0.90$$

where *NA* and *NN* are the number of amines and nitro groups. The dominating term in the equation is *ASA* and its appearance is in agreement with similar results on the importance of molecular size for log *P* prediction indicated in previous studies. To some extent it is surprising that fundamental MD parameters, such as solute–solvent Coulomb and Lennard–Jones energies, were not significant in the equation and were substituted with more simple parameters such as *NA* and *NN*.

QikProp does not perform MD simulations but calculates required parameters from supplied 3D structures of molecules. For example, it computes atomic charges using the semi-empirical CM1p method, *ASA* is calculated using a 1.4 Å probe radius [43]. The recent parameters of QikProp were optimized using 500 drugs and related heterocyclic compounds and the model calculated an r^2 of 0.93 and an RMSE of 0.49 for more than 400 drug-like compounds [42].

15.2.5 MLP Methods

The molecular forces underlying log *P*, in particular H-bonds, are also important for binding of small molecules to their biological targets. Effects due to solvation/desolvation of water molecules are quite difficult to describe either with quantum mechanical or MD calculations, since both approaches mainly cover electrostatic interactions. The MLP methods were proposed to overcome these limitations. They determine the distribution of an empirically introduced lipophilicity potential of molecules, which accumulates all effects determining log *P*. The MLP can be used to approximate the interactive forces governing not only separation of molecules in the lipid phase, but also interactions between bioactive molecules and receptors.

15.2.5.1 Early Methods of MLP Calculations

Considering the success of fragment methods, which apply additive models for log *P* prediction, one can assume that additive approaches may also satisfactory work for MLP. Indeed, similar to the Generalized Born model, one can consider fragments of molecules as centers of some potential functions and use an empirically defined distance function *f*() to calculate the MLP value by:

$$\mathrm{MLP}(\mathbf{x}) = \sum_i p_i f(|\mathbf{x}_i - \mathbf{x}|) \tag{20}$$

where summation is done over all fragments $i = 1, \ldots, n$ in the molecule, p_i are lipophilic constants of the fragment i and $|\mathbf{x}_i - \mathbf{x}|$ is the distance from the fragment i to the point of interest.

The MLP potential defined by Eq. (20) was introduced in 80th simultaneously by several groups [44–46]. Dubost et al. [46] used a hyperbolic distance function and calculated p_i using the fragmental system of Rekker [47]. The same function was also applied by Furet et al. [44] who used the fragmental system of Ghose and Crippen [48]. The MLP approach introduced by Fauchere et al. [45] used an exponential distance function. Two approaches for MLP calculations will be discussed in more details.

15.2.5.2 Hydrophobic Interactions (HINT)

The HINT method [49–51] refers to the idea of Abraham and Leo [52] that hydrophobic fragment constants, reduced to atomic values, could be used to evaluate interactions between small and large molecules. The interaction energy term b_{ij} scores atom–atom interactions (i,j) within or between the molecules using the following equation:

$$b_{ij} = a_i S_i a_j S_j T_{ij} R_{ij} + r_{ij} \tag{21}$$

where R_{ij} is an exponential function $f(|\mathbf{x}_i - \mathbf{x}|)$ from Eq. (20), a_i are the hydrophobic atom constants, S_i is the accessible surface area (estimated using the H_2O probe [50]), $T_{ij}=\{1,-1\}$ is a logic function describing the character of interacting pairs (attraction or repulsion) and r_{ij} describes a penalty term to flag van der Waals violations. The a_i contributions are calculated by an adaptation of the CLOGP method [52, 53], i.e. by calculating the hydrophobic atom constants following summation of CLOGP group contributions. Later on, the authors used Eq. (21) to derive a non-Newtonian HINT "natural" force field [51] that can be used to optimize molecular structures and to analyze protein–protein and protein–ligand complexes. The HINT software has been successfully used to calculate and view hydrophobic fields, estimate pK_a values of protein residues, and calculate free energy of binding between protein subunits and between substrates and enzymes [50, 54].

15.2.5.3 Calculated Lipophilicity Potential (CLIP)

The CLIP approach was developed by the group of Testa [55]. It is based on the atomic lipophilic system of Broto and Moreau [56] and uses a modified exponential distance function of $e^{-d/2}$, which differs from the e^{-d} function of Fauchere et al. [45]. In addition the authors restricted the distance function at 4 Å to avoid influence of too distant elements. The most recent version of CLIP uses a Fermi-like distance function, which does not need any cutoff values [57]. Another implementation of the approach is available in the VEGA software provided by Pedretti et al. [58].

Both HINT and CLIP allow to calculate whole-molecule log P values by integrating the MLP over the surface, e.g. by calculating ΣMLP, ΣMLP+, ΣMPL– corresponding to a total sum of all MLP values and its "lipophilic" (positive MLP values) and "hydrophilic" (negative) parts, respectively. Calculated in such a way log P values are sensitive to intramolecular factors, such as proximity between polar groups and, especially, molecular conformations and are known as "virtual" log P values [55]. The dependence of log P on molecular conformations was directly confirmed by recent experimental studies of clenbuterol rotamers. The nuclear magnetic resonance-based measurements indicated a difference of 0.73 log units in the log P values for different conformers [59].

15.2.6 Log *P* Prediction Using Lattice Energies

One of the first studies to predict log *P* by using potential energy fields calculated using the GRID and CoMFA approaches was done by Kim [60]. The author investigated H^+, CH_3 and H_2O probes, and calculated the best models using the "hydrophobic probe" H_2O for relatively small series (20 or less compounds each) of furans, carbamates, pyridines and pyrazines. A similar study was performed by Waller [61] who predicted a small series of 24 polyhalogenated compounds. Recently, Caron and Ermondi [62] used a new version of Cruciani's software, VolSurf [63], to predict the octanol–water and alkane–water partition coefficients for 152 compounds with $r^2 = 0.77$, $q^2 = 0.72$, SDEP = 0.60 for octanol–water and $r^2 = 0.76$, $q^2 = 0.71$, SDEP = 0.85 for alkane–water.

15.3 Methods Based on Topological Descriptors

A large number of log *P* calculation methods are based on topological descriptors. One of their main advantages is speed. Indeed, these algorithms can be hundreds of thousand times faster than the resource-demanding algorithms based on *ab initio* or MD calculations. Speed renders these methods an important tool for predicting large datasets, in particular for screening virtual combinatorial libraries.

15.3.1 MLOGP

One of the first but yet successful and well known method, MLOGP, was developed by Moriguchi et al. [64]. MLOGP uses the sum of lipophilic (carbons and halogens) and hydrophilic atoms (nitrogens and oxygens) as two basic descriptors. These two descriptors were able to explain 73% of the variance in the experimental log *P* values for 1230 compounds. The use of 11 correction factors covered 91% of the variance. Due to the simplicity of implementation, the MLOGP method was widely used as a calculation and reference approach for many years.

15.3.2 Graph Molecular Connectivity

Molecular formulas are traditionally represented by their two-dimensional (2D) graphs. Graph analysis allows us to introduce graph-theoretical invariants that are well related to physicochemical parameters with relevance for lipophilicity [65, 66], such as their size, surface area, polarizability and dipolarity. Niemi et al. [67] divided molecules in 14 groups according to the number of H-bonds per atom and calculated regression coefficients for each group using graph-theoretical descriptors. Within each group the authors used 50%/50% training/test set partition to avoid chance correlation. The resulting set of best-subset equations calculated an RMSE of 0.64 for more than 4000 compounds for the joined training/test set. The results for the training/test set protocol, i.e. when models developed with training sets were used to predict test set molecules, calculated RMSE = 0.85. The particular poor prediction results were obtained for complex molecules. In another study Devillers developed the AUTOLOGP approach [68] using 35 2D autocorrelation descriptors [65, 66] (n = 7200, RMSE = 0.37, r = 0.97).

15.3.2.1 **TLOGP**

Junghans and Pretsch [69] used topological descriptors of uniform length [70] representing n-dimensional vectors. The vectors showed the occurrence of atom pairs in relative distances of 1 to n bonds. The heteroatoms or atoms with specified hybridization were painted to produce colored derivatives of the graph for which the path counting was repeated. In the 3D version of this graph the corresponding normalized interatomic distances were summed up for each path length. The 3D structures were calculated using CORINA [71]. The authors built two classes of models: global and local. The global model included all molecules and local models were calculated following clustering of molecules for each cluster. The authors noticed that the quality of prediction depended on similarity to training set compounds. In case of low similarity the accuracy of local models significantly decreased and the authors performed predictions using the global model.

15.3.3
Methods Based on Electrotopological State (E-state) Descriptors

The E-state indices [72, 73] were developed to cover both topological and valence states of atoms. These indices were successfully used to build correlations between the structure and activity for different physicochemical and biological properties [72]. New applications of this methodology are also extensively reviewed in Chapter 4. Several articles by different authors demonstrated the applicability of these indices for lipophilicity predictions [74–83].

15.3.3.1 **VLOGP**

Gombar and Enslein [75, 76] used E-state descriptors and other topological indices for a better description of molecular symmetry and shape. The model was based on 8686 compounds from the BioByte StarList [10]. The final 363-variable equation was derived using 6675 compounds and a set of 2011 (25%) compounds which were outside of the so-called optimal prediction space (OPS) were set aside during the model development. A standard error of 0.2 log unit and r^2 of 0.98 were calculated for compounds inside the OPS. After creation of a model, the range-based cutoffs are used to determine whether the query molecule is inside or outside of the space of the training set of molecules. For a test set of 113 compounds the predictive ability of the method for 29 compounds outside the OPS was about 5 times lower compared to the molecules inside [75]. Thus, the use of the OPS allowed to discriminate "bad" from "good" predictions. In our testing we used the recent version of this program provided by Accelrys Software Inc. [84].

15.3.3.2 ALOGPS

A series of studies with increasing numbers of molecules was done by Tetko et al. [77–83]. The first study [78] was rather a toy problem and was performed using a small set of 345 drug-like compounds. Only 32 atom-type E-state descriptors were used to calculate linear and neural network models, which had very similar performance (q^2 = 0.83–0.84, $RMSE_{LOO}$ = 0.69–0.71). The second study used 1754 molecules by including the set of Klopman [85]. The authors extended the basic set of E-state indices to better cover the surrounding of nitrogens and oxygens. In the next study [79] an increased set of 12 908 molecules from the PHYSPROP database [37] demanded the additional use of bond-type E-state descriptors to build a predictive model. The neural networks produced significantly better results (r^2 = 0.95, RMSE = 0.39) compared to the linear regression (r^2 = 0.89, RMSE = 0.61) using 75 descriptors. The most recent version, ALOGPS 2.1 [80–82], was developed using the associative neural network method [86, 87] (n = 12908, RMSE = 0.35). Similar to Niemi et al. [67] the authors noticed a decrease in performance of their and several other methods as a function of the number of nonhydrogen atoms in the molecules, i.e. their size and complexity.

15.3.3.3 CSlogP

Two programs to predict lipophilicity of molecules were developed by Parham et al. [74]. The authors used an ensemble of 10 neural networks that were applied to the initial training set of 12942 organic compounds. The compounds used in their study were taken from the PHYSPROP [37] as well as from some private sources. Their model was developed using 107 E-state indices that were selected from the original set of 224 MolconnZ indices using the Interactive Analysis software. The prediction accuracy of their model, r^2 = 0.96, MAE = 0.31, was calculated using a test set of 1258 molecules. Later on the authors developed a new model, CSlogP, based on a set of 16893 compounds compiled from LOGKOW [88], CLOGP BioByte StarList [10] and PHYSPROP [37]. The authors selected a set of 519 topological and E-state descriptors. The final model calculated r^2 = 0.89, MAE = 0.43.

15.3.3.4 A_S+logP

The 2D model was built from a wide array of descriptors, including also E-state indices, by Simulations Plus [89]. The model is based on the associative neural network ensembles [86, 87] constructed from n = 9658 compounds selected from the BioByte StarList [10] of ion-corrected experimental log P values. The model produced MAE = 0.24, r^2 = 0.96 (R. Fraczkiewicz, personal communication).

15.4
Prediction Power of Property-based Approaches

Since the final purpose of any log P model is to predict new data, the benchmarking of different models allows us to compare their advantages and disadvantages. The methods were benchmarked using a dataset of 284 molecules selected from the book of Avdeef [90]. This dataset mainly contains experimental measurements from Avdeef as well as some data from literature. The dataset is available for download at http://www.vcclab.org/datasets. See Tab. 15.1.

15.4.1 Datasets Quality and Consistence

TLOGP does not predict molecules used in the training set but it rather reports an experimental value for the compound. 79 molecules were used in the training set of TLOGP. The difference between values reported by TLOGP and the values of the same molecules in our benchmarking set gave an MAE of 0.18, $n=79$. Warfarin (CAS-RN: 81-81-2) had the largest difference in $\log P$ values between the TLOGP training set ($\log P=2.70$) and the current benchmarking set ($\log P=3.54$).

About 80% (234/284) of the molecules in the benchmarking dataset were also in the BioByte StarList [10]. The differences between the benchmarking and Star-list $\log P$ values contributed an MAE of 0.19, $n=234$. The largest difference was observed for labetalol (CAS-RN: 36894-69-6), which had a $\log P$ of 3.09 in the BioByte StarList and 1.33 in the benchmarking dataset. An MAE of 0.2 thus can be considered as an estimation of the accuracy of the experimental measurements, i.e. the lower limit of error for these data. The BioByte StarList [10] and its subsets were considered as the "gold standard" for $\log P$ values and were intensively used to develop many $\log P$ prediction methods, including those considered in this article. Contrary to the fact that the benchmarking molecules that were not found in this set may correspond to relatively new chemical classes of molecules. Since the results for the "star" set and the entire set were practically identical, we decided to consider only two datasets: molecules from the "star" set ($n=234$), and molecules from the remaining "nostar" set ($n=50$).

15.4.2 Background Models

For comparing models, it is useful to have one or several background models that could provide an estimation of an upper limit of errors of a model. Two such models were used. The first model was the arithmetic average model (AAM). The molecules in the benchmarking set have a mean $\log P$ of 2.1 that was similar to 1.97, the value averaged over all molecules in the PHYSPROP database [37]. Either of these values could be used as an estimation of an *a priori* $\log P$ value of any molecule in the benchmarking set, i.e. as a model which always predicts only one value for any given structure. We will refer to this "model" as the AAM according to [91]. The AAM provided an MAE=1.46–1.47 using $\log P$ of 2.1 and 1.97, respectively. The second model used for comparison was MLOGP [64]. This model, based only on 13 one-dimensional descriptors, allows better understanding how far the "advanced approaches" stand from this very basic model developed more than 15 years ago.

The statistical difference in performance of models was estimated with a bootstrap test using 10000 replicas (see details in Ref. [91]). The significance level of $p<0.05$ was used. For each dataset all methods were classified in four categories.

Tab. 15.1 Software packages used in benchmarking.

Name	*Provider*	*Internet/e-mail*	*Structure*	*Reference*
A_S+logP	Simulations Plus Inc., USA	http://www.simulations-plus.com	2D	89
ABSOLV[1], LSER[2]	Pharma Algorithms, Lithuania/Canada	http://www.ap-algorithms.com	2D	8, 9
ALOGPS[3]	Virtual Computational Chemistry Laboratory, Germany	http://www.vcclab.org	2D	79, 87, 98
CLIP	University of Geneva, Switzerland	pierre-alain.carrupt@pharm.unige.ch	3D/CORINA	55, 57
COSMOFrag[3]	COSMOlogic GmbH & Co. KG, Germany	http://www.cosmologic.de	2D	35, 36
CSlogP	ChemSilico LLC, USA	http://www.chemsilico.com	2D	
GBLOGP	Max Totrov, USA	max@molsoft.com	3D/ICM	31
HINT	EduSoft LC, USA	http://www.edusoft-lc.com	3D/CORINA	49–51
LSER UFZ[4]	Helmholtz Centre for Environmental Research – UFZ, Germany	ralph.kuehne@ufz.de	2D	12
MLOGP			2D	64
QikProp[5]	Schrödinger LLC, USA	http://www.schrodinger.com	3D/CORINA	3, 42
QLOGP	University of Miami, USA	PBuchwald@med.miami.edu	3D/CORINA	28, 30
QuantlogP	Quantum Pharmaceuticals, Russia	http://q-pharm.com	3D/CORINA	40
SLIPPER	Institute of Physiologically Active Compounds, Russia	http://camd.ipac.ac.ru	2D	13, 14
SPARC	University of Georgia, USA	http://ibmlc2.chem.uga.edu	2D	16–18
TLOGP	Upstream Solutions, Switzerland	http://www.upstream.ch	2D	69
VEGA	University of Milan, Italy	http://www.ddl.unimi.it	3D/CORINA	58
VLOGP	Accelrys Software Inc., USA	http://www.accelrys.com	2D	84

1 LSER Eq. (5).
2 LSER Eq. (3).
3 Calculated at the Virtual Computational Chemistry Laboratory (VCCLAB) site [98].
4 LSER Eq. (4).
5 Calculated using an internal server at Pfizer Inc.

Methods that produced results not significantly different from the method with the minimal MAE received rank I, i.e. all methods with rank I had a similar "best" performance. After determination of this group, the remaining methods with results significantly better than those of MLOGP or AAM received rank II and III, respectively. Methods that had an MAE non-significantly different from (or even worse than) AAM received rank IV.

Several methods, e.g. CLIP, VEGA and HINT, required 3D structures of molecules. For these methods 3D structures were generated using Corina [71], while 3D structures for GBLOGP were generated by MMFF94 force field geometry optimization as implemented in ICM [92].

15.4.3 Benchmarking Results

Table 15.2 lists models ranked by their predictive performance of the "star" dataset. The lowest MAE were calculated using the A_S+logP and ALOGPS methods, which had statistically indistinguishable results for the prediction of both datasets. Moreover, their results were statistically significantly different compared to MLOGP. A_S+logP and ALOGPS resemble in their design; both programs were developed using E-state indices [72, 73] and associative neural networks [86, 87].

A group of eight methods (rank II) showed significantly higher prediction ability compared to MLOGP for the "star" set. The group includes VLOGP and CSlogP that were developed using 2D descriptors such as E-state indices. QikProp and QuantlogP are based on quantum mechanical calculations and another two methods, SLIPPER and ABSOLV, are based on empirical treatment of H-bonding. VLOGP program calculated MAE of 0.48 ($n=257$) and MAE of 1.51 ($n=27$) for molecules within and out of its OPS ($p<0.05$). This difference in the performance of the program confirmed previous conclusions that exclusion of unreliably predicted molecules can dramatically increase the prediction accuracy of the VLOGP method [75, 76].

For all methods, the prediction performance for the "nostar" set was on average by 0.4 log units lower. Four programs, VLOGP, SLIPPER, CSlogP and ABSOLV, had rank I, i.e. they had statistically indistinguishable performance at $p<0.05$ to the results of the algorithms with the lowest MAE. However, the results of these programs were also statistically not significantly different compared to those of MLOGP.

A considerable number of methods showed results not significantly different from the AAM model, i.e. these methods failed to provide predictive models.

15.4.4 Pitfalls of the Benchmarking

15.4.4.1 Do We Compare Methods or Their Implementations?

Our analysis included three LSER-like models. Two of them were provided by Pharma Algorithms and differed only with respect to the used regression coeffi-

Tab. 15.2 Performance of algorithms.

Method	*Star (234)*		*Non-Star (50)*		*Zwitterions (18)*		*Other (266)*
	MAE	*Rank*[1]	*MAE*	*Rank*	*MAE*	*AE*	*MAE*
A_S+logP	0.33	I	0.7	I	0.4	−0.01	0.4
ALOGPS[2]	0.39	I	0.7	I	0.64	−0.51	0.44
VLOGP[3]	0.50 (0.41)	II	0.95 (0.84)	I, III	0.87 (0.69)	−0.8 (−0.62)	0.56 (0.47)
SLIPPER	0.58	II	0.91	I, III	1.2	−1.14	0.6
QikProp	0.58	II	1.01	III	0.83	−0.48	0.64
CSlogP	0.61	II	0.95	I, III	0.54	−0.06	0.68
TLOGP[4]	0.64	II	1.01	III	1.26	−0.97	0.69
ABSOLV	0.65	II	0.94	I, III	1.98	−1.97	0.61
QuantlogP	0.7	II	1.03	III	1.91	−1.9	0.68
QLOGP	0.72	II	1.19	III	0.9	−0.24	0.79
VEGA[5]	0.8	III	1.07	III	1.53	0.95	0.8
CLIP[6]	0.82	III	1.27	III	1.3	−0.95	0.87
LSER	0.87	III	1.26	III	2.32	−2.31	0.84
MLOGP	0.93	III	1.12	III	1.64	−1.51	0.92
SPARC[7,8]	0.93	III	1.17	III	0.72	0.06	0.99
COSMOFrag[2]	1.13	III	1.38	IV	2.48	−2.47	1.09
LSER UFZ[7]	1.19	IV	2.15	IV	2.32	−1.75	1.29
GBLOGP[6]	1.25	IV	1.76	IV	2.51	2.46	1.26
HINT[5]	1.27	IV	1.87	IV	1.25	0.54	1.39
AAM	1.37	IV	1.87	IV	2.96		1.36

1 See definition of ranks in Section 15.4.2.
2 log *P* values were calculated at the http://www.vcclab.org site.
3 VLOGP results for 257 molecules inside the OPS are shown in parentheses.
4 90 molecules (including 79 with experimental values in TLOGP and 11 molecules that were not calculated) were excluded.
5 Ionized forms were used to estimate log *P*.
6 14 molecules were not calculated.
7 Zidovudine was not calculated.
8 log $D^{7.4}$ values assuming an ionic strength of 0.1 were calculated for zwitterions.

cients in the LSER equation (see Section 15.2.1). However, even this presumably minor difference resulted in significantly different performance of both models ($p<0.05$) for the whole set of molecules. The third LSER approach was implemented by the group of Schüürman [12] according to publications by Platts [9, 93] and used the original regression coefficients from Eq. (4). It provided statistically significantly worth results ($p<0.05$) compared to both LSER approaches implemented by Pharma Algorithms. There is also a difference in results calculated using the original CLIP method and its VEGA implementation. Thus the variations in performance of different implementations of the same approach (e.g. LSER) should not be neglected when drawing conclusions in comparative studies.

15.4.4.2 **Overlap in the Training and Benchmarking Sets**

A serious complication for comparing methods performances is due to differences in their training sets. For example, in our previous study ALOGPS [79] and KowWIN [94] provided higher prediction accuracy (RMSE=0.44–0.46) compared to CLOGP [53] (RMSE=0.62) for prediction of compounds that were missed in the BioByte StarList [10]. However, when the ALOGPS program was redeveloped using a subset of molecules from the BioByte StarList, its prediction accuracy for the remaining molecules was similar to that of CLOGP (RMSE=0.57). In other studies a 2- to 10-fold increase in the prediction accuracy of ALOGPS was demonstrated following extension of the training set in the LIBRARY mode [80–83].

A similar impact of the size of the training set can be expected in the current study. A_S+logP, ALOGPS and CSlogP were developed using neural networks and large databases of compounds with experimental values that were at least partially included in the BioByte StarList [10]. Therefore the observed performance of these methods for the "star" subset may correspond to "data memorizing" by the neural network. This is also true for several other top-ranked methods. For example, SLIPPER makes its prediction using a similarity search in a large database of experimental values. VLOGP was developed using compounds from the BioByte StarList. A newer version of QikProp was recently developed using a large dataset of diverse drug-like compounds. Fitting parameters of ABSOLV were recalculated using a large database of molecules with experimental log *P* values. On the other hand, methods, which are based on theoretically derived models, such as GBLOGP and COSMOfrag, do not have the ability to "memorize" the data and their generalization abilities were, presumably, estimated more rigorously. Thus, it is difficult to assess the degree of fitting/generalization in different models. A proper estimation of method performance can only be performed using data that have never been used in the method development, e.g. using in-house data of pharmaceutical companies [81, 82, 95, 96]. The "non-star" set, which does not include molecules from the BioByte StarList, provides presumably a more realistic estimation of the prediction ability of methods. However, still it is possible that "non-star" molecules (or their analogues) were used as training sets of some methods.

CLIP, VEGA and HINT, were included in the comparison only for illustrative purposes. These methods were not developed to predict log *P* on their own and may require calibration within each specific series of compounds [55].

15.4.4.3 **Zwitterions**

Our dataset contained 18 zwitterions, which exist in water as ionic species at any pH, as either the cationic or anionic group is charged. Thus, for such compounds only distribution coefficients, log *D* (see Chapter 16), rather than partition coefficients, log *P*, can be measured. Several programs, such as A_S+logP, ALOGPS, CSlogP, VLOGP and QikProp, were presumably parameterized on datasets with "zwitterionic log *D*" values and produced similar results with and without zwitterions (Tab. 15.2). QLOGP used a correction factor of $-0.73{*}3=-2.2$ units for zwitterions (P. Buchwald, personal communication) and also gave similar results for both subsets. The performance of the SPARC model [MAE = 2.33, average error (AE) = −2.3] dramatically increased (MAE = 0.72, AE = 0.06) when results for log *D* at pH 7.4 and ionic strength of 0.1 were used (Tab. 15.2). The use of ionized forms of zwitterions improved results for VEGA (MAE = 1.53, AE = 0.95 versus MAE = 2.09, AE = −2.07) and HINT (MAE = 1.25, AE = 0.54 versus MAE = 3.25, AE = −3.24). The omission of zwitterions only slightly rearranged the ranks of the methods. Still, zwitterions and compounds charged at any pH could bias comparison of log *P* methods if they represent a considerable fraction of the benchmarking dataset. The nature of such molecules and difficulties to identify their genuine log *P* values raise a question whether they should be used in benchmarking of log *P* programs. However, the identification and exclusion of such compounds is non-trivial in general case.

15.4.4.4 Tautomers and Aromaticity

Molecules in real solvents can exist in one or more tautomeric forms. The use of different tautomers in calculations can lead to significant variation in the estimated log *P* values (Fig. 15.1). Accurate prediction of the dominant tautomer requires *ab initio* calculations. Due to speed limitations such calculations are not feasible for virtual screening and prediction of large compound collections. Moreover, the interpretation of the results can also be difficult. For example, the lacton-lactim (Fig. 15.1B) is the stable form of maleic hydrazide in the gas phase but the difference between this and the dilacton form (Fig. 15.1C and D) disappears in solution

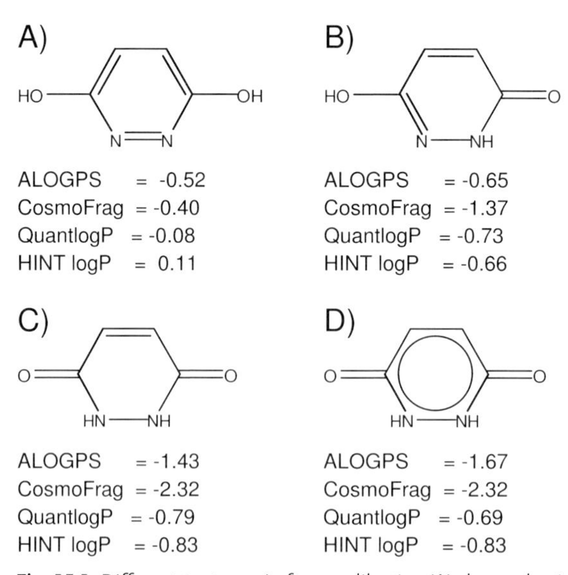

Fig. 15.1 Different tautomeric forms, dilactim (A), lacton-lactim (B) and dilacton (C and D), of maleic hydrazide (CAS-RN: 123-33-1), log P_{exp} = −0.56 (−0.84 in the BioByte StarList [10]).

[97]. The ratios of the tautomeric forms in water and octanol can be also different, thus further complicating accurate prediction of log *P*.

Other difficulties arise from the incorrect treatment of aromaticity, e.g. the tautomeric form C can be also represented as an aromatic compound. However, ALOGPS, for example, does not consider this ring as an aromatic one. Inconsistently defined aromaticity lowers the prediction performance (Fig. 15.1D). The use of SDF files, which do not explicitly define aromaticity solves this problem. All these factors are not limiting when the data are prepared with the same coding scheme and consistency. However, these issues are very important for method application and benchmarking.

15.5
Conclusions

Prediction of log *P* based on descriptors derived from molecular structure has received considerable interest during the last years. Despite a great variety of developed methods, all of them indicate importance of molecular size and H-bonding for an accurate prediction of this property. These parameters are in good agreement with those selected as the most important in LSER and other empirical methods. The methods based on time-demanding calculations, such as MD and *ab initio* calculation, and optimization of molecular structures in octanol and water phases started to appear. However, there is still a need to prove their advantage over more simple approaches. The A_S+logP and ALOGPS methods, based on 2D structure representation, provided the most accurate prediction of 278 molecules from the benchmarking dataset. Since most of these molecules were also presumably used as the training set of some of the programs, one should treat the observed ranking of the analyzed approaches with caution. From another side, a number of methods could not provide significantly better results compared to the 15-year-old MLOGP program and even over the AAM, which does not require any calculation. The pitfalls of benchmarking studies arising from sensitivity of methods to different tautomeric forms, different treatments of aromaticity were also discussed.

Acknowledgments

The authors want to thank P. Buchwald (University of Miami, USA) for providing QLOGP calculations, P.-A. Carrupt (University of Geneva, Switzerland) for CLIP calculations, R. Fraczkiewicz (SimulationsPlus, USA) for A_S+logP calculations, D. Haney (EduSoft, USA) for HINT calculations, P. Japertas (Pharma Algorithms, Canada) for calculation using ABSOLV and original LSER equation, R. Kühne (Helmholtz Centre for Environmental Research, Germany) for calculations using LSER equation, H. McClelland (Accelrys Software, Inc., USA) for VLOGP calculations, O. Raevsky (Institute of Physiologically Active Compounds, Russia) for

SLIPPER calculations, A. Saravanaraj (University of Georgia, USA) for SPARC calculations, M. Totrov (MolSoft, USA) for GBLOGP calculations, G. Vistoli (University of Milan) for VEGA calculations, and J. Votano (ChemSilico, USA) for CSlogP calculations.

I.V.T. thanks A. Klamt (COSMOlogic, Germany) for the possibility to use their algorithm as demo version at the Virtual Computational Chemistry Laboratory site. The authors also thank A. Avdeef (Pion, USA), R. Mannhold (University of Düsseldorf, Germany) and C. Ostermann (NYCOMED Pharma, Germany) for preparation and verification of the benchmarking dataset used in this study. We also thank M. Medlin (BioByte) for providing a demo version of Bio-Loom to access CLOGP 5.0.

References

1 Sangster, J. Octanol–water partition coefficients of simple organic compounds. *J. Phys. Chem. Ref. Data* **1989**, *18*, 1111–1229.

2 Born, M. Volumen und Hydrationwärme der Ionen. *Z. Phys.* **1920**, *1*, 45–48.

3 Duffy, E. M., Jorgensen, W. L. Prediction of properties from simulations: free energies of solvation in hexadecane, octanol, and water. *J. Am. Chem. Soc.* **2000**, *122*, 2878–2888.

4 Kamlet, M. J., Doherty, R. M., Carr, P., Abraham, M. H., Marcus, Y., Taft, R. W. Linear solvation energy relationships. 46. An improved equation for correlation and prediction of octanol–water partition coefficients of organic non-electrolytes (including strong hydrogen bond donor solutes). *J. Phys. Chem.* **1988**, *92*, 5244–5255.

5 Kamlet, M. J., Abboud, J.-L. M., Abraham, M. H., Taft, R. W. Linear solvation energy relationships. 23. A comprehensive collection of the solvatochromic parameters, π^*, α, and β, and some methods for simplifying the generalized solvatochromic equation. *J. Org. Chem.* **1983**, *48*, 2877–2887.

6 Abraham, M. H. Scales of solute hydrogen-bonding: their construction and application to physicochemical and biochemical processes. *Chem. Soc. Rev.* **1993**, *22*, 73–83.

7 Abraham, M. H., Chadha, H. S., Whiting, G. S., Mitchell, R. C. Hydrogen bonding. 32. An analysis of water–octanol and water–alkane partitioning and the Δlog *P* parameter of Seiler. *J. Pharm. Sci.* **1994**, *83*, 1085–1100.

8 Platts, J. A., Butina, D., Abraham, M. H., Hersey, A. Estimation of molecular linear free energy relation descriptors using a group contribution approach. *J. Chem. Inf. Comput. Sci.* **1999**, *39*, 835–845.

9 Platts, J. A., Abraham, M. H., Butina, D., Hersey, A. Estimation of molecular linear free energy relationship descriptors by a group contribution approach. 2. Prediction of partition coefficients. *J. Chem. Inf. Comput. Sci.* **2000**, *40*, 71–80.

10 Leo, A. J. *The MedChem Database*, BioByte Corp. and Pomona College, Daylight Chemical Information Systems, Mission Viejo, CA, **2003**.

11 ABSOLV, http://www.ap-algorithms.com/absolv.htm.

12 Schüürmann, G., Ebert, R. U., Kühne, R. Prediction of physicochemical properties of organic compounds from 2D molecular structure – fragment methods vs. LFER models. *Chimia* **2006**, *60*, 691–698.

13 Raevsky, O. A., Trepalin, S. V., Trepalina, H. P., Gerasimenko, V. A., Raevskaja, O. E. SLIPPER-2001 – software for predicting molecular properties on the basis of physicochemical descriptors and structural similarity. *J. Chem. Inf. Comput. Sci.* **2002**, *42*, 540–549.

14 Raevsky, O. A. Molecular lipophilicity calculations of chemically heterogeneous

chemicals and drugs on the basis of structural similarity and physicochemical parameters. *SAR QSAR Environ. Res.* **2001**, *12*, 367–381.

15 Raevsky, O. A. Quantification of non-covalent interactions on the basis of the thermodynamic hydrogen bond parameters. *J. Phys. Org. Chem.* **1997**, *10*, 405–413.

16 Hilal, S. H., Carreira, L. A., Karickhoff, S. W., Melton, C. M. Estimation of gas–liquid chromatographic retention times from molecular structure. *J. Chromatogr. A* **1994**, *662*, 269–280.

17 Hilal, S. H., Carreira, L. A., Karickhoff, S. W., Melton, C. M. Estimation of electron affinity based on structure activity relationships. *Quant. Struct.-Activ. Rel.* **1993**, *12*, 389–396.

18 Karickhoff, S. W., McDaniel, V. K., Melton, C., Vellino, A. N., Nute, D. E., Carreira, L. A. Predicting chemical reactivity by computer. *Environ. Toxicol. Chem.* **1991**, *10*, 1405–1416.

19 Hammett, L. P. *Physical Organic Chemistry*, 2 edn., McGraw-Hill, New York, **1970**.

20 Dewar, M. J. S. *The Molecular Orbital Theory for Organic Chemistry*, McGraw-Hill, New York, **1969**.

21 Hilal, S. H., Karickhoff, S. W., Carreira, L. A. Prediction of the solubility, activity coefficient and liquid/liquid partition coefficient of organic compounds. *QSAR Comb. Sci.* **2004**, *23*, 709–720.

22 Rogers, K. S., Cammarata, A. A molecular orbital description of the partitioning of aromatic compounds between polar and nonpolar phases. *Biochim. Biophys. Acta* **1969**, *193*, 22–29.

23 Klopman, G., Iroff, L. D. Calculation of partition coefficients by the charge density method. *J. Comput. Chem.* **1981**, *2*, 157–160.

24 Bodor, N., Gabanyi, Z., Wong, C. A new method for the estimation of partition coefficient. *J. Am. Chem. Soc.* **1989**, *111*, 3783–3786.

25 Bodor, N., Huang, M. J. An extended version of a novel method for the estimation of partition coefficients. *J. Pharm. Sci.* **1992**, *81*, 272–281.

26 Breindl, A., Beck, B., Clark, T., Glen, R. C. Prediction of the *n*-octanol/water partition coefficient, log *P*, using a combination of semiempirical MO-calculations and a neural network. *J. Mol. Model.* **1997**, *3*, 142–155.

27 Beck, B., Breindl, A., Clark, T. QM/NN QSPR models with error estimation: vapor pressure and log *P*. *J. Chem. Inf. Comput. Sci.* **2000**, *40*, 1046–1051.

28 Bodor, N., Buchwald, P. Molecular size based approach to estimate partition properties for organic solutes. *J. Phys. Chem. B* **1997**, *101*, 3404–3412.

29 Dewar, M. J. S., Zoebisch, E. G., Healy, E. F., Stewart, J. J. P. AM1: a new general purpose quantum mechanical molecular model. *J. Am. Chem. Soc.* **1985**, *107*, 3902–3909.

30 Buchwald, P., Bodor, N. Octanol–water partition of nonzwitterionic peptides: predictive power of a molecular size-based model. *Proteins* **1998**, *30*, 86–99.

31 Totrov, M. Accurate and efficient generalized Born model based on solvent accessibility: derivation and application for log *P* octanol/water prediction and flexible peptide docking. *J. Comput. Chem.* **2004**, *25*, 609–619.

32 Moriguchi, I., Hirono, S., Nakagome, I., Hirano, H. Comparison of reliability of log *P* values for drugs calculated by several methods. *Chem. Pharm. Bull.* **1994**, *42*, 976–978.

33 Klamt, A., Schüürmann, G. COSMO: a new approach to dielectric screening in solvents with explicit expressions for the screening energy and its gradient. *J. Chem. Soc. Perkin Trans.* **1993**, *2*, 799–805.

34 Eckert, F., Klamt, A. Fast solvent screening via quantum chemistry: COSMO-RS approach. *Aiche J.* **2002**, *48*, 369–385.

35 Klamt, A., Eckert, F. COSMO-RS: a novel way from quantum chemistry to free energy, solubility and general QSAR-descriptors for partitioning. In *Rational Approaches to Drug Design*, Höltje, H.-D., Sippl, W. (eds.), Prous, Barcelona, **2001**, pp. 195–205.

36 Hornig, M., Klamt, A. COSMOfrag: a novel tool for high-throughput ADME property prediction and similarity screening based on quantum chemistry. *J. Chem. Inf. Model.* **2005**, *45*, 1169–1177.

37 The Physical Properties Database (PHYSPROP), Syracuse Research Corp., http://www.syrres.com.

38 Haeberlein, M., Brinck, T. Prediction of water–octanol partition coefficients using theoretical descriptors derived from the molecular surface area and the electrostatic potential. *J. Chem. Soc. Perkin Trans. 2* **1997**, 289–294.

39 Chuman, H., Mori, A., Tanaka, H., Yamagami, C., Fujita, T. Analyses of the partition coefficient, log P, using *ab initio* MO parameter and accessible surface area of solute molecules. *J. Pharm. Sci.* **2004**, *93*, 2681–2697.

40 Fedichev, P. O., Men'shikov, L. I., Long-range order and interactions of macroscopic objects in polar liquids. *ArXiv Condensed Matter e-prints* 0601129, 2006.

41 Quantum Pharmaceuticals, Moscow, Russia, http://q-pharm.com.

42 QikProp, Schrödinger LLC, http://www.schrodinger.com.

43 Jorgensen, W. L., Duffy, E. M. Prediction of drug solubility from structure. *Adv. Drug. Deliv. Rev.* **2002**, *54*, 355–366.

44 Furet, P., Sele, A., Cohen, N. C. 3D molecular lipophilicity potential profiles: a new tool in molecular modeling. *J. Mol. Graph.* **1988**, *6*, 182–189.

45 Fauchère, J. L., Quarendon, P., Kaetterer, L. Estimating and representing hydrophobicity potential. *J. Mol. Graph.* **1988**, *6*, 203–206.

46 Audry, E., Dallet, P., Langlois, M. H., Colleter, J. C., Dubost, J. P. Quantitative structure affinity relationships in a series of alpha-2 adrenergic amines using the molecular lipophilicity potential. *Prog. Clin. Biol. Res.* **1989**, *291*, 63–66.

47 Rekker, R. F., de Kort, H. M. The hydrophobic fragmental constant; an extension to a 1000 data point set. *Eur. J. Med. Chem.* **1979**, *14*, 479–488.

48 Ghose, A. K., Crippen, G. M. Atomic physicochemical parameters for three-dimensional structure-directed quantitative structure–activity relationships. I. Partition coefficients as a measure of hydrophobicity. *J. Comp. Chem.* **1986**, *7*, 565–677.

49 Kellogg, G. E., Semus, S. F., Abraham, D. J. HINT: a new method of empirical hydrophobic field calculation for CoMFA *J. Comput.-Aided Mol. Des.* **1991**, *5*, 454–552.

50 Kellogg, G. E., Abraham, D. J. Hydrophobicity: is Log$P_{o/w}$ more than the sum of its parts? *Eur. J. Med. Chem.* **2000**, *35*, 651–661.

51 Kellogg, G. E., Burnett, J. C., Abraham, D. J. Very empirical treatment of solvation and entropy: a force field derived from Log $P_{o/w}$. *J. Comput.-Aided Mol. Des.* **2001**, *15*, 381–393.

52 Abraham, D. J., Leo, A. J. Extension of the fragment method to calculate amino acid zwitterion and side chain partition coefficients. *Proteins* **1987**, *2*, 130–152.

53 Leo, A. J., Hoekman, D. Calculating log P_{oct} with no missing fragments; the problem of estimating new interaction parameters. *Perspect. Drug Discov. Des.* **2000**, *18*, 19–38.

54 Amadasi, A., Spyrakis, F., Cozzini, P., Abraham, D. J., Kellogg, G. E., Mozzarelli, A. Mapping the energetics of water–protein and water–ligand interactions with the "natural" HINT force field: predictive tools for characterizing the roles of water in biomolecules. *J. Mol. Biol.* **2006**, *358*, 289–309.

55 Gaillard, P., Carrupt, P. A., Testa, B., Boudon, A. Molecular lipophilicity potential, a tool in 3D QSAR: Method and applications. *J. Comput.-Aided Mol. Des.* **1994**, *8*, 83–96.

56 Broto, P., Moreau, G., Vandycke, C. Molecular structures, perception, autocorrelation descriptor and SAR studies; system of atomic contributions for the calculation of the octanol–water partition coefficient. *Eur. J. Med. Chem.* **1984**, *19*, 71–78.

57 Testa, B., Carrupt, P. A., Gaillard, P., Billois, F., Weber, P. Lipophilicity in molecular modeling. *Pharm. Res.* **1996**, *13*, 335–343.

58 Pedretti, A., Villa, L., Vistoli, G. VEGA: a versatile program to convert, handle and visualize molecular structure on Windows-based PCs. *J. Mol. Graph. Model.* **2002**, *21*, 47–49.

59 Kraszni, M., Banyai, I., Noszal, B. Determination of conformer-specific partition coefficients in octanol/water systems. *J. Med. Chem.* **2003**, *46*, 2241–2245.

60 Kim, K. H. Calculation of hydrophobic parameters directly from three-dimensional structures using comparative molecular field analysis. *J. Comput.-Aided Mol. Des.* **1995**, *9*, 308–318.

61 Waller, C. L. A three dimensional technique for the calculation of octanol-water partition coefficients. *Quant. Struct.-Act. Relat.* **1994**, *13*, 172–176.

62 Caron, G., Ermondi, G. Calculating virtual $\log P$ in the alkane/water system ($\log P^{\mathrm{N}}_{\mathrm{alk}}$) and its derived parameters deltalog $\log P^{\mathrm{N}}_{\mathrm{oct-alk}}$ and $\log D^{\mathrm{pH}}_{\mathrm{alk}}$. *J. Med. Chem.* **2005**, *48*, 3269–3279.

63 Cruciani, G., Pastor, M., Guba, W. VolSurf: a new tool for the pharmacokinetic optimization of lead compounds. *Eur. J. Pharm. Sci.* **2000**, *11*, S29–S39.

64 Moriguchi, I., Hirono, S., Liu, Q., Nakagome, I., Matsushita, Y. Simple method of calculating octanol/water partition coefficient. *Chem. Pharm. Bull.* **1992**, *40*, 127–130.

65 Devillers, J., Balaban, A. T. *Topological indices and Related Descriptors In QSAR and QSPR*, Gordon & Breach, Amsterdam, **1999**.

66 Todeschini, R., Consonni, V. *Handbook of Molecular Descriptors (Methods and Principles in Medicinal Chemistry)*, Wiley-VCH, Weinheim, **2000**.

67 Niemi, G. J., Basak, S. C., Veith, G. D., Grunwald, G. D. Prediction of octanol water partition coefficient (K_{ow}) using algorithmically-derived variables. *Environ. Toxicol. Chem.* **1992**, *11*, 893–900.

68 Devillers, J., Domine, D., Guillon, C., Karcher, W. Simulating lipophilicity of organic molecules with a back-propagation neural network. *J. Pharm. Sci.* **1998**, *87*, 1086–1090.

69 Junghans, M., Pretsch, E. Estimation of partition coefficients of organic compounds: local database modeling with uniform-length structure descriptors. *Fresenius J. Anal. Chem.* **1997**, *359*, 88–92.

70 Baumann, K., Clerc, J. T. Computer-assisted IR spectra prediction – linked similarity searches for structures and spectra. *Anal. Chim. Acta* **1997**, *348*, 327–343.

71 Sadowski, J., Gasteiger, J., Klebe, G. Comparison of automatic three-dimensional model builders using 639 X-ray structures. *J. Chem. Inf. Comput. Sci.* **1994**, *34*, 1000–1008.

72 Kier, L. B., Hall, L. H. *Molecular Structure Description: The Electrotopological State*, Academic Press, London, **1999**.

73 Hall, L. H., Kier, L. B. Electrotopological state indices for atom types – a novel combination of electronic, topological, and valence state information. *J. Chem. Inf. Comput. Sci.* **1995**, *35*, 1039–1045.

74 Parham, M. E., Hall, L. H., Kier, L. B. High quality of property predictions by Molconn-Z and artificial neural network modeling. *Abstr. Papers Am. Chem. Soc.* **2000**, *220*, U288.

75 Gombar, V. K., Enslein, K. Assessment of *n*-octanol/water partition coefficient: when is the assessment reliable? *J. Chem. Inf. Comput. Sci.* **1996**, *36*, 1127–1134.

76 Gombar, V. K. Reliable assessment of $\log P$ of compounds of pharmaceutical relevance. *SAR QSAR Environ. Res.* **1999**, *10*, 371–380.

77 Huuskonen, J. J., Livingstone, D. J., Tetko, I. V. Neural network modeling for estimation of partition coefficient based on atom-type electrotopological state indices. *J. Chem. Inf. Comput. Sci.* **2000**, *40*, 947–955.

78 Huuskonen, J. J., Villa, A. E., Tetko, I. V. Prediction of partition coefficient based on atom-type electrotopological state indices. *J. Pharm. Sci.* **1999**, *88*, 229–233.

79 Tetko, I. V., Tanchuk, V. Y., Villa, A. E. Prediction of *n*-octanol/water partition coefficients from PHYSPROP database using artificial neural networks and E-state indices. *J. Chem. Inf. Comput. Sci.* **2001**, *41*, 1407–1421.

80 Tetko, I. V., Tanchuk, V. Y. Application of associative neural networks for prediction of lipophilicity in ALOGPS 2.1 program. *J. Chem. Inf. Comput. Sci.* **2002**, *42*, 1136–1145.

81 Tetko, I. V., Poda, G. I. Application of ALOGPS 2.1 to predict log *D* distribution coefficient for Pfizer proprietary compounds. *J. Med. Chem.* **2004**, *47*, 5601–5604.

82 Tetko, I. V., Bruneau, P. Application of ALOGPS to predict 1-octanol/water distribution coefficients, log *P*, and log *D*, of AstraZeneca in-house database. *J. Pharm. Sci.* **2004**, *93*, 3103–3110.

83 Tetko, I. V., Livingstone, D. J. Rule-based systems to predict lipophilicity. In *ADME/Tox Approaches*, Van de Waterbeemd, H., Testa, B. (vol. eds.), Vol. 5 in *Comprehensive Medicinal Chemistry*, 2nd edn., Taylor, J. B., Triggle, D. J. (eds.), Elsevier, Oxford, **2006**, pp. 649–668.

84 VLOGP (v 3.1) model from the Toxicity Prediction (TOPKAT) protocol in Discovery Studio 1.7, http://www.accelrys.com/products/topkat.

85 Klopman, G., Li, J.-Y., Wang, S., Dimayuga, M. Computer automated log *P* calculations based on an extended group contribution approach. *J. Chem. Inf. Comput. Sci.* **1994**, *34*, 752–781.

86 Tetko, I. V. Neural network studies. 4. Introduction to associative neural networks. *J. Chem. Inf. Comput. Sci.* **2002**, *42*, 717–728.

87 Tetko, I. V. Associative neural network. *Neural Process. Lett.* **2002**, *16*, 187–199.

88 A databank of evaluated octanol–water partition coefficients (LOGKOW), http://logkow.cisti.nrc.ca/logkow.

89 ADMET Predictor, Simulations Plus Inc., http://www.simulations-plus.com.

90 Avdeef, A. *Absorption and Drug Development. Solubility, Permeability and Charge State*. Wiley-Interscience, Hoboken, **2003**.

91 Tetko, I. V., Solov'ev, V. P., Antonov, A. V., Yao, X., Doucet, J. P., Fan, B., Hoonakker, F., Fourches, D., Jost, P., Lachiche, N., Varnek, A. Benchmarking of linear and nonlinear approaches for quantitative structure–property relationship studies of metal complexation with ionophores. *J. Chem. Inf. Model.* **2006**, *46*, 808–819.

92 ICM (Internal Coordinate Mechanics) software, MolSoft LLC, http://www.molsoft.com/icm_pro.html.

93 Platts, J. A., Abraham, M. H., Zhao, Y. H., Hersey, A., Ijaz, L., Butina, D. Correlation and prediction of a large blood–brain distribution data set – an LFER study. *Eur. J. Med. Chem.* **2001**, *36*, 719–730.

94 Meylan, W. M., Howard, P. H. Atom/fragment contribution method for estimating octanol–water partition coefficients. *J. Pharm. Sci.* **1995**, *84*, 83–92.

95 Tetko, I. V., Bruneau, P., Mewes, H. W., Rohrer, D. C., Poda, G. I. Can we estimate the accuracy of ADME-Tox predictions? *Drug Discov. Today* **2006**, *11*, 700–707.

96 Walker, M. J. Training ACD/LogP with experimental data. *QSAR Comb. Sci.* **2004**, *23*, 515–520.

97 Katrusiak, A. Polymorphism of maleic hydrazide. I. *Acta Crystallogr. B* **2001**, *57*, 697–704.

98 Tetko, I. V., Gasteiger, J., Todeschini, R., Mauri, A., Livingstone, D., Ertl, P., Palyulin, V. A., Radchenko, E. V., Zefirov, N. S., Makarenko, A. S., Tanchuk, V. Y., Prokopenko, V. V. Virtual Computational Chemistry Laboratory – design and description. *J. Comput.-Aided Mol. Des.* **2005**, *19*, 453–463.

16
The Good, the Bad and the Ugly of Distribution Coefficients: Current Status, Views and Outlook

Franco Lombardo, Bernard Faller, Marina Shalaeva, Igor Tetko, and Suzanne Tilton

Abbreviations

ADMET	absorption, distribution, metabolism, excretion and toxicity
BBB	blood–brain barrier
CHI	chromatographic hydrophobicity index
CRO	contract research organization
DMSO	dimethylsulfoxide
EPC	egg phosphatidylcholine liposomes
HPLC	high-performance liquid chromatography
IAM	immobilized artificial membrane
MAE	mean absolute error
MEEKC	microemulsion electrokinetic chromatography
NCE	new chemical entity
ODS	octadecylsilane
PK	pharmacokinetics
RP	reversed-phase
RMSE	root mean square error
QSAR	quantitative structure–activity relationship
QSPR	quantitative structure–property relationship

Symbols

$\log P_{oct}$	octanol–water partition coefficient
$\log D_{oct}$	octanol–water distribution coefficient
pK_a	ionization constant
$\log k'$	logarithm of the capacity factor
$\log k_w$	logarithm of the capacity factor extrapolated for a hypothetical 100% water eluent

Molecular Drug Properties. Measurement and Prediction. R. Mannhold (Ed.)

ISBN: 978-3-527-31755-4

16.1 Log *D* and Log *P*

16.1.1 Definitions and Equations

A general definition of log *P* and log *D*, in its simplest form, can be given as the logarithm of the ratio (*P* or *D*) of the concentration of species of interest (the "drug" in a pharmaceutical context) in each phase, assuming the phases are immiscible and well separated prior to analysis. *P* is defined as the *partition coefficient*, whereas *D* is the *distribution coefficient*. However, the simplest form does not reveal some of the intricacies of the determination and use of these parameters, and further explanation is necessary.

For example, the definition of the two phases is necessary even though they generally represent water (or buffer) and octanol, mutually saturated prior to the determination of *P*. The usual symbol for this parameter then becomes log P_{oct} and a three-letter subscript (e.g. cyc for cyclohexane, chl for chloroform) is generally applied to any organic water-immiscible solvent used as the organic phase, with the implicit assumption that water (or buffer) constitutes the aqueous phase. By far, the octanol–water (or buffer) system is the most used and reported in the now vast and ever-growing array of absorption, distribution, metabolism and excretion (ADME), quantitative structure–activity relationship (QSAR), and quantitative structure–property relationship (QSPR) applications of "lipophilicity" expressed as log P_{oct}. This stems from the choice, performed over 40 years ago by Corwin Hansch [1], of the octanol–water system as a relatively simple and accessible experimental "model" of much more complicated phenomena, related to the partition of a solute within biological systems comprised of aqueous and lipophilic ("fat-like") environments. It follows that, for the vast majority of scientists in the field, the octanol–water (or buffer) system is the system of choice, and the most utilized and referenced. The growing set of computational approaches aimed at predicting log P_{oct} and log D_{oct} is the direct result of the popularity of that choice, and the consequent availability of such data (see Chapters 14 and 15).

We should then define what the "species of interest" or the "drug" represent as this is of paramount importance for the application and an understanding of these parameters. Log P_{oct}, or any other partition coefficient, is a thermodynamic property of the solute or "drug" which does not depend on the pH of the aqueous phase and it is determined well below saturation concentration of the drug in either phase, i.e. P_{oct} represents the ratio of the neutral form of the drug in each phase, yielding no information on its ionization state. It is worth mentioning, at this point, that mutual saturation of the phases should be achieved prior to experiment and this is not a trivial aspect as it may impact significantly on the result of the determination. Nor it is a rapid process that can be achieved in 1–2 h while equilibrating the drug between phases. Water-saturated octanol is approximately 2 M in water, while octanol-saturated water is approximately 1×10^{-3} M in octanol, but it takes quite some time to achieve such equilibrium between phases. The excellent and timeless article by Dearden [2] discusses and sets forth the requirements and good practices for these determinations. The equation describing P_{oct} can be written as:

$$P_{oct} = \frac{[\text{Drug}^{N}]_{oct}}{[\text{Drug}^{N}]_{w}} \tag{1}$$

where Drug^{N} represents the neutral form of the drug.

D_{oct}, on the other hand, refers to the "distribution" of all species (neutral or ionized) related to the compound of interest and is, by its nature, invariably less than or equal to P_{oct}, the two values being coincident when all of the drug is in its neutral form and thus $P_{oct} = D_{oct}$. The definition of D_{oct}, most commonly expressed as its logarithm, then requires the definition of the pH of the aqueous phase used for the determination and, generally, this is accomplished by adding a superscript reflecting the pH value, i.e. $\log D_{oct}^{7.4}$ for a buffer at pH 7.4. Due to the physiological importance of the latter pH value, most determinations and applications adopt this parameter. However, other values are encountered in the literature, such as determinations at pH 6.5 or 6.8, e.g. for applications to the prediction of intestinal absorption [3]. In general, for a monoprotic acid or base in the octanol–buffer system, the difference between the partition and distribution coefficient of the completely ionized drug, can be generally expected to be between 3 and 4 units, as shown by several examples and common observations [4, 5]. The equation describing the distribution coefficient D in a generic organic-buffer system, and at an unspecified pH can be generally written as:

$$D = D^{N} + D_{1}^{I} + \ldots + D_{n}^{I} \tag{2}$$

Where D^{N} is the distribution coefficient of the neutral form of the drug and D_{n}^{I} represents the distribution coefficient of each of the n ionized species.

The relationships between the two quantities for a generic monoprotic base can be written as:

$$D^{N} = \frac{P^{N}}{(1 + 10^{(pK_a - \text{pH})})} \tag{3}$$

and:

$$D^{I} = \frac{P^{I} \times 10^{(pK_a - \text{pH})}}{(1 + 10^{(pK_a - \text{pH})})} \tag{4}$$

where, in the symbolism used by Caron et al. [4], P^{N} represents the partition of the neutral species and P^{I} represents the partition of the completely ionized species. This yields D^{N} and D^{I}, respectively, upon correction for the fraction of each present at a specified pH.

Lastly, some authors may refer to $\log D$ as $\log P'$ with the "prime" superscript indicating that the determination does not refer to the ratio of neutral species and it is, rather, a $\log D$ determination. However, such symbolism should not be used, especially since it may lead to confusion among the users of these parameters.

On the subject of applications of such experimental parameters, the questions of whether an ionized molecule may or may not partition into octanol and whether bulk-phase partitioning is in turn adequate to describe bilayer partitioning are important and pertinent ones, and are addressed in Section 16.3.

In this section it will suffice to say that ion-pairing, including the drug's ability to alter its conformation and thus intramolecularly mask charges, may dramatically change the outcome of the solute (drug) partition between octanol and water. The practice of using back-calculations of $\log P_{oct}$ based on a $\log D_{oct}$ determination at a given pH and the molecule pK_a, often performed for water-solubility reasons, is based on the assumption that only the neutral species can partition into octanol. In other words, it relies on the assumption that the concentration of the ionized species, in octanol, is negligible. Such an assumption, depending upon the nature and concentration of the counterions and drug, may yield significantly inaccurate results and it may have to be verified case by case.

16.1.2
Is There Life After Octanol?

The question posed by the title of this section may at first be answered with a "no" and thus yield the undoubtedly shortest section of the present, and possibly any, chapter while saving the authors, as well as the readers, time and effort.

Nevertheless, and even though we have stated above that by far the most used and modeled system is in fact represented by the octanol–water pair of solvents, the question deserves some clarification and discussion.

Hartmann and Schmitt [6], in a recent review article, provide a comparison of technologies that attempt to go beyond octanol–water systems, aimed at reproducing or substituting the information encoded by the octanol–water system, and/or to find lipophilicity parameters that may be better suited to reproduce biological phenomena and, in particular, ADME aspects. These authors roughly classify the approaches in five main areas: solvent–water partitioning, chromatographic approaches, artificial membranes, electrokinetic approaches, and partitioning between lipid and water phases. Several chapters (Chapters 2, 3, 12 and 13) in this book discuss in detail these approaches and we will limit our discussion to bulk-phase systems and, in particular, to the so-called "critical quartet" defined by Leahy et al. [7], and to the $\Delta\log P$ parameter, defined by Seiler [8], with a few examples and references.

One of the likely reasons octanol–water gained such a widespread use is its amphiprotic character, i.e. its ability to serve as an H-bond donor and acceptor while, for example, alkane–water systems are inert in that sense, from the point of view, of course, of the organic phase. The other two systems of the critical quartet are represented by chloroform–water (proton donor) and propylene glycol dipelargonate–water, the latter organic phase being an H-bond acceptor, as in biological membranes. These systems, collectively, encode all important H-bonding properties and others, such as dichloroethane–, dibutyl ether– and *o*-nitrophenyl octyl ether–water [9, 10] have been proposed. However, we are not aware of any of these systems being routinely or at all used in industrial physicochemical characterization laboratories, whether in a high-throughput mode or not. On the other hand, artificial membrane, chromatographic, automated bulk phase partitioning (octanol–water) and electrokinetic systems have gained acceptance and application in industrial laboratories, with the first two probably accounting for the vast majority of the approaches taken.

The cyclohexane–water (or generally the alkane–water) system deserves further mention having been successfully used in a seminal paper on blood–brain barrier (BBB) permeation prediction by Young et al. [11]. These authors synthesized series of H_2 histamine receptor antagonists and determined their partition coefficients in octanol–water, chloroform–water and cyclohexane–water systems. Although significant linear relationships between the partition coefficients and the logarithm of the brain–blood concentration ratio of the drugs ($\log BB$) were found, using only a limited number of compounds ($n=6$), for $\log P_{cyc}$ and $\log P_{chl}$ (but not for $\log P_{oct}$), the best correlation was found using the $\Delta\log P$ parameter and expressed as:

$$\Delta \log P_{oct-cyc} = \log P_{oct} - \log P_{cyc} \tag{5}$$

This approach also resulted in a very significant correlation (69% of variance) in a retrospective analysis of 20 compounds. The authors speculated that $\log P_{oct}$ might account for plasma protein binding in blood, which may limit the amount of the free drug able to permeate the BBB, while $\log P_{cyc}$ might reflect the partitioning in nonpolar regions of the brain. An alternative explanation is that the difference between the two partition coefficients brings out the H-bond donor character of the solute, all other parameters (e.g. volume, dipolarity/polarizability, H-bond acceptor character) essentially canceling out as expressed in the solvation parameters analysis of Abraham et al. [12]. The high propensity for H-bonding was interpreted as a detrimental factor for permeation, increasing the tendency of a drug to remain in the blood compartment rather than passing through and into a fairly nonpolar environment.

The difficulties and low-throughput nature of the experimental dual determination, especially in alkane–water systems, the development of other techniques more amenable to automation, as well as more refined computational approaches for octanol–water systems, all have contributed to limit the use of the alkane–water system as a “second” bulk-phase system. However, efforts have been devoted to the development of $\log P_{alk}$ (alkane) computational prediction methods by Rekker et al. [13] as well as Caron and Ermondi [14].

At any rate, due to its wide popularity and the vast availability of literature and in-house data, whether computed or experimentally determined, octanol–water partition or distribution data remain by far the most utilized single parameter for ADME, QSPR and QSAR predictions.

16.1.3 Log *P* or Log *D*?

The determination of $\log P_{oct}$ is far from being trivial whether potentiometric, shake-flask, chromatographic or other techniques are used and, often, this value is derived from the back-calculation discussed above, using $\log D_{oct}^{pH}$ and pK_a, which of course has to be known. This is often done in shake-flask determinations so that there is appreciable aqueous solubility (and thus partition) in the aqueous phase for highly lipophilic drugs and it may also be accompanied by a variation of the phase ratio, in favor of the phase where the compound is expected to be less soluble, to avoid saturation phenomena.

The danger of calculating $\log P$ from $\log D$ determinations lies primarily in the assumption that only the neutral form of the drug could partition into octanol and we have briefly illustrated, and we will discuss in Section 16.3, that this is not necessarily true and, in fact, may often be an incorrect assumption.

In the majority of cases, excluding the very low pH environment of the stomach, the determination of $\log D$ at pH 7.4 and/or at pH 6.8, as a model of the lipophilicity of a drug in the systemic compartment or at intestinal pH, respectively, should yield the most desirable value for prediction and modeling. However, all predictions and hypotheses should be mindful of the caveat expressed and, especially, of the limitations of using a bulk property to model much more complex interactions with biological systems. Thus, we would favor the determination of a "native" $\log D$ value since often the back-calculation of the value for the neutral form is not realistic given the polyprotic nature of many drugs. When determination is not achievable because of very high lipophilicity, especially for neutral compounds at the pH of interest, potentiometric techniques (when the compound is ionizable, with a cosolvent used for pK_a determination) or chromatographic approaches are used, the latter with extrapolation to a "zero" cosolvent amount [15]. To our knowledge this is a much more acceptable approach and widely used in the industrial pharmaceutical laboratories.

Another question may be asked: why is there a plethora of $\log P_{oct}$ prediction packages, with more of them continually appearing in the literature, but only a few available models for $\log D$ prediction, to our knowledge? One answer may be that most of the data, for a great many number of solutes available throughout the vast $\log P_{oct}$ literature, are indeed for neutral compounds. Furthermore, and even with good-quality $\log D_{oct}$ data, the presence of charges, with the complications represented by ion-pairing and conformational changes, turns modeling $\log D$ into a nontrivial exercise and requires, obviously, the calculation of fairly accurate pK_a values. In our experience, the performance of predictive software packages is well above an accuracy of 0.5 log D or pK_a units, and it could reach several units depending on the class of compounds, nature of substituents and conformational flexibility present in the molecule, even for neutral molecules. We discuss computational aspects in more detail in Section 16.4.

We also note that many ADME, QSAR or QSPR models, based on experimental or computed parameters, use a combination of log P and partial charges and/or fraction ionized at a given pH, as independent variables, rather than the potentially more "physiological" log $D^{7.4}$ or log $D^{6.8}$ values. This tendency may reflect a perceived "superiority" and "accuracy" of the log P values, whether computed or experimentally determined, and may also be reflected by the nature of the data stored observed among different industrial settings.

16.1.4
ADME Applications

A thorough review, even with a very superficial mention of the vast literature on ADME, QSPR and QSAR applications involving the use of lipophilicity, would be a daunting task and it is far beyond the scope of this chapter. However, it is no surprise that three international conferences, in recent years, have been specifically dedicated to this topic, and that the development of newer and faster screening methods, in some cases seeking to produce alternative lipophilicity parameters to the classical or nonclassical log P_{oct} determinations are still an active area of interest [6].

Smith et al. [16] have used a tabular approach to describe the importance and impact of lipophilicity on physicochemical properties and ADME behavior of drugs, the latter two being inextricably related. In a quali-quantitative approach they reported a powerful and yet succinct way of describing the "response" of, for example, clearance to an increase in log P (or log D) by using symbols such as "+" and "–" for a direct or inverse correlation of a given property or pharmacokinetic (PK) parameters with lipophilicity, and they used repeated symbols to describe the strength of the correlation. Another classification, based on log D interval can be found in the work of Comer [17].

It is generally accepted, in very broad terms, that a high lipophilicity compound will likely increase metabolic liabilities, while compounds that are hydrophilic (and of fairly low molecular weight) such as varenicline [18] will tend to be excreted unchanged by the kidney. Compounds that are very hydrophilic tend not to pass the BBB, which may be a positive aspect for non-central nervous system compounds, although they may have a solubility advantage useful for absorption. However, compounds that have a high level of hydrophilicity may offer, at the same time, concern in terms of their passive diffusion through enterocytes thus resulting in intestinal absorption issue, due to permeability, and the "balance" of solubility and permeability, for passive diffusion, has to be considered.

We also note, in closing, that lipophilicity is a key parameter of the well-known "Rule-of-5" [19] and it represents probably the single most broadly used parameter in these efforts whether the approach is experimental or *in silico* [20].

16.2
Issues and Automation in the Determination of Log *D*

16.2.1
Shake-Flask Method

As chronicled by Dearden [21], the association of compound lipophilicity with membrane penetration was first implied by Overton and Meyer more than a century ago. To enhance this understanding, lipophilicity measurements were initially performed using a variety of lipid phases [22], while the comprehensive review by Hansch et al. [23], with extensive data from literature and their own measurements, lent further support to the now accepted wide use of the octanol–water solvent system.

A detailed description of octanol–water distribution coefficient measurements by shake-flask can be found in publications by Dearden [2] and Hansch [24]. The method usually involves the following: solubilization of the compound in a mixture of mutually presaturated buffered water and octanol, agitation until equilibrium has been reached, careful separation of octanol and aqueous phases, and direct measurement of the solute concentration in both phases. Although seemingly simple, the method has a number of caveats making it inappropriate for some compounds.

Gocan et al. [25] summarized the drawbacks of shake-flask determinations in a recent review and identified the following issues:

- The formation of extremely stable emulsions for some compounds prevent the complete separation of the octanol and water phases, and, therefore, an accurate measurement of the analyte concentration cannot be made.
- The measurement of extreme log *P*/log *D* values (generally accepted to be below −3 and above 4) requires a disproportionate volume ratio of the aqueous and solvent phases, making it difficult to sample a reasonable amount for the concentration determination and automated sampling almost impossible.
- The sample concentration must be lower than the critical micelle concentration.
- The sample concentration must be set below the aqueous solubility limit.
- Some compounds, particularly lipophilic compounds, can adhere to the surface of the vessel.
- Some compounds can act as surfactants that concentrate at the interface between two phases or form foams.
- Compounds must have high purity.

Some of the complications listed above could be flagged and, sometimes, remedied in a manual shake-flask experiment, but that is unlikely to be the case in automated "shake-vial" procedures, especially if performed in a 96-well plate setting. Nevertheless, the demands of modern pharmaceutical discovery operations emphasize high-throughput measurements, low compound consumption and method versatility to accommodate diverse compounds. Several reviews and critical evaluations of a variety of methods for lipophilicity determination adapted to these demands are available [17, 23, 25–28].

Numerous automated "shake-flask" systems have been developed and reported. Recently, Hitzel et al. [29] developed a modified "shake-flask" procedure using a 96-well plate and a Beckman Biomek 2000 robotic liquid handler. The method performance has been validated on a test set of 30 compounds containing mostly drugs in the $\log D$ range from −2 to 4. The shaking time sufficient for equilibration to be reached in the 96-well plate was established at 30 min. It was also noted that in order to achieve good reproducibility, the compounds with negative $\log D$ values should be first solubilized in buffer solution, rather than in octanol. Sampling of each phase was done by inserting the autosampler needle to an appropriate depth, thus avoiding the phase separation step. The analyses were performed by a generic fast-gradient high-performance liquid chromatography (HPLC) method using a HP1090 system with a diode array detector.

An automated $\log P$ workstation using a shake-flask method and robotic liquid handling in 96-well plate format is commercially available [30]. The system is equipped with a diode-array spectrophotometer and equimolar nitrogen detector. Mass spectrometric detection could be incorporated into the detection system as well, allowing for a greater variety of detection methods. That is advantageous in a pharmaceutical discovery setting where it is challenging to quantify novel compounds with structural diversity and purity issues by any single detection method. While the system is optimized to handle dimethylsulfoxide (DMSO)-solubilized samples commonly found in industry, the required sample amount of approximately 130 μL of 30 mM DMSO solution is quite high.

16.2.2
Potentiometric Method

The potentiometric method for $\log P$ determination has been correlated with the shake-flask method [31, 32] and is available in a commercial instrument (GlpKa; Sirius Analytical Instruments, UK). This method has some advantages over the shake-flask, including experimental time, no separation step, less analytical burden and determination of ion partitioning. The potentiometric $\log P$ is characterized by comparing an aqueous pK_a to an apparent pK_a measured in the two-phase system (generally octanol–water) using difference curve analysis [33]. Therefore, the method is appropriate only for ionizable compounds with accurately determined aqueous pK_a values. However, knowledge of the aqueous pK_a allows a $\log D$ calculation at any pH. Further, when measuring ampholytic drugs it is not always possible to characterize the lipophilicity by a single shake-flask or HPLC $\log D$ value, due to overlapping acid–base protonation. In this case the potentiometric method may be used to deduce the true partition coefficient [34]. The potentiometric method suffers from the same compound solubility challenge faced by shake-flask and many other ADMET measurements. Precipitation during the titration effectively introduces an uncharacterized third phase into the system which can lead to erroneous values. While the difference curve analysis can highlight the onset of precipitate, it is not always possible to detect its presence in a log *P* titration. The potentiometric method is not considered high throughput and has been utilized more often for in-depth compound analysis to correlate physicochemical properties with PK results [35]. It has been applied to biphasic systems beyond octanol–water, including 1, 2-dichloroethane–water (isotropic system) and liposomes (anisotropic system) [36]. Recently the potentiometric method was used to investigate partition coefficients in the *o*-nitrophenyl octyl ether–water system [10]. This system retains a similar partitioning mechanism to 1, 2-dichloroethane, yet may be favorable because of improved properties such as lower solubility in water.

16.2.3 Chromatographic Methods

Advances in understanding solute interactions in liquid–liquid systems in a non-equilibrium environment brought reversed-phase (RP)-HPLC into the forefront of lipophilicity determination. The development and manufacturing of rigid, reproducible and well-characterized stationary phases and columns, as well as the accessibility and high level of automation of modern HPLC systems, have made RP-HPLC the method of choice for many laboratories.

The comprehensive review by Gocan et al. [25] focused specifically on lipophilicity measurements by liquid chromatography, including reversed phase, thin-layer, micellar, RP-ion-pair and countercurrent chromatography.

Valko et al. [37] developed a fast-gradient RP-HPLC method for the determination of a chromatographic hydrophobicity index (CHI). An octadecylsilane (ODS) column and 50 mM aqueous ammonium acetate (pH 7.4) mobile phase with acetonitrile as an organic modifier (0–100%) were used. The system calibration and quality control were performed periodically by measuring retention for 10 standards unionized at pH 7.4. The CHI could then be used as an independent measure of hydrophobicity. In addition, its correlation with linear free-energy parameters explained some molecular descriptors, including H-bond basicity/acidity and dipolarity/polarizability. It is noted [27] that there are significant differences between CHI values and octanol–water log *D* values.

Poole and Poole [26] examined suitability of various separation methods for the estimation of log *P* and log *D* using a solvation parameter model. RP-HPLC systems with different stationary phases and mobile-phase compositions along with variables such as $\log k'$ versus $\log k_w$ (the logarithm of the capacity factor and of its extrapolated value for a hypothetical 100% water eluent, respectively) were analyzed. The authors demonstrated that intermolecular interactions contributing to retention in RP-HPLC are similar, but not identical, in character to those responsible for the octanol–water partition coefficient. The contribution from H-bonding interactions was the most different. Only three separation systems were identified as possible models for octanol–water partition systems, based on the comparison of their solvatochromic parameters. In particular, the Supelcosil LC-ABZ+ column with methanol–water mobile phase was found suitable for estimating log P.

The most recent systematic analysis of retention mechanisms in relation to lipophilicity determination on three popular base-deactivated RP-HPLC stationary phases is presented in the work of Stella et al. [38]. These authors studied the correlation of log k_w and log P using solvatochromic analysis parameters and evaluated the intermolecular interaction forces underlying the partitioning mechanisms on three columns i.e. LC-ABZ+, Discovery RP-Amide C_{16} and Zorbax Extend C_{18}. An optimized set of 80 simple neutral and basic compounds with known solvatochromic parameters was selected. The results confirmed that the principal factors governing retention are van der Waals volume and H-bond acceptor basicity of the solutes, which are also the governing factors of partitioning in the octanol–water system. As expected, good correlation of log k_w and log P was achieved on all columns for neutral compounds. The factors influencing basic solutes in the buffered mobile phase in the presence of organic solvent were briefly discussed. The pK_a of basic compounds as well as the buffered mobile phase were influenced by the organic solvent and depended on its proportion in the solution. The results indicated that the stationary phases and mobile phases studied were not suitably optimized for the estimation of lipophilicity of basic compounds.

These studies confirmed results of Lombardo et al. describing an industrial-strength, high-throughput RP-HPLC method for the determination of log P [39] and log $D^{7.4}$ values [15]. The lipophilicity values obtained by this method were named Elog P and Elog D, respectively. The Supelcosil LC-ABZ column was used in this method together with a mobile phase consisting of 20 mM MOPS buffer at pH 7.4 and methanol with octanol added. The addition of octanol, proposed earlier by Minick et al. [40], provided saturation of the stationary liquid phase and appearance of a quasi-immobilized octanol stationary phase. The column performance was monitored periodically by ten standard compounds with lipophilicity ranging from –0.44 to 6.10 and sensitivity to variation in HPLC conditions.

For each compound the log k' values were obtained by isocratic methods at three different methanol: water ratios and the log k'_w value was derived by extrapolation to 100% water.

The obtained log k'_w correlated well with log $D^{7.4}$ values reported in the literature:

$$\log D_{\mathrm{oct}} = 1.1267(\pm 0.0233)\log k'_w + 0.2075(\pm 0.0430) \quad (6)$$

$$n = 90,\ r^2 = 0.964,\ s = 0.309,\ F = 2339,\ q^2 = 0.962$$

A more recent test set of 163 drugs [41] yielded a comparable correlation:

$$\log D_{\mathrm{oct}} = 1.084(\pm 0.020)\log k'_w + 0.200(\pm 0.040) \quad (7)$$

$$n = 163,\ r^2 = 0.949,\ s = 0.369,\ F = 3000,\ q^2 = 0.948$$

The major drawback of the Elog D method is that it cannot be applied to acidic compounds ionized at pH 7.4. However, on practical grounds, the latter represent a minor fraction of compounds in pharmaceutical libraries and structural alerts could be developed which are used to filter acidic compounds prior to the determination of Elog D in an automated fashion.

Lipophilicity determinations using HPLC on octanol-coated columns of a commercially available instrument from Sirius Analytical Instruments were reported [17]. A proprietary stationary phase dynamically coated with octanol and an octanol-saturated mobile phase were used for log $D^{7.4}$ measurements. Depending on the expected log D value each compound was measured using one of three methods optimized for the following log D ranges: 1.0–2.0, 1.5–3.5 and 3.0–4.5 [25]. The column coating uniformity should be carefully monitored and the column replaced about every 300 injections, precluding the system from being used in rugged pharmaceutical screening approaches.

The immobilized artificial membrane (IAM) chromatography method introduced by Pidgeon and Venkataram [42] was similar to other RP-HPLC methods except for the column, which was a silica resin modified by covalently attached phospholipid-like groups. The method was intended for the prediction of drug distribution in biological systems directly from HPLC measurements and has been widely reviewed [43, 44]. Recently, Lazaro et al. [45] applied the solvation parameter model to characterize an IAM column and compare it to common C_{18} columns in their ability to model biological processes. Using IAM.PC.DD2 and X-Terra MSC_{18} and RP_{18} columns and a set of 17–56 compounds they showed that, contrary to common belief, IAM was not always the best choice to model a biological drug distribution system. For example, human skin permeation was better modeled by C_{18} columns, while blood–brain or tissue-blood distribution could not be modeled well by either column. On the other hand, the octanol–water partition was predicted fairly well by the IAM system.

16.2.4 Electrophoretic Methods

Microemulsion electrokinetic chromatography (MEEKC), a variation of the capillary electrophoresis method, was successfully applied for measuring log P [46–48]. It was demonstrated by linear free energy relationship analysis [49] that a microemulsion composed of 1.44% (w/w) sodium dodecyl sulfate, 6.49% (w/w) *n*-butanol and 0.82% (w/w) heptane in borate-phosphate buffer, pH 7.0, was an exact model for water–octanol partition. However, the method was only suitable for estimating log P of neutral compounds, as additional electrophoretic and electrostatic interactions of ionized compounds caused changes in retention not related to partition mechanisms. The system could be optimized for determinations at lower and higher pH to accommodate for a wider range of weakly acidic and basic compounds, e.g. pH 3 and 10 [48]. However, compounds must be neutral in the appropriate range and therefore must be sorted based on pK_a predictions. Recently, the MEEKC method was transferred to a commercially available multiplexed capillary electrophoresis system from CombiSep [50, 51], providing increased throughput and a greater degree of measurement automation.

16.2.5 IAMs

High-throughput measurements of octanol–water partition coefficients have also been achieved through the use of microtiter plates. Faller et al. [52] reported on an OCT-PAMPA assay based on diffusion of compounds between two aqueous compartments of a 96-well plate separated by a thin octanol layer. Unlike some chromatographic methods, the method was suitable for neutral, acidic and basic compounds. The octanol–water partition coefficients were derived from a calibration curve of apparent permeability of the neutral species ($\log Pa_N$). Calculated or measured pK_a values were used to determine at which aqueous compartment pH the compound is more than 90% neutral, and as such some ampholytes and zwitterions were less accurately measured due to lack of a full neutral fraction throughout the pH range 2–11. The use of a liquid chromatography-mass spectroscopy detection system enhanced the dynamic range from −2 to 8, although the extremes may be affected by the flatter relationship between $\log Pa_N$ and $\log P$ at the low end and solubility at the high end. When compared to reference $\log P_{oct}$ values, this high-throughput method offered a standard deviation of 0.31 over the predicted CLOGP standard deviation of 0.59 for 20 generic drugs.

16.2.6 Applications Perspective

While several methods exist for $\log P/D$ measurement, the timing and degree of application are less defined. The role of lipophilicity in describing *in vivo* properties such as biological permeation [53, 54], binding [55, 56] and other important PK parameters [3, 57, 58] has received much attention. It is also noteworthy to mention the very recent work of Valko et al. on the prediction of volume of distribution using immobilized human serum albumin and IAM columns [59], even though these authors do not use a "direct" $\log P/\log D$ determination. The contribution of lipophilicity is understood well enough to drive medicinal chemistry efforts toward an appropriate physicochemical space, guided by $\log P$ as well as its components of molecular size and hydrogen bonding capacity [19, 60, 61]. *In silico* models for $\log P/D$ are necessary for this volume of compounds, especially when compounds are not yet synthesized. However, as new chemical entities are made and profiled for their ADME/toxicity (ADMET) properties, the measurement of $\log P/D$ becomes significant.

Lipophilicity correlations with ADMET properties are often promising when marketed or well-characterized compounds are utilized in the training set. Even when structurally diverse sets of marketed drugs are chosen, the common denominator is that most of those compounds have optimized physicochemical properties. The error in calculated log P/D values for newly synthesized compounds is likely to be higher; the error then propagates into the correlation and yields inconsistencies, e.g. with *in vivo* results [27, 62].

Satisfactory correlations may be obtained more often for a structurally similar series of compounds where intermittent experimental log P/D values are obtained [56, 63]. In our experience, correlations with specific ADMET properties such as protein binding or relationships between calculated and experimental log P have been established with specific drug discovery programs without over-burdening the profiling resources. The number of compounds measured in a series may depend on the degree of structural modification. Van de Waterbeemd [64] described a series of β-blockers, Ca^{+2} channel antagonists and peptidic renin inhibitors, and gave examples of interrelationships between lipophilicity and PK properties. A number of properties, including *in vitro* potency and unbound volume of distribution, show excellent correlations to lipophilicity with nine to 10 compounds. Once established, such correlations may also serve to reduce PK laboratory resources by prioritizing new compounds with small structural modifications.

Not only does the practice of "spot checking" compounds for experimental measurement of log P/D free resources, it also allows room for higher quality results. The successful correlations reported in the literature rely on carefully determined results by measurement experts. In-house relationships should be strengthened by the same practice. Alternatively, outsourced measurements are available and are being applied more frequently to ADMET profiling strategies. It is good practice to check the quality of the contract research organization (CRO) performing the measurements by comparing their results to in-house measurements on new chemical entities (NCEs), if possible, as marketed drugs are not representative of the measurement challenges that the CRO will face with new discovery programs.

A CRO may also allow for the in-house introduction of specialized lipophilic scales by transferring routine measurements. While the octanol–water scale is widely applied, it may be advantageous to utilize alternative scales for specific QSAR models. Solvent systems such as alkane or chloroform and biomimetic stationary phases on HPLC columns have both been advocated. Seydel [65] recently reviewed the suitability of various systems to describe partitioning into membranes. Through several examples, he concludes that drug–membrane interaction as it relates to transport, distribution and efficacy cannot be well characterized by partition coefficients in bulk solvents alone, including octanol. However, octanol–water partition coefficients will persist in valuable databases and decades of QSAR studies.

Lipophilicity is one of the oldest physicochemical parameters and retains a steady presence in understanding ADMET and PK. While explorations of fully computational models for these parameters are continually being pursued and applied especially at early screening stages, the measurement of log P/D continues to be crucial for NCEs.

16.3
pH-partition Theory and Ion-pairing

16.3.1
General Aspects and Foundation of the pH-partition Theory

The pH-partition theory or nonionic permeability hypothesis was first described by Jacobs in 1940 [66]. According to this concept, only neutral, preferably nonpolar compounds are able to cross biological membranes. The transcellular permeability pH-profile is then essentially characterized by the membrane partition coefficient and the pKa of the compound. The simplest quantitative description of membrane permeation is given by:

$$\log P_{\mathrm{m}} = \log D_{\mathrm{mem}} + \log(D/h) \tag{8}$$

where P_{m} is the membrane permeability, D_{mem} the membrane–water distribution coefficient, h the membrane thickness and D the diffusion coefficient.

$$D_{\mathrm{mem}} = \frac{P_{\mathrm{HA}}}{1+10^{(\mathrm{pH}-\mathrm{p}K_{\mathrm{a}})}} \quad \text{for an acid} \tag{9}$$

or:

$$D_{\mathrm{mem}} = \frac{P_{\mathrm{B}}}{1+10^{(\mathrm{p}K_{\mathrm{a}}-\mathrm{pH})}} \quad \text{for an base} \tag{10}$$

Seventy years later, this theory largely holds true, although periodically challenged [67, 68]. Observation of transmembrane permeability of ionic species was initially explained by the formation of neutral ion-pair [69, 70]. A comprehensive review of the physicochemical properties influencing permeation has been written by Mälkiä et al. [5]. The reality is that, despite many studies, the effect of ionization on permeation is still a matter of discussion and active research. In contrast, it became clear that bulk-phase partitioning measurements are not adequate to describe bilayer partitioning [71–73].

16.3.2
Ion-pairing: *In Vitro* and *In Vivo* Implications

16.3.2.1 Ion-pairing *In Vitro*

The experimental approaches used to characterize ion-pair partitioning are cyclic voltammetry and potentiometric titration. Cyclic voltammetry is overall more powerful, but requires special instrumentation which is not commercially available as a ready-to-use set-up. For this reason the potentiometric titration technique has been more widely used.

The measurement of log P_n and log P_i by dual-phase potentiometric titration has been described by Avdeef [33]. Briefly, the method is based on the shift in the apparent pK_a upon addition of the partition solvent. In absence of ion-pair extraction, the apparent pK_a (p$_o K_a$) is related to the aqueous pK_a by the equation below:

$$p_oK_a = pK_a + \log\left(P_{HA}\frac{V_o}{V_w} + 1\right) \quad (11)$$

where V_o is the volume of organic phase, V_w the volume of the aqueous phase and P_{HA} the partition coefficient. In the presence of ion-pair extraction, Eq. (11) becomes:

$$p_oK_a = pK_a + \log\left[\frac{1+(V_o/V_w)P_{HA}}{1+(V_o/V_w)P_A^-}\right] \quad (12)$$

The change in the apparent pK_a (p$_o K_a$) with varying water/organic phase ratios is an effective way to characterize ion-pair extraction (Fig. 16.1).

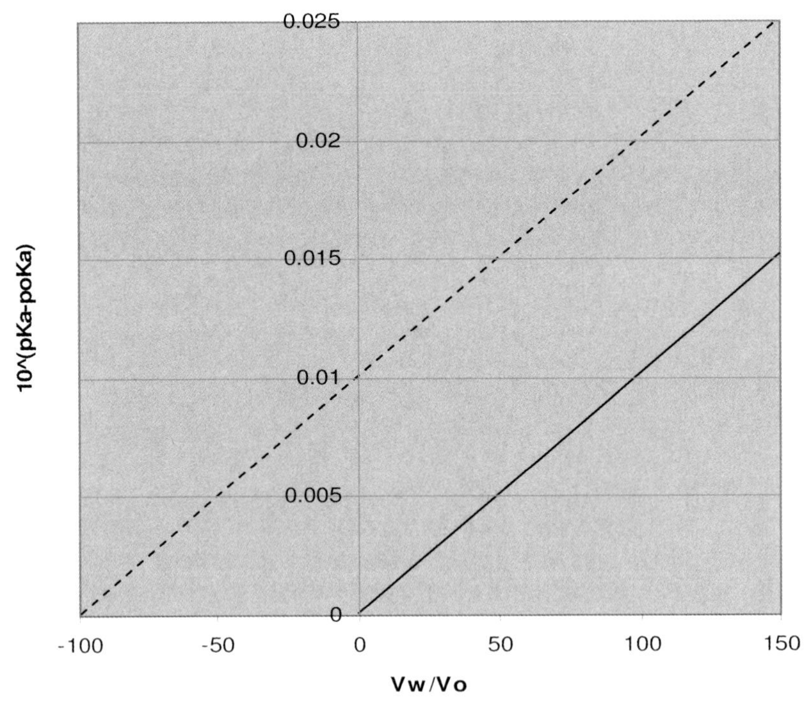

Fig. 16.1 Theoretical variation of the pK_a shift with in increasing water/octanol ratio for a compound with a log P_{HA} of 4.0 and logP_{A^-} of 0 (solid line) or 2 (dashed line). The intercept on the x-axis corresponds to minus P_{A^-}, the slope gives P_{HA} and the y-axis intercept gives P_{A^-}/P_{HA}.

As V_o/V_w increases, p_oK_a reaches a limit p_oK_a lim $= pK_{a,oct}$, which some people refer to "Scherrer pK_a" in reference to the work of Scherrer who pioneered the concept of maximal p_oK_a and introduced the concept of pK_a in lipids [69]. In the absence of ion-pairing p_oK_a lim $= pK_a + \log P_{HA}$, while in the presence of ion-pairing $p_oK_a \lim = pK_a + \log(P_{HA}/P_{A}^{-})$.

However, the potentiometric titration approach has been challenged [74] for the following reasons: (i) potentiometric titration only measures the apparent partition coefficient of the ion because the Galvani potential difference between the two phases is not taken into account and (ii) even in the presence of a supporting electrolyte, it has been shown that the Galvani potential can change significantly during the course of the titration in the presence of lipophilic ions [75]. When this happens, the value $\log P_i$ (or p_oK_a) is no longer constant during the titration and Eq. (12) needs to be handled with caution. Cyclic voltammetry requires the use of a polarizable interface and initially most studies were done with 1,2-dichloroethane as a partition solvent. More recently, voltammetry with octanol as a partition solvent has been described [76]. Ion formation influences the lipophilicity pH profile of ionizable drugs, as illustrated for a typical monoprotic acid in Fig. 16.2.

The $\log D$ value at a single pH is often influenced by ion partitioning, especially at neutral pH where $\log D$ may be used in quantitative structure–permeation relationships. Therefore, the lipophilicity profile may be essential to interpreting PK

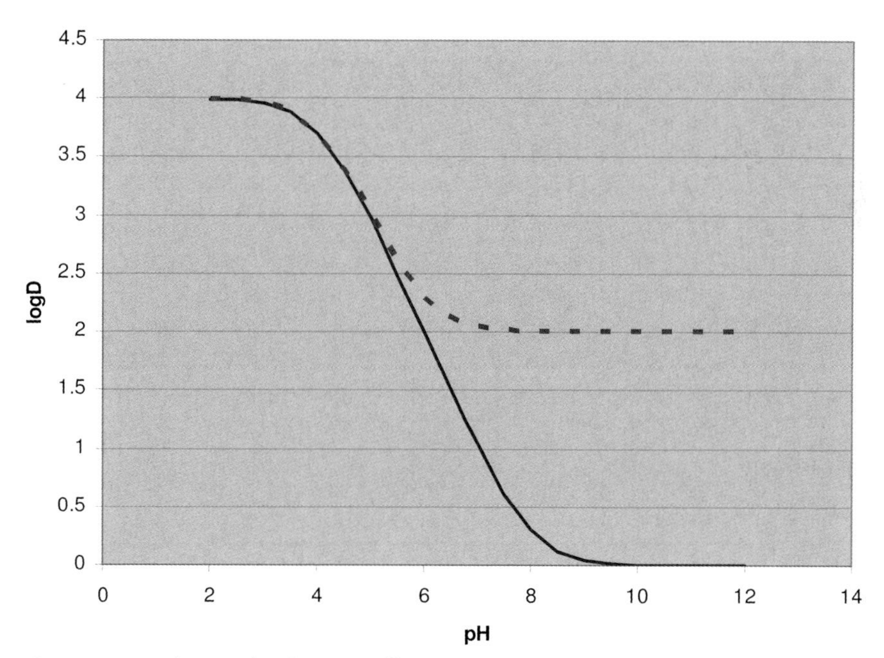

Fig. 16.2 Octanol–water distribution coefficient versus pH for a weak acid with $pK_a = 4$, $\log P_{HA} = 4$ and $\log P_i = 0$ (solid line) or 2 (dashed line).

or other properties. Lipophilicity profiles for ampholytes deserve special attention, as they may be bell-shaped or U-shaped depending on their properties [77].

The distribution of the ionic species is determined by the molecular properties of the compound, but also by the nature and the concentration of the counterions present in the media [78]. For example, the influence of $[Na^+]$ on the transport kinetics of warfarin through an octanol membrane has been reported [79].

However, as stated above, the partition coefficients measured by the shake-flask method or by potentiometric titration can be influenced by the potential difference between the two phases, and are therefore apparent values which depend on the experimental conditions (phase volume ratio, nature and concentrations of all ions in the solutions). In particular, it has been shown that the difference between the apparent and the standard $\log P_i$ depends on the phase volume ratio and that this relationship itself depends on the lipophilicity of the ion [80]. In theory, the most relevant case for *in vivo* extrapolation is when $V_o/V_w << 1$ as it corresponds to the phase ratio encountered by a drug as it distributes within the body. The measurement of apparent $\log P_i$ values does not allow to differentiate between ion-pairing effect and partitioning of the ions due to the Galvani potential difference, and it has been shown that the apparent lipophilicity of a number of quaternary ion drugs is not due to ion-pair partitioning as initially thought [80].

Ion-pair extraction is moderate in partition solvents like octanol, very low with aprotic solvents (hexadecane, cyclohexane) and very significant with phospholipid bilayers like liposomes where strong surface ion-pairs can be formed [73]. Numerous studies with liposomes as a partitioning solvent have been performed in the 1990s [72, 81, 82], however, no clear advantage has been found over bulk octanol for the prediction of drug absorption. The partition coefficient of the neutral species usually correlates well with octanol–water $\log P$. From the work by Avdeef [73] one gets $\log P_{DOPCliposome} = 0.71 \log P_{oct} + 1.06$ ($r^2 = 0.82$). From our own experience with egg phosphatidylcholine liposomes (EPC) a similar relationship was found: $\log P_{EPC} = 0.74 \log P_{oct} + 0.83$ ($r^2 = 0.96$). It is interesting to note that the slope is lower than 1, which means that lipophilic compounds partition less in liposomes and, as the intercept is higher than 0, hydrophilic compounds partition more in liposomes than in bulk octanol. There is a consensus to say that $\log P_{ion}$ are quite different in octanol and liposomes, due to the ionic interactions with the phospholipids head groups.

16.3.2.2 Ion-pairing *In Vivo*

Although there is a lot of experimental evidence showing that ions influence drug distribution *in vitro*, its physiological relevance remains uncertain. The most striking evidence for an *in vivo* effect is perhaps the PK differences of the calcium channel blockers nimodipine and amlodipine. Nimodipine is characterized by a rapid onset and a short duration, whereas amlodipine has a slow onset and a long duration of action. The longer duration of action of the latter is thought to be due to the ionic interactions between the drug and the polar head groups of the phospholipids [83, 84]. The evidence of oral absorption enhancement of poorly permeable compounds through ion-pair formation remains tenuous. This is perhaps because the phase volume ratios used *in vitro* have relatively little *in vivo* relevance. A recent publication and corresponding response [85, 86] provide an example of the ongoing debate surrounding ion-pairing.

In concluding this section we hope we have shown that there is a clear need for more elaborated studies to understand how lipophilic ions interact with biological membranes – an aspect that may look deceptively simple, but which is not yet completely understood.

16.4 Computational Approaches

While there are plenty of methods to predict 1-octanol–water partition coefficients, log P (see Chapters 14 and 15), the number of approaches to predict 1-octanol–water distribution coefficients is rather limited. This is due to a lower availability of log D data and, in general, higher computational complexity of this property compared to that of log P. The approaches to predict log D can be roughly classified into two major categories: (i) calculation of log D at an arbitrary pH and (ii) calculation of log D at a fixed pH.

16.4.1 Methods to Predict Log D at Arbitrary pH

These are more general, and they can perform log D calculation at any pH and ionic strength. The distribution coefficient for monoprotic base given by Eq. (2) can be simplified to Eq. (3) assuming that only the neutral form of a molecule will partition into the organic phase and thus D^{I} is zero.

For compounds with $N>1$ ionizable groups, the problem becomes more complicated. Formally, for such compounds one can use a combination of Eq. (3) where pK_a will be ionization constants of each group by assuming sequential ionization of the groups using macroionization constants. However, this simplification does not correspond to the physics of ionization. Indeed, during the ionization not one, but several centers can be ionized simultaneously. This provides, in practice, a serious computational issue to predict the macroconstants and to allocate them to one or another ionizable center [87]. The equilibria between several microspecies are regulated by microconstants as illustrated in Fig. 16.3. The number of possible microspecies and thus microconstants increases exponentially with the number of ionizable centers. The experimental determination of microconstants is very difficult and it becomes practically impossible in cases which have several protonation sites [87]. All these factors dramatically contribute to the problem of pK_a and thus log D calculation. A detailed review of methods to predict pK_a can be found elsewhere [88, 89].

The ionic partition in the octanol phase could invalidate Eq. (3) for pK_a − pH > 3 log units, when partitioning of ionic species starts to play a significant contribution. The partition of ionic species in the octanol phase was recently considered

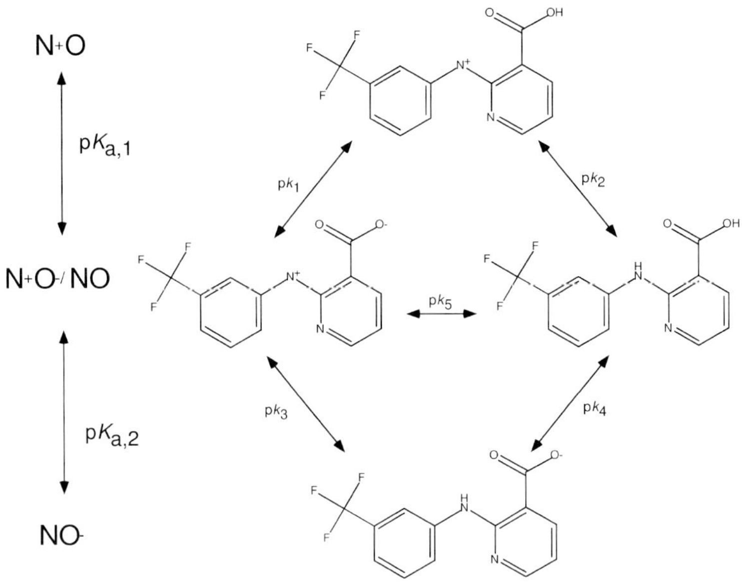

Fig. 16.3 ionization scheme of niflumic acid (CAS-RN: 4394-00-7). The macro- ($pK_{a,i}$) and microprotonation (pk_i) schemes of the compound are shown. The log *P* value is determined by partition of neutral species (dubbed as NO) only. The microconstants and partition coefficients of each microspecies as well as the nature and the concentration of the counterions present in the media are required to calculate log *D* at arbitrary pH. Measurement of log *P* and prediction of log *D* values are difficult and challenging problems for experimental and theoretical approaches, respectively. Contrary to that, measurement of log *D* value at fixed pH can be done using standard methods (see Section 16.2).

by Zhao and Abraham [90]. The authors classified 544 cations on nine functional groups, and observed linear regression between log *P* and partition of single ions within each group. The authors concluded that it is possible to predict partition coefficients for additional cations by classifying them into one of the groups and comparing with log *P* values for the corresponding neutral analogue.

Equation (2) for monoprotic base can be rewritten using a definition of the "Scherrer pK_a" $pK_{a,oct}$ (see Section 16.3.2.1) as:

$$\log D = \log P - \log(1 + 10^{pK_a - pH}) + \log(1 + 10^{pK_{a,oct} - pH}) \quad (13)$$

It can be simplified for three regions of pH as:

$$\log D \approx \begin{cases} \log P, & pH > pK_a \\ \log P - (pK_a - pH), & pK_{a,oct} < pH < pK_a \\ \log P - (pK_a - pK_{a,oct}), & pH < pK_{a,oct} \end{cases} \quad (14)$$

where the last equation corresponds to the distribution coefficient of completely ionized species, $\log P^I$. The difference $\log P - \log P^I$ ($\log P^{diff}$) is proportional to the difference in ionization constants, $pK_a - pK_{a,oct}$, of a compound in water and octanol [91].

The partition of ionic pair depends on ionic strength and may play a significant role for partition of a compound between water and lipid phase in the gastrointestinal tract. Indeed, the ionic strength in the jejunum was estimated to be about 0.14 M and it is maintained at a constant level, probably by means of water and ion secretion [92]. The model of Csizmadia et al. [87] considered partitioning of both ionic species and ion-pairing. In order to make calculations feasible, the authors had to make several simplifications. First, they generalized the observation of Schwarrenbach [93] that the partitioning of dissociated phenols correlated with the strength of the electron-withdrawing or -donating effects of the substituents on the aromatic ring. Since pK_a was known to correlate with the charge delocalization, the authors used the protonation constant to estimate the partition of ion-pairs in the octanol phase. Second, the authors noticed that difference in partition of neutral and ionic species, $\log P^{diff}$, was approximately constant for acids. As a result of these simplifications, the authors derived several empirical equations for monovalent ions and zwitterionic species, which were implemented in the PrologD program [87]. This study demonstrated the complexity of $\log D$ calculations and a need for a large amount of experimental data to correctly estimate the values of constants used in the equations in addition to $\log P$ and pK_a values. Methods to calculate $\log D$ using predicted $\log P$ and pK_a values are available from several commercial providers (e.g. Ref. [94–99]).

16.4.2
Methods to Predict Log *D* at Fixed pH

The recognition of the importance of lipophilicity for drug discovery as well as development of automated and fairly high-throughput methods for $\log P$ measurements (Section 16.2) resulted in the generation of a large amount of experimental data within the pharmaceutical industry. A rugged set-up for such measurements should include strict control of experimental conditions, i.e. of pH and ionic strength of solution. Thus the data would consistently have very high quality.

The availability of large amounts of in-house measurements allowed for an easy benchmarking of methods for $\log D$ prediction. The calculated results in most cases were disappointing and large experimental errors [mean absolute error (MAE) > 1] were reported in several studies [100–103]. These results may be explained by "clustered" sets of compounds used to develop methods and different test sets of compounds measured in pharmaceutical companies. Stouch [104] discussed these issues in an informative article which, albeit not particularly focused on lipophilicity, is nevertheless of interest in the context of *in silico* ADME and QSPR modeling. Some companies approached this problem and provided tools to incorporate new data in their models. For example, DeWitte et al. reported a "user training" feature of their $\log D$ program [105]. The experimental testing of this feature, however, demonstrated only "a modest increase in accuracy of the model, with the r^2 value of the prediction improving in the test set from 0.316 to 0.527" using Merck datasets [106]. The other limitations on memory and speed of calculations of the method were also reported.

Considering the failure of available log *D* methods to predict in-house data and taking into account that such data are usually generated just for a few fixed pH values, a number of companies started to elaborate in-house methods for log *D* prediction at fixed pH. Up to date several companies have reported development of such methods. For example, Cerep has developed methods to predict log *D* at pH 7.4 and 6.5 included in their BioPrint package [107], but details of their method are not published. HQSAR Tripos descriptors were used by Bayer to develop log *D* models at pH 2.3 and 7.5 using 70 000 ($q^2 = 0.76$, $STD_{cv} = 0.60$) and 7000 ($q^2 = 0.83$, $STD_{cv} = 0.67$) compounds, respectively [108]; however, again, no details of the approach were provided.

The log *D* values for 11 283 measurements performed using the shake-flask method at pH 7.4 were used by Merck [109]. The authors studied the performance of model prediction accuracy as a function of a similarity of training and test set compounds. Several types of molecular descriptors and several machine-learning methods were investigated. One of the best models reported was built using random forest [110] and regular atom pair descriptors [111]. The model developed with 4000 compounds calculated root mean square error (RMSE) = 0.90 for the remaining test set compounds.

The aforementioned studies were usually performed by employees of the companies. The absence of public repositories of such data has made it difficult for academic users to develop similar models. A possible solution to this problem was proposed by Tetko et al. [100, 101], where the authors successfully developed log *D* models for fixed pH by correcting the prediction errors of the log *P* model. Both studies were based on the ALOGPS program [112], using the LIBRARY mode (see Chapter 15). By doing the LIBRARY correction the authors assumed that the errors to predict log *D* values when using a global log *P* model were likely the same for similar molecules. The success of this assumption strongly depended on the correct definition of the similarity of molecules that was introduced elsewhere [113]. The LIBRARY mode calculated highly predictive models for the Pfizer [100] (RMSE = 0.69, $N = 17861$) and the AstraZeneca [101] (RMSE = 0.70, $N = 8081$) datasets. The same approach was directly used to improve predictions of the log *D* model by Bruneau and McElroy [114]. In their study the LIBRARY correction decreased RMSE of 0.58 to RMSE of 0.45 for $\log D_{7.4}$ of 11 461 test set molecules measured at several AstraZeneca sites.

16.4.3 Issues and Needs

16.4.3.1 Log *D* Models in ADMET Prediction

Actually, why do we need log *D* models? Why can't we use just log *P* models? One of the main requirements for prediction of octanol–water coefficients is to optimize bioavailability of chemical compounds. During the absorption process the drug passes through the different compartments of the gastrointestinal tract, which have different pH values. A compound with ionizable groups will have different degrees of ionization in each compartment and thus different chances to be absorbed in each compartment. Therefore, one can expect that the use of log *D* rather than log *P* values can provide better models for absorption of chemical compounds. This was shown in a number of studies, where log *D* values were also successfully used as one of the parameters to predict human volume of distribution [57], bioavailability [3], plasma protein binding [115], BBB [116] and other properties. Therefore, the determination of log *D* is important for the drug discovery process and cannot be easily substituted with log *P* measurements only.

16.4.3.2 Applicability Domain of Models

In order to be applied the models should be predictive. Unfortunately, the models frequently fail and demonstrate significantly lower prediction ability compared to the estimated one, when they are applied to new unseen data [100–103, 106]. One of the main reasons for such failures can be the lack of available experimental data and difficulties in calculating log *D*, as discussed in Section 16.4.2. Another problem of low prediction ability of log *D* models can be attributed to different chemical diversity of molecules in the in-house databases compared to the training sets used to develop the programs.

The models developed for log *D* prediction usually aim at being global ones. This, however, does not work on practice. Sheridan et al. [109] noticed that the accuracy of log *D* prediction of molecules decreased approximately 2–3 times (RMSE = 0.75 versus 1.5–2) as the similarity of the test molecule to the molecules in the training set (using Dice definition with the atom pair descriptors) changed from 1 to 0 (most to least similar). Thus, if a test set molecule had a very similar molecule in the training set, it was possible to accurately predict its log *D* value. A detailed overview of state of the art methods to access the same problem was published elsewhere [117].

However, there is still a strong need to develop new methods that will be able to quantitatively or at least qualitatively estimate the prediction accuracy of log *D* models. Such models will allow the computational chemist to distinguish reliable versus nonreliable predictions and to decide whether the available model is sufficiently accurate or whether experimental measurements should be provided. For example, when applying ALOGPS in the LIBRARY model it was possible to predict more than 50% and 30% compounds with an accuracy of MAE < 0.35 for Pfizer and AstraZeneca collections, respectively [117]. This precision approximately corresponds to the experimental accuracy, $s = 0.4$, of potentiometric lipophilicity determinations [15]. Thus, depending on the required precision, one could skip experimental measurements for some of the accurately predicted compounds.

A low accuracy of models for prediction of log *D* at any pH would not encourage the use of these models for practical applications in industry. Thus, it is likely that the methods for log *D* prediction at fixed pH that are developed in house by pharmaceutical companies will dominate in industry. However, log *D* measurements can also be used to dramatically enhance the quality of the approaches for log *D* prediction at arbitrary pH and can lead to the development of new theories and methodology.

16.5 Some Concluding Remarks: The Good, the Bad and the Ugly

We have tried to cover some of the important aspects of the determination and use of log *P* and log *D* parameters. Far from being exhaustive, this chapter attempted to offer some considerations and perspective in a field where, after 40 years from its beginning at the hand of Corwin Hansch et al., there does not seem to be much alternative to the balance of forces encoded by the octanol–water system to model lipophilicity.

It is difficult to unravel whether this is due to the reasonably good initial choice of octanol and water to mimic the forces encoded in the passage of solutes through hydrophilic and lipophilic environments in biological systems of varying complexity. It could be that the vast array of data available (the "good"), the tendency of most scientists in the field, with notable exceptions, to use the octanol–water system on virtually every attempt at ADME and QSAR modeling (the "bad"), and the experimental difficulties associated with other systems (the "ugly") may all share any blame that may be assessed.

The net result, at any rate, is that there has not been a strong need, apparently, to develop alternative systems and/or these systems have not gained wide popularity and use within the industrial and academic communities in the field. When this has been attempted, no clear winner has emerged when consideration was given to advantages and disadvantages of potential alternatives [6] and to an increase in the complexity of data analysis and automation.

The question of "how" and "what" log *P*/log *D* values do we use in our daily work in medicinal chemistry, PK and metabolism, toxicology, and environmental applications, is an important one. The plethora of available computational approaches is, of course, a very strong "anchor" to the use of log *P*/log *D* values, which is difficult to set aside, but the practice may be to decrease the number of measured values and use routinely a computed value as a surrogate. However, for each class of compounds, no matter how refined and accurate the computational method might be, significant deviations from the calculated values might be expected and, in our direct experience, from week to week and plate to plate, experimental values differing by 1–2 units from computed ones were not infrequent. This, in itself, is "bad", but it may become "ugly" if there is no comparative experimental data to establish whether or not the computational method is reasonably good in predicting the lipophilicity of a particular molecular scaffold. The "good" practice should be to determine at least a few values for representative compounds and continue monitoring the performance of the computational method with additional determinations alongside the medicinal chemistry work, especially when significant modifications have been made to the molecular scaffold.

At the same time, an educational effort should accompany the use of log P/log D by putting the data in proper perspective for the user and undertaking an analysis of "failures" to gain potential new insight on when the answer is the answer. This is, of course, a paradigm to ensure quality and successful application of data, not limited to log P/log D.

References

1 Fujita, T., Iwasa, J., Hansch, C. A new substituent constant, *P*, derived from partition coefficients. *J. Am. Chem. Soc.* **1964**, *86*, 5175–5180.

2 Dearden, J. C., Bresnen, G. M. The measurement of partition coefficients. *Quant. Struct.-Act. Relat.* **1988**, *7*, 133–144.

3 Yoshida, F., Topliss, J. G. QSAR model for drug human oral bioavailability. *J. Med. Chem.* **2000**, *43*, 2575–2585.

4 Caron, G., Ermondi, G., Scherrer, R. A. Lipophilicity, polarity, and hydrophobicity. In *ADME/Tox Approaches*, Van de Waterbeemd, H., Testa, B. (vol. eds.), Vol. 5 in *Comprehensive Medicinal Chemistry*, 2nd edn., Taylor, J. B., Triggle, D. J. (eds.), Elsevier, Oxford, **2006**, pp. 425–452.

5 Malkia, A., Murtomaki, L., Urtti, A., Kontturi, K. Drug permeation in biomembranes: *in vitro* and *in silico* prediction and influence of physicochemical properties. *Eur. J. Pharm. Sci.* **2004**, *23*, 13–47.

6 Hartmann, T., Schmitt, J. Lipophilicity – beyond octanol/water: a short comparison of modern technologies. *Drug Discov. Today Technol.* **2004**, *1*, 431–439.

7 Leahy, D. E., Morris, J. J., Taylor, P. J., Wait, A. R. Membranes and their models: towards a rational choice of partitioning system. In *QSAR: Rational Approaches to the Design of Bioactive Compounds*, Silipo, C., Vittoria, A. (eds.), Elsevier, Amsterdam, **1991**, Vol. 16, pp. 75–82.

8 Seiler, P. Interconversion of lipophilicities from hydrocarbon/water systems into the octanol/water system. *Eur. J. Med. Chem.* **1974**, *9*, 473–479.

9 Reymond, F., Steyaert, G., Carrupt, P. A., Testa, B., Girault, H. H. 10. Mechanism of transfer of a basic drug across the water/1,2-dichloroethane interface: the case of quinidine. *Helv. Chim. Acta* **1996**, *79*, 101–117.

10 Liu, X., Bouchard, G., Müller, N., Galland, A., Girault, H., Testa, B., Carrupt, P. A. Solvatochromic analysis of partition coefficients in the *o*-nitrophenyl octyl ether (*o*-NPOE)/water system. *Helv. Chim. Acta* **2003**, *86*, 3533–3547.

11 Young, R. C., Mitchell, R. C., Brown, T. H., Ganellin, C. R., Griffiths, R., Jones, M., Rana, K. K., Saunders, D., Smith, I. R., Sore, N. E., Wilks, T. J. Development of a new physicochemical model for brain penetration and its application to the design of centrally acting H2 receptor histamine antagonists. *J. Med. Chem.* **1988**, *31*, 656–671.

12 Abraham, M. H., Chadha, H. S., Whiting, G. S., Mitchell, R. C. Hydrogen bonding. 32. An analysis of water–octanol and water–alkane partitioning and the delta log *P* parameter of Seiler. *J. Pharm. Sci.* **1994**, *83*, 1085–1100.

13 Rekker, R. F., Mannhold, R., Bijloo, G., De Vries, G., Dross, K. The lipophilic behaviour of organic compounds: 2. The development of an aliphatic hydrocarbon/water fragmental system via interconnection with octanol–water partitioning data. *Quant. Struct.-Activ. Rel.* **1998**, *17*, 537–548.

14 Caron, G., Ermondi, G. Calculating virtual log *P* in the alkane/water system ($\log P_{alk}^{N}$) and its derived parameters deltalog $\log P_{oct-alk}^{N}$ and $\log P_{alk}^{pH}$. *J. Med. Chem.* **2005**, *48*, 3269–3279.

15 Lombardo, F., Shalaeva, M. Y., Tupper, K. A., Gao, F. Elog D_{oct}: a tool for lipophilicity determination in drug discovery. 2. Basic and neutral compounds. *J. Med. Chem.* **2001**, *44*, 2490–2497.

16 Smith, D. A., Van de Waterbeemd, H., Walker, D. K. *Pharmacokinetics and Metabolism In Drug Design (Methods and Principles in Medicinal Chemistry)*, Wiley-VCH, Weinheim, **2001**.

17 Comer, J. E. A. High-throughput measurements of log D and pK_a. In *Drug Bioavailability (Methods and Principles in Medicinal Chemistry)*, Van de Waterbeemd, H., Lennernäs, H., Artursson, P. (eds.), Wiley-VCH, Weinheim, **2003**, Vol. 18, pp. 21–45.

18 Obach, R. S., Reed-Hagen, A. E., Krueger, S. S., Obach, B. J., O'Connell, T. N., Zandi, K. S., Miller, S., Coe, J. W. Metabolism and disposition of varenicline, a selective alpha4beta2 acetylcholine receptor partial agonist, *in vivo* and *in vitro*. *Drug Metab. Dispos.* **2006**, *34*, 121–130.

19 Lipinski, C. A., Lombardo, F., Dominy, B. W., Feeney, P. J. Experimental and computational approaches to estimate solubility and permeability in drug discovery and development settings. *Adv. Drug. Deliv. Rev.* **1997**, *23*, 3–25.

20 Lombardo, F., Gifford, E., Shalaeva, M. Y. *In silico* ADME prediction: data, models, facts and myths. *Minirev. Med. Chem.* **2003**, *3*, 861–875.

21 Dearden, J. C. Partitioning and lipophilicity in quantitative structure–activity relationships. *Environ. Health Perspect.* **1985**, *61*, 203–228.

22 Tute, M. S. Lipophilicity: a history. In *Lipophilicity in Drug Action and Toxicology (Methods and Principles in Medicinal Chemistory)*, Pliska, V., Testa, B., Van de Waterbeemd, H. (eds.), Wiley-VCH, Weinheim, **1996**, pp. 7–26.

23 Leo, A., Hansch, C., Elkins, D. Partition coefficients and their uses. *Chem. Rev.* **1971**, *61*, 525–616.

24 Hansch, C., Leo, A., Hoekman, D. *Hydrophobic, Electronic, and Steric Constants.* American Chemical Society, Washington, **1995**, Vol. 2, p. 368.

25 Gocan, S., Cimpan, G., Comer, J. Lipophilicity measurements by liquid chromatography. *Adv. Chromatogr.* **2006**, *44*, 79–176.

26 Poole, S. K., Poole, C. F. Separation methods for estimating octanol–water partition coefficients. *J. Chromatogr. B* **2003**, *797*, 3–19.

27 Valko, K., Reynolds, D. P. High-throughput physicochemical and *in vitro* ADMET screening: a role in pharmaceutical profiling. *Am. J. Drug Deliv.* **2005**, *3*, 83–100.

28 Valko, K. Application of high-performance liquid chromatography based measurements of lipophilicity to model biological distribution. *J. Chromatogr. A* **2004**, *1037*, 299–310.

29 Hitzel, L., Watt, A. P., Locker, K. L. An increased throughput method for the determination of partition coefficients. *Pharm. Res.* **2000**, *17*, 1389–1395.

30 Analiza Inc., http://analiza.com (July 12, 2007).

31 Slater, B., McCormack, A., Avdeef, A., Comer, J. E. pH-metric log P. 4. Comparison of partition coefficients determined by HPLC and potentiometric methods to literature values. *J. Pharm. Sci.* **1994**, *83*, 1280–1283.

32 Takacs-Novak, K., Avdeef, A. Interlaboratory study of log P determination by shake-flask and potentiometric methods. *J. Pharm. Biomed. Anal.* **1996**, *14*, 1405–1413.

33 Avdeef, A. pH-metric log P. Part 1. Difference plots for determining ion-pair octanol–water partition coefficients of multiprotic substances. *Quant. Struct.-Activ. Relat.* **1992**, *11*, 510–517.

34 Takacs-Novak, K., Avdeef, A., Box, K. J., Podanyi, B., Szasz, G. Determination of protonation macro- and microconstants and octanol/water partition coefficient of the antiinflammatory drug niflumic acid. *J. Pharm. Biomed. Anal.* **1994**, *12*, 1369–1377.

35 Deak, K., Takacs-Novak, K., Tihanyi, K., Noszal, B. Physico-chemical profiling of antidepressive sertraline: solubility, ionisation, lipophilicity. *Med. Chem.* **2006**, *2*, 385–389.

36 Caron, G., Ermondi, G., Damiano, A., Novaroli, L., Tsinman, O., Ruelle, J. A., Avdeef, A. Ionization, lipophilicity, and molecular modeling to investigate permeability and other biological properties of amlodipine. *Bioorg. Med. Chem.* **2004**, *12*, 6107–6118.

37 Valko, K., Bevan, C., Reynolds, D. Chromatographic hydrophobicity index by fast-gradient RP HPLC: A high-throughput alternative to log *P* log *D*. *Anal. Chem.* **1997**, *69*, 2022–2029.

38 Stella, C., Galland, A., Liu, X., Testa, B., Rudaz, S., Veuthey, J. L., Carrupt, P. A. Novel RPLC stationary phases for lipophilicity measurement: solvatochromic analysis of retention mechanisms for neutral and basic compounds. *J. Sep. Sci.* **2005**, *28*, 2350–2362.

39 Lombardo, F., Shalaeva, M. Y., Tupper, K. A., Gao, F., Abraham, M. H. Elog P_{oct}: a tool for lipophilicity determination in drug discovery. *J. Med. Chem.* **2000**, *43*, 2922–2928.

40 Minick, D. J., Frenz, J. H., Patrick, M. A., Brent, D. A. A comprehensive method for determining hydrophobicity constants by reversed-phase high-performance liquid chromatography. *J. Med. Chem.* **1988**, *31*, 1923–1933.

41 Lombardo, F., Shalaeva, M. Y., Bissett, B. D., Chistokhodova, N. Elog $D^{7.4}$ 20 000 compounds later: refinements, observations and applications. In *Pharmacokinetic Profiling in Drug Research: Biological, Physicochemical, and Computational Strategies*, Testa, B., Krämer, S., Wunderli-Allenspach, H., Folkers, G. (eds.), Wiley-VCH, Weinheim, **2006**, pp. 187–201.

42 Pidgeon, C., Venkataram, U. V. Immobilized artificial membrane chromatography: supports composed of membrane lipids. *Anal. Biochem.* **1989**, *176*, 36–47.

43 Taillardat-Bertschinger, A., Carrupt, P. A., Barbato, F., Testa, B. Immobilized artificial membrane HPLC in drug research. *J. Med. Chem.* **2003**, *46*, 655–665.

44 Taillardat-Bertschinger, A., Barbato, F., Quercia, M. T., Carrupt, P. A., Reist, M., La Rotonda, M. I., Testa, B. Structural properties governing retention mechanisms on immobilized artificial membrane (IAM) HPLC columns. *Helv. Chim. Acta* **2002**, *85*, 519–532.

45 Lazaro, E., Rafols, C., Abraham, M. H., Roses, M. Chromatographic estimation of drug disposition properties by means of immobilized artificial membranes (IAM) and C_{18} columns. *J. Med. Chem.* **2006**, *49*, 4861–4870.

46 Ishihama, Y., Oda, Y., Uchikawa, K., Asakawa, N. Evaluation of solute hydrophobicity by microemulsion electrokinetic chromatography. *Anal. Chem.* **1995**, *67*, 1588–1595.

47 Gluck, S. J., Benko, M. H., Hallberg, R. K., Steele, K. P. Indirect determination of octanol–water partition coefficients by microemulsion electrokinetic chromatography. *J. Chromatogr. A* **1996**, *744*, 141–146.

48 Poole, S. K., Durham, D., Kibbey, C. Rapid method for estimating the octanol–water partition coefficient (log P_{ow}) by microemulsion electrokinetic chromatography. *J. Chromatogr. B* **2000**, *745*, 117–126.

49 Abraham, M. H., Treiner, C., Roses, M., Rafols, C., Ishihama, Y. Linear free energy relationship analysis of microemulsion electrokinetic chromatographic determination of lipophilicity. *J. Chromatogr. A* **1996**, *752*, 243–249.

50 Wehmeyer, K. R., Tu, J., Jin, Y., King, S., Stella, M., Stanton, D. T., Strasburg, R., Kenseth, J., Wong, K. S. The application of multiplexed microemulsion electrokinetic chromatography for the rapid determination of log P_{ow} values for neutral and basic compounds. *LC GC North Am.* **2003**, *21*, 1078–1088.

51 Wong, K. S., Kenseth, J., Strasburg, R. Validation and long-term assessment of an approach for the high throughput determination of lipophilicity (log P_{ow}) values using multiplexed, absorbance-based capillary electrophoresis. *J. Pharm. Sci.* **2004**, *93*, 916–931.

52 Faller, B., Grimm, H. P., Loeuillet-Ritzler, F., Arnold, S., Briand, X. High-throughput lipophilicity measurement with immobilized artificial membranes. *J. Med. Chem.* **2005**, *48*, 2571–2576.

53 Cross, S. E., Magnusson, B. M., Winckle, G., Anissimov, Y., Roberts, M. S. Determination of the effect of lipophilicity on the *in vitro* permeability and tissue reservoir characteristics of topically applied solutes in human skin layers. *J. Invest. Dermatol.* **2003**, *120*, 759–764.

54 Luco, J. M., Marchevsky, E. QSAR studies on blood–brain barrier permeation. *Curr. Comput.-Aided Drug Des.* **2006**, *2*, 31–55.

55 Lewis, D. F. V., Jacobs, M. N., Dickins, M. Compound lipophilicity for substrate binding to human P450s in drug metabolism. *Drug Discov. Today* **2004**, *9*, 530–537.

56 Lewis, D. F., Lake, B. G., Ito, Y., Anzenbacher, P. Quantitative structure–activity relationships (QSARs) within cytochromes P450 2B (CYP2B) subfamily enzymes: the importance of lipophilicity for binding and metabolism. *Drug Metab. Drug Interact.* **2006**, *21*, 213–231.

57 Lombardo, F., Obach, R. S., Shalaeva, M. Y., Gao, F. Prediction of human volume of distribution values for neutral and basic drugs. 2. Extended data set and leave-class-out statistics. *J. Med. Chem.* **2004**, *47*, 1242–1250.

58 Lombardo, F., Obach, R. S., Dicapua, F. M., Bakken, G. A., Lu, J., Potter, D. M., Gao, F., Miller, M. D., Zhang, Y. A hybrid mixture discriminant analysis–random forest computational model for the prediction of volume of distribution of drugs in human. *J. Med. Chem.* **2006**, *49*, 2262–2267.

59 Hollosy, F., Valko, K., Hersey, A., Nunhuck, S., Keri, G., Bevan, C. Estimation of volume of distribution in humans from high throughput HPLC-based measurements of human serum albumin binding and immobilized artificial membrane partitioning. *J. Med. Chem.* **2006**, *49*, 6958–6971.

60 Remko, M., Swart, M., Bickelhaupt, F. M. Theoretical study of structure, pK_a, lipophilicity, solubility, absorption, and polar surface area of some centrally acting antihypertensives. *Bioorg. Med. Chem.* **2006**, *14*, 1715–1728.

61 Yalkowsky, S. H., Johnson, J. L., Sanghvi, T., Machatha, S. G. A "rule of unity" for human intestinal absorption. *Pharm. Res.* **2006**, *23*, 2475–2481.

62 Testa, B., Crivori, P., Reist, M., Carrupt, P. A. The influence of lipophilicity on the pharmacokinetic behavior of drugs: concepts and examples. *Perspect. Drug Discov. Des.* **2000**, *19*, 179–211.

63 Chou, C. H., Rowland, M. Relationship between lipophilicity and protein binding of a homologous series of barbiturates. *Chin. Pharm. J.* **2002**, *54*, 87–94.

64 Van de Waterbeemd, H., Smith, D. A., Jones, B. C. Lipophilicity in PK design: Methyl, Ethyl, futile. *J. Comput.-Aided Mol. Des.* **2001**, *15*, 273–286.

65 Seydel, J. K., Wiese, M. Octanol–water partitioning versus Partitioning into membranes. In *Drug–Membrane Interactions (Methods and Principles in Medicinal Chemistry)*, Seydel, J. K., Wiese, M. (eds.), Wiley-VCH, Weinheim, **2002**, Vol. 15, pp. 35–50.

66 Jacobs, M. H. Some aspects of cell permeability to weak electrolytes. *Cold Spring Harb. Symp. Quant. Biol.* **1940**, *8*, 30–39.

67 Palm, K., Luthman, K., Ros, J., Grasjo, J., Artursson, P. Effect of molecular charge on intestinal epithelial drug transport: pH-dependent transport of cationic drugs. *J. Pharmacol. Exp. Ther.* **1999**, *291*, 435–443.

68 Artursson, P., Palm, K., Luthman, K. Caco-2 monolayers in experimental and theoretical predictions of drug transport. *Adv. Drug. Deliv. Rev.* **2001**, *46*, 27–43.

69 Scherrer, R. A. The treatment of ionizable compounds in quantitative structure–activity studies with special consideration to ion partitioning. In *Pesticide Synthesis Through Rational Approaches (ACS Symp. Ser. 255)*, Magee, P. S., Kohn, G. K., Menn, J. J. (eds.), American Chemical Society, Washington, DC, **1984**, pp. 225–246.

70 Takacs-Novak, K., Szasz, G. Ion-pair partition of quarternary ammonium drugs: the influence of counter ions of different lipophilicity, size, and flexibility. *Pharm. Res.* **1999**, *16*, 1633–1638.

71 Wimley, W. C., White, S. H. Membrane partitioning: distinguishing bilayer effects from the hydrophobic effect. *Biochemistry* **1993**, *32*, 6307–12.

72 Kramer, S. D., Wunderli-Allenspach, H. The pH-dependence in the partitioning behaviour of (*RS*)-[^{3}H]propranolol between MDCK cell lipid vesicles and buffer. *Pharm. Res.* **1996**, *13*, 1851–1855.

73 Avdeef, A., Box, K. J., Comer, J. E., Hibbert, C., Tam, K. Y. pH-metric log *P* 10. Determination of liposomal membrane–water partition coefficients of ionizable drugs. *Pharm. Res.* **1998**, *15*, 209–215.

74 Bouchard, G., Carrupt, P. A., Testa, B., Gobry, V., Girault, H. H. Lipophilicity and solvation of anionic drugs. *Chemistry* **2002**, *8*, 3478–3484.

75 Reymond, F., Gobry, V., Bouchard, G., Girault, H. H. Electrochemical aspects of drug partitioning. In *Pharmacokinetic Optimization in Drug Research: Biological, Physicochemical, and Computational Strategies*, Testa, B., Van de Waterbeemd, H., Folkers, G., Guy, R. (eds.), Wiley-VCH, Weinheim, **2001**, pp. 327–350.

76 Bouchard, G., Galland, A., Carrupt, P. A., Gulaboski, R., Mirceski, V., Scholz, F., Girault, H. H. Standard partition coefficients of anionic drugs in the *n*-octanol/water system determined by voltammetry at three-phase electrodes. *Phys. Chem. Chem. Phys.* **2003**, *5*, 3748–3751.

77 Pagliara, A., Carrupt, P. A., Caron, G., Gaillard, P., Testa, B. Lipophilicity profiles of ampholytes. *Chem. Rev.* **1997**, *97*, 3385–3400.

78 Austin, R. P., Barton, P., Davis, A. M., Manners, C. N., Stansfield, M. C. The effect of ionic strength on liposome-buffer and 1-octanol-buffer distribution coefficients. *J. Pharm. Sci.* **1998**, *87*, 599–607.

79 Cools, A. A., Janssen, L. H. Influence of sodium ion-pair formation on transport kinetics of warfarin through octanol-impregnated membranes. *J. Pharm. Pharmacol.* **1983**, *35*, 689–691.

80 Bouchard, G., Carrupt, P. A., Testa, B., Gobry, V., Girault, H. H. The apparent lipophilicity of quaternary ammonium ions is influenced by Galvani potential difference, not ion-pairing: a cyclic voltammetry study. *Pharm. Res.* **2001**, *18*, 702–708.

81 Pauletti, G. M., Wunderli-Allenspach, H. Partition coefficients *in vitro*: Artificial membranes as a standardized distribution model. *Eur. J. Pharm. Sci.* **1994**, *1*, 273–282.

82 Austin, R. P., Davis, A. M., Manners, C. N. Partitioning of ionizing molecules between aqueous buffers and phospholipid vesicles. *J. Pharm. Sci.* **1995**, *84*, 1180–1183.

83 Mason, R. P., Rhodes, D. G., Herbette, L. G. Reevaluating equilibrium and kinetic binding parameters for lipophilic drugs based on a structural model for drug interaction with biological membranes. *J. Med. Chem.* **1991**, *34*, 869–877.

84 Herbette, L. G., Rhodes, D. G., Mason, R. P. New approaches to drug design and delivery based on drug–membrane interactions. *Drug Des. Deliv.* **1991**, *7*, 75–118.

85 Thomae, A. V., Wunderli-Allenspach, H., Krämer, S. D. Permeation of aromatic carboxylic acids across lipid bilayers: the pH-partition hypothesis revisited. *Biophys. J.* **2005**, *89*, 1802–1811.

86 Saparov, S. M., Antonenko, Y. N., Pohl, P. A new model of weak acid permeation through membranes revisited: does Overton still rule? *Biophys. J.* **2006**, *90*, L86–88.

87 Csizmadia, F., Tsantili-Kakoulidou, A., Panderi, I., Darvas, F. Prediction of distribution coefficient from structure. 1. Estimation method. *J. Pharm. Sci.* **1997**, *86*, 865–871.

88 Wan, H., Ulander, J. High-throughput pK_a screening and prediction amenable for ADME profiling. *Expert Opin. Drug Metab. Toxicol.* **2006**, *2*, 139–155.

89 Fraczkiewicz, R. *In silico* prediction of ionization. In *ADME/Tox Approaches*, Van de Waterbeemd, H., Testa, B. (vol.

eds.), Vol. 5 in *Comprehensive Medicinal Chemistry*, 2nd edn., Taylor, J. B., Triggle, D. J. (eds.), Elsevier, Oxford, **2006**, pp. 603–626.

90 Zhao, Y. H., Abraham, M. H. Octanol/water partition of ionic species, including 544 cations. *J. Org. Chem.* **2005**, *70*, 2633–2640.

91 Scherrer, R. A. Biolipid pK_a values and the lipophilicity of ampholytes and ion pairs. In *Pharmacokinetic Optimization in Drug Research: Biological, Physicochemical, and Computational Strategies*, Testa, B., Van de Waterbeemd, H., Folkers, G., Guy, R. (eds.), Wiley-VCH, Weinheim, **2001**, pp. 351–381.

92 Lindahl, A., Ungell, A. L., Knutson, L., Lennernäs, H. Characterization of fluids from the stomach and proximal jejunum in men and women. *Pharm. Res.* **1997**, *14*, 497–502.

93 Schwarrenbach, R. P., Stierli, R., Folsom, B. R., Zeyer, J. Compound properties relevant for assessing the environmental partitioning of nitrophenols. *Environ. Sci. Technol.* **1988**, *22*, 83–92.

94 CompuDrug International Inc., http://www.compudrug.com (July 12, 2007).

95 SLIPPER log *D*, http://www.timtec.net/software/slipper.htm (July 12, 2007).

96 Advanced Chemistry Development Inc., http://www.acdlabs.com (July 12, 2007).

97 Pharma Algorithms, http://www.ap-algorithms.com (July 12, 2007).

98 ChemSilico LLC, http://www.chemsilico.com (July 12, 2007).

99 ADMET Predictor, Simulations Plus Inc., http://www.simulations-plus.com (July 12, 2007).

100 Tetko, I. V., Poda, G. I. Application of ALOGPS 2.1 to predict log *D* distribution coefficient for Pfizer proprietary compounds. *J. Med. Chem.* **2004**, *47*, 5601–5604.

101 Tetko, I. V., Bruneau, P. Application of ALOGPS to predict 1-octanol/water distribution coefficients, log *P*, and log *D*, of AstraZeneca in-house database. *J. Pharm. Sci.* **2004**, *93*, 3103–3110.

102 Lombardo, F., Shalaeva, M. Y., Bissett, B. D., Chistokhodova, N. Elog $D^{7.4}$ 20000 compounds later: refinements, observations and applications. In *Proc. Log P 2004: The Third Lipophilicity Symposium*, Folkers, G., Krämer, S., Testa, B., Wunderli-Allespach, H. (eds.), ETH, Zurich, **2004**, p. L-22.

103 Morris, J. J., Bruneau, P. Prediction of physicochemical properties. In *Virtual Screening for Bioactive Molecules*, Bohm, H. G., Schneider, G. (eds.), Wiley-VCH, Weinheim, Vol. 10, **2000**, pp. 33–58.

104 Stouch, T. R., Kenyon, J. R., Johnson, S. R., Chen, X. Q., Doweyko, A., Li, Y. *In silico* ADME/Tox: why models fail. *J. Comput.-Aided Mol. Des.* **2003**, *17*, 83–92.

105 DeWitte, R., Gorohov, F., Kolovanov, E. Using targeted measurements to improve the accuracy of physical property prediction. Presented at *ADMET-1 Conference 2004*, San Diego, **2004**.

106 Walker, M. J. Training ACD/Log*P* with experimental data. *QSAR Comb. Sci.* **2004**, *23*, 515–520.

107 Cerep Inc., http://www.cerep.com.

108 Goller, A. H., Hennemann, M., Keldenich, J., Clark, T. *In silico* prediction of buffer solubility based on quantum-mechanical and HQSAR- and topology-based descriptors. *J. Chem. Inf. Model.* **2006**, *46*, 648–658.

109 Sheridan, R. P., Feuston, B. P., Maiorov, V. N., Kearsley, S. K. Similarity to molecules in the training set is a good discriminator for prediction accuracy in QSAR. *J. Chem. Inf. Comput. Sci.* **2004**, *44*, 1912–1928.

110 Breiman, L. Random forests. *Mach. Learn.* **2001**, *45*, 5–32.

111 Carhart, R. E., Smith, D. H., Ventkataraghavan, R. Atom pairs as molecular features in structure–activity studies: definition and application. *J. Chem. Inf. Comput. Sci.* **1985**, *25*, 64–73.

112 Tetko, I. V., Tanchuk, V. Y. Application of associative neural networks for prediction of lipophilicity in ALOGPS 2.1 program. *J. Chem. Inf. Comput. Sci.* **2002**, *42*, 1136–1145.

113 Tetko, I. V. Neural network studies. 4. Introduction to associative neural networks. *J. Chem. Inf. Comput. Sci.* **2002**, *42*, 717–728.

114 Bruneau, P., McElroy, N. R. Log $D^{7.4}$ modeling using Bayesian regularized neural networks. Assessment and correction of the errors of prediction. *J. Chem. Inf. Model.* **2006**, *46*, 1379–1387.

115 Yamazaki, K., Kanaoka, M. Computational prediction of the plasma protein-binding percent of diverse pharmaceutical compounds. *J. Pharm. Sci.* **2004**, *93*, 1480–1494.

116 Liu, X., Tu, M., Kelly, R. S., Chen, C., Smith, B. J. Development of a computational approach to predict blood–brain barrier permeability. *Drug Metab. Dispos.* **2004**, *32*, 132–139.

117 Tetko, I. V., Bruneau, P., Mewes, H. W., Rohrer, D. C., Poda, G. I. Can we estimate the accuracy of ADME-Tox predictions? *Drug Discov. Today* **2006**, *11*, 700–707.

Part VI

Drug- and Lead-likeness

17
Properties Guiding Drug- and Lead-likeness

Sorel Muresan and Jens Sadowski

Abbreviations

3D	three-dimensional
ACD	Available Chemicals Directory
ADMET	adsorption, distribution, metabolism, excretion and toxicity
Caco-2	adenocarcinoma cell line derived from human colon
ClogP	calculated log *P*
CNS	central nervous system
CMC	MDL Comprehensive Medicinal Chemistry
MDDR	MDL Drug Data Report
HBA	H-bond acceptor
HBD	H-bond donor
HTS	high-throughput screening
MW	molecular weight
MPS	multiple parallel synthesis
PDR	Physicians' Desk Reference
PSA	polar surface area
RTB	rotatable bonds
SAR	structure–activity relationship
WDI	World Drug Index

Symbols

LE	ligand efficiency
log *P*	octanol–water partition coefficient

17.1
Introduction

The identification of suitable lead molecules is a crucial process with important implications for success in drug development. There is a growing common under-

Molecular Drug Properties. Measurement and Prediction. R. Mannhold (Ed.)

ISBN: 978-3-527-31755-4

standing that typical lead molecules differ in many aspects of their property profile from typical drug candidates after a successful lead optimization. The reason is that lead molecules, as starting points for lead optimization, have to leave room for both increasing biological potency, and for achieving favorable metabolic and pharmacokinetic profiles. In a simplified manner, leads have to be smaller and simpler than typical drugs. There is emerging consensus in the pharmaceutical industry how to profile leads according to rather simple physicochemical properties which indirectly ensure this.

In this chapter we present the most commonly used set of properties and make recommendations for proper ranges of properties. In addition to property profiles, we will briefly touch upon drug-likeness as a classification problem. The enormous increase in both the number of available compounds for hit finding and in the throughput of hit finding-technologies such as combinatorial chemistry ("combichem") and high-throughput screening (HTS), have led to an explosion of the number of primary hits. As a consequence we will focus on *in silico* methods for lead and drug profiling. Finally, an application example will take us through a complete compound selection exercise in order to show how to put everything together.

17.2
Properties of Leads and Drugs

17.2.1
Simple Molecular Properties

HTS of compound libraries is currently the major source of novel hits in drug discovery. The development of new screening technologies and robust screening assays combined with combinatorial chemistry and multiple parallel synthesis (MPS) enabled rapid access to a large chemical space to identify novel active compounds. However, it was soon realized that screening vast numbers of chemicals as produced by combinatorial chemistry did not necessarily increase the number of leads nor did it generate the best chemical starting points. The screening library size (in the range of multimillion compounds for top pharmaceutical companies) and diversity (as defined, for example, by the number of chemical clusters or pharmacophore coverage), although important, are not enough to provide high quality hits and leads for further optimization [1]. Combinatorial chemistry and MPS output needs to be shaped by medicinal chemistry knowledge [such as privileged structures and pharmacophores for certain targets or target classes, and adsorption, distribution, metabolism, excretion and toxicity (ADMET) issues] in order to restrict the large synthetically accessible chemical space to the much smaller medicinal space.

Lead-likeness and drug-likeness were introduced to describe high-quality medicinally relevant compounds and a great deal of effort was devoted over the last 10 years to define the boundaries in terms of molecular properties for leads and drugs. Both lead-likeness and drug-likeness are complex properties, and they suggest, above all, the potential of a molecule to reach the status of a lead and a drug, respectively.

Drug-likeness refers to the right balance of properties that one finds in marketed drugs or compounds in advanced clinical trials including efficacy, safety and pharmacokinetic/pharmacodynamic profiles [2, 3], and one can think of drug-like molecules as similar to existing drugs in terms of scaffolds, functional groups and physicochemical properties. To extract molecular properties related to drug-likeness one can use available drug databases such as the MDL Comprehensive Medicinal Chemistry (CMC) [4], MDL Drug Data Report (MDDR) [5], World Drug Index (WDI) [6], Physicians' Desk Reference (PDR) [7], SPRESI [8] and GVKBIO Drug Database [9]. MDL's Available Chemicals Directory (ACD) [10] is commonly used as the nondrug counterpart.

Lipinski et al. at Pfizer [11] analyzed the distribution of physicochemical properties of 2245 drugs from the WDI that have entered clinical trials after excluding natural products and actively transported molecules. They proposed the "Rule-of-5" to indicate that poor absorption or permeation is more likely when:

Molecular weight (MW)	>500
H-bond acceptors (HBA)	$n>10$
H-bond donors (HBD)	$n>5$
Calculated log P (ClogP)	>5

Oxygen (O) and nitrogen (N) atoms count as HBA and -NH or -OH groups add to the number of HBD. Compounds that violate any two of the "Rule-of-5" conditions ($MW \leq 500$, $HBA \leq 10$, $HBD \leq 5$, $ClogP \leq 5$) are unlikely to be oral drugs.

A modified Lipinski rule was proposed by Congreve et al. [12] following the analysis of a diverse set of 40 fragment hits against a range of targets. The molecular property cutoffs for the "Rule-of-3" [$MW \leq 300$, $HBA \leq 3$, $HBD \leq 3$, $ClogP \leq 3$, number of rotatable bonds $(RTB) \leq 3$, $PSA \leq 60$] reflect the smaller molecules and can be used to design fragment libraries for fragment-based lead generation.

Ghose et al. [13] analyzed a larger set of 6304 drug-like compounds from CMC and determined qualifying ranges which cover more than 80% of the compounds in the set for the following molecular properties: ALOG P (–0.4 to 5.6, average value 2.52), MW (160–480, average value 357), molar refractivity (40–130, average value 97) and number of atoms (20–70, average value 48).

Oprea [14] performed a Pareto analysis on several datasets and concluded that 70% of the drug compounds contain zero to two HBD, two to eight RTB and one to four rings. Similar distributions for MW, ClogP, HBD and HBA were found on filtered ACD and MDDR datasets. As a consequence 80% of both ACD and MDDR passed the "Rule-of-5". Additional parameters related to molecular complexity such as the number of nonterminal RTB, the number of rigid bonds and the number of rings can be used to distinguish between drugs and nondrugs.

Molecular polar surface area (PSA; in Å^2) or normalized to the total surface area (%PSA) is another descriptor that has been used extensively to model drug-likeness especially when it comes to describing permeability and bioavailability of compounds [15–17]. In its simplest form PSA is defined as the van der Waals surface area (Å^2) of all nitrogen and oxygen atoms, and the polar hydrogens bonded to these heteroatoms and can be calculated from the three-dimensional (3D) molecular structure (multiple low-energy conformations have to be sampled). To overcome the need of generating 3D structures Ertl et al. [18] introduced topological PSA – a method based on contributions of polar fragments determined by least-square fitting with single conformer 3D PSA for drug-like structures from WDI.

Kelder et al. [19] have shown that PSA can be used to model oral absorption and brain penetration of drugs that are transported by the transcellular route. A good correlation was found between brain penetration and PSA (n=45, r=0.917). From analyzing a set of 2366 central nervous system (CNS) and non-CNS oral drugs that have reached at least phase II clinical trials it was concluded that orally active drugs that are transported passively by the transcellular route should have PSA < 120 Å^2. In addition, different PSA distributions were found for CNS and non-CNS drugs.

Deconinck et al. [17] used Classification And Regression Trees (CART) to classify 141 drug-like compounds in absorption classes and revealed the high importance of PSA and log P for the predictive models. Palm et al. [20] studied the correlation between PSA and intestinal drug absorption of six β-adrenoreceptors antagonists and reported very good correlations with cell permeability (Caco-2 cells, r^2=0.99; rat ileum, r^2=0.92). Molecules with PSA ≤ 60 Å^2 will exhibit high and almost complete intestinal absorbance, whereas molecules with a PSA ≥ 140 Å^2 exhibit poor intestinal absorbance. Similar PSA cutoffs were found by Egan et al. [15] using a pattern recognition method based on PSA and AlogP98 to describe passive intestinal absorption. The model validated on known orally available drugs and drug-like molecules with Caco-2 cell permeability data demonstrated good predictive power (74–92% depending on the dataset and criterion used).

Leads are key starting points in drug discovery projects. A lead compound should be amenable for further medicinal chemistry optimization and as such it should display simple chemical features, membership to a well-defined structure–activity relationship (SAR) series, favorable patent situation and good ADME properties [21]. Detailed milestone criteria for leads as used by AstraZeneca-Charnwood to guide the hit-to-lead process have been published [22]. The lead-like paradigm was introduced by Teague, Davis and Oprea from AstraZeneca [1]. They analyzed 18 pairs of leads and drugs by looking at various properties and found statistically significant differences in their values. A follow-up study was performed on a larger set of 96 lead-drug pairs [21]. The general trend was an increase in molecular complexity in going from lead to drug (an increase in MW, ClogP, HBD and HBA). A similar study was done at GlaxoSmithKline on a set of 450 lead-drug pairs [23]. Their analysis also indicated that, when compared with their corresponding drugs, leads have lower MW, lower ClogP, fewer aromatic rings and fewer HBA. No differences were observed for HBD.

The generic term of drug-likeness implies a number of other properties [24] such as aqueous solubility, metabolism, blood–brain barrier penetration and oral absorption which are covered by other chapters in this book.

17.2.2 Chemical Filters

Compounds unstable under the screening conditions or having functional groups reactive towards proteins tend to give false HTS hits (false positives) in biochemical screens. Rishton [25] discussed simple chemical guidelines to filter out unwanted chemicals and extended the lead-likeness concept by including chemical properties. Good lead compounds should not contain chemically reactive functional groups [26], "promiscuous inhibitors" [27], "frequent hitters" [28] and warheads [25]. Chemical filters have been also developed for specific screens. For example, assays based on fluorimetric or colorimetric detections are highly sensible to fluorescent and colored compounds.

Several compilations of "unwanted" chemical (sub)structures have been published and the chemical filters typically include [22, 25, 29–32]:

- Metals and isotopes, atoms other than C, H, N, O, P, S, F, Cl, Br, I.
- Substructural counts to flag compounds difficult to optimize (e.g. $NO_2>1$, $Cl>3$ or $F>6$ or sum of $Cl+Br+I>3$, structures containing no heteroatoms, structures without rings).
- Unstable compounds under the screening conditions (esters, anhydrides, thiols, heteroatom–heteroatom acyclic single bonds).
- Reactive structures that interfere with the biochemical assay (aldehydes, acyl-halides, sulfonyl-halides, Michael acceptors, epoxides, aziridines, oximes, N-oxides).
- Substructures commonly found in dyes (chromophores), fluorescent compounds, pesticides (polyhalogenated alkanes, cycloalkanes).
- Toxicophores [33] (polycyclic aromatic and polycyclic planar systems, nitro- and amino-aromatics).
- Carcinogenic and mutagenic substructures [34].

Many of these unwanted functionalities have been collected based on chemists feedback from hit identification and lead optimization projects, and by looking at compounds not considered good starting points for optimization by medicinal chemistry or difficult to synthesize [35]. However, one could say that "beauty is in the eye of the beholder" and selecting attractive chemical starting points depends upon the experience and prejudice of individual chemists. An interesting study at Pharmacia in which 13 chemists reviewed about 22 000 compounds in a compound acquisition program showed that medicinal chemists were inconsistent in the compounds they reject [36]. Furthermore, it was found that individual medicinal chemists do not consistently reject the same compound.

These unwanted functional groups can be easily encoded as SMARTS [37] to be used as structural alerts for HTS compound prioritization or compound acquisition. OpenEye's Filter [38] is an excellent example of such a filtering tool.

17.2.3 Correlated Properties

We have discussed now a number of important molecular properties which are used to profile lead and drug molecules. In many cases, certain combinations of these properties are correlated to some extent within a series of compounds. In particular, the size-related properties of MW, PSA, and log *P* show this tendency. One should be aware of this phenomenon and it should be taken into account when interpreting the underlying SAR data. However, there is no strong correlation between these three properties in general. When looking at a random subset of 10 000 compounds from GVKBIO [9], we find that MW and log *P* are correlated with $r=0.32$, log *P* and PSA with $r=0.35$, and MW and PSA with $r=0.61$.

Correlations of simple properties within compound series become more important when comparing them to biological activity. In the ligand binding efficiency part, we will discuss how to monitor the relation between size and potency in a compact manner. Here, we comment on correlations of calculated properties and biological activities in series of compounds. In many cases, simple physicochemical parameters, such as lipophilicity, have an influence on potency and other properties. It is important to know about possible correlations in order to understand related SAR series and in order to predict to what extent a lead series can be optimized without ending up in areas of unfavorable properties. Often, a direct correlation between potency and a physicochemical property can uncover trends in the SAR within a series of compounds. If, for example, potency is highly correlated with lipophilicity, it is very likely that the SAR within the series shows a strong influence from typical lipophilic substituents. Moreover, knowing these trends can steer further optimization towards desirable property profiles.

In the following example from the literature [39], the correlation with lipophilicity is studied for 47 compounds in two series which have been designed for dopamine D_2 receptor affinity. As seen in Fig. 17.1, there is no clear relation between D_2 activity and lipophilicity ($r=0.11$).

In addition, for all compounds the binding affinity towards the σ receptor (a counter-target) was determined. Here exists a significant correlation with $r=0.85$, see Fig. 17.2.

In this case, there are two rather simple, although important, conclusions to be drawn. For the main target, D_2, lipophilicity is not a driver for potency. Thus, making the compounds more potent does not impose an implicit risk to make them more lipophilic. For the counter-target, the σ receptor, lipophilicity is strongly correlated to potency. This opens up for the simple hypothesis for separating D_2 and σ activity by designing less lipophilic compounds.

In another example from the literature [40], we see a strong correlation between potency and PSA. In this case a series of 53 CCR5 receptor agonists have been

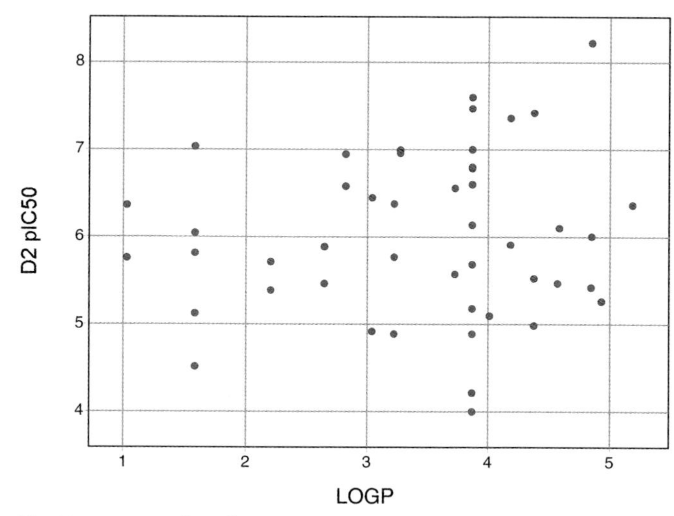

Fig. 17.1 Scatter plot of D_2 pIC_{50} versus $\log P$.

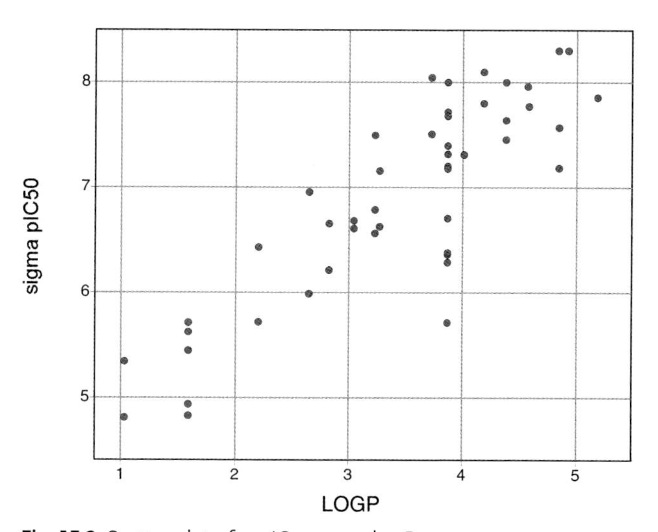

Fig. 17.2 Scatter plot of σ pIC_{50} versus $\log P$.

measured in a CCR5 binding assay. The scatter plot presented in Fig. 17.3 shows a strong correlation of the pIC_{50} values with PSA. Since the target is inhibition of HIV replication, higher PSA should have a positive effect on inhibiting the compounds from entering the brain. Thus, correlation of potency with PSA in this case might be an advantage.

There are a number of such relevant correlations to be always considered. Based on both relevance for lead profiling and *in silico* availability of molecular properties,

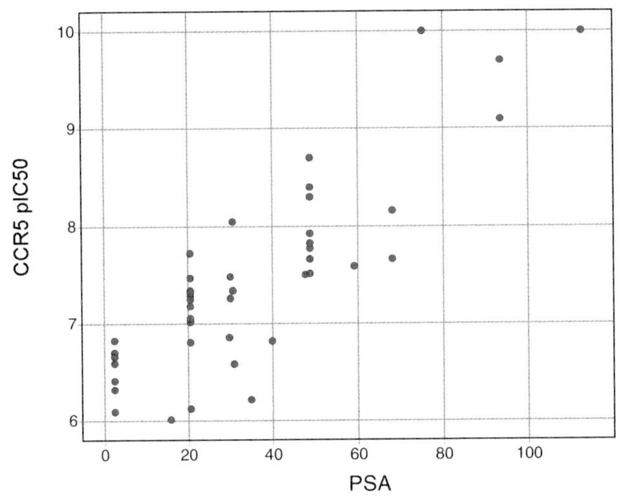

Fig. 17.3 Scatter plot of CCR5 pIC_{50} versus PSA.

the following list of potential correlations of potency with simple properties is recommended to be checked: log *P*, PSA, MW, HBD, HBA and solubility.

In addition, one should check other measured properties and their relevant correlations to other properties as:

- Solubility versus lipophilicity and pK_a.
- Clearance versus lipophilicity and solubility.
- Permeability versus solubility.

17.2.4
Property Trends and Property Ranges

Analysis of simple molecular properties for leads and drugs indicates a steady increase in molecular complexity when progressing compounds from lead to drug. This can be directly linked to the high attrition in late development stages due to poor bioavailability and pharmacokinetics. To tackle the problem a number of researchers have studied the changes in molecular profiles for lead-drug pairs, and the distribution of calculated molecular properties for drugs and leads. As a result, specific guidelines for lead-like and drug-like compounds have been suggested to help medicinal and computational chemists in their drug-hunting tasks. The molecular property cutoffs proposed in literature are slightly different from one study to another depending mostly on the dataset used to derive them. There is, nevertheless, a general consensus regarding the utility of such metrics in library design, HTS data analysis and compound acquisition to name just a few major applications.

A systematic analysis of drug-lead pairs performed by Oprea et al. [21] revealed a right shift in property medians between drugs and leads: an increase in MW

Tab. 17.1 Changes in simple molecular properties between leads and drugs according to Oprea et al. [21].

Property	*Lead*	*Drug*	*Difference*
MW	315	384	69
ClogP	2.54	2.11	0.43
HBA	4	5	1
HBD	2	2	0
RTB	–	–	2
Rings	–	–	1

(69 Da) and lipophilicity (0.43 log units), two additional RTB, one additional ring, one additional HBA and no change in HBD (see Table 17.1). Similar trends were reported by Hann et al. [23] on a larger set of lead-drug pairs. In contrast, Proudfoot [41] reported small differences between leads and drugs for a set of small molecules drugs launched in 2000. These studies indicate that the lead-like chemical space is somewhat smaller than the drug-like chemical space and the following property ranges were recommended for lead-like compounds [42]: MW$\leq$460, $-4\leq$ClogP$\leq$4.2, log $S_w\leq-5$, RTB $\leq$10, rings$\leq$4, HBD$\leq$5, HBA$\leq$9.

Smaller and more soluble leads as starting points for lead optimization is also supported by several studies on the analysis of property distributions for compounds in clinical development (phase I, II and III, and launched) which indicate that on average compounds that are progressed from one phase to the next one have lower MW, lower log *P* and higher solubility [43–45]. Property data for marketed drugs and subsets of oral drugs reported by different research groups are presented in Table 17.2. We have also included statistics for the GVKBIO Drug Database containing all US Food and Drug Administration-approved drugs, and other drugs extracted from standard books, online sources and various pharmacological journals.

The molecular descriptors and the chemical filters discussed throughout this chapter are fast to compute and can be readily integrated in cheminformatic platforms to support medicinal chemists in making decisions for library design, HTS compound prioritization or compound acquisition. At AstraZeneca we have developed CLASS [46], a cheminformatic tool supported by database technologies, for tracking and profiling chemical libraries. It is intended to guide medicinal chemists during the "plan–make–test" of library design processes and to encourage them to adhere to standard criteria required for all compounds in our compound collection [22]. It also provides an easy way to check the novelty of the proposed libraries against internal and external compound collections deposited in the system, and to explore large virtual libraries. Selecting and prioritizing compounds based on molecular properties and chemical filters needs to be done cautiously as different targets may demand different property profiles [47]. Therefore, training and awareness is an integral part of an efficient exploitation of such a tool. Similar

Tab. 17.2 Molecular property data for sets of marketed drugs from different sources.

Property	*Marketed oral*[1]	*Marketed oral*[2]	*Launched drugs*[3]	*Marketed drugs*[4]
Compounds	594	1193	884	1759
MW	337	344	338	368 (329)
CLOGP	2.5	2.3	2.5	2.1 (2.3)
HBD	2.1	1.8	1	2.4 (2)
HBA	4.9	5.5	4	6.1 (5)
PSA	–	78	122	90 (69)
RTB	5.9	5.4	6	5.7 (5)
Rings	–	2.6	–	2.9 (3)

1 Data from Wenlock et al. [44]. Mean values for marketed oral drugs extracted from PDR.
2 Data from Vieth et al. [70]. Mean values for a compilation of marketed oral drugs from various sources.
3 Data from Blake [45]. Median values for compounds classified as launched drugs.
4 This work. Mean (median) values for marketed drugs from GVKBIO Drug Database.

molecular property calculators and library profiling tools have been implemented by other pharmaceutical companies [29, 48, 49], compound suppliers [32, 50, 51] or in academia [31, 52, 53].

17.2.5
Ligand Efficiency

At the early stages of drug discovery, active-to-hit and hit-to-lead, potency is one of the main drivers for progressing compounds. Analysis of simple molecular properties for leads and drugs indicates a steady increase for the mean MW and log *P*, whereas the opposite trend was found for oral drugs progressing from clinical candidates to marketed drugs [44]. Therefore, improving potency at the expense of increasing molecular complexity may subsequently hamper further optimization due to unacceptable drug metabolism and pharmacokinetics profiles.

Kuntz et al. [54] surveyed dissociation constants and IC_{50} values of 160 ligand complexes and showed that, for strong-binding ligands of up to 15 nonhydrogen atoms, the maximum free energy of binding is approximately 1.5 $kcal\,mol^{-1}$ per nonhydrogen atom. The majority of medicinal chemistry compounds have efficiencies far below the observed maximal affinity per atom. Following this work researchers at Pfizer [55] introduced ligand efficiency (*LE*) as a measure for the quality of hits and leads, a concept that has rapidly gained acceptance in the pharmaceutical industry. Ligand efficiency is defined as the average binding energy contributed per nonhydrogen atom in the molecule:

$$LE = -\Delta G/N = -RT \ln K_d/N \qquad (1)$$

where ΔG is the free energy of binding of the ligand for a specific target and N is the number of nonhydrogen atoms (heavy atoms). For comparison purposes K_d

can be substituted by IC_{50} or even the extrapolated IC_{50} values from percentage inhibition data.

In the same paper Hopkins et al. [55] suggested a minimum *LE* value for compounds obeying Lipinski's rule. An analysis of Pfizer's screening collection indicated a mean MW of 13.3 for heavy atoms. A compound with MW of 500 and affinity of 10 nM would have 38 heavy atoms and a ligand efficiency of 0.27 $kcal\,mol^{-1}$ per heavy atom. A compound with ligand efficiency of 0.36 $kcal\,mol^{-1}$ per heavy atom would require only 30 atoms (MW of 405) to bind with $K_d = 10\,nM$. Hence, optimizing hits or leads with the highest ligand efficiencies rather than the highest potencies is recommended and a 0.3 *LE* cut-off can be used as a guideline.

As an example, Fig. 17.4 presents the ligand efficiency of 107 compounds active against human 5-hydroxytryptamine 1A receptor (data extracted from GVKBIO). Compounds **A** and **B** have different potencies (IC_{50}**A** = 12.2 μM and IC_{50}**B** = 1.0 μM) and properties (MW**A** = 244, HA**A** = 18, CLOGP**A** = 2.8 and MW**B** = 294, HA**B** = 22, CLOGP**B** = 4.1), but display a similar *LE* of 0.37.

The idea of ligand efficiency was extended by considering other molecular descriptors. Abad-Zapatero and Metz [56] introduced a percentage efficiency index, a binding efficiency index and a surface-binding efficiency index by normalizing

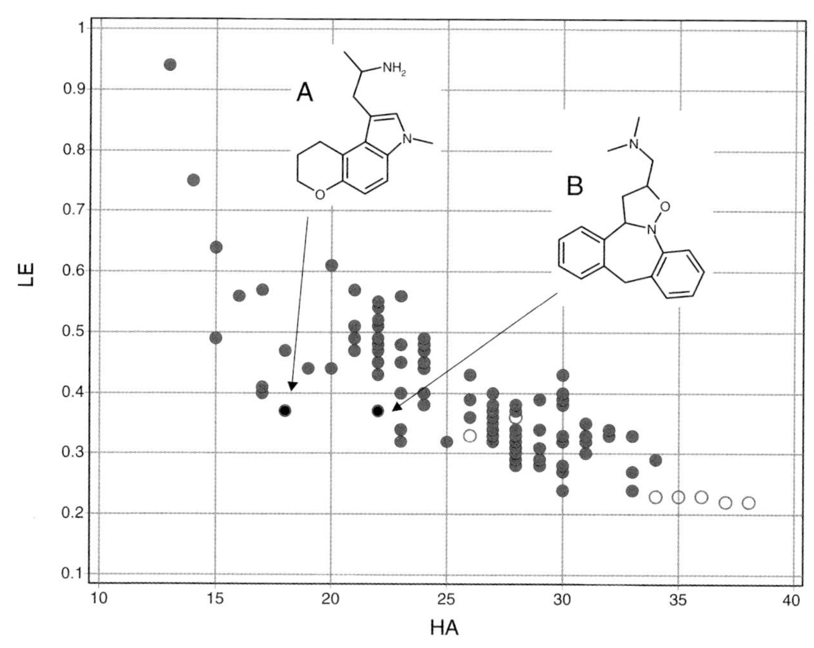

Fig. 17.4 Ligand efficiency for a set of 107 compounds active against human 5-hydroxytryptamine 1A receptor. Exemplified compounds **A** and **B** both have *LE* of 0.37, but different potencies and molecular sizes (**A**: IC_{50} = 12.2 μM, MW = 244, HA = 18; **B**: IC_{50} = 1.0 μM, MW = 294, HA = 22). Open circles indicate compounds with RO5 ≥ 1 (at least one parameter from Lipinski's "Rule-of-5" is out of range). Activity data extracted from GVKBIO.

the binding affinity by MW and PSA, respectively. Burrows and Griffen at AstraZeneca [57] showed that potency (pIC_{50}) per heavy atom, potency per unit Clog D, and potency difference to human serum protein binding (log K_{app}) can discriminate between hits that have been successfully developed into candidate drugs and those which have not. The following ranges are recommended for good hits/leads:

$$pIC_{50}/HA > 0.2$$

$$pIC_{50} - C\log D > 2$$

$$pIC_{50} - \log K_{app} > 1$$

Combining molecular properties with potency provides a simple yet powerful overview of a screening dataset and can be used to quickly identify ligand-efficient lead-like compounds. In Fig. 17.5, a set of 429 compounds active against 5-hydroxytryptamine 1A receptor are displayed in a PSA–ClogP plot with the size of the circle related to ligand efficiency. One can easily spot the efficient binders in area of favorable properties.

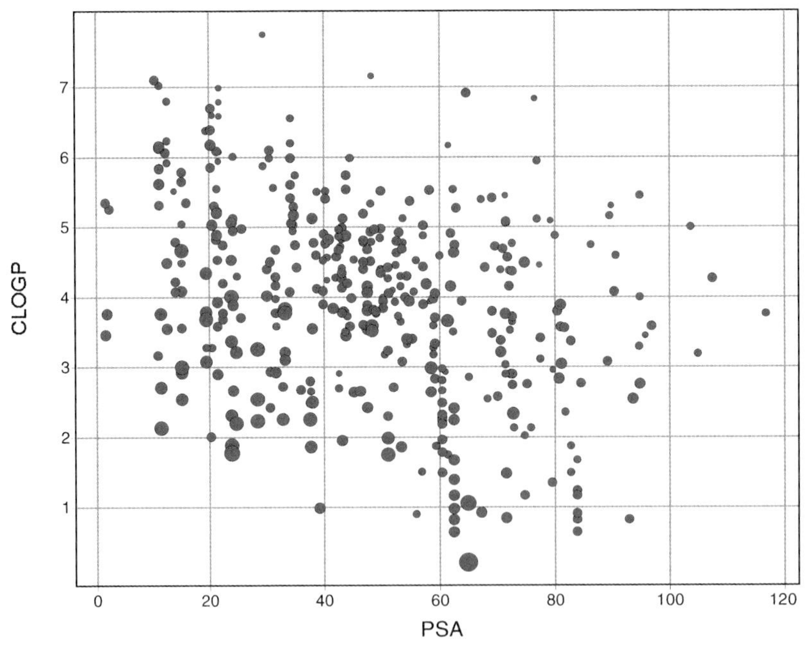

Fig. 17.5 Plot of ClogP versus PSA for a set of 429 compounds active against 5-hydroxytryptamine 1A receptor. Compounds represented by larger circles have higher ligand efficiency. Such visualization is useful to identify ligand-efficient lead-like compounds. Activity data extracted from GVKBIO.

Ligand efficiency normalizes affinity, and it can be used to compare the efficiency of binding across compound series and across targets. Easy to calculate and interpret, it provides a useful metric for medicinal chemists to identify weak binders with potential of being progressed into a potent clinical candidate. This is particularly useful in fragment-based screening where low-MW compounds (100–300 Da) commonly exhibit weak binding (100–10 μM).

17.3 Drug-likeness as a Classification Problem

Another way to tackle drug-likeness is to treat it as a classification problem, i.e. drugs versus nondrugs. When characterizing leads and drugs, we often use the properties discussed in the previous sections in order to assess directly parameters which steer towards favorable metabolic and pharmacokinetic profiles, i.e. make them drug-like. In addition, we often discard compounds with chemical features known to make them instable, toxic or otherwise unwanted. Apart from that, when involving experienced medicinal chemists in ranking and selecting compounds, there seems to be an extra element of intuition or "gut-feeling" which cannot be expressed easily by property ranges or chemical features. Sadowski and Kubinyi [58] and Ajay et al. [59] were the first to describe an astonishingly easy way to separate drugs from nondrugs in a very general manner. Following these publications, other researcher have repeated these studies in various ways (different datasets, different descriptors and different statistical methods) and came to similar results. For a recent example, see the work of Müller et al. [60] and references cited therein. The general concept consists of the following three steps:

(i) Select representative datasets of drugs and nondrugs.
(ii) Describe the molecules by using suitable descriptors.
(iii) Obtain a predictive model.

This will be illustrated by following the route described by Sadowski and Kubinyi [58] (see the original publication for references and more details). The most crucial part is the selection of representative sets of molecules for both the drug and the non-drug classes. The WDI was used to obtain typical drug molecules, whereas the ACD was used as source for typical nondrugs.

Both classes of compounds were then filtered in order to get rid of obviously unsuitable compounds, which violated MW cutoffs or contained otherwise unwanted chemical features such as reactive groups. This all was done in order to remove explicit chemical features which could otherwise, in a trivial manner, discriminate between drugs and nondrugs. For example, there will not be any drug molecules containing acid chloride, but this is of course a typical class of reagent from the ACD. The final datasets contained about 170 000 nondrugs and about 40 000 drugs.

After all the careful filtering, there was neither a single molecular property from the profiles discussed above which could effectively discriminate the two classes nor a predictive model obtained from all those property descriptors. Only after deriving a descriptor which takes into account the chemical structure of the molecules in both classes could a neural network model be trained to separate the two classes. In this case the molecular descriptor was the counts of the Ghose–Crippen atom types [61], a set of atom types originally developed for predicting log *P*.

The neural network was a feedforward backpropagation net with the atom type counts as input values and the predicted class value (0 for nondrug and 1 for drug) as output. It was trained by using 5000 drugs and 5000 nondrugs as a training set. The individual compounds were selected randomly. The model obtained turned out to be highly predictive as illustrated in Fig. 17.6. The predicted value is a continuous score between 0 (nondrug) and 1 (drug). Figure 17.6 shows the percentage of structures from the remaining datasets which fall into certain intervals of the score in separate graphs for nondrugs (dashed line) and drugs (continuous line). One can see clearly a very good separation between the two classes. Using a symmetric cutoff score value of 0.5, 83% of the ACD compounds and 77% of the WDI compounds were correctly classified. These figures were obtained from the remaining compounds which were not in the training set; therefore the model is considered to be highly predictive.

In summary, such simple classification schemes for drug-likeness can, in a very fast and robust manner, help to enrich compound selections with drug-like molecules. These filters are very general and cannot be interpreted any further. Thus, they are seen rather as a complement to the more in-depth profiling of leads and drugs by using molecular properties and identifying trends in compound series.

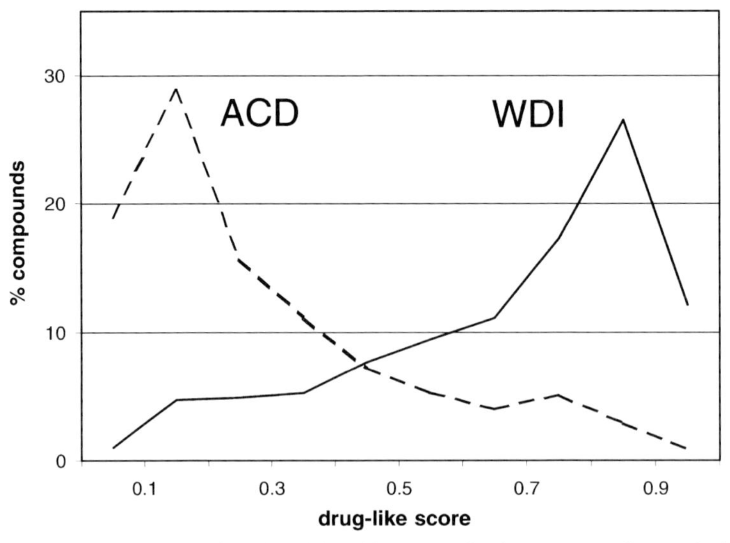

Fig. 17.6 Distribution of predicted drug-like scores for the test sets of ACD (dashed line) and WDI (continuous line) compounds.

17.4 Application Example: Compound Acquisition

The corporate compound collection is an important asset of any pharmaceutical company. Having access to a large chemistry base not only provides hits and leads for new targets by using the current HTS technology, but also significantly accelerates the lead optimization process. Compound collections larger than 1 million are routinely screened nowadays. These collections are replenished on a regular basis to refill some of the depleted compounds or to complement the in-house chemistry with novel compounds. Commercially and freely available databases are an important source of screening compounds, and many of them are readily accessible on the Internet (for a comprehensive list see, e.g. http://www.bioscreening.com/index/Companies).

Scientists within compound management groups face the challenge to select the "best" compounds from this huge pool of available chemicals. At AstraZeneca criteria used for compound selection include:

- Compound availability and price.
- Purity.
- Lead-like and drug-like filters.
- Diversity and similarity with internal and external compounds.
- Availability of small clusters of compounds (close analogs).

Here we provide a simple example of a compound acquisition process guided by lead-like and drug-like criteria. We have randomly selected 10 000 unique compounds from the GVKBIO database (published in journals and having measured IC_{50} or K_i values against at least one target) and designated them as Collection A. Another set of 100 000 unique compounds, Collection B, was randomly selected from an internally compiled set of compounds from external vendors. We aim to double Collection A by selecting the "best" compounds from Collection B (we assume availability and quality to be similar for all compounds in Collection B).

Usually one starts by preprocessing Collection B in order to produce normalized structures that can be used for direct one-to-one comparisons. This steps includes removing small fragments (adducts, counterions, water) and neutralizing remaining charges as well as generating canonical tautomers, and producing a unique list of structures. The links to the suppliers are kept to be used later to chose the best price per compound or to minimize the number of suppliers that cover the final selection.

Collection B is then filtered with user-defined drug-like and lead-like criteria. In this example AstraZeneca acquisition filters are used. The next step is to identify identical compounds and close analogs with Collection A by using, for example, molecular hashcodes [62] and structural fingerprints [63]. Classification techniques, such as neural networks, can also be used to assess "in-house likeness" and rapidly compare large compound collections [64]. At this point we might decide to replenish certain compounds or expand biologically interesting singletons from Collection A. Otherwise identical compounds and close neighbors are removed. Additionally one can include other external databases in this step to ensure novelty such as GVKBIO or PubChem [65]. The remaining set can be considered as containing potentially interesting compounds and the final selection can be diversity or knowledge-based driven. The final set is made available to medicinal chemists who will accept or reject compounds or clusters of compounds. This is also a good opportunity to enhance the chemical filters if unwanted structures pass through the filtering process. The process is summarized in Fig. 17.7.

If, after applying lead-like criteria (property and chemical filters), there are still too many compounds left one can use diversity/similarity-based techniques to select the final set. For the example above, 64% of the starting compounds survive the compound acquisition process (around 64 000 compounds in 11 000 clusters), see Fig. 17.8.

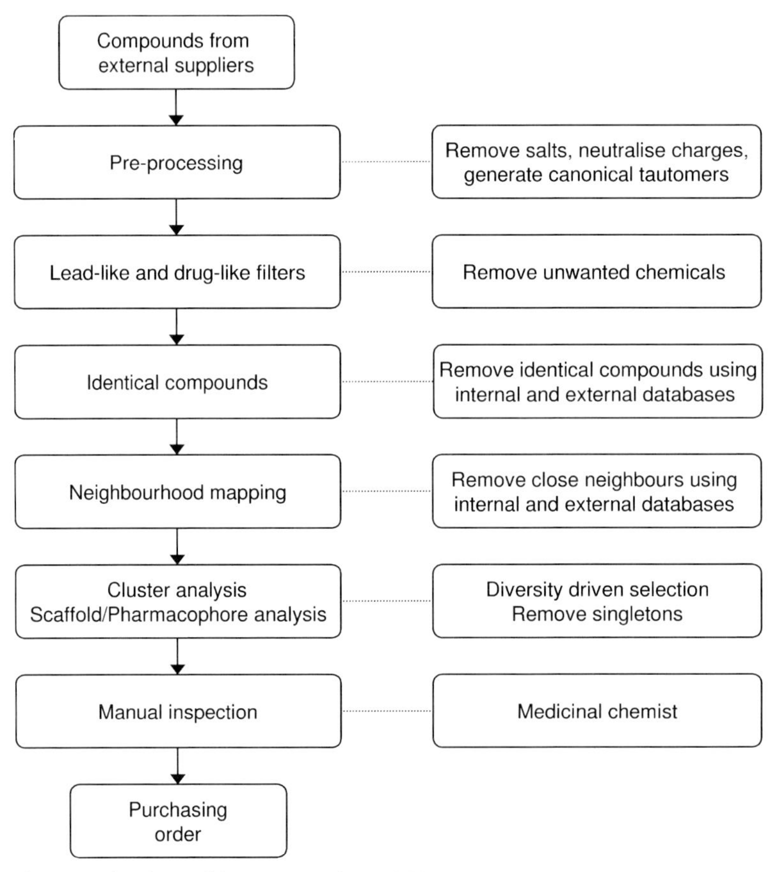

Fig. 17.7 Flowchart of the compound acquisition process.

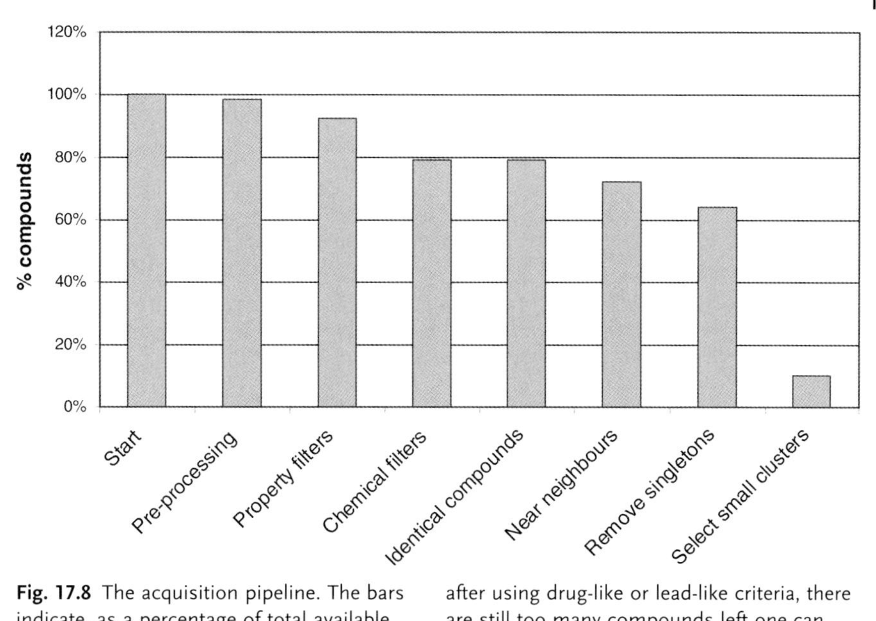

Fig. 17.8 The acquisition pipeline. The bars indicate, as a percentage of total available compounds, the survival rate at each step. If, after using drug-like or lead-like criteria, there are still too many compounds left one can apply diversity-based selection techniques.

Hence there are multiple solutions for the final set of 10 000 compounds. The final selection can be diversity driven using for example cluster analysis based on multiple fingerprints [63], hole filling strategies by using scaffold/ring analysis (LeadScope [66], SARVision [66]) or pharmacophore analysis [67, 68]. For a review of computational approaches to diversity and similarity-based selections, see the paper of Mason and Hermsmeier [69] and the references therein.

17.5 Conclusions

We have discussed simple calculated compound properties and related aspects in the context of drug-likeness and lead-likeness. Careful inspection of property ranges, the presence or absence of specific chemical features as well as their correlation with each other and with biological potency are of great importance for success in selecting starting points for lead generation and in guiding chemical optimization. A number of important concepts such as property ranges, chemical substructure filters, ligand efficiency, and drug-likeness as a classification problem have been discussed, and some of them have finally been demonstrated in an example of how to select compounds for acquisition.

References

1 Teague, S. J., Davis, A. M., Leeson, P. D., Oprea, T. The design of leadlike combinatorial libraries. *Angew. Chem. Int. Ed.* **1999**, *38*, 3743–3748.
2 Ajay. Predicting drug-likeness: why and how? *Curr. Top. Med. Chem.* **2002**, *2*, 1273–1286.
3 Matter, H., Baringhaus, K.-H., Naumann, T., Klabunde, T., Pirard, B. Computational approaches towards the rational design of drug-like compound libraries. *Comb. Chem. High-Throughput Screen.* **2001**, *4*, 453–475.
4 MDL Comprehensive Medicinal Chemistry provided by MDL Information Systems Inc., San Leandro, CA, USA, http://www.mdl.com/products/knowledge/medicinal_chem.
5 MDL Drug Data Report provided by MDL Information Systems Inc., San Leandro, CA, USA, http://www.mdl.com/products/knowledge/drug_data_report.
6 WDI: Derwent World Drug Index provided by Thomson Scientific, http://scientific.thomson.com/products/dwpi.
7 Physicians' Desk Reference distributed by Thomson Scientific, http://scientific.thomson.com/products/pdr.
8 SPRESI database, distributed by Daylight Chemical Information Systems Inc., Mission Viejo, CA, USA, http://www.daylight.com.
9 GVKBIO database, http://www.gvkbio.com.
10 MDL Available Chemicals Directory provided by MDL Information Systems Inc., San Leandro, CA, USA, http://www.mdl.com/products/experiment/available_chem_dir.
11 Lipinski, C. A., Lombardo, F., Dominy, B. W., Feeney, P. J. Experimental and computational approaches to estimate solubility and permeability in drug discovery and development settings. *Adv. Drug Deliv. Rev.* **1997**, *23*, 3–25.
12 Congreve, M., Carr, R., Murray, C., Jhoti, H. A rule of three for fragment-based lead discovery? *Drug Discov. Today* **2003**, *8*, 876–877.
13 Ghose, A. K., Viswanadhan, V. N., Wendoloski, J. J. A knowledge-based approach in designing combinatorial or medicinal chemistry libraries for drug discovery. 1. A qualitative and quantitative characterization of known drug databases. *J. Comb. Chem.* **1999**, *1*, 55–68.
14 Oprea, T. I. Property distribution of drug-related chemical databases. *J. Comput.-Aided Mol. Des.* **2000**, *14*, 251–264.
15 Egan, W. J., Merz, K. M., Baldwin, J. J. Prediction of drug absorption using multivariate statistics. *J. Med. Chem.* **2000**, *43*, 3867–3877.
16 Martin, Y. C. A Bioavailability Score. *J. Med. Chem.* **2005**, *48*, 3164–3170.
17 Deconinck, E., Hancock, T., Coomans, D., Massart, D. L., Heyden, Y. V. Classification of drugs in absorption classes using the Classification And Regression Trees (CART) methodology. *J. Pharm. Biomed. Analysis* **2005**, *39*, 91–103.
18 Ertl, P., Rohde, B., Selzer, P. Fast calculation of molecular polar surface area as a sum of fragment-based contributions and its application to the prediction of drug transport properties. *J. Med. Chem.* **2000**, *43*, 3714–3717.
19 Kelder, J., Grootenhuis, P. D. J., Bayada, D. M., Delbressine, L. P. C., Ploemen, J.-P. Polar molecular surface as a dominating determinant for oral absorption and brain penetration of drugs. *Pharm. Res.* **1999**, *V16*, 1514–1519.
20 Palm, K., Luthman, K., Ungell, A.-L., Strandlund, G., Artursson, P. Correlation of drug absorption with molecular surface properties. *J. Pharm. Sci.* **1996**, *85*, 32–39.
21 Oprea, T. I., Davis, A. M., Teague, S. J., Leeson, P. D. Is there a difference between leads and drugs? A historical perspective. *J. Chem. Inf. Comput. Sci.* **2001**, *41*, 1308–1315.
22 Davis, A. M., Keeling, D. J., Steele, J., Tomkinson, N. P., Tinker, A. C. Components of successful lead generation. *Curr. Top. Med. Chem.* **2005**, *5*, 421–439.
23 Hann, M. M., Leach, A. R., Harper, G. Molecular complexity and its impact on the probability of finding leads for drug discovery. *J. Chem. Inf. Comput. Sci.* **2001**, *41*, 856–864.

24 Walters, W. P., Murcko, M. A. Prediction of "drug-likeness". *Adv. Drug Deliv. Rev.* **2002**, *54*, 255–271.

25 Rishton, G. M. Reactive compounds and *in vitro* false positives in HTS. *Drug Discov. Today* **1997**, *2*, 382–384.

26 Rishton, G. M. Nonleadlikeness and leadlikeness in biochemical screening. *Drug Discov. Today* **2003**, *8*, 86–96.

27 McGovern, S. L., Caselli, E., Grigorieff, N., Shoichet, B. K. A common mechanism underlying promiscuous inhibitors from virtual and high-throughput screening. *J. Med. Chem.* **2002**, *45*, 1712–1722.

28 Roche, O., Schneider, P., Zuegge, J., Guba, W., Kansy, M., Alanine, A., Bleicher, K., Danel, F., Gutknecht, E. M., Rogers-Evans, M., Neidhart, W., Stalder, H., Dillon, M., Sjögren, E., Fotouhi, N., Gillespie, P., Goodnow, R., Harris, W., Jones, P., Taniguchi, M., Tsujii, S., von der Saal, W., Zimmermann, G. Schneider, G. Development of a virtual screening method for identification of "frequent hitters" in compound libraries. *J. Med. Chem.* **2002**, *45*, 137–142.

29 Charifson, P. S., Walters, W. P. Filtering databases and chemical libraries. *J. Comput.-Aided Mol. Des.* **2002**, *16*, 311–323.

30 Baurin, N., Baker, R., Richardson, C., Chen, I., Foloppe, N., Potter, A., Jordan, A., Roughley, S., Parratt, M., Greany, P., Morley, D., Hubbard, R. E. Drug-like annotation and duplicate analysis of a 23-supplier chemical database totalling 2.7 million compounds. *J. Chem. Inf. Comput. Sci.* **2004**, *44*, 643–651.

31 Olah, M. M., Bologa, C. G., Oprea, T. I. Strategies for compound selection. *Curr. Drug Discov. Technol.* **2004**, *1*, 211–220.

32 Verheij, H. J. Leadlikeness and structural diversity of synthetic screening libraries. *Mol. Diversity* **2006**, *V10*, 377–388.

33 Kazius, J., McGuire, R., Bursi, R. Derivation and validation of toxicophores for mutagenicity prediction. *J. Med. Chem.* **2005**, *48*, 312–320.

34 Ashby, J., Tennant, R. W. Chemical structure, *Salmonella* mutagenicity and extent of carcinogencity as indicators of genotoxic carcinogenesis among 222 chemicals tested in rodents by the US NCI/NTP. *Mutat. Res.* **1988**, *204*, 17–115.

35 Takaoka, Y., Endo, Y., Yamanobe, S., Kakinuma, H., Okubo, T., Shimazaki, Y., Ota, T., Sumiya, S., Yoshikawa, K. Development of a method for evaluating drug-likeness and ease of synthesis using a data set in which compounds are assigned scores based on chemists' intuition. *J. Chem. Inf. Comput. Sci.* **2003**, *43*, 1269–1275.

36 Lajiness, M. S., Maggiora, G. M., Shanmugasundaram, V. Assessment of the consistency of medicinal chemists in reviewing sets of compounds. *J. Med. Chem.* **2004**, *47*, 4891–4896.

37 James, C. A., Weininger, D., Delaney, J. *Daylight Theory Manual*, Daylight Chemical Information Systems, Mission Viejo, CA, USA, http://www.daylight.com/dayhtml/doc/theory/index.html.

38 Takaoka, Y., Endo, Y., Yamanobe, S., Kakinuma, H., Okubo, T., Shimazaki, Y., Ota, T., Sumiya, S. Filter 2.0, OpenEye Scientific Software Inc., Santa Fe, NM, USA, http://www.eyesopen.com.

39 Wikström, H., Andersson, B., Elebring, T., Svensson, K., Carlsson, A., Largent, B. *N*-Substituted 1,2,3,4,4a,5,6,10b-octahydrobenzo[*f*]quinolines and 3-phenylpiperidines: effects on central dopamine and sigma-receptors. *J. Med. Chem.* **1987**, *30*, 2169–2174.

40 Hale, J. J., Budhu, R. J., Holson, E. B., Finke, P. E., Oates, B., Mills, S. G., MacCoss, M., Gould, S. L., DeMartino, J. A., Springer, M. S., Siciliano, S., Malkowitz, L., Schleif, W. A., Hazuda, D., Miller, M., Kessler, J., Danzeisen, R., Holmes, K., Lineberger, J., Carella, A., Carver, G., Emini, E. 1,3,4-Trisubstituted pyrrolidine CCR5 receptor antagonists. Part 2: lead optimization affording selective, orally bioavailable compounds with potent anti-HIV activity. *Bioorg. Med. Chem. Lett.* **2001**, *11*, 2741–2745.

41 Proudfoot, J. R. Drugs, leads, and drug-likeness: an analysis of some recently launched drugs. *Bioorg. Med. Chem. Lett.* **2002**, *12*, 1647–1650.

42 Hann, M. M., Oprea, T. I. Pursuing the leadlikeness concept in pharmaceutical research. *Curr. Opin. Chem. Biol.* **2004**, *8*, 255–263.

43 Oprea, T. I. Current trends in lead discovery: are we looking for the appropriate properties? *J. Comput.-Aided Mol. Des.* **2002**, *16*, 325–334.

44 Wenlock, M. C., Austin, R. P., Barton, P., Davis, A. M., Leeson, P. D. A comparison of physiochemical property profiles of development and marketed oral drugs. *J. Med. Chem.* **2003**, *46*, 1250–1256.

45 Blake, J. F. Identification and evaluation of molecular properties related to preclinical optimization and clinical fate. *Med. Chem.* **2005**, *1*, 649–655.

46 Xie, P., Muresan, S., Li, J. CLASS – a high-throughput cheminformatic tool for chemical library profiling. Presented at *ChemAxon's User Group Meeting 2005*, Budapest Hungary, http://www.chemaxon.com/forum/download382.ppt.

47 Vieth, M., Sutherland, J. J. Dependence of molecular properties on proteomic family for marketed oral drugs. *J. Med. Chem.* **2006**, *49*, 3451–3453.

48 Walters, P. W., Stahl, M. T., Murcko, M. A. Virtual screening – an overview. *Drug Discov. Today* **1998**, *3*, 160–178.

49 Martin, Y. C. What works and what does not: lessons from experience in a pharmaceutical company. *QSAR Comb. Sci.* **2006**, *25*, 1192–1200.

50 ChemNavigator.com Inc., http://www.chemnavigator.com.

51 Maybridge, Tintagel, UK, http://www.maybridge.com.

52 Irwin, J. J., Shoichet, B. K. ZINC – a free database of commercially available compounds for virtual screening. *J. Chem. Inf. Model.* **2005**, *45*, 177–182.

53 Monge, A., Arrault, A., Marot, C., Morin-Allory, L. Managing, profiling and analyzing a library of 2.6 million compounds gathered from 32 chemical providers. *Mol. Diversity* **2006**, *10*, 389–403.

54 Kuntz, I. D., Chen, K., Sharp, K. A., Kollman, P. A. The maximal affinity of ligands. *Proc. Natl Acad. Sci. USA* **1999**, *96*, 9997–10002.

55 Hopkins, A. L., Groom, C. R., Alex, A. Ligand efficiency: a useful metric for lead selection. *Drug Discov. Today* **2004**, *9*, 430–431.

56 Abad-Zapatero, C., Metz, J. T. Ligand efficiency indices as guideposts for drug discovery. *Drug Discov. Today* **2005**, *10*, 464–469.

57 Burrows, J. Lead generation by fragment screening. Presented at SCIpharm 2006 International Pharmaceutical Industry Conference, Edinburgh, **2006**.

58 Sadowski, J., Kubinyi, H. A scoring scheme for discriminating between drugs and nondrugs. *J. Med. Chem.* **1998**, *41*, 3325–3329.

59 Ajay; Walters, W. P., Murcko, M. A. Can we learn to distinguish between drug-like and nondrug-like molecules. *J. Med. Chem.* **1998**, *41*, 3314–3324.

60 Müller, K.-R., Ratsch, G., Sonnenburg, S., Mika, S., Grimm, M., Heinrich, N. Classifying "drug-likeness" with kernel-based learning methods. *J. Chem. Inf. Model.* **2005**, *45*, 249–253.

61 Ghose, A. K., Viswanadhan, V. N., Wendoloski, J. J. Prediction of hydrophobic (lipophilic) properties of small organic molecules using fragmental methods: an analysis of ALOGP and CLOGP methods. *J. Phys. Chem. A* **1998**, *102*, 3762–3772.

62 Ihlenfeldt, W. D., Gasteiger, J. Hash codes for the identification and classification of molecular structure elements. *J. Comput. Chem.* **1994**, *15*, 793–813.

63 Kogej, T., Engkvist, O., Blomberg, N., Muresan, S. Multifingerprint based similarity searches for targeted class compound selection. *J. Chem. Inf. Model.* **2006**, *46*, 1201–1213.

64 Muresan, S., Sadowski, J. "In-house likeness": comparison of large compound collections using artificial neural networks. *J. Chem. Inf. Model.* **2005**, *45*, 888–893.

65 PubChem provides information on the biological activities of small molecules. It is a component of the NIH's Molecular Libraries Roadmap Initiative, http://pubchem.ncbi.nlm.nih.gov.

66 Roberts, G., Myatt, G. J., Johnson, W. P., Cross, K. P., Blower, P. E. LeadScope: software for exploring large sets of screening data. *J. Chem. Inf. Comput. Sci.* **2000**, *40*, 1302–1314.

67 Beno, B. R., Mason, J. S. The design of combinatorial libraries using properties and 3D pharmacophore fingerprints. *Drug Discov. Today* **2001**, *6*, 251–258.

68 McGregor, M. J., Muskal, S. M. Pharmacophore fingerprinting. 2. Application to primary library design. *J. Chem. Inf. Model.* **2000**, *40*, 117–125.

69 Mason, J. S., Hermsmeier, M. A. Diversity assessment. *Curr. Opin. Chem. Biol.* **1999**, *3*, 342–349.

70 Vieth, M., Siegel, M. G., Higgs, R. E., Watson, I. A., Robertson, D. H., Savin, K. A., Durst, G. L., Hipskind, P. A. Characteristic physical properties and structural fragments of marketed oral drugs. *J. Med. Chem.* **2004**, *47*, 224–232.

Index

Molecular Drug Properties. Measurement and Prediction. R. Mannhold (Ed.)

ISBN: 978-3-527-31755-4

d

e